Biochemical Thermodynamics:
Applications of Mathematica

**METHODS OF
BIOCHEMICAL ANALYSIS**

Volume 48

Biochemical Thermodynamics: Applications of Mathematica

Robert A. Alberty

Massachusetts Institute of Technology
Cambridge, MA

WILEY-INTERSCIENCE

A JOHN WILEY & SONS, INC., PUBLICATION

Published by John Wiley & Sons, Inc., Hoboken, New Jersey.
Published simultaneously in Canada.

For general information on our other products and services or for technical support, please contact our Customer Care Department within the United States at (800) 762-2974, outside the United States at (317) 572-3993 or fax (317) 572-4002.

Wiley also publishes its books in a variety of electronic formats. Some content that appears in print may not be available in electronic format. For information about Wiley products, visit our web site at www.wiley.com.

Library of Congress Cataloging-in-Publication Data is available.

Alberty, Robert A.
   Biochemical Thermodynamics: Applications of Mathematica

   ISBN-13 978-0-471-75798-6
   ISBN-10 0-471-75798-5

Printed in the United States of America.

10 9 8 7 6 5 4 3 2 1

Contents

Preface

This book is about calculations on the thermodynamics of biochemical reactions that is based on Legendre transforms of the Gibbs energy that bring in the pH, pMg, and concentrations of coenzymes and ligands as independent variables. Chemical reactions are studied under the constraints of constant temperature and constant pressure, but biochemical reactions are studied under the addional constraints of pH and, perhaps, pMg or free concentrations of other metal ions. In considering systems of biochemical reactions, it may be useful to constrain concentrations of various coenzymes. In considering macromolecule-ligand binding it may be useful to constrain the ligand concentration. For example, the binding of oxygen by hemoglobin can be treated at specified concentrations of molecular oxygen. As more intensive variables are specified, more thermodynamic properties of a system are defined, and the equations that represent thermodynamic properties as functions of independent variables become more complicated. Since more independent variables are involved in biochemical thermodynamics than in chemical thermodynamics, the equations for calculating properties become more complicated. Thanks to the development of mathematical applications for personal computers, these complicated calculations can be carried out much more easily. Thermodynamic calculations on such systems require the use of computers, and *Mathematica* is a very convenient application because of its symbolic capabilities, its calculation of partial derivatives, and its facilities for construction of databases, tables, and plots. The fact that *Mathematica* can be used to derive these very complicated functions and take partial derivatives to obtain other properties makes it possible to make calculations that would previously have been considered impractical. Therefore, this book has been written in *Mathematica*. *Mathematica* makes it possible to intermingle text with calculations, as illustrated by a number of recent books. In this book, the calculations of biochemical thermodynamics are described in words, equations are derived by use of *Mathematica*, and then evaluated for specified values of independent variables. All the *Mathematica* calculations are shown, and so the data, equations and programs can be used for the calculation of properties of reactants at other specified values of independent variables.

The main question that thermodynamics deals with is the direction of spontaneous change when a system is initially in a specified state. The state of a biochemical reaction system is described by specifying concentrations of reactants, temperature, pH, ionic strength, and concentrations of free metal ions that are bound by reactants. Concentrations of coenzymes and ligands like molecular oxygen can also be specified. Thermodynamics is important in biochemistry because it can tell us whether a given enzyme-catalyzed reaction or ligand binding will go to the right or to the left under specified conditions. It can also give us the equilibrium composition. Enzymes determine the reactions that are catalyzed in a given system and their rates, but enzymes do not determine the directions of reactions or the amount of energy that is stored, transferred, or is required to synthesize a needed reactant. In a cell, certain reactions are needed to store energy and other reactions are needed to use this energy for necessary purposes of life. If we want to understand how energy is stored and used, we need to know the apparent equilibrium constants K' of the reactions involved under the ambient conditions and also heats of reaction. Other biochemical reactions, like the binding of oxygen by hemoglobin do not require enzyme catalysis, but the equilibrium extents of binding reactions and the influence of other ligands are determined by thermodynamics. With knowledge of pKs, a measurement of the apparent equilibrium constant K' of a biochemical reaction at a single pH and ionic strength makes it possible to calculate K' as a function of pH and ionic strength. When enthalpies of reaction are available and enthalpies of acid dissociation are known, a measurement of K' at a single temperature, pH, and ionic strength makes it possible to calculate K' and other transformed thermodynamic properties as functions of temperature, pH, and ionic strength.

The apparent equilibrium constants of about 500 enzyme-catalyzed reactions have been determined under various sets of conditions, and enthalpy changes have been measured calorimetrically for some of these reactions or can be calculated from the effect of temperature on the apparent equilibrium constant. In principle these data can be used to calculate standard Gibbs energies of formation and standard enthalpies of formation for the species of about 1000 reactants. The current number of known species matrices is 199. For 94 of these reactants, the $\Delta_f H°$ of all species are known. Further analysis of existing experimental data and new measurements will make it possible to extend the current database BasicBiochemData3. The most efficient way to store thermodynamic information on enzyme-catalyzed reactions is to store data on species because then apparent equilibrium constants and other transformed thermodynamic properties of reactants and reactions can be calculated for specified conditions. Such a database makes it possible to calculate apparent equilibrium constants and transformed thermodynamic properties for many more reactions than it takes to make the data table. An even larger data set can be based on analogies because of similarities in the underlying chemistry of some reactions. Such a database can be used to calculate apparent equilibrium constants that are too large to measure directly. The number of apparent equilibrium constants that can be calculated from a database increases exponentially with the number of reactants in the data base. The calculation of species properties from different enzyme-catalyzed reactions reveals inconsistencies between different equilibrium and calorimetric experiments.

Apparent equilibrium constants cannot be determined experimentally on reactions that go nearly completion. Calorimetric measurements of enthalpies of reaction do not have this problem, Proteins may be reactants in enzyme-catalyzed reactions. When apparent equilibrium constants can be measured on reactions involving proteins, the thermodynamic properties of the reaction site in the protein can be calculated.

It is assumed that the reader has had some introduction to thermodynamics at the level of an undergraduate course in physical chemistry. My previous book "Thermodynamics of Biochemical Reactions," Wiley, Hoboken, NJ (2003) provides a more complete introduction to the structure of thermodynamics and its relation to statistical mechanics. This successor book is needed because more recent research has clarified the structure of biochemical thermodynamics and opened up new possibilities for learning about the flow of energy in living things. Three aspects of these calculations are as follows:

1. Experimental data on enzyme-catalyzed reactions are in the form of apparent equilibrium constants K', heats of reaction, and pKs (and corresponding heats of dissociation), but the most efficient way to store the thermodynamic properties of biochemical reactions is by means of small matrices that give standard Gibbs energies of formation, standard enthalpies of formation, charge numbers, and numbers of hydrogen atoms in each species of a reactant. The bridge between treating enzyme-catalyzed reactions in terms of species and in terms of reactants, like ATP, which is a sum of species, is provided by the Legendre transform $G' = G - n_c(\text{H})\mu(\text{H}^+)$, where G' is the transformed Gibbs energy of the system, G is the Gibbs energy of the system, $n_c(\text{H})$ is the amount of the hydrogen component in the system (total amount of hydrogen atoms), and μ (H^+) is the specified chemical potential of hydrogen ions, which is determined by the pH. The standard transformed Gibbs energy of a biochemical reaction is given by $\Delta_r G'° = -RT\ln K'$. The dependence of thermodynamic properties on ionic strength can be calculated using the extended Debye-Hückel equation, which involves a temperature-dependent parameter. When a reactant consists of pseudoisomers that are at equilibrium at a specified pH, isomer group thermodynamics has to be used to calculate the standard transformed Gibbs energy of formation $\Delta_f G'°$ of the reactant. This process leads to functions of temperature, pH, and ionic strength that are too complicated to be written out by hand, but *Mathematica* can be used to derive these functions and to calculate the standard transformed enthalpy of formation, standard transformed entropy of formation, average number of hydrogen ions bound, and other thermodynamic properties by taking partial derivatives.

2. Going from the experimental thermodynamic properties K' and transformed enthalpies of reaction to properties of species involves the concept of the inverse Legendre transform ($G = G' + n_c(\text{H})\mu(\text{H}^+)$). Computer programs can be written to go from the experimental properties directly to the standard Gibbs energies of formation and standard enthalpies of formation of the species involved in a reactant. These programs are more complicated than the programs using properties of species to derive the standard transformed thermodynamic properties of reactants.

3. Equilibrium compositions of systems of chemical reactions or systems of enzyme-catalyzed reactions can only be calculated by iterative methods, like the Newton-Raphson method, and so computer programs are required. These computer programs involve matrix operations for going back and forth between conservation matrices and stoichiometric number matrices. A more global view of biochemical equilibria can be obtained by specifying steady-state concentrations of coenzymes. These are referred to as calculations at the third level to distinguish them from the first level (chemical thermodynamic calculations in terms of species) and the second level (biochemical thermodynamic calculations at specified pH in terms of reactants).

In *Mathematica* reactants need to named with words starting with lower case letters because words starting with capital letters refer to operations. Also the names of reactants need to be as short as convenient and cannot involve spaces, subscripts, superscripts, hyphens, dots or other symbols that are *Mathematica* operations. Therefore, ATP is referred to as atp both in the text and in computer programs. Most of these abbreviated names will be recognized immediately, but a glossary of names is provided in the Appendix.

The Appendix contains a copy of the *Mathematica* notebook BasicBiochemData3.nb, Tables of Transformed Thermodynamic Properties, the Glossary of Names of Reactants, the Glossary of Symbols for Thermodynamic Properties, a List of *Mathematica* programs, and Sources of Biochemical Thermodynamic Information on the Web. The *Mathematica* package BasicBiochemData3.m , which is also available at

$$http : // library.wolfram.com / infocenter / MathSource / 5704$$

contains all of the species data at 298.15 K and zero ionic strength. It also contains functions of pH and ionic strength for the standard transformed Gibbs energies of formation of 199 reactants at 298.15 K; these functions are named atp, adp,... The functions are also given for the average number of hydrogen atoms in the reactant at 298.15 K as functions of the pH and ionic strength; these functions are named atpNH, adpNH,... Since $\Delta_f H°$ values are known for all species of 94 reactants at 298.15 K and zero ionic strength, the functions of temperature, pH, and ionic strength are given for these 94 reactants for the following transformed thermodynamic properties: $\Delta_f G'°$ (named atpGT, adpGT,...), $\Delta_f H'°$ (named atpHT, adpHT,...), $\Delta_f S'°$ (named atpST, adpST,...), and $\overline{N}_H$ (named atpNHT, adpNHT,...).

Since functions of pH and ionic strength for $\Delta_f G'°$ and $\overline{N}_H$ are known for 199 reactants at 298.15 K, $\Delta_r G'°$ and $\Delta_r N_H$ are calculated in Chapter 12 for 229 enzyme-catalyzed reactions as functions of pH and ionic strength. Since $\Delta_f G°$ and $\Delta_f H°$ are known for all the species of 94 reactants, functions of temperature, pH, and ionic strength that yield $\Delta_f G'°$, $\Delta_f H'°$, $\Delta_f S'°$, and $\overline{N}_H$ for 90 enzyme-catalyzed reactions are given in Chapter 13.

It is not necessary to be a programmer in order to use the programs and procedures illustrated in this book. Names of reactants, temperatures, pHs, and ionic strengths are readily changed in using the various programs.

The CD at the back of the book contains the whole book in *Mathematica*. It can be downloaded into a personal computer with *Mathematica* installed, but it can be read on a computer with *MathReader*, which is freely available from Wolfram Research, Inc. (100 Trade Center Drive, Champaign, IL 61820-7237, and www.wolfram.com). A chapter can be downloaded into a personal computer as a notebook. The following chapters do not require that BasicBiochemData3 be loaded: Chapters 2, 3, 5, 6, and 14. Chapters 1, 4, 7, 8, 9, 10, 11, 12, 13, and 15 need BasicBiochemData3 to be loaded by typing <<BiochemThermo`BasicBiochemData3` (see Use of *Mathematica*).

I am indebted to Irwin Oppenheim for my introduction to Legendre transforms. I am indebted to Robert N. Goldberg for many helpful discussions of biochemical thermodynamics. I am indebted to the National Institutes of Health for support of the research on which this book is based (5-RO1-0948358). At Wiley I am indebted to my Editor Darla Henderson and Editorial Assistant Christine Moore.

Robert A. Alberty
Cambridge, Massachusetts

Use of Mathematica

Even if you are not familiar with *Mathematica* (Wolfram Research, Inc. 100 Trade Center Drive, Champaign, IL 61820-7237 and www.wolfram.com), you should be able to read this book. The concepts and calculations in biochemical thermodynamics are explained in words in the textual parts of the book. And the results of calculations are discussed in words. When *Mathematica* is used to make tables and figures, explanatory titles are given. Names of biochemical reactants in *Mathematica* call up mathematical functions of temperature, pH, and ionic strength or mathematical functions of pH and ionic strength at $T = 298.15$ K (25 °C). Since mames of mathematical functions in *Mathematica* have to start with lowercase letters and cannot contain dots, dashes, or spaces, short names like atp are used for reactants. There is a complete list of these short names and the corresponding scientific names in the Appendix 3.

BasicBiochemData3 contains functions of pH and ionic strength that give standard transformed thermodynamic properties of 199 reactants for which values of $\Delta_f\, G°$ are known for all species that have significant concentrations in the pH 5 to pH 9 range. Using ATP as an example, these two functions are as follows:

atp: This function of pH and ionic strength yields the standard transformed Gibbs energy of formation of ATP at 298.15 K.
atpNH: This function of pH and ionic strength yields the average number of hydrogen atoms in ATP at 298.15 K.

For 94 of these reactants $\Delta_f\, G°$ and $\Delta_f\, H°$ are known for all species that have significant concentrations in the pH 5 to pH 9 range. Using ATP as an example, these four functions are as follows:

atpGT: This function of T, pH, and ionic strength yields the standard transformed Gibbs energy of formation of ATP.
atpHT: This function of T, pH, and ionic strength yields the standard transformed enthalpy of formation of ATP.
atpST: This function of T, pH, and ionic strength yields the standard transformed entropy of formation of ATP.
atpNHT: This function of T, pH, and ionic strength yields the average number of hydrogen atoms in ATP.

Since *Mathematica* is a high level language that uses words like Integrate and abbreviations like D for differentiate, you can see what mathematical operations are involved in a program. Everything that is involved in making these calculations is shown. When a calculation is made, a semicolon is often put at the end of the input so that the result is not shown immediately. When calculations are performed in your computer, the semicolon can be deleted to see what the result is, but semicolons are used in the book to save space.

The CD contains the whole book in *Mathematica*. When you put the CD in your computer, you will see a list of chapters and appendices. If *Mathematica* is installed in your computer, you can click on a chapter or appendix, and it will come up on the screen. If it is Chapter 2, 3, 5, 6, or 14, all the calculations can be run by using Kernel/Evaluation/Evaluate Notebook. This will take a few seconds or a couple of minutes depending on the chapter. When you load a new chapter, you can also run a cell at a time in the order they are in the chapter. The values of arguments in programs can be changed to make calculations at different temperatures, pHs, and ionic strengths. Each chapter should be opened in a fresh workspace.

If you want to run programs in Chapters 1, 4, 7, 8, 9, 10, 11, 12, 13, or 15, BasicBiochemData3 is needed (see Appendix 1). BasicBiochemData3.nb and BasicBiochemData3.m are also available at

http : // library.wolfram.com / infocenter / MathSource / 5704

Note that there is a BasicBiochemicalData3.m file in the CD that is not in the printed book. When the notebook BasicBiochemData3.nb was made, a package version BasicBiochemData3.m was made automatically. The package consists of the *Mathematica* input without the text. The package BasicBiochemData3.m needs to be installed in your personal computer as described in the following *Instructions for the use of the package BasicBiochemData3.m*. When it is installed, it possible to load it into a workspace by use of <<BiochemThermo`BasicBiochemData3`. The value of a mathematical function can be calculated using ReplaceAll (/.x->). For example, the standard transformed Gibbs energy of formation of ATP in kJ mol^{-1} at

298.15 K, pH 7, and ionic strength 0.25 M can be calculated by typing atp/.pH->7/.is->.25. This value can also be obtained by typing atpGT/.t->298.15/.pH->7/.is->.25.

In calculations on enzyme-catalyzed reactions, one of the ways a reaction equation can be entered is in the form atp+h2o+de==adp+pi. Note that hydrogen ions are never shown in a reaction equation at specified pH. Other programs may require that the reaction equation be written in the form ec3x6x1x3=adp+pi-(atp+h2o), where the name of the reaction is the EC (Enzyme Nomenclature) number with decimal points replaced by x's. It is especially simple to change the ranges of independent variables in tables and figures.

When a change is made in a chapter, the chapter can be saved in your computer, but the version in the CD cannot be changed. If you do not want to save the whole chapter, you can copy a calculation and paste it into a new notebook, where it can be saved where you want it in your computer. When you make a new notebook it needs to contain the programs that are used.

Mathematica provides for several styles of programing: functional programming, procedural programing, rule-based programming, and recursion. Almost all of the programs in this book are examples of functional programming. This is a style that is quite distinct from what is available in traditional computer languages. A functional program is a mathematical function and the inputs to the program are the arguments of the function. When the program is run, the function is applied to its arguments. The way of writing a functional program looks like a mathematical equation. For example, a function of x and y is written as functionname[x_,y_]:=body, where x and y are arguments. The body of the function can be a single expression or a series of expressions. The := is referred to as a delayed assignment. When the program is typed in, nothing is returned. But when the name of the function is typed in with values for the arguments, the program returns the result of the calculation. For example,

```
In[260]:=
        square[x_] := x^2

In[261]:=
        square[5]

Out[261]=
        25
```

Instructions for the use of the package BasicBiochemData3.m

When the 199 small matrices of species data or the 774 functions are needed, the command
<<BiochemThermo`BasicBiochemData3`
is used to make this information available. It is necessary for the user of this book to put BasicBiochemData3.m in their computer so that
<<BiochemThermo`BasicBiochemData3`
will work. To load this package properly, it is first necessary to create a folder named BiochemThermo and put BasicBiochemData3.m into this folder. Then it is necessary to find out where to put this folder as follows: Open a *Mathematica* session and evaluate $UserBaseDirectory. On my Mac computer, this yields /Users/robertalberty/Library/*Mathematica*. In this *Mathematica* file, you will find Applications. Put the folder BiochemThermo (with BasicBiochemData3.m in it) into this Applications directory. Now this package becomes available whenever you load it using
<<BiochemThermo`BasicBiochemData3`

When this package is loaded, all the species data on 199 reactants and the 774 functions under this package will be available for calculations. Each chapter should be opened in a fresh workspace. A whole chapter can be run by use of Kernel/Evaluation/Evaluate Notebook. The properties for adenosine triphosphate are named atpsp, atp, atpNH, atpGT, atpHT, atpST, and atpNHT. These functions are all protected; that is, none of them can be changed without unprotecting them. These thermodynamic values are based on the usual conventions of chemical thermodynamic tables that $\Delta_f G° = \Delta_f H° = 0$ for elements in defined reference states and for $H^+ (a=1)$. Additional canventions are that $\Delta_f G° = \Delta_f H° = 0$ for glutathione$_{ox}^{2-}$, NAD$_{ox}^{-1}$, NADP$_{ox}^{3-}$, retinal^0, thioredoxin$_{ox}^0$, and ubiquinone$_{ox}^0$.

Sources of Biochemical Thermodynamic Information on the Web

In the lists of references in the chapters, some have URLs (Uniform Resource Locator). These URLs in the CD that contains this book are active in the sense that if you click on them in a computer connected with the Web, the data source will come up on your screen. These URLs are all given in one place in Appendix 6, which includes a short description of their content.

A number of books have been written to help people get started in *Mathematica*: Three of these are
B. F. Torrence and E. A. Torrence, The Student's Introduction to *Mathematica*, Cambridge University Press, 1999.
P. Wellin, R. Gaylord, and S. Kumin, An Introduction to Programming with *Mathematica*, Cambridge University Press, 2005.
R. J. Silbey, R. A. Alberty, and M. Bawendi, Solutions Manual to Accompany Physical Chemistry, Wiley, Hoboken, NJ, 2005.

Biochemical Thermodynamics:
Applications of Mathematica

Chapter 1 Thermodynamics of the Dissociation of Weak Acids

1.1 Basic Chemical Thermodynamics for Dilute Aqueous Solutions

The first law of thermodynamics introduces a thermodynamic property U of a system that is referred to as the **internal energy**. The change in the internal energy in a change in state of a system is given by $\Delta U = q + w$, where q is the heat flow into the system and w is the work done on the system. The second law of thermodynamics introduces the **entropy** S, and it has two parts. The change in entropy in a reversible change from one state of a system to another is given by $\Delta S = q/T$, where T is the thermodynamic temperature. According to the second part of the second law, when a change takes place spontaneously in an isolated system, ΔS is greater than zero. This is very important because it provides a way to calculate whether a specified change in state can take place in a system on the basis of thermodynamic measurements. The entropy provides the criterion for spontaneous change and equilibrium at specified internal energy and volume: $(dS)_{U,V} \geq 0$. Thus the entropy of an isolated system increases to a maximum in a spontaneous change. These conclusions apply to systems consisting of phases that are uniform in composition and do not have gradients of temperature or concentration in them. According to the third law of thermodynamics, the entropy of a pure crystalline substance is equal to zero at absolute zero. The molar entropy of a crystalline substance at room temperature can be determined by making heat capacity measurements down to close to absolute zero.

The **enthalpy** H of a system is defined by

$$H = U + PV \tag{1.1-1}$$

The enthalpy provides the criterion for spontaneous change and equilibrium at specified internal energy and pressure:

$(dH)_{U,P} \leq 0$. Thus the enthalpy of a system decreases to a minimum in a spontaneous change at specified U and V. Gibbs (1) introduced what we now refer to as the **Gibbs energy** G that he defined with

$$G = U + PV - TS = H - TS \tag{1.1-2}$$

The Gibbs energy is especially useful in considering chemical reactions because it provides the criterion for spontaneous change and equilibrium at specified temperature and pressure: $(dG)_{T,P} \leq 0$. It decreases to a minimum in a system at specified temperature and pressure when a spontaneous change occurs. Equations 1.1-1 and 1.1-2 are called **Legendre transforms**. Note that Legendre transform 1.1-1 introduces P as an independent variable, and Legendre transform 1.1-2 introduces T as an independent variable. The **transformed Gibbs energy** G' at specified pH will be introduced in Chapter 3. The transformed Gibbs energy provides the criterion for spontaneous change and equilibrium when temperature, pressure and pH are independent variables: $(dG')_{T,P,pH} \leq 0$. More information on Legendre transforms is given in Chapter 3. More information on introductory thermodynamics is given in textbooks on physical chemistry (2) and in "Thermodynamics of Biochemical Reactions" (3).

1.2 Database BasicBiochemData3

A long time ago chemists realized that the most efficient way to store thermodynamic data on chemical reactions is by making tables of standard thermodynamic properties of species. The NBS Tables of Chemical Thermodynamic Properties (4) gives $\Delta_f G°$, $\Delta_f H°$, and $S_m°$ for species at 298.15 K at xero ionic strength. Since the standard molar entropy $S_m°$ is not available for many species of biochemical interest, the standard entropies of formation $\Delta_f S°$ are used. This property of a species is calculated by using

$$\Delta_f G° = \Delta_f H° - T\Delta_f S° \tag{1.2-1}$$

Because of this relation $\Delta_f S°$ is redundant, but it is of interest because the factors determining the entropy are quite different from those determining the enthalpy. The standard thermodynamic properties of some of the species of interest in biochemistry are available in the NBS Tables, but these tables are limited to inorganic species and C_1 and C_2. When the equilibrium constant K has been determined for a chemical reaction of biochemical interest and the standard Gibbs energies of formation are available in the NBS Tables for all the species but one, the standard Gibbs energy of that one species can be calculated from the value for K (see next section). When $\Delta_f H°$ is available for all the species but one, the standard enthalpy of formation of that one species can also be calculated. When the properties are unknown for two species, it may be useful to assign $\Delta_f G° = 0$ and $\Delta_f H° = 0$ to one of the species. This was done in 1992 for adenosine so that $\Delta_f G°$ and $\Delta_f H°$ for all the species in the ATP series could be calculated (5). But when $\Delta_f G°$ and $\Delta_f H°$ became available for adenosine in dilute aqueous solution in 2001 (6), it became possible to calculate the properties of all species in the ATP series with respect to the elements.

A database on the thermodynamic properties of species of biochemical interest has been developed in *Mathematica (7)* as a package. In this package, BasicBiochemData3 (8), small matrices for 199 reactants (sums of species) contain the data at 298.15 K and zero ionic strength. There is a row in the matrix for each species that gives $\{\Delta_f G°, \Delta_f H°, z, N_H\}$. $\Delta_f G°$ and $\Delta_f H°$ are given in kJ mol^{-1}. N_H is the number of hydrogen atoms in the species. N_H is not needed in this chapter, but it will play a key role in Chapter 3. The package BasicBiochemData3 exists in two forms: (1) as a Mathemaica notebook BasicBiochemData3.nb, and as (2) a *Mathematica* package with the name BasicBiochemData3.m. The notebook form, which contains only the *Mathematica* input, is the first item in the Appendix of this book. The package form is loaded as follows:

```
In[2]:= Off[General::spell1];
        Off[General::spell];
```

```
In[4]:=
 << BiochemThermo`BasicBiochemData3`
```

Now the species matrices can be obtained by typing in the name of the reactant with the suffix sp. The species properties for the reactants in atp + h2o = adp + pi are given by

```
In[5]:=
atpsp
```

Out[5]= {{-2768.1, -3619.21, -4, 12}, {-2811.48, -3612.91, -3, 13}, {-2838.18, -3627.91, -2, 14}}

```
In[6]:=
adpsp
```

Out[6]= {{-1906.13, -2626.54, -3, 12}, {-1947.1, -2620.94, -2, 13}, {-1971.98, -2638.54, -1, 14}}

```
In[7]:=
pisp
```

Out[7]= {{-1096.1, -1299., -2, 1}, {-1137.3, -1302.6, -1, 2}}

```
In[8]:=
h2osp
```

Out[8]= {{-237.19, -285.83, 0, 2}}

The names of these data matrices start with a lower case letter because capital letters are used for *Mathematica* operations. The names cannot contain - or a period because these are mathematical operations. pKs less than 3 and more than 10 are generally ignored in the database because their effects are not significant in the range pH 5 to pH 9.

The reason for using matrices in *Mathematica* is that the individual properties can be retrieved. $\Delta_f G°(\text{ATP}^{4-})$ is given by

```
In[9]:=
atpsp[[1, 1]]
```

Out[9]= -2768.1

$\Delta_f H°(\text{ATP}^{4-})$ is given by

```
In[10]:=
atpsp[[1, 2]]
```

Out[10]= -3619.21

$\Delta_f G°(\text{HATP}^{3-})$ is given by

```
In[11]:=
atpsp[[2, 1]]
```

Out[11]= -2811.48

1.3 Treatment of Activity Coefficients

The dissociations of weak acids play important roles in the thermodynamics of enzyme-catalyzed reactions, and so they are discussed first. Their dissociation constants determine the effects of pH on apparent equilibrium constants. The National Bureau of Standards Tables of Chemical Thermodynamic Properties (4) give the properties of species in water at 298.15 K and zero ionic strength, but biochemists are concerned with properties at ionic strengths in the physiological range. The ionic strength is defined by $I = (1/2)\sum z_j^2 c_j$, where z_j is the charge number of the jth species and c_j is its molar concentration. The **Gibbs energy of formation** $\Delta_f G_j$ of species j in an aqueous solution is given by

$$\Delta_f G_j = \Delta_f G_j{}^\circ + RT\ln\gamma_j c_j = \Delta_f G_j{}^\circ + RT\ln\gamma_j + RT\ln c_j \tag{1.3-1}$$

where $\Delta_f G_j{}^\circ$ is the standard Gibbs energy of formation of ion j at zero ionic strength and γ_j is its **activity coefficient**. The standard Gibbs energy $\Delta_f G_j{}^\circ$ of formation of species j is relative to the elements it contains, each taken as $\Delta_f G^\circ$(element) = 0 for a reference form. Neutral species in water are not significantly affected by the ionic strength, but activity coefficients γ_j of ions in the physiological range of ionic strengths can be represented by the extended Debye-Hückel equation (9):

$$\ln\gamma_j = -\alpha z_j^2 I^{1/2} / (1 + BI^{1/2}) \text{ or } \gamma_j = \exp\{-\alpha z_j^2 I^{1/2} / (1 + BI^{1/2})\} \tag{1.3-2}$$

where the **Debye-Hückel constant** α is a function of temperature, as shown in the next section, and $B = 1.6 \text{ L}^{1/2} \text{ mol}^{-1/2}$ is an empirical constant that is taken to be independent of temperature.

At 298.15 K, the coefficient α in the Debye-Hückel equation is 1.17582 $\text{kg}^{1/2} \text{ mol}^{-1/3}$ (10). The original Debye-Hückel equation does not have the $(1 + BI^{1/2})$ in the denominator. It is a limiting law, which means it is approached as the ionic strength approaches zero. The empirical term in the denominator leads to estimates at higher ionic strength, but should not be used above about 0.35 M. The exact thermodynamics of aqueous solutions containing ions is very complicated (11), but we will be concerned here with what might be called practical calculations. Substituting equation 1.3-2 into equation 1.3-1 yields

$$\Delta_f G_j = \Delta_f G_j{}^\circ - RT\alpha z_j^2 I^{1/2} / (1 + BI^{1/2}) + RT\ln c_j \tag{1.3-3}$$

We will actually use this equation in the form

$$\Delta_f G_j(I) = \Delta_f G_j{}^\circ(I) + RT\ln c_j \tag{1.3-4}$$

where the **standard Gibbs energy of formation** of species j at ionic strength I is given by

$$\Delta_f G_j{}^\circ(I) = \Delta_f G_j{}^\circ(I=0) - RT\alpha z_j^2 I^{1/2} / (1 + BI^{1/2}) \tag{1.3-5}$$

Thus we will consider the standard Gibbs energy of formation of species j, $\Delta_f G_j{}^\circ(I)$, to be a function of ionic strength. A table of standard formation properties of species can be prepared for a specified ionic strength and can be used at that ionic strength without having to deal with activity coefficients.

The **standard enthalpy of formation** of species j at ionic strength I is obtained by applying the Gibbs-Helmholtz equation: $H = -T^2\{\partial(G/T)/\partial T\}$ to equation 1.3-5.

$$\Delta_f H_j{}^\circ(I) = \Delta_f H_j{}^\circ(I=0) - RT^2(\partial\alpha/\partial T)z_j^2 I^{1/2} / (1 + BI^{1/2}) \tag{1.3-6}$$

Table 1.1 gives the coefficients in the Debye-Hückel equation, and in the equations for $\Delta_f G_j{}^\circ(I)$ and $\Delta_f H_j{}^\circ(I)$ that have been calculated by Clarke and Glew (10).

Table 1.1 Debye-Hückel constant α in $kg^{1/2}$ $mol^{-1/2}$ and the coefficients of the ionic strength terms for Δ_f $G°$ in kJ $mol^{-3/2}$ $kg^{1/2}$, Δ_f $H°$ in kJ $mol^{-3/2}$ $kg^{1/2}$ and Δ_f $S°$ in J $mol^{-3/2}$ $kg^{1/2}$ K^{-1}

```
In[12]:=
line1 = {0, 1.12938, 2.56494, 1.075, 13.3258};
line2 = {10, 1.14717, 2.70073, 1.213, 13.8221};
line3 = {20, 1.16589, 2.84196, 1.3845, 14.4174};
line4 = {25, 1.17582, 2.91482, 1.4775, 14.7319};
line5 = {30, 1.18599, 2.98934, 1.5775, 15.0646};
line6 = {40, 1.20732, 3.14349, 1.800, 16.1057};
TableForm[{line1, line2, line3, line4, line5, line6},
 TableHeadings → {None, {"t/°C ", "α ", "RTα", "RT² ∂α/∂T ", "R(α+T∂α/∂T)"}}]
```

Out[18]//TableForm=

t/°C	α	RTα	RT$^2 \partial\alpha/\partial$T	R(α+T$\partial\alpha/\partial$T)
0	1.12938	2.56494	1.075	13.3258
10	1.14717	2.70073	1.213	13.8221
20	1.16589	2.84196	1.3845	14.4174
25	1.17582	2.91482	1.4775	14.7319
30	1.18599	2.98934	1.5775	15.0646
40	1.20732	3.14349	1.8	16.1057

Clarke and Glew did not give the entropy coefficient, but this is discussed in Chapter 15 and in papers on temperature effects (12,13).

The **equilibrium constant** K_k for reaction k between species at a specified ionic strength can be calculated using

$$\Delta_r G_k° = -RT\ln K_k = \sum_{j=1}^{N} v_{jk} \Delta_f G_j° \qquad (1.3\text{-}7)$$

where v_{jk} is the stoichiometric number for species j in reaction k and N is the number of species involved in chemical reaction k. Whenever a thermodynamic property of a reaction is given, the chemical reaction must be specified because there are different ways to write chemical equations, and so it is important to know the way a chemical equation is written. The equilibrium constant K_k is written in terms of concentrations in biochemical thermodynamics because equation 1.3-4 is written in terms of concentrations. Thus for the dissociation of a weak acid, HA = H$^+$ + A$^-$, the dissociation constant is given by

$$K_k = [\text{H}^+][\text{A}^-]/[\text{HA}] = 10^{-\text{pH}} [\text{A}^-]/[\text{HA}] = 10^{-pK_k} \qquad (1.3\text{-}8)$$

where $pK_k = -\log K_k$ and pH $= -\log[\text{H}^+]$. This is not the pH obtained with a pH meter, but the required adjustment is calculated in the next section.

1.4 Calculation of pH = -log [H⁺]

The symbol pH_c can be used to distinguish $pH_c = -\log[H^+]$ from $pH_a = -\log\{\gamma(H^+)[H^+]\}$ obtained with a glass electrode. Equation 1.3-2 can be used to show that (12)

$$pH_a - pH_c = \alpha I^{1/2} / \ln(10)(1 + 1.6\,I^{1/2}) \tag{1.4-1}$$

This is the adjustment to be subtracted from the pH measured with a pH meter to yield the pH used in equation 1.3-7. pH_c is lower than pH_a because the ion atmosphere of H^+ reduces its activity.

To calculate the adjustments shown in equation 1.4-1 as a function of temperature, we need the temperature depedence of the Debye-Hückel constant α. Clarke and Glew (10) have provided tables that show that

```
In[19]:=
α=1.10708 - 0.00154508*t + 5.95584*10^-6*t^2
```

$$Out[19]= \ 1.10708 - 0.00154508\,t + 5.95584 \times 10^{-6}\,t^2$$

where t is the temperature in Kelvins. The dark type is input in *Mathematica* (7), and the next line is output from *Mathematica*. A lower case t is used here because capital letters in *Mathematica* are used for operations. In *Mathematica* an asterix is used as the multiply sign and ^ is used to indicate an exponent. The value of α at a specified temperature can be calculated by use of the ReplaceAll operation (/.t->).

```
In[20]:=
α /. t → 298.15
```

$$Out[20]= \ 1.17585$$

The units of α are $kg^{1/2}\,mol^{-1/2}$.

In *Mathematica* the right hand side of equation 1.4-1 is given by

```
In[21]:=
α * is·5 / (Log[10] * (1 + 1.6 * is^.5))
```

$$Out[21]= \ \frac{is^{0.5}\,(1.10708 - 0.00154508\,t + 5.95584 \times 10^{-6}\,t^2)}{(1 + 1.6\,is^{0.5})\,Log[10]}$$

where **is** is the symbol for ionic strength and **Log[10]** is ln10. The pH adjustment at 298.15 K, pH 7.00, and ionic strength 0.25 M can be calculated by use of the ReplaceAll operation:

```
In[22]:=
α * is·5 / (Log[10] * (1 + 1.6 * is^.5)) /. t -> 298.15 /. pH → 7.0 /. is → .25
```

$$Out[22]= \ 0.141851$$

Equation 1.4-1 yields the following table.

Table 1.2 Adjustments (as functions of ionic strength and Celsius temperature) to be subtracted from pH_a measured with a pH meter to obtain $pH_c = -\log[H^+]$.

```
In[23]:=
PaddedForm[TableForm[α*is^.5/(Log[10]*(1+1.6*is^.5))/.is->{0,.05,.1,.15,.2,.25}/.t->{283.15,29
8.15,313.15},TableHeadings->{{"0","0.05","0.10","0.15","0.20","0.25"},{"10 C","25 C","40
C"}}],{3,2}]
```

```
Out[23]//PaddedForm=
                  10 C        25 C        40 C
        0         0.00
        0.05      0.08        0.08        0.09
        0.10      0.10        0.11        0.11
        0.15      0.12        0.12        0.13
        0.20      0.13        0.13        0.14
        0.25      0.14        0.14        0.15
```

1.5 Calculation of Activity Coefficients for Ionic Species with Different Charge Numbers

The activity coefficients γ of ions in water at 298.15 K are given as a function of ionic strength by equation 1.3-2:

```
In[24]:=
γ = Exp[-1.17582 * z^2 * is^.5 / (1 + 1.6 * is^.5)];
```

Note that when a semicolon is put at the end of input, *Mathematica* does not print output. The effect of ionic strength on the activity coefficient of an ion is very sensitive to the charge number. At 298.15 K and 0.25 M ionic strength, the activity coefficients of ions with 1, 2, 3, and 4 charges are given by

```
In[25]:=
γ /. is → .25 /. z → {1, 2, 3, 4}
```

```
Out[25]= {0.72136, 0.270775, 0.0528895, 0.0053757}
```

Figure 1.1 gives the activity coefficients at 298.15 K for ions with charge numbers of 1, 2, 3, and 4 as a function of ionic strength.

```
In[26]:=
Plot[Evaluate[γ /. z → {1, 2, 3, 4}], {is, 0, .4}, AxesLabel → {"I/M", "γ"}];
```

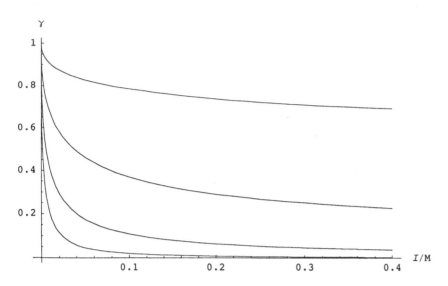

Figure 1.1 Activity coefficients as a function of ionic strength at 298.15 K in water for ions with charge numbers 1, 2, 3, and 4.

The standard Gibbs energy of formation $\Delta_f G_j\,^\circ(I)$ of ion j in aqueous solution is given by equation 1.3-5, where $RT\alpha = 2.91482$ kJ mol$^{-1/2}$ kg$^{1/2}$ at 298.15 K. The *Mathematica* expression for the adjustment from $I = 0$ to I is given by

```
In[27]:=
 stdgibbse = -2.91482 * z^2 * is^.5 / (1 + 1.6 * is^.5);
```

The effect of ionic strength on the standard Gibbs energy of formation of an ion is quite sensitive to the charge number. The effects on $\Delta_f G_j\,^\circ$ for $z = 1, 2, 3$, and 4 are given by

```
In[28]:=
 stdgibbse /. is -> .25 /. z → {1, 2, 3, 4}
```

```
Out[28]= {-0.809672, -3.23869, -7.28705, -12.9548}
```

where the adjustments of the standard Gibbs energies are given in kJ mol^{-1}. Figure 1.2 gives the standard Gibbs energies of formation at 298.15 K as functions of ionic strength for ions with charge numbers of 1, 2, 3, and 4.

```
In[29]:=
 Plot[Evaluate[stdgibbse /. z → {1, 2, 3, 4}],
   {is, 0, .4}, AxesLabel → {"I/M", "Δ_fG_j°"}, AxesOrigin → {0, -15}];
```

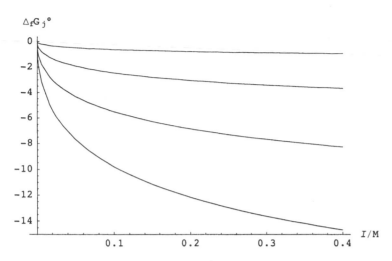

Figure 1.2 Adjustments of $\Delta_f G°$ in kJ mol^{-1} as a function of ionic strength at 298.15 K in water for ions with charge numbers 1, 2, 3, and 4.

1.6 pKs of Weak Acids at Various Ionic Strengths

Package BasicBiochemData3 can be used to calculate pKs for weak acids at 298.15 K and desired ionic strengths. This database contains 60 reactants that have a total of 82 pKs. In the program **calcpK**, it is necessary to give the number of the pK. pKs are numbered 1, 2, 3,... from the highest to the lowest. This program is used here to calculate the pKs for all the weak acids in the database at five ionic strengths. The logarithm of the acid dissociation $K(298.15 \text{ K},I)$ is given as a function of ionic strength by the following equation:

$$\ln K(298.15 \text{ K},I) = \ln K(298.15 \text{ K},I=0) + \alpha \, I^{1/2} \sum v_j \, z_j^2/(1 + 1.6 \, I^{1/2}) \tag{1.6-1}$$

Note that $\alpha = 1.17582 \text{ kg}^{1/2} \text{ mol}^{-1/2}$ at 298.15 K in Clarke and Glew (10). The following program (8) uses this equation to obtain pKs at 298.15 K at desired ionic strengths.

```
In[30]:=
calcpK[speciesmat_, no_, is_] :=
  Module[{lnkzero, sigmanuzsq, lnK}, (*Calculates pKs for a weak acid at 298.15 K at specified
    ionic strengths (is) when the number no of the pK is specified. pKs are numbered 1,
    2,3,... from the highest pK to the lowest pK,
    but the highest pK for a weak acid may be omitted if it is outside
    of the range 5 t0 9. For H3PO4,pK1=calcpK[pisp,1,{0}]=7.22.*)
  lnkzero = (speciesmat[[no + 1, 1]] - speciesmat[[no, 1]]) / (8.31451 * 0.29815);
  sigmanuzsq = speciesmat[[no, 3]]^2 - speciesmat[[no + 1, 3]]^2 + 1;
  lnK = lnkzero + (1.17582 * is^0.5 * sigmanuzsq) / (1 + 1.6 * is^0.5);
  N[-(lnK / Log[10])]]
```

This is the first *Mathematica* program in this book, and so it is important to observe its structure. In the **Module** the first list in {...} gives the names of expressions to be kept within the program. The purpose and operation of the program are described in (*...*). The first line of the program calculates $\ln K(I=0)$. The second line calculates $\sum v_j z_j^2$. The third line calculates $\ln K$, and the last line calculates $pK = -\ln K/\log(10)$. The pKs of atp at 298.15 K and ionic strengths of 0, 0.05, 0.10, 0.15, and 0.25 M are calculated using

```
In[31]:=
 calcpK[atpsp, 1, {0, .05, .1, .15, .25}]

Out[31]= {7.5998, 6.92702, 6.74198, 6.62294, 6.46502}
```

```
In[32]:=
calcpK[atpsp, 2, {0, .05, .1, .15, .25}]

Out[32]= {4.67761, 4.17303, 4.03424, 3.94497, 3.82652}
```

When a program is to be used many times, it should have just one argument so that it can be applied to a list of species properties using **Map** in *Mathematica*. This can be done with the program **calcpK298is** (14). This program derives a list of functions of ionic strength that yield the successive p*K*s of a reactant at 298.15 K.

```
In[33]:=
calcpK298is[speciesmat_] :=
 Module[{glist, hlist, zlist, nHlist, glistis, ghydionis}, (*This program derives the functions
     of ionic strength that yields the pKs at 298.15 K for weak acids.  The first
     function of ionic strength is for the acid with the fewest hydrogen atoms.  The
     program has a single argument so that it can be used with Map.  The functions
     can be evaluated by use of calcpK298is[atpsp]/.is→{0,.1,.25}, for example.*)
  {glist, hlist, zlist, nHlist} = Transpose[speciesmat];
 glistis =
    Table[glist[[i]] - 2.91482 * zlist[[i]]^2 * is^.5 / (1 + 1.6 * is^.5), {i, 1, Length[zlist]}];
  ghydionis = -2.91482 * is^.5 / (1 + 1.6 * is^.5);
  Table[((glistis[[i - 1]] - glistis[[i]] + ghydionis) / (8.31451 * .29815 * Log[10])),
  {i, 2, Length[zlist]}]]
```

In the second line of the program, **Length[zlist]** is used to calculate the number of p*K*s the reactant has. The *Mathematica* operation **Table** makes a list of values. The ionic strength dependencies of the two p*K*s of atp can be calculated as follows:

```
In[34]:=
TableForm[calcpK298is[atpsp] /. is → {0, .05, .1, .15, .25},
  TableHeadings → {{"atp pK1", "atp pK2"}, {"I=0", "I=0.05", "I=0.10", "I=0.15", "I=0.25"}}]
```

```
Out[34]//TableForm=
              I=0         I=0.05      I=0.10      I=0.15      I=0.25
    atp pK1   7.5998      6.92702     6.74198     6.62295     6.46502
    atp pK2   4.67761     4.17303     4.03424     3.94497     3.82653
```

BasicBiochemData3 contains data on the following 60 weak acids at 298.15 K and zero ionic strength.

```
In[35]:=
listreactantsdata = {acetatesp, acetylphossp, adeninesp, adenosinesp, adpsp, ammoniasp, ampsp,
    arabinose5phossp, atpsp, bpgsp, citratesp, citrateisosp, co2totsp, coAsp, cysteineLsp,
    deoxyribose1phossp, deoxyribose5phossp, deoxyadenosinesp, deoxyampsp, deoxyadpsp,
    deoxyatpsp, dihydroxyacetonephossp, fructose6phossp, fructose16phossp, fumaratesp,
    galactose1phossp, galactose6phossp, gluconolactone6phossp, glucose6phossp, glucose1phossp,
    glutathioneredsp, glyceraldehydephossp, glycerol3phossp, h2saqsp, idpsp, impsp,
    inosinesp, itpsp, malatesp, malylcoAsp, mannitol1phossp, mannose1phossp, methylmaleatesp,
    methylmalonylcoAsp, nicotinamideribonucleotidesp, oxalatesp, pepsp, phosphoglycerate2sp,
    phosphoglycerate3sp, phosphoserinesp, pisp, prppsp, ppisp, ribose1phossp, ribose5phossp,
    ribulose5phossp, sorbitol6phossp, succinatesp, succinylcoAsp, thioredoxinredsp};
```

Map is used to apply **calcpK298is** to these species matrices. The functions of ionic strength for the 60 reactants are joined together by use of **Join**. Then the functions are evaluated at five ionic strengths.

```
In[36]:=
Join[Map[calcpK298is, listreactantsdata]] /. is → {0, .05, .1, .15, .25};
```

Before making a table it is necessary to **Flatten** the matrix of pKs. This makes a table of pKs at $I = 0$, 0.05, 0.10, 0.15, and 0.25 M for 82 pKs.

In[37]:=
```
Dimensions[Flatten[Join[Map[calcpK298is, listreactantsdata]] /. is → {0, .05, .1, .15, .25}, 1]]
```

Out[37]= {82, 5}

The names of the rows are given by the following list.

In[38]:=
```
names = {"acetate pK1", "acetylphos pK1", "acetylphos pK2", "adenine pK1", "adenosine pK1",
    "adp pK1", "adp pK2", "ammonia pK1", "amp pK1", "amp pK2", "arabinose5phos pK1", "atp pK1",
    "atp pK2", "bpg pK1", "citrate pK1", "citrate pK2", "citrateiso pK1", "citrateiso pK2",
    "co2tot pK1", "co2tot pK2", "coA pK1", "cysteineL pK1", "deoxyribose1phos pK1",
    "deoxyribose5phos pK1", "deoxyadenosine pK1", "deoxyamp pK1", "deoxyamp pK2",
    "deoxyadp pK1", "deoxyadp pK2", "deoxyatp pK1", "deoxyatp pK2", "dihydroxyacetonephos pK1",
    "fructose6phos pK1", "fructose16phos pK1", "fructose16phos pK2", "fumarate pK1",
    "fumarate pK2", "galactose1phos pK1", "galactose6phos pK1", "gluconolactone6phos pK1",
    "glucose6phos pK1", "glucose1phos pK1", "glutathionered pK1", "glyceraldehydephos pK1",
    "glycerol3phos pK1", "h2saq pK1", "h2saq pK2", "idp pK1", "idp pK2", "imp pK1", "imp pK2",
    "inosine pK1", "itp pK1", "itp pK2", "malate pK1", "malylcoA pK1", "mannitol1phos pK1",
    "mannose1phos pK1", "methylmaleate pK1", "methylmalonylcoA pK1", "nmn pK1", "oxalate pK1",
    "pep pK1", "phosphoglycerate2 pK1", "phosphoglycerate3 pK1", "phosphoserine pK1", "pi pK1",
    "prpp pK1", "prpp pK2", "ppi pK1", "ppi pK2", "ppi pK3", "ppi pK4", "ribose1phos pK1",
    "ribose5phos pK1", "ribulose5phos pK1", "sorbitol6phos pK1", "succinate pK1",
    "succinate pK2", "succinylcoA pK1", "thioredoxinred pK1", "thioredoxinred pK2"};
```

Table 1.3 p*K*s of weak acids in water at 298.15 K at ionic strengths 0, 0.05, 0.10, 0.15, and 0.25 M.

```
In[39]:=
PaddedForm[
  TableForm[Flatten[Join[Map[calcpK298is, listreactantsdata]] /. is → {0, .05, .1, .15, .25}, 1],
    TableHeadings → {names, {"I=0", "I=0.05", "I=0.1", "I=0.15", "I=0.25"}},
    TableSpacing → {1, 1}], {3, 2}]

Out[39]//PaddedForm=
```

	I=0	I=0.05	I=0.1	I=0.15	I=0.25
acetate pK1	4.75	4.59	4.54	4.51	4.47
acetylphos pK1	8.69	8.35	8.26	8.20	8.12
acetylphos pK2	5.11	4.94	4.90	4.87	4.83
adenine pK1	4.20	4.20	4.20	4.20	4.20
adenosine pK1	3.47	3.47	3.47	3.47	3.47
adp pK1	7.18	6.67	6.53	6.44	6.33
adp pK2	4.36	4.02	3.93	3.87	3.79
ammonia pK1	9.25	9.25	9.25	9.25	9.25
amp pK1	6.73	6.39	6.30	6.24	6.16
amp pK2	3.99	3.82	3.77	3.74	3.71
arabinose5phos pK1	6.69	6.35	6.26	6.20	6.12
atp pK1	7.60	6.93	6.74	6.62	6.47
atp pK2	4.68	4.17	4.03	3.94	3.83
bpg pK1	7.96	7.29	7.10	6.98	6.83
citrate pK1	6.39	5.89	5.75	5.66	5.54
citrate pK2	4.76	4.42	4.33	4.27	4.19
citrateiso pK1	6.40	5.90	5.76	5.67	5.55
citrateiso pK2	4.71	4.38	4.28	4.22	4.15
co2tot pK1	10.30	9.99	9.90	9.84	9.76
co2tot pK2	6.37	6.20	6.15	6.12	6.08
coA pK1	8.38	8.21	8.16	8.14	8.10
cysteineL pK1	8.38	8.21	8.16	8.13	8.09
deoxyribose1phos pK1	6.69	6.35	6.26	6.20	6.12
deoxyribose5phos pK1	6.69	6.35	6.26	6.20	6.12
deoxyadenosine pK1	3.47	3.47	3.47	3.47	3.47
deoxyamp pK1	6.73	6.39	6.30	6.24	6.16
deoxyamp pK2	3.99	3.82	3.77	3.74	3.71
deoxyadp pK1	7.18	6.67	6.53	6.44	6.33
deoxyadp pK2	4.36	4.02	3.93	3.87	3.79
deoxyatp pK1	7.60	6.93	6.74	6.62	6.47
deoxyatp pK2	4.68	4.17	4.03	3.94	3.83
dihydroxyacetonephos pK1	5.70	5.36	5.27	5.21	5.13
fructose6phos pK1	6.27	5.94	5.84	5.78	5.70
fructose16phos pK1	6.65	5.98	5.79	5.67	5.52
fructose16phos pK2	6.05	5.54	5.41	5.32	5.20
fumarate pK1	4.60	4.27	4.17	4.11	4.03
fumarate pK2	3.09	2.93	2.88	2.85	2.81
galactose1phos pK1	6.15	5.81	5.72	5.66	5.58
galactose6phos pK1	6.44	6.10	6.01	5.95	5.87
gluconolactone6phos pK1	6.42	6.08	5.99	5.93	5.85
glucose6phos pK1	6.42	6.08	5.99	5.93	5.85
glucose1phos pK1	6.50	6.16	6.07	6.01	5.93
glutathionered pK1	8.34	8.00	7.91	7.85	7.77
glyceraldehydephos pK1	5.70	5.36	5.27	5.21	5.13

glycerol3phos pK1	6.67	6.33	6.24	6.18	6.10
h2saq pK1	12.90	12.60	12.50	12.40	12.30
h2saq pK2	6.99	6.82	6.78	6.75	6.71
idp pK1	9.56	8.89	8.70	8.58	8.43
idp pK2	7.18	6.67	6.54	6.45	6.33
imp pK1	9.63	9.13	8.99	8.90	8.78
imp pK2	6.73	6.39	6.30	6.24	6.16
inosine pK1	8.96	8.79	8.74	8.71	8.67
itp pK1	10.10	9.25	9.02	8.87	8.67
itp pK2	7.60	6.93	6.74	6.62	6.47
malate pK1	5.26	4.92	4.83	4.77	4.69
malylcoA pK1	4.21	4.04	4.00	3.97	3.93
mannitol1phos pK1	6.50	6.16	6.07	6.01	5.93
mannose1phos pK1	6.44	6.10	6.01	5.95	5.87
methylmaleate pK1	6.27	5.93	5.84	5.78	5.70
methylmalonylcoA pK1	4.21	4.04	4.00	3.97	3.93
nmn pK1	6.44	6.10	6.01	5.95	5.87
oxalate pK1	4.28	3.94	3.85	3.79	3.71
pep pK1	7.00	6.50	6.36	6.27	6.15
phosphoglycerate2 pK1	7.64	7.14	7.00	6.91	6.79
phosphoglycerate3 pK1	7.53	7.03	6.89	6.80	6.68
phosphoserine pK1	6.44	6.10	6.01	5.95	5.87
pi pK1	7.22	6.88	6.79	6.73	6.65
prpp pK1	7.18	6.34	6.11	5.96	5.76
prpp pK2	6.69	6.02	5.83	5.71	5.56
ppi pK1	9.46	8.79	8.60	8.48	8.33
ppi pK2	6.72	6.21	6.08	5.99	5.87
ppi pK3	2.26	1.92	1.83	1.77	1.69
ppi pK4	0.83	0.66	0.62	0.59	0.55
ribose1phos pK1	6.69	6.35	6.26	6.20	6.12
ribose5phos pK1	6.69	6.35	6.26	6.20	6.12
ribulose5phos pK1	6.69	6.35	6.26	6.20	6.12
sorbitol6phos pK1	6.42	6.08	5.99	5.93	5.85
succinate pK1	5.64	5.30	5.21	5.15	5.07
succinate pK2	4.21	4.04	3.99	3.96	3.92
succinylcoA pK1	4.21	4.04	4.00	3.97	3.93
thioredoxinred pK1	8.64	8.30	8.21	8.15	8.07
thioredoxinred pK2	8.05	7.88	7.83	7.80	7.76

This table can be helpful in estimating the pKs of other weak acids from their structures. In using this table it is important to remember that -log$[H^+]$ is used in the expression for the acid dissociation constant in terms of pH. To obtain pKs based on $-\log\{\gamma(H^+)[H^+]\}$, add 0, 0.08, 0.11, 0.12, and 0.14 at ionic strengths of 0, 0.05, 0.10, 0.15, and 0.25 M, respectively, at 298.15 K as indicated by Table 1.3. **PaddedForm** rounds the output to two figures to the right of the decimal point. There is a list of full names of reactants in the Appendix of this book. The reactants bpg, nmn, pep, and prpp are bisphosphoglycerate, nicotinamidemononucleotide, phosphoenolpyruvate, and 5-phosphoribosyl-alpha-pyrophosphate, respectively.

Note that, except for ammonia, adenine, and adenosine, the pKs always decrease as the ionic strength increases; in other words, the acids become stronger as the ionic strength increases. For a weak acid dissociation represented by HA = H^+ + A^- increasing the ionic strength stabilizes H^+ + A^- more than HA. The pK shift is greater when HA is an ion. There is no shift in pK with ionic strength for weak acids like the ammonium ion because there is a single charge on each side of the dissociation equation.

1.7 p*K*s of Weak Acids at Various Ionic Strengths and Temperatures

When $\Delta_f H°$ values are known for all the species of a weak acid at 298.15 K in addition to $\Delta_f G°$, the p*K*s at tempera-tures other than 298.15 K can be calculated on the assumption that $\Delta_f H°$ is constant over the range of temperature. If $\Delta_f H°$ is independent of temperature, $\Delta_f S°$ is also independent of temperature. BasicBiochemData3 contains data on 27 weak acids for which $\Delta_f H°$ values are known. In order to calculate p*K*s at temperatures other than 298.15 K, we need to do two things: (1) Express the standard Gibbs energies of formation of species as functions of temperature. (2) Express the coefficient of the ionic strength term as a function of temperature. When the standard enthalpies of formation of species at $I = 0$ are independent of temperature, their standard Gibbs energies of formation are given by

$$\Delta_f G_j°(T) = \frac{T}{298.15} \Delta_f G_j°(298.15 \text{ K}) + (1 - \frac{T}{298.15}) \Delta_f H_j°(298.15 \text{ K}) \tag{1.7-1}$$

The coefficient $RT\alpha$ in equation 1.2-5 is given as a function of temperature by

$$RT\alpha = 9.20483 \times 10^{-3} T - 1.28467 \times 10^{-5} T^2 + 4.95199 \times 10^{-8} T^3 \tag{1.7-2}$$

The program **calcpKT** (8) can be used to calculate the p*K* of a weak acid at a series of temperatures.

```
In[40]:=
calcpKT[speciesmat_, n_, is_, t_] :=
  Module[{basicspeciesG, acidicspeciesG, coeff, basicspeciesI, acidicspeciesI, hydionI},
    (*Calculates pKs for a weak acid at temperatures (t) in the range 273.15-
      313.15 K and ionic strengths (is) in the rangw 0-0.35 M when the number (n) of the
      pK is specified. pKs are numbered 1,2,3,... from the highest pK to the lowest pK,
    but the highest pK for a weak acid may be omitted if it is outside of the range
    5 to 9. The first step is to calculate the standard Gibbs energies of formation at
    zero ionic strength as a function of temperature.  The second step is to adjust these
    values to the desired ionic strength.  A list of temperatures can be used.  For example,
    pK1=calcpKT[atpsp,1,0,{273.15,298.15,313.15}]={7.50,7.60,7,65}.*)
    basicspeciesG = (t / 298.15) * speciesmat[[n, 1]] + (1 - t / 298.15) * speciesmat[[n, 2]];
    acidicspeciesG = (t / 298.15) * speciesmat[[n + 1, 1]] + (1 - t / 298.15) * speciesmat[[n + 1, 2]];
    coeff = (9.20483 * 10^-3) * t - (1.28467 * 10^-5) * t^2 + (4.95199 * 10^-8) * t^3;
    basicspeciesI = basicspeciesG - coeff * speciesmat[[n, 3]]^2 * is^.5 / (1 + 1.6 * is^.5);
    acidicspeciesI = acidicspeciesG - coeff * speciesmat[[n + 1, 3]]^2 * is^.5 / (1 + 1.6 * is^.5);
    hydionI = -coeff * is^.5 / (1 + 1.6 * is^.5);
    (1 / (8.31451 * (t / 1000) * Log[10])) * (hydionI + basicspeciesI - acidicspeciesI)]
```

The two p*K*s of atp are each calculated at three ionic strengths and three temperatures as follows:

```
In[41]:=
atp10 = calcpKT[atpsp, 1, 0, {273.15, 298.15, 313.15}];
```

```
In[42]:=
atp11 = calcpKT[atpsp, 1, .10, {273.15, 298.15, 313.14}];
```

```
In[43]:=
atp125 = calcpKT[atpsp, 1, .25, {273.15, 298.15, 313.14}];
```

```
In[44]:=
atp20 = calcpKT[atpsp, 2, 0, {273.15, 298.15, 313.15}];
```

```
In[45]:=
atp21 = calcpKT[atpsp, 2, .10, {273.15, 298.15, 313.14}];
```

```
In[46]:=
 atp225 = calcpKT[atpsp, 2, .25, {273.15, 298.15, 313.14}];

In[47]:=
 TableForm[{atp10, atp11, atp125, atp20, atp21, atp225}]

Out[47]//TableForm=
        7.49879      7.5998       7.65267
        6.67482      6.74196      6.77187
        6.4088       6.465        6.48751
        4.91813      4.67761      4.55173
        4.30015      4.03423      3.89124
        4.10063      3.82651      3.67797
```

It is more convenient to write the program **calcpKTfn** that has a single argument so that **Map** can be used to derive functions of temperature and ionic strength for a list of reactants.

```
In[48]:=
calcpKTfn[speciesmat_] :=
 Module[{glist, hlist, zlist, nHlist, coeff, speciesGT, spceiesGTis, hydionis},
   (*Derives the function of temperature, pH,and ionic strength that gives the pKs for
     a weak acid.  pKs are numbered 1,2,3,... from the highest pK to the lowest pK,
    but the highest pK for a weak acid may be omitted if it is outside of the range
    5 to 9. The first step is to calculate the standard Gibbs energies of formation at
    zero ionic strength as a function of temperature.  The second step is to adjust
    these values to the desired ionic strength. The output is a list of functions,
    with as many functions as pKs.  The third step is to make a table of the pKs.  For example,
    calcpKTfn[atpsp]/.t→{273.15,298.15,313.15}/.is→{0,.1,.25}*)
   {glist, hlist, zlist, nHlist} = Transpose[speciesmat];
   (*Calculate functions of temperature for the Gibbs energies of all species.*)
   speciesGT = (t / 298.15) * glist + (1 - t / 298.15) * hlist;
   (*Adjust these functions of
     temperature to make them functions of ionic strength as well.*)
   coeff = (9.20483 * 10^-3) * t - (1.28467 * 10^-5) * t^2 + (4.95199 * 10^-8) * t^3;
   speciesGTis = speciesGT - coeff * zlist^2 * is^.5 / (1 + 1.6 * is^.5);
   hydionis = -coeff * is^.5 / (1 + 1.6 * is^.5);
   (*Make a list of the Gibbs energies
     of dissociation for all weak acids and convert them to pKs.*)
   Table[((speciesGTis[[i - 1]] - speciesGTis[[i]] + hydionis) / (8.31451 * (t / 1000) * Log[10])),
    {i, 2, Length[zlist]}]]
```

This program can be used to calculate pK_1 and pK_2 for atp at three temperatures and five ionic strengths.

```
In[49]:=
 calcpKTfn[atpsp] /. t → {273.15, 298.15, 313.15} /. is → {0, .05, .1, .15, .25}

Out[49]= {{{7.49879, 6.85256, 6.67482, 6.56049, 6.4088},
           {7.5990, 6.92701, 6.74196, 6.62292, 6.465},
           {7.65267, 6.96189, 6.77189, 6.64967, 6.48753}},
          {{4.91813, 4.43346, 4.30015, 4.2144, 4.10063}, {4.67761, 4.17301, 4.03423,
             3.94495, 3.82651}, {4.55173, 4.03365, 3.89115, 3.79949, 3.67788}}}

In[50]:=
 PaddedForm[TableForm[calcpKTfn[atpsp] /. t → {273.15, 298.15, 313.15} /. is → {0, .05, .1, .15, .25}
   TableHeadings → {None, {"         273.15 K", "  298.15 K", "313.15 K"},
     {"I=0", "I=0.05", "I=0.10", "I=0.15", "I=0.25"}}], {3, 2}]
```

```
Out[50]//PaddedForm=
            273.15 K          298.15 K          313.15 K
    I=0       7.50              7.60              7.65
    I=0.05    6.85              6.93              6.96
    I=0.10    6.67              6.74              6.77
    I=0.15    6.56              6.62              6.65
    I=0.25    6.41              6.46              6.49

    I=0       4.92              4.68              4.55
    I=0.05    4.43              4.17              4.03
    I=0.10    4.30              4.03              3.89
    I=0.15    4.21              3.94              3.80
    I=0.25    4.10              3.83              3.68
```

Note that the pKs for the acid dissociations at 298.15 K agree with the values in Table 1.3 This calculation can be repeated for the following 27 reactants for which $\Delta_f H°$ is known for all species. **Map** can be used to calculate all these pKs in one step.

```
In[51]:=
listHreactants = {acetatesp, adeninesp, adenosinesp, adpsp, ammoniasp, ampsp,
    atpsp, citratesp, co2totsp, fructose16phossp, fructose6phossp, fumaratesp,
    glucose6phossp, glycerol3phossp, h2saqsp, idpsp, impsp, inosinesp, itpsp, malatesp,
    mannose6phossp, pepsp, pisp, ppisp, ribose1phossp, ribose5phossp, succinatesp};
```

Table 1.4 p*K*s of weak acids as functions of temperature and ionic strength.

```
In[52]:=
PaddedForm[
  TableForm[Flatten[Join[Map[calcpKTfn, listHreactants]] /. t → {273.15, 298.15, 313.15} /.
    is → {0, .05, .1, .15, .25}, 1],
   TableHeadings → {{"acetate pK1", "adenine pK1", "adenosine pK1", "adp pK1", "adp pK2",
     "ammonia pK1", "amp pK1", "amp pK2", "atp pK1", "atp pK2", "citrate pK1",
     "citrate pK2", "co2tot pK1", "co2tot pK2", "fructose16phos pK1", "fructose16phos pK2",
     "fructose6phos pK1", "fumarate pK1", "fumarate pK2", "glucose6phos pK1",
     "glycerol3phos pK1", "h2saq pK1", "h2saq pK2", "idp pK1", "idp pK2", "imp pK1", "imp pK2",
     "inosine pK1", "itp pK1", "itp pK2", "malate pK1", "mannose6phos pK1", "pep pK1", "pi pK1",
     "ppi pK1", "ppi pK2", "ppip pK3", "ppip pK4", "ribose1phosp K1", "ribose5phos pK1",
     "succinate pK1", "succinate pK2"}, {"        273.15 K", "  298.15 K", "313.15 K"},
    {"I=0", "I=0.05", "I=0.10", "I=0.15", "I=0.25"}}], {4, 2}]

Out[52]//PaddedForm=
```

		273.15 K	298.15 K	313.15 K
	I=0	4.75	4.75	4.76
	I=0.05	4.59	4.59	4.58
acetate pK1	I=0.10	4.54	4.54	4.54
	I=0.15	4.52	4.51	4.51
	I=0.25	4.48	4.47	4.47
	I=0	4.52	4.20	4.03
	I=0.05	4.52	4.20	4.03
adenine pK1	I=0.10	4.52	4.20	4.03
	I=0.15	4.52	4.20	4.03
	I=0.25	4.52	4.20	4.03
	I=0	3.73	3.47	3.33
	I=0.05	3.73	3.47	3.33
adenosine pK1	I=0.10	3.73	3.47	3.33
	I=0.15	3.73	3.47	3.33
	I=0.25	3.73	3.47	3.33
	I=0	7.09	7.18	7.22
	I=0.05	6.60	6.67	6.71
adp pK1	I=0.10	6.47	6.53	6.56
	I=0.15	6.38	6.44	6.47
	I=0.25	6.27	6.33	6.35
	I=0	4.64	4.36	4.21
	I=0.05	4.32	4.02	3.87
adp pK2	I=0.10	4.23	3.93	3.77
	I=0.15	4.17	3.87	3.71
	I=0.25	4.10	3.79	3.63
	I=0	10.09	9.25	8.81
	I=0.05	10.09	9.25	8.81
ammonia pK1	I=0.10	10.09	9.25	8.81
	I=0.15	10.09	9.25	8.81
	I=0.25	10.09	9.25	8.81
	I=0	6.64	6.73	6.77
	I=0.05	6.32	6.39	6.43
amp pK1	I=0.10	6.23	6.30	6.33
	I=0.15	6.17	6.24	6.27
	I=0.25	6.10	6.16	6.19
	I=0	4.28	3.99	3.84
	I=0.05	4.12	3.82	3.66
amp pK2	I=0.10	4.07	3.77	3.62
	I=0.15	4.04	3.74	3.59
	I=0.25	4.01	3.71	3.55
	I=0	7.50	7.60	7.65
	I=0.05	6.85	6.93	6.96
atp pK1	I=0.10	6.67	6.74	6.77
	I=0.15	6.56	6.62	6.65
	I=0.25	6.41	6.46	6.49

atp pK2	I=0	4.92	4.68	4.55
	I=0.05	4.43	4.17	4.03
	I=0.10	4.30	4.03	3.89
	I=0.15	4.21	3.94	3.80
	I=0.25	4.10	3.83	3.68
citrate pK1	I=0	6.45	6.39	6.36
	I=0.05	5.96	5.89	5.85
	I=0.10	5.83	5.75	5.70
	I=0.15	5.74	5.66	5.61
	I=0.25	5.63	5.54	5.49
citrate pK2	I=0	4.80	4.76	4.74
	I=0.05	4.47	4.42	4.39
	I=0.10	4.38	4.33	4.30
	I=0.15	4.33	4.27	4.23
	I=0.25	4.25	4.19	4.15
co2tot pK1	I=0	10.57	10.33	10.20
	I=0.05	10.24	9.99	9.86
	I=0.10	10.16	9.90	9.76
	I=0.15	10.10	9.84	9.70
	I=0.25	10.02	9.76	9.62
co2tot pK2	I=0	6.49	6.37	6.30
	I=0.05	6.33	6.20	6.13
	I=0.10	6.28	6.15	6.08
	I=0.15	6.25	6.12	6.05
	I=0.25	6.22	6.08	6.01
fructose16phos pK1	I=0	6.62	6.65	6.67
	I=0.05	5.98	5.98	5.97
	I=0.10	5.80	5.79	5.78
	I=0.15	5.68	5.67	5.66
	I=0.25	5.53	5.52	5.50
fructose16phos pK2	I=0	6.02	6.05	6.06
	I=0.05	5.54	5.54	5.55
	I=0.10	5.40	5.41	5.40
	I=0.15	5.32	5.32	5.31
	I=0.25	5.20	5.20	5.19
fructose6phos pK1	I=0	6.24	6.27	6.29
	I=0.05	5.92	5.94	5.94
	I=0.10	5.83	5.84	5.85
	I=0.15	5.77	5.78	5.79
	I=0.25	5.70	5.70	5.70
fumarate pK1	I=0	4.56	4.60	4.63
	I=0.05	4.23	4.27	4.28
	I=0.10	4.14	4.17	4.19
	I=0.15	4.09	4.11	4.13
	I=0.25	4.01	4.03	4.04
fumarate pK2	I=0	3.10	3.09	3.09
	I=0.05	2.94	2.93	2.92
	I=0.10	2.89	2.88	2.87
	I=0.15	2.87	2.85	2.84
	I=0.25	2.83	2.81	2.80
glucose6phos pK1	I=0	6.39	6.42	6.44
	I=0.05	6.07	6.08	6.09
	I=0.10	5.98	5.99	6.00
	I=0.15	5.92	5.93	5.93
	I=0.25	5.85	5.85	5.85
glycerol3phos pK1	I=0	6.64	6.67	6.69
	I=0.05	6.32	6.33	6.34
	I=0.10	6.23	6.24	6.25
	I=0.15	6.17	6.18	6.18
	I=0.25	6.10	6.10	6.10
h2saq pK1	I=0	13.73	12.92	12.49
	I=0.05	13.40	12.58	12.14
	I=0.10	13.32	12.49	12.05
	I=0.15	13.26	12.43	11.99
	I=0.25	13.18	12.35	11.91

h2saq pK2	I=0	7.35	6.99	6.81
	I=0.05	7.18	6.82	6.63
	I=0.10	7.14	6.78	6.59
	I=0.15	7.11	6.75	6.56
	I=0.25	7.07	6.71	6.52
idp pK1	I=0	10.02	9.56	9.32
	I=0.05	9.37	8.89	8.63
	I=0.10	9.20	8.70	8.44
	I=0.15	9.08	8.58	8.32
	I=0.25	8.93	8.43	8.15
idp pK2	I=0	7.09	7.18	7.23
	I=0.05	6.60	6.67	6.71
	I=0.10	6.47	6.54	6.57
	I=0.15	6.39	6.45	6.47
	I=0.25	6.27	6.33	6.35
imp pK1	I=0	10.20	9.63	9.34
	I=0.05	9.71	9.13	8.82
	I=0.10	9.58	8.99	8.68
	I=0.15	9.49	8.90	8.59
	I=0.25	9.38	8.78	8.47
imp pK2	I=0	6.64	6.73	6.78
	I=0.05	6.32	6.39	6.43
	I=0.10	6.23	6.30	6.34
	I=0.15	6.18	6.24	6.27
	I=0.25	6.10	6.16	6.19
inosine pK1	I=0	9.39	8.96	8.73
	I=0.05	9.23	8.79	8.56
	I=0.10	9.19	8.74	8.51
	I=0.15	9.16	8.71	8.48
	I=0.25	9.12	8.67	8.44
itp pK1	I=0	10.47	10.09	9.89
	I=0.05	9.67	9.25	9.02
	I=0.10	9.44	9.02	8.79
	I=0.15	9.30	8.87	8.63
	I=0.25	9.11	8.67	8.43
itp pK2	I=0	7.50	7.60	7.65
	I=0.05	6.85	6.93	6.96
	I=0.10	6.67	6.74	6.77
	I=0.15	6.56	6.62	6.65
	I=0.25	6.41	6.46	6.49
malate pK1	I=0	5.26	5.26	5.26
	I=0.05	4.94	4.92	4.91
	I=0.10	4.85	4.83	4.82
	I=0.15	4.79	4.77	4.76
	I=0.25	4.72	4.69	4.68
mannose6phos pK1	I=0	6.41	6.44	6.46
	I=0.05	6.09	6.10	6.11
	I=0.10	6.00	6.01	6.01
	I=0.15	5.94	5.95	5.95
	I=0.25	5.87	5.87	5.87
pep pK1	I=0	6.97	7.00	7.02
	I=0.05	6.49	6.50	6.50
	I=0.10	6.35	6.36	6.36
	I=0.15	6.27	6.27	6.26
	I=0.25	6.15	6.15	6.14
pi pK1	I=0	7.28	7.22	7.19
	I=0.05	6.95	6.88	6.84
	I=0.10	6.86	6.79	6.75
	I=0.15	6.81	6.73	6.69
	I=0.25	6.73	6.65	6.61
ppi pK1	I=0	9.48	9.46	9.45
	I=0.05	8.84	8.79	8.76
	I=0.10	8.66	8.60	8.57
	I=0.15	8.54	8.48	8.45
	I=0.25	8.39	8.33	8.28

	I=0	6.73	6.72	6.71
	I=0.05	6.24	6.21	6.20
ppi pK2	I=0.10	6.11	6.08	6.05
	I=0.15	6.02	5.99	5.96
	I=0.25	5.91	5.87	5.84
	I=0	2.18	2.26	2.30
	I=0.05	1.86	1.92	1.96
ppip pK3	I=0.10	1.77	1.83	1.86
	I=0.15	1.71	1.77	1.80
	I=0.25	1.63	1.69	1.72
	I=0	0.68	0.83	0.91
	I=0.05	0.52	0.66	0.73
ppip pK4	I=0.10	0.48	0.62	0.69
	I=0.15	0.45	0.59	0.66
	I=0.25	0.41	0.55	0.62
	I=0	6.51	6.69	6.78
	I=0.05	6.18	6.35	6.44
ribose1phosp K1	I=0.10	6.10	6.26	6.34
	I=0.15	6.04	6.20	6.28
	I=0.25	5.96	6.12	6.20
	I=0	6.51	6.69	6.78
	I=0.05	6.18	6.35	6.44
ribose5phos pK1	I=0.10	6.10	6.26	6.34
	I=0.15	6.04	6.20	6.28
	I=0.25	5.96	6.12	6.20
	I=0	5.64	5.64	5.64
	I=0.05	5.32	5.30	5.29
succinate pK1	I=0.10	5.23	5.21	5.20
	I=0.15	5.17	5.15	5.13
	I=0.25	5.10	5.07	5.05
	I=0	4.26	4.21	4.18
	I=0.05	4.10	4.04	4.01
succinate pK2	I=0.10	4.06	3.99	3.96
	I=0.15	4.03	3.96	3.93
	I=0.25	3.99	3.92	3.89

In using these pKs it is important to remember that the acid dissociation constants are written in terms of $[H^+]$, rather than $\gamma(H^+)[H^+]$.

1.8 Standard Thermodynamic Properties for Acid Dissociations

The standard Gibbs energy of a chemical reaction is determined by the standard enthalpy of reaction, standard entropy of reaction, and temperature, as shown by

$$\Delta_r G_k° = \Delta_r H_k° - T\Delta_r S_k° \tag{1.8-1}$$

Therefore it is of interest to consider the relative importances of $\Delta_r H_k°$ and $\Delta_r S_k°$ in acid dissociation. This can be done for the 27 reactants in BasicBiochemData3 for which $\Delta_f G_j°(298.15\ K, I=0)$ and $\Delta_f H_j°(298.15\ K, I=0)$ are known for all species with significant concentrations in the pH range 5 to 9. Calculations are made only at 298.15 K because $\Delta_f H_j°(T, I=0)$ and $\Delta_f S_j°(T, I=0)$ do not change much in the range 273.15 K to 313.15 K. The program **calcpK** given Section 1.6 is readily modified to yield $\{\Delta_f G°_k(298.15\ K, I), \Delta_f H°_k(298.15\ K, I), \Delta_f S°_k(298.15\ K, I)\}$.

These three standard properties can be calculated from the species matrices in BasicBiochemData3. The following program first calculates $lnK(298.15\ K, I=0)$ and then adjusts for the ionic strength according to equation 1.6-1. $\Delta_f G_k°$ is calculated using equation 1.3-7. $\Delta_r H_k°$ is calculated using equation 1.3-6, and $\Delta_r S_k°$ is calculated using equation 1.8-1. The program **calcGHSdiss** (14) requires specification of the number of the pK, starting with the highest pK as number 1.

```
In[53]:=
calcGHSdiss[speciesmat_, no_, is_] := Module[{lnkzero, sigmanuzsq, lnK, dGI, dHzero, dHI, dSI},
  (*Calculates {dGI,dHI,dSI} for a weak acid at 298.15 K at specified ionic
    strengths (is) when the number no of the pK is specified. pKs are numbered 1,2,
    3,... from the highest pK to the lowest pK,but the highest pK for a weak acid
    may be omitted if it is outside of the range 5 t0 9. The Gibbs energy and enthalpy
    are given in kJ mol^-1, and the entropy is given in J K^-1 mol^-1.  For H3PO4,
    pK1=calcGHSdiss[pisp,1,{0}]={{41.2},{3.6},{-126.111}}.*)
  lnkzero = (speciesmat[[no + 1, 1]] - speciesmat[[no, 1]]) / (8.31451 * 0.29815);
  sigmanuzsq = speciesmat[[no, 3]] ^ 2 - speciesmat[[no + 1, 3]] ^ 2 + 1;
  lnK = lnkzero + (1.17582 * is ^ 0.5 * sigmanuzsq) / (1 + 1.6 * is ^ 0.5);
  (*Calculate the Gibbs energy of acid dissociation.*)
  dGI = -8.31451 * .29815 * lnK;
  (*Calculate the enthalpy of dissociation.*)
  dHzero = speciesmat[[no, 2]] - speciesmat[[no + 1, 2]];
  sigmanuzsq = speciesmat[[no, 3]] ^ 2 - speciesmat[[no + 1, 3]] ^ 2 + 1;
  dHI = dHzero - (1.4775 * is ^ 0.5 * sigmanuzsq) / (1 + 1.6 * is ^ 0.5);
  (*Calculate the entropy of dissociation.*)
  dSI = (dHI - dGI) / .29815;
  Transpose[{dGI, dHI, dSI}]]
```

For example, the standard thermodynamic properties for pK_1 and pK_2 of atp at 298.15 K and three ionic strengths can be calculated as follows:

```
In[54]:=
PaddedForm[TableForm[calcGHSdiss[atpsp, 1, {0, .1, .25}], TableHeadings →
  {{"I=0", "I=0.10", "I=0.25"}, {"ΔᵣG°/kJ mol⁻¹", "ΔᵣH°/kJ mol⁻¹", "ΔᵣS°/J K⁻¹ mol⁻¹"}}], {5, 2}]
```

```
Out[54]//PaddedForm=
```

	$\Delta_r G°/kJ\ mol^{-1}$	$\Delta_r H°/kJ\ mol^{-1}$	$\Delta_r S°/J\ K^{-1}\ mol^{-1}$
I=0	43.38	-6.30	-166.63
I=0.10	38.48	-8.78	-158.53
I=0.25	36.90	-9.58	-155.91

```
In[55]:=
PaddedForm[TableForm[calcGHSdiss[atpsp, 2, {0, .1, .25}], TableHeadings →
  {{"I=0", "I=0.10", "I=0.25"}, {"ΔᵣG°/kJ mol⁻¹", "ΔᵣH°/kJ mol⁻¹", "ΔᵣS°/J K⁻¹ mol⁻¹"}}], {5, 2}]
```

```
Out[55]//PaddedForm=
```

	$\Delta_r G°/kJ\ mol^{-1}$	$\Delta_r H°/kJ\ mol^{-1}$	$\Delta_r S°/J\ K^{-1}\ mol^{-1}$
I=0	26.70	15.00	-39.24
I=0.10	23.03	13.14	-33.17
I=0.25	21.84	12.54	-31.21

However, it is more convenient to have a program that has a single argument and calculates the properties for all the acid dissociations of a reactant. The following program derives functions of ionic strength at 298.15 K.

```
In[56]:=
calcGHSdissfn[speciesmat_] := Module[
   {glist, hlist, zlist, nHlist, glistis, ghydionis, gibbs, hlistis, hhydionis, enthalpy, entropy},
   (*This program derives the functions of ionic strength that yield {G,H,S} functions
      of ionic strength at 298.15 K for weak acids.  The first function of ionic
      strength is for the acid with the fewest hydrogen atoms.  The program has
      a single argument so that it can be used with Map.  The functions can be
      evaluated by use of calcGHSdissfn[atpsp]/.is→{0,.1,.25}, for example.*)
   {glist, hlist, zlist, nHlist} = Transpose[speciesmat];
 glistis =
   Table[glist[[i]] - 2.91482 * zlist[[i]]^2 * is^.5 / (1 + 1.6 * is^.5), {i, 1, Length[zlist]}];
   ghydionis = -2.91482 * is^.5 / (1 + 1.6 * is^.5);
   gibbs = Table[((glistis[[i - 1]] - glistis[[i]] + ghydionis)), {i, 2, Length[zlist]}];
   hlistis =
   Table[hlist[[i]] - 1.4775 * zlist[[i]]^2 * is^.5 / (1 + 1.6 * is^.5), {i, 1, Length[zlist]}];
   hhydionis = -1.4775 * is^.5 / (1 + 1.6 * is^.5);
   enthalpy = Table[((hlistis[[i - 1]] - hlistis[[i]] + hhydionis)), {i, 2, Length[zlist]}];
   entropy = (enthalpy - gibbs) / .29815;
   Transpose[{gibbs, enthalpy, entropy}]]]
```

The use of this program is illustrated by applying it to atp.

```
In[57]:=
PaddedForm[TableForm[calcGHSdissfn[atpsp] /. is → {0, .1, .25},
   TableHeadings → {{"atp pK1", "atp pK2"}, {"            Δ_rG°/kJ mol^-1",
      "   Δ_rH°/kJ mol^-1", "  Δ_rS°/J K^-1 mol^-1"}, {"I=0", "I=0.10", "I=0.25"}}], {5, 2}]
```

```
Out[57]//PaddedForm=
```

		$\Delta_r G°/kJ\ mol^{-1}$	$\Delta_r H°/kJ\ mol^{-1}$	$\Delta_r S°/J\ K^{-1}\ mol^{-1}$
atp pK1	I=0	43.38	−6.30	−166.63
	I=0.10	38.48	−8.78	−158.53
	I=0.25	36.90	−9.58	−155.91
atp pK2	I=0	26.70	15.00	−39.24
	I=0.10	23.03	13.14	−33.17
	I=0.25	21.84	12.54	−31.21

Now we use **Map** to apply this program to all 27 reactants for which $\Delta_f G_j°(298.15\ K, I=0)$ and $\Delta_f H_j°(298.15\ K, I=0)$ are known for all species.

```
In[58]:=
listHreactants = {acetatesp, adeninesp, adenosinesp, adpsp, ammoniasp, ampsp,
   atpsp, citratesp, co2totsp, fructose16phossp, fructose6phossp, fumaratesp,
   glucose6phossp, glycerol3phossp, h2saqsp, idpsp, impsp, inosinesp, itpsp, malatesp,
   mannose6phossp, pepsp, pisp, ppisp, ribose1phossp, ribose5phossp, succinatesp};
```

Table 1.5 $\Delta_r G°$, $\Delta_r H°$, and $\Delta_r S°$ for acid dissociations at 298.15 K and five ionic strengths

```
In[59]:=
PaddedForm[
 TableForm[Flatten[Join[Map[calcGHSdissfn, listHreactants]]] /. t → {273.15, 298.15, 313.15} /.
   is → {0, .05, .1, .15, .25}, 1],
  TableHeadings → {{"acetate pK1", "adenine pK1", "adenosine pK1", "adp pK1",
    "adp pK2", "ammonia pK1", "amp pK1", "amp pK2", "atp pK1", "atp pK2",
    "citrate pK1", "citrate pK2", "co2tot pK1", "co2tot pK2", "fructose16phos pK1",
    "fructose16phos pK2", "fructose6phos pK1", "fumarate pK1", "fumarate pK2",
    "glucose6phos pK1", "glycerol3phos pK1", "h2saq pK1", "h2saq pK2", "idp pK1",
    "idp pK2", "imp pK1", "imp pK2", "inosine pK1", "itp pK1", "itp pK2", "malate pK1",
    "mannose6phos pK1", "pep pK1", "pi pK1", "ppi pK1", "ppi pK2", "ppip pK3",
    "ppip pK4", "ribose1phosp K1", "ribose5phos pK1", "succinate pK1", "succinate pK2"},
   {"          Δ_rG°/kJ mol⁻¹", "   Δ_rH°/kJ mol⁻¹", "Δ_rS°/J K⁻¹mol⁻¹"},
   {"I=0", "I=0.05", "I=0.10", "I=0.15", "I=0.25"}}, TableSpacing → {1, 1}], {5, 2}]
```

Out[59]//PaddedForm=

		$\Delta_r G°/\text{kJ mol}^{-1}$	$\Delta_r H°/\text{kJ mol}^{-1}$	$\Delta_r S°/\text{J K}^{-1}\text{mol}^{-1}$
	I=0	27.14	-0.25	-91.87
	I=0.05	26.18	-0.74	-90.28
acetate pK1	I=0.10	25.92	-0.87	-89.84
	I=0.15	25.75	-0.96	-89.56
	I=0.25	25.52	-1.07	-89.19
	I=0	23.97	20.10	-12.98
	I=0.05	23.97	20.10	-12.98
adenine pK1	I=0.10	23.97	20.10	-12.98
	I=0.15	23.97	20.10	-12.98
	I=0.25	23.97	20.10	-12.98
	I=0	19.78	16.40	-11.34
	I=0.05	19.78	16.40	-11.34
adenosine pK1	I=0.10	19.78	16.40	-11.34
	I=0.15	19.78	16.40	-11.34
	I=0.25	19.78	16.40	-11.34
	I=0	40.97	-5.60	-156.20
	I=0.05	38.09	-7.06	-151.43
adp pK1	I=0.10	37.30	-7.46	-150.12
	I=0.15	36.79	-7.72	-149.28
	I=0.25	36.11	-8.06	-148.16
	I=0	24.88	17.60	-24.42
	I=0.05	22.96	16.63	-21.24
adp pK2	I=0.10	22.43	16.36	-20.37
	I=0.15	22.09	16.19	-19.81
	I=0.25	21.64	15.96	-19.06
	I=0	52.81	52.22	-1.98
	I=0.05	52.81	52.22	-1.98
ammonia pK1	I=0.10	52.81	52.22	-1.98
	I=0.15	52.81	52.22	-1.98
	I=0.25	52.81	52.22	-1.98
	I=0	38.41	-5.40	-146.94
	I=0.05	36.49	-6.37	-143.76
amp pK1	I=0.10	35.96	-6.64	-142.89
	I=0.15	35.62	-6.81	-142.33
	I=0.25	35.17	-7.04	-141.58
	I=0	22.77	18.10	-15.66
	I=0.05	21.81	17.61	-14.08
amp pK2	I=0.10	21.55	17.48	-13.64
	I=0.15	21.38	17.39	-13.36
	I=0.25	21.15	17.28	-12.99
	I=0	43.38	-6.30	-166.63
	I=0.05	39.54	-8.25	-160.28
atp pK1	I=0.10	38.48	-8.78	-158.53
	I=0.15	37.80	-9.13	-157.41
	I=0.25	36.90	-9.58	-155.91

atp pK2	I=0	26.70	15.00	-39.24
	I=0.05	23.82	13.54	-34.48
	I=0.10	23.03	13.14	-33.17
	I=0.15	22.52	12.88	-32.33
	I=0.25	21.84	12.54	-31.21
citrate pK1	I=0	36.49	3.35	-111.15
	I=0.05	33.61	1.89	-106.39
	I=0.10	32.82	1.49	-105.08
	I=0.15	32.31	1.23	-104.24
	I=0.25	31.63	0.89	-103.12
citrate pK2	I=0	27.15	2.42	-82.94
	I=0.05	25.23	1.45	-79.77
	I=0.10	24.70	1.18	-78.90
	I=0.15	24.36	1.01	-78.33
	I=0.25	23.91	0.78	-77.59
co2tot pK1	I=0	58.96	14.85	-147.95
	I=0.05	57.04	13.88	-144.77
	I=0.10	56.51	13.61	-143.90
	I=0.15	56.17	13.44	-143.33
	I=0.25	55.72	13.21	-142.59
co2tot pK2	I=0	36.34	7.64	-96.26
	I=0.05	35.38	7.15	-94.67
	I=0.10	35.12	7.02	-94.24
	I=0.15	34.95	6.93	-93.95
	I=0.25	34.72	6.82	-93.58
fructose16phos pK1	I=0	37.96	-1.80	-133.36
	I=0.05	34.12	-3.75	-127.00
	I=0.10	33.06	-4.28	-125.26
	I=0.15	32.38	-4.63	-124.13
	I=0.25	31.48	-5.08	-122.64
fructose16phos pK2	I=0	34.53	-1.80	-121.85
	I=0.05	31.65	-3.26	-117.09
	I=0.10	30.86	-3.66	-115.78
	I=0.15	30.35	-3.92	-114.93
	I=0.25	29.67	-4.26	-113.82
fructose6phos pK1	I=0	35.80	-1.80	-126.11
	I=0.05	33.88	-2.77	-122.94
	I=0.10	33.35	-3.04	-122.06
	I=0.15	33.01	-3.21	-121.50
	I=0.25	32.56	-3.44	-120.75
fumarate pK1	I=0	26.27	-2.93	-97.94
	I=0.05	24.35	-3.90	-94.76
	I=0.10	23.82	-4.17	-93.89
	I=0.15	23.48	-4.34	-93.33
	I=0.25	23.03	-4.57	-92.58
fumarate pK2	I=0	17.66	0.42	-57.82
	I=0.05	16.70	-0.07	-56.24
	I=0.10	16.44	-0.20	-55.80
	I=0.15	16.27	-0.29	-55.52
	I=0.25	16.04	-0.40	-55.15
glucose6phos pK1	I=0	36.65	-1.80	-128.96
	I=0.05	34.73	-2.77	-125.79
	I=0.10	34.20	-3.04	-124.91
	I=0.15	33.86	-3.21	-124.35
	I=0.25	33.41	-3.44	-123.61
glycerol3phos pK1	I=0	38.08	-1.80	-133.76
	I=0.05	36.16	-2.77	-130.58
	I=0.10	35.63	-3.04	-129.71
	I=0.15	35.29	-3.21	-129.15
	I=0.25	34.84	-3.44	-128.40
h2saq pK1	I=0	73.72	50.70	-77.21
	I=0.05	71.80	49.73	-74.03
	I=0.10	71.27	49.46	-73.16
	I=0.15	70.93	49.29	-72.60
	I=0.25	70.48	49.06	-71.85

	I=0	39.91	22.10	-59.74
	I=0.05	38.95	21.61	-58.15
h2saq pK2	I=0.10	38.69	21.48	-57.71
	I=0.15	38.52	21.39	-57.43
	I=0.25	38.29	21.28	-57.06
	I=0	54.57	28.70	-86.77
	I=0.05	50.73	26.75	-80.42
idp pK1	I=0.10	49.67	26.22	-78.67
	I=0.15	48.99	25.87	-77.55
	I=0.25	48.09	25.42	-76.06
	I=0	40.98	-5.60	-156.23
	I=0.05	38.10	-7.06	-151.47
idp pK2	I=0.10	37.31	-7.46	-150.16
	I=0.15	36.80	-7.72	-149.31
	I=0.25	36.12	-8.06	-148.20
	I=0	54.99	35.10	-66.71
	I=0.05	52.11	33.64	-61.95
imp pK1	I=0.10	51.32	33.24	-60.64
	I=0.15	50.81	32.98	-59.79
	I=0.25	50.13	32.64	-58.68
	I=0	38.42	-5.40	-146.97
	I=0.05	36.50	-6.37	-143.80
imp pK2	I=0.10	35.97	-6.64	-142.92
	I=0.15	35.63	-6.81	-142.36
	I=0.25	35.18	-7.04	-141.62
	I=0	51.13	27.10	-80.60
	I=0.05	50.17	26.61	-79.01
inosine pK1	I=0.10	49.91	26.48	-78.57
	I=0.15	49.74	26.39	-78.29
	I=0.25	49.51	26.28	-77.92
	I=0	57.59	23.95	-112.83
	I=0.05	52.79	21.52	-104.89
itp pK1	I=0.10	51.47	20.85	-102.71
	I=0.15	50.62	20.42	-101.30
	I=0.25	49.49	19.85	-99.44
	I=0	43.38	-6.35	-166.80
	I=0.05	39.54	-8.30	-160.44
itp pK2	I=0.10	38.48	-8.83	-158.70
	I=0.15	37.80	-9.18	-157.57
	I=0.25	36.90	-9.63	-156.08
	I=0	30.02	0.16	-100.15
	I=0.05	28.10	-0.81	-96.98
malate pK1	I=0.10	27.57	-1.08	-96.10
	I=0.15	27.23	-1.25	-95.54
	I=0.25	26.78	-1.48	-94.79
	I=0	36.76	-1.81	-129.36
	I=0.05	34.84	-2.78	-126.19
mannose6phos pK1	I=0.10	34.31	-3.05	-125.32
	I=0.15	33.97	-3.22	-124.75
	I=0.25	33.52	-3.45	-124.01
	I=0	39.96	-1.80	-140.06
	I=0.05	37.08	-3.26	-135.30
pep pK1	I=0.10	36.29	-3.66	-133.99
	I=0.15	35.78	-3.92	-133.15
	I=0.25	35.10	-4.26	-132.03
	I=0	41.20	3.60	-126.11
	I=0.05	39.28	2.63	-122.94
pi pK1	I=0.10	38.75	2.36	-122.06
	I=0.15	38.41	2.19	-121.50
	I=0.25	37.96	1.96	-120.75
	I=0	54.00	1.40	-176.42
	I=0.05	50.16	-0.55	-170.07
ppi pK1	I=0.10	49.10	-1.08	-168.32
	I=0.15	48.42	-1.43	-167.20
	I=0.25	47.52	-1.88	-165.71

	I=0	38.35	0.50	-126.95
	I=0.05	35.47	-0.96	-122.19
ppi pK2	I=0.10	34.68	-1.36	-120.88
	I=0.15	34.17	-1.62	-120.03
	I=0.25	33.49	-1.96	-118.91
	I=0	12.90	-5.00	-60.04
	I=0.05	10.98	-5.97	-56.86
ppip pK3	I=0.10	10.45	-6.24	-55.99
	I=0.15	10.11	-6.41	-55.43
	I=0.25	9.66	-6.64	-54.68
	I=0	4.74	-9.20	-46.75
	I=0.05	3.78	-9.69	-45.17
ppip pK4	I=0.10	3.52	-9.82	-44.73
	I=0.15	3.35	-9.91	-44.45
	I=0.25	3.12	-10.02	-44.08
	I=0	38.18	-11.30	-165.96
	I=0.05	36.26	-12.27	-162.78
ribose1phosp K1	I=0.10	35.73	-12.54	-161.91
	I=0.15	35.39	-12.71	-161.35
	I=0.25	34.94	-12.94	-160.60
	I=0	38.18	-11.30	-165.96
	I=0.05	36.26	-12.27	-162.78
ribose5phos pK1	I=0.10	35.73	-12.54	-161.91
	I=0.15	35.39	-12.71	-161.35
	I=0.25	34.94	-12.94	-160.60
	I=0	32.18	0.16	-107.40
	I=0.05	30.26	-0.81	-104.22
succinate pK1	I=0.10	29.73	-1.08	-103.35
	I=0.15	29.39	-1.25	-102.78
	I=0.25	28.94	-1.48	-102.04
	I=0	24.02	3.36	-69.29
	I=0.05	23.06	2.87	-67.71
succinate pK2	I=0.10	22.80	2.74	-67.27
	I=0.15	22.63	2.65	-66.99
	I=0.25	22.40	2.54	-66.62

The standard entropies of dissociation are always negative because the products have a lower entropy than the weak acid. The lower entropy of the products is a result of the orientation of water molecules around the ions that are produced by the dissociation. $\Delta_r G°$ decreases when the ionic strength is raised, or remains constant for ammonia, adenine, and adenosine (HA^+ acids). $\Delta_r S°$ always decreases when the ionic strength is raised. $\Delta_r H°$ decreases when the ionic strength is raised, except for two cases. The ionic strength effects can be attributed to the shielding of charged groups that reduces the degree of hydration. The most negative $\Delta_r S°$ are observed when the charged groups are adjoining as in $Hatp^{3-} = H^+ + atp^{4-}$. The standard entropy of dissociation of ammonia is quite small (-2 $J K^{-1}$ mol^{-1}) because there is a single positive ion on both sides of the reaction. The same comment applies to the adenine dissociation in atp, although the effect is not so striking. When $\Delta_r H°$ is negative, heat is evolved and raising the temperature weakens the acid. When $\Delta_r H°$ is positive, heat is absorbed and raising the temperature strengthens the acid.

1.9 Discussion

This chapter has been about chemical thermodynamics in the sense that all the calculations have dealt with species. In the next chapter we will see that in considering biochemical reaction systems, it is advantageous to deal with sums of species (like, $Hatp^{3-}$ and atp^{4-}), which are referred to as reactants. In the third chapter, we will see that the discussion of the thermodynamics of enzyme-catalyzed reactions at specified pH involves thermodynamic properties that are different from $\Delta_r G°$, $\Delta_r H°$, and $\Delta_r S°$. These calculations may be very complicated, and so it is convenient to use *Mathematica*. In *Mathematica*, related programs and databases can be put in packages, and we have seen an example of that with BasicBiochemData3. Akers and Goldberg (15) wrote a package BioEqCalc that brings together programs for making calculations on activity coefficients, effects of ionic strength, etc. The predecessor to BasicBiochemData3 is BasicBiochemData2 (16) that gives more explanatory material and programs.

References

1. J. W. Gibbs, The Scientific Papers of J. Willard Gibbs,Vol. 1, Thermodynamics, Dover, New York, 1961.
2. R. J. Silbey, R. A. Alberty, and M. Bawendi, Physical Chemistry, Wiley, Hoboken, NJ, 2005.
3. R. A. Alberty, Thermodynamics of Biochemical Reactions, Wiley, Hoboken, NJ, 2003.
4. D. D. Wagman, W. H. Evans, V. B. Parker, R. H. Schumm, I. Halow, S. M. Bailey, K. L. Churney, and R. L. Nuttall, The NBS tables of chemical thermodynamic properties, J. Phys. Chem. Ref. Data, 11, Supplement 2 (1982).
5. R. A. Alberty and R. N. Goldberg, Calculation of thermodynamic formation properties for the ATP series at specified pH and pMg, Biochem. 31, 10610-10615 (1992).
6. J. Boerio-Goates, M. R. Francis, R. N. Goldberg, M. A. V. Ribeiro da Silva, M. D. M. C. Ribeiro da Silva, and Y. Tewari, Thermochemistry of adenosine, J. Chem. Thermo. 33, 929-947 (2001).
7. S. Wolfram, The *Mathematica* Book, Third Ed., Cambridge University Press, 1999.
8. R. A. Alberty, BasicBiochemData3, 2005.

   http : // library.wolfram.com / infocenter / MathSource / 5704

9. R. N. Goldberg and Y. Tewari, Thermodynamic and transport properties of carbohydrates and their monophosphates: The pentoses and hexoses, J. Phys. Chem. Ref. Data 18, 809-880 (1989).
10. E. C. W. Clarke and D. N. Glew, Evaluation of Debye-Hückel limiting slopes for water between 0 and 50 C, Chem. Soc. 1, 76,1911 (1980).
11. K. S. Pitzer, Activity Coefficients in Electrolyte Solutions, CRC Press, Boca Raton, FL, 1991.
12. R. A. Alberty, Effect of temperature on standard transformed Gibbs energies of formation of reactants at specified pH and ionic strength and apparent equilibrium constants of biochemical reactions, J. Phys. Chem. 105 B, 7865-7870 (2001). (Supplementary Information is available.)
13. R. A. Alberty, Standard molar entropies, standard entropies of formation, and standard transformed entropies of formation in the thermodynamics of enzyme-catalyzed reactions, J. Chem. Thermodyn., in press.
14. R. A. Alberty, Thermodynamic properties of weak acids involved in enzyme-catalyzed reactions, in preparation.
15. D. L. Akers and R. N. Goldberg, BioEqCalc: A package for performing equilibrium calculations on biochemical reactions, Mathematica J., 8, 86-113 (2001).

   http://www.mathematica-journal.com/issue/v8i1/

16. R. A. Alberty, BasicBiochemData2: Data and Programs for Biochemical Thermodynamics (2003).

   http : // library.wolfram.com / infocenter / MathSource / 797

Chapter 2 Introduction to Apparent Equilibrium Constants

2.1 Average Binding of Hydrogen Ions by ATP

Two types of equilibrium constant expressions are needed in biochemistry. Enzyme-catalyzed reactions can be described in terms of chemical reactions of species, and this is especially important in discussing mechanisms and the detailed consideration of **chemical reactions**. On the other hand, in discussing metabolism a more global view of enzyme-catalyzed reactions is needed. In this more global view it is convenient to consider ATP as an entity with its own thermodynamic properties. As a first example, we consider the binding of hydrogen ions by ATP. Since ATP is made up of three species in the physiological pH range, its concentration is given by

$$[ATP] = [ATP^{4-}] + [HATP^{3-}] + [H_2ATP^{2-}] \qquad (2.1-1)$$

When the acid dissociations are at equilibrium, substituting the expressions for the two acid dissociation constants of ATP yields

$$[ATP] = [ATP^{4-}]\left(1 + \frac{[H^+]}{K_{1\,ATP}} + \frac{[H^+]^2}{K_{1\,ATP}\,K_{2\,ATP}}\right) \qquad (2.1-2)$$

$$= [ATP^{4-}](1 + 10^{pK1ATP-pH} + 10^{pK1ATP+pK2ATP-2\,pH})$$

where the factor multiplying $[ATP^{4-}]$ is referred to as a **binding polynomial** (1,2,3) and is often represented by P.   This equation shows that the equilibrium mole fraction r of ATP^{4-} in ATP is given by

$$r(ATP^{4-}) = \frac{[ATP^{4-}]}{[ATP]} = \frac{1}{1+10^{pK1ATP-pH}+10^{pK1ATP+pK2ATP-2\,pH}}$$ (2.1-3)

The mole fractions of the other two species are given by

$$r(HATP^{3-}) = \frac{[HATP^{3-}]}{[ATP]} = \frac{10^{pK1ATP-pH}}{1+10^{pK1ATP-pH}+10^{pK1ATP+pK2ATP-2\,pH}}$$ (2.1-4)

$$r(H_2ATP^{4-}) = \frac{[H_2ATP^{4-}]}{[ATP]} = \frac{10^{pK1ATP+pK2ATP-2\,pH}}{1+10^{pK1ATP-pH}+10^{pK1ATP+pK2ATP-2\,pH}}$$ (2.1-5)

These equilibrium mole fractions can be plotted versus pH at 298.15 K and 0.25 M ionic strength by using the pKs for ATP. The binding polynomial for ATP is represented as follows in *Mathematica*:

```
In[2]:=  Off[General::spell1];
         Off[General::spell];
```

$In[4]:=$ **patp = 1 + $10^{pK1ATP-pH}$ + $10^{pK1ATP+pK2ATP-2\,pH}$** ;

The equilibrium mole fractions of the three species of ATP are given in *Mathematica* by

$In[5]:=$ **r1 = 1 / patp;**

$In[6]:=$ **r2 = ($10^{pK1ATP-pH}$) / patp;**

$In[7]:=$ **r3 = $10^{-2\,pH+pK1ATP+pK2ATP}$ / patp;**

The two pKs for ATP at 298.15 K and 0.25 M strength are given in Table 1.3. They are put into this notebook as follows:

$In[8]:=$ **pK1ATP = 6.47;**

$In[9]:=$ **pK2ATP = 3.83;**

The equilibrium mole fractions of these three species can be plotted as a function of pH.

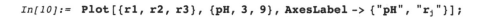

$In[10]:=$ **Plot[{r1, r2, r3}, {pH, 3, 9}, AxesLabel -> {"pH", "r$_j$"}];**

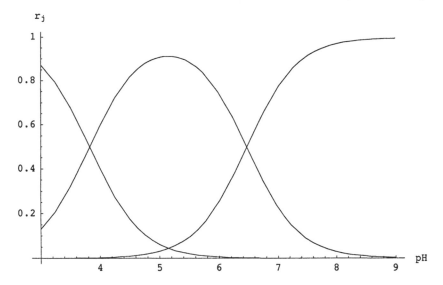

Figure 2.1 Equilibrium mole fractions of H_2ATP^{2-}, $HATP^{3-}$, and ATP^{4-} at 298.15 K and 0.25 M ionic strength as functions of pH.

The average number $\overline{N}_H$ of hydrogen ions bound by ATP beyond those bound at pH 9 is given as a function of pH by $\overline{N}_H = r_2 + 2r_3$.

$In[11]:=$ **$\overline{N}_H$ = (10$^{pK1ATP-pH}$ + 2 * 10$^{-2\,pH+pK1ATP+pK2ATP}$) / patp;**

In *Mathematica* it is very easy to plot $\overline{N}_H$ as a function of pH.

$In[12]:=$ **Plot[$\overline{N}_H$, {pH, 3, 9}, AxesLabel -> {"pH", "$\overline{N}_H$"}];**

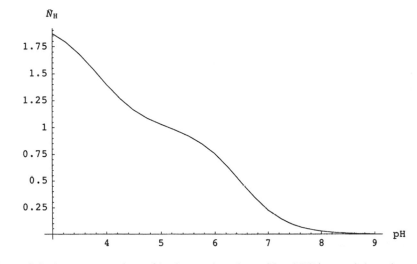

Figure 2.2 Average number of hydrogen ions bound by ATP beyond those bound above pH 9 at 298.15 K and 0.25 M ionic strength.

There is another way this plot can be obtained, and that is by taking the derivative of $\ln P$ with respect to pH (1,2,3).

$$\bar{N}_H = \frac{-1}{\ln(10)} \frac{d\ln P}{dpH} \qquad (2.1\text{-}6)$$

It will be clear later how this equation appears logically, but it is evident here that taking the derivative of $\ln P$ with respect to pH eliminates the first term in P and puts a 2 in front of the third term. In *Mathematica* $\bar{N}_H$ is represented by

```
In[13]:=  N̄ₕ = - (1 / Log[10]) * D[Log[p], pH];
```

In *Mathematica*, the differentiation operator is D. Plotting this expression for $\bar{N}_H$ also yields Figure 2.2.

2.2 Apparent Equilibrium Constant for the Hydrolysis of ATP as a Function of pH

The hydrolysis of ATP to ADP is represented by the **biochemical reaction equation**

$$ATP + H_2O = ADP + P_i \qquad (2.2\text{-}1)$$

This represents the reaction at a specified pH, and so it does not balance hydrogen atoms or electric charge, but it does balance atoms of all other elements. The expression for the apparent equilibrium constant K' is given by

$$K' = \frac{[ADP][P_i]}{[ATP]} \qquad (2.2\text{-}2)$$

K' is referred to as an **apparent equilibrium constant** because it is generally a function of pH, while equilibrium constants of chemical reactions are independent of pH. There is no term in the equilibrium expression for the activity of water because it is equal to unity for dilute aqueous solutions. The terminology used here is that recommended by the IUPAC-IUB Committee chaired by Wadso (4). In the next chapter we will find it necessary to use the terminology recommended by the IUPAC-IUBMB committee chaired by Alberty (5).

Equation 2.2-2 can be written in terms of the concentrations of the various species.

$$K' = \frac{([ADP^{3-}]+[HADP^{2-}]+[H_2 ADP^-])\,([HPO_4^{2-}]+[H_2 PO_4^{2-}])}{([ATP^{4-}]+[HATP^{3-}]+[H_2 ATP^{2-}])} \qquad (2.2\text{-}3)$$

This equation can be rearranged by introducing the expressions for the acid dissociations.

$$K' = \frac{K_{ref}}{[H^+]} \frac{(1+10^{pK1ADP-pH}+10^{pK1ADP+pK2ADP-2\,pH})\,(1+10^{pK1Pi-pH})}{(1+10^{pK1ATP-pH}+10^{pK1ATP+pK2ATP-2\,pH})} \qquad (2.2\text{-}4)$$

where

$$K_{ref} = \frac{[ADP^{3-}][HPO_4^{2-}][H^+]}{[ATP^{4-}]} \qquad (2.2\text{-}5)$$

is the equilibrium constant for the **chemical reference reaction**:

$$ATP^{4-} + H_2O = ADP^{3-} + HPO_4^{2-} + H^+ \qquad (2.2\text{-}6)$$

Note that an equilibrium constant must be dimensionless so that $-RT\ln K$ can be calculated. Strictly speaking the right-hand side of equation 2.2-5 should be divided by $(c^0)^2$, where $c^0 = 1$ mol L^{-1}, but this factor will be omited as a simplification.

Equation 2.2-4 can be written in terms the binding polynomials for ATP, ADP, and P$_i$. P(ATP) has been given earlier, and the other two binding polynomials are given by

$In[14]:=$ **padp = 1 + 10$^{\text{pK1ADP-pH}}$ + 10$^{\text{pK1ADP+pK2ADP-2 pH}}$;**

$In[15]:=$ **ppi = 1 + 10$^{\text{pK1Pi-pH}}$;**

The pKs in these equations (6) at 298.15 K and 0.25 M ionic strength are given by Table 1.3.

$In[16]:=$ **pK1ADP = 6.33;**
        pK2ADP = 3.79;
        pK1Pi = 6.65;

Since $K_{ref} = 0.222$ (6) at 298.15 K and ionic strength 0.25 M, the apparent equilibrium constant for the hydrolysis of ATP to ADP at 298.15 K and 0.25 M ionic strength is given by (see equation 2.2-4)

$In[19]:=$ **kapp = 0.222 * padp * ppi / (patp * 10$^{\text{-pH}}$);**

At pHs 5, 6, 7, 8, and 9 and 0.25 M ionic strength the apparent equilibrium constants for the hydrolysis of ATP to ADP are given by

$In[20]:=$ **kapp /. pH → {5, 6, 7, 8, 9}**

$Out[20]=$ $\{739088., 963030., 3.00983 \times 10^6, 2.30084 \times 10^7, 2.22811 \times 10^8\}$

$-RT\ln K'$ at 298.15 K and 0.25 M ionic strength can be plotted in kJ mol^{-1} versus pH.

$In[21]:=$ **Plot[-8.31451 * .29815 * Log[kapp], {pH, 5, 9}, AxesLabel → {"pH", "-RTlnK'"}];**

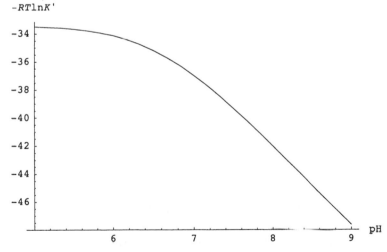

Figure 2.3 $-RT\ln K'$ in kJ mol^{-1} at 298.15 K and 0.25 M ionic strength as a function of pH. A factor of 10^3 has been put in the denominator so that the ordinate is in kJ mol^{-1}.

2.3 Change in the Binding of Hydrogen Ions in the Hydrolysis of ATP

Since equation 2.1-6 applies to each reactant, we expect that the change in the binding of hydrogen ions $\bar{N}_H$ in the hydrolysis of ATP to ADP will be given by

$$\Delta_r N_H = \frac{-1}{\ln(10)} \frac{d\ln K'}{dpH}$$
(2.3-1)

In *Mathematica* the change in binding of hydrogen ions in a biochemical reaction is given by

In[22]:= **(-1 / Log[10]) * D[Log[kapp], pH];**

This is a rather complicated function of pH that can be examined by removing the semicolon. The changes in moles of hydrogen ion per mole of reaction at 298.15 K and 0.25 M ionic strength and pHs 5, 6, 7, 8, and 9 are given by

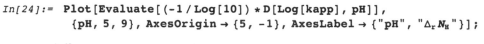

In[23]:= **(-1 / Log[10]) * D[Log[kapp], pH] /. pH → {5, 6, 7, 8, 9}**

Out[23]= {-0.0390838, -0.249292, -0.743044, -0.964977, -0.996362}

Since there is a decrease in the binding of hydrogen ions, hydrogen ions are produced in the hydrolysis of ATP. The change in binding with pH can be plotted.

In[24]:= **Plot[Evaluate[(-1 / Log[10]) * D[Log[kapp], pH]],**
**   {pH, 5, 9}, AxesOrigin → {5, -1}, AxesLabel → {"pH", "$\Delta_r N_H$"}];**

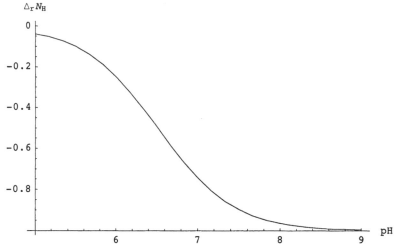

Figure 2.4 Change in the binding of hydrogen ions in the hydrolysis of ATP to ADP at 298.15 K and 0.25 M ionic strength.

The operation **Evaluate** is needed so that the derivative is calculated before the plotting is attempted.

2.4 Average Binding of Hydrogen Ions by ATP as a Function of pH and pMg

When magnesium ions are present, the binding of hydrogen ions by ATP is affected (8). The complex ions of ATP are $MgATP^{-2}$, $MgHATP^{-1}$, and $Mg_2 ATP$. A number of studies of the binding of magnesium ions by ATP have been made. The most thorough evaluation of these data has been made by Goldberg and Tewari (6) and their values were used by Alberty and Goldberg (7). The binding polynomial for ATP is given as a function of pH and pMg by

In[25]:= **pmgatp = 1 + 10^{pK1ATP-pH} + 10^{pK1ATP+pK2ATP-2 pH} +**
10^{pK3ATP-pMg} + 10^{pK1ATP+pK4ATP-pH-pMg} + 10^{pK3ATP+pK5ATP-2*pMg} ;

The dissociation constants for the three complex ions at 298.15 K and 0.25 M ionic strength are 0.0001229, 0.01181, and 0.02785. The pKs can be calculated as follows:

In[26]:= **-Log[10, {.0001229, .01181, .02785}]**

Out[26]= {3.91045, 1.92775, 1.55517}

The pKs are input into *Mathematica* as follows:

In[27]:= **pK3ATP = 3.91;**
pK4ATP = 1.93;
pK5ATP = 1.55;

The binding of hydrogen ions $\bar{N}_{HMg}$ when magnesium ions are present can be calculated using equation 2.1-6, where the derivative is now a partial derivative because the binding is affected by both pH and pMg.

In[30]:= **$\bar{N}_{HMg}$ = -(1 / Log[10]) * D[Log[pmgatp], pH];**

In[31]:= **Plot3D[$\bar{N}_{HMg}$, {pH, 3, 9}, {pMg, 1, 6}, AxesLabel → {"pH", "pMg", "$\bar{N}_{HMg}$"}];**

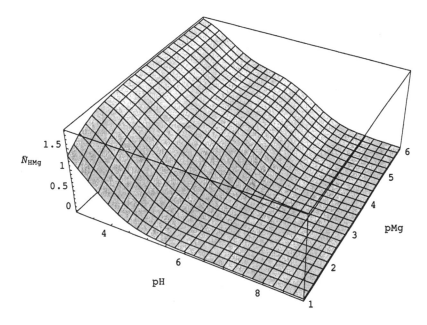

Figure 2.5 Average binding of hydrogen ions $\bar{N}_{HMg}$ by ATP at 298.15 K and 0.25 M ionic strength in the presence of magnesium ions.

2.5 Average Binding of Magnesium Ions by ATP as a Function of pH and pMg and Linked Functions

The average binding of magnesium ions $\bar{N}_{\mathrm{MgH}}$ by ATP is given by

$In[32]:=$ $\bar{N}_{\mathrm{MgH}}$ = - (1 / Log[10]) * D[Log[pmgatp], pMg];

$In[33]:=$ Plot3D[$\bar{N}_{\mathrm{MgH}}$, {pH, 3, 9}, {pMg, 1, 6}, AxesLabel → {"pH", "pMg", "$\bar{N}_{\mathrm{MgH}}$"}];

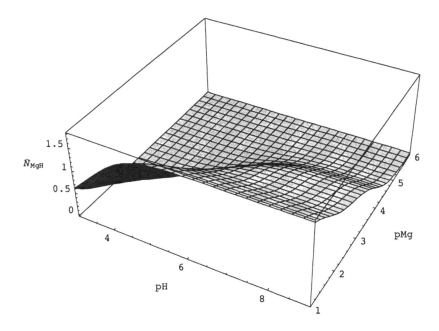

Figure 2.6 Average binding of magnesium ions $\bar{N}_{\mathrm{MgH}}$ by ATP at 298.15 K and 0.25 M ionic strength.

Figures 2.5 and 2.6 are related in a remarkable way, in that at each pH and pMg the slope of Figure 2.5 in the pMg direction is equal to the slope of Figure 2.6 in the pH direction. The slope of Figure 2.5 in the pMg direction is given by $\partial \bar{N}_{\mathrm{HMg}} / \partial \mathrm{pMg}$

$In[34]:=$ Plot3D[Evaluate[- (1 / Log[10]) * D[Log[pmgatp], pH, pMg]],
          {pH, 3, 9}, {pMg, 1, 6}, AxesLabel → {"pH", "pMg", "$\partial \bar{N}_{\mathrm{HMg}} / \partial \mathrm{pMg}$"}];

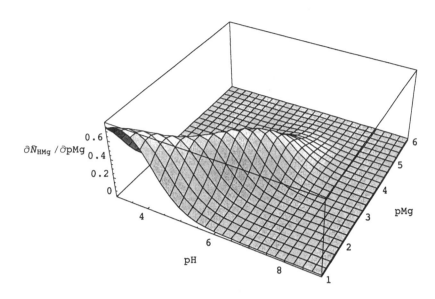

Figure 2.7 $\partial \bar{N}_{HMg} / \partial \text{pMg}$ for ATP at 298.15 K and 0.25 M ionic strength.

The calculation of $\partial \bar{N}_{MgH} / \partial \text{pH}$ yields Figure 2.7 again. This shows that

$$\frac{\partial \bar{N}_H}{\partial \text{pMg}} = \frac{\partial \bar{N}_{Mg}}{\partial \text{pH}} \tag{2.5-1}$$

where the first partial derivative is taken at constant pH and the second partial derivative is taken at constant pMg. In thermodynamics this is referred to as a **Maxwell relation**. Since the effect of pMg on the binding of hydrogen ions is equal to the effect of pH on the binding of magnesium ions, these are called **reciprocal effects**. The bindings of hydrogen ions and magnesium ions are **linked functions**. Substituting the equations that define $\bar{N}_{HMg}$ and $\bar{N}_{MgH}$ into equation 2.5-1 yields

$$\frac{\partial \ln (\text{pmgatp})}{\partial \text{pH} \, \partial \text{pMg}} = \frac{\partial \ln (\text{pmgatp})}{\partial \text{pMg} \, \partial \text{pH}} \tag{2.5-2}$$

which makes it clear why equation 2.5-1 is true.

2.6 Apparent Equilibrium Constant for the Hydrolysis of ATP as a Function of pH and pMg

In order to study the effects of pH and pMg on the apparent equilibrium constant for the hydrolysis of ATP to ADP, the binding polynomials with magnesium terms for ADP and P_i are needed.

$In[35]:=$ **pmgadp =**
$$\mathbf{1 + 10^{pK1ADP-pH} + 10^{pK1ADP+pK2ADP-2\,pH} + 10^{pK3ADP-pMg} + 10^{pK1ADP+pK4ADP-pH-pMg}} \mathbf{;}$$

$In[36]:=$ **pmgpi = $1 + 10^{pK1Pi-pH} + 10^{pK2Pi-pMg}$;**

The dissociation constants for $MgADP^{-1}$, MgHADP, and $MgHPO_4$ are 0.00113, 0.0431, and 0.0266 at 298.15 K and 0.25 M ionic strength, and the pKs can be calculated as follows:

In[37]:= **-Log[10, {.00113, .0431, .0266}]**

Out[37]= {2.94692, 1.36552, 1.57512}

The p*K*s are entered into *Mathematica* as follows:

In[38]:= **pK3ADP = 2.95;**
            pK4ADP = 1.37;
            pK2Pi = 1.58;

The apparent equilibriun constant for the hydrolysis of ATP to ADP at 298.15 K and 0.25 M ionic strength is given by

In[41]:= **kappmg = 0.222 * pmgadp * pmgpi / (pmgatp * 10^{-pH});**

In[42]:= **Plot3D[-8.31451 * .29815 * Log[kappmg], {pH, 3, 9},**
            {pMg, 1, 6}, AxesLabel → {"pH", "pMg", "-RTlnK '"}];

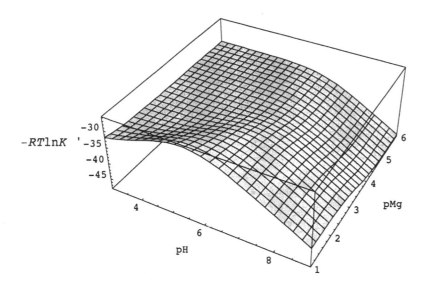

Figure 2.8 -*RT*ln*K* ' for the reaction ATP + H$_2$O = ADP + P$_i$ at 298.15 K and 0.25 M ionic strength as a function of pH and pMg.

2.7 Change in the Binding of Hydrogen Ions in the Hydrolysis of ATP as a Function of pH and pMg

The change in the binding of hydrogen ions in the hydrolysis of ATP to ADP when magnesium ions are present can be calculated using

In[43]:= Δ**N**$_{\text{HMg}}$ **= - (1 / Log[10]) * D[Log[kappmg], pH];**

In[44]:= **Plot3D[ΔN$_{\text{HMg}}$, {pH, 3, 9}, {pMg, 1, 6}, AxesLabel → {"pH", "pMg", "ΔN$_{\text{HMg}}$"}];**

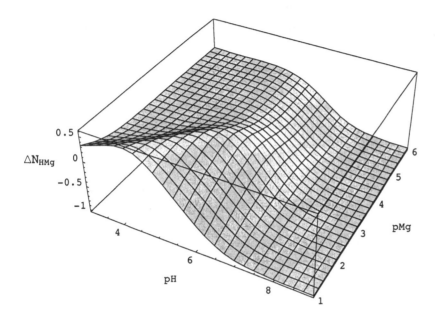

Figure 2.9 Change in the binding of hydrogen ions in the hydrolysis of ATP at 298.15 K and 0.25 M ionic strength as a function of pH and pMg.

2.8 Change in the Binding of Magnesium Ions in the Hydrolysis of ATP as a Function of pH and pMg and Linked Functions

The change in the binding of magnesium ions in the hydrolysis of ATP to ADP can be calculated using

In[45]:= Δ**N**$_{\text{MgH}}$ **= - (1 / Log[10]) * D[Log[kappmg], pMg];**

In[46]:= **Plot3D[ΔN$_{\text{MgH}}$, {pH, 3, 9}, {pMg, 1, 6}, AxesLabel → {"pH", "pMg", "ΔN$_{\text{MgH}}$"}];**

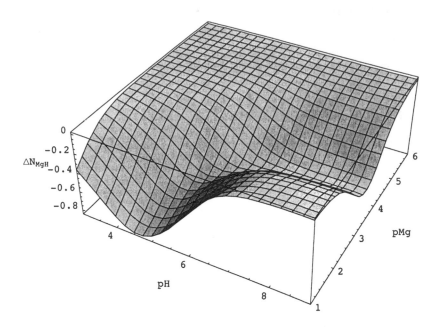

Figure 2.10 Change in the binding of magnesium ions in the hydrolysis of ATP at 298.15 K and 0.25 M ionic strength as a function of pH andf ionic strength.

As in the binding of hydrogen ions and magnesium ions by ATP, Figures 2.9 and 2.10 are related in a remarkable way, in that at each pH and pMg the slope of Figure 2.9 in the pMg direction is equal to the slope of Figure 2.10 in the pH direction. The slope of Figure 2.9 in the pMg direction is given by Figure 2.11:

```
In[47]:= Plot3D[Evaluate[D[ΔN_HMg, pMg]], {pH, 3, 9}, {pMg, 1, 6},
         AxesLabel → {"pH", "pMg", "∂ΔN_HMg /∂pMg"}, PlotRange -> {-0.4, 0.6}];
```

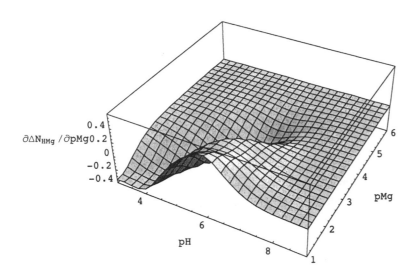

Figure 2.11 $\partial \Delta N_{HMg} / \partial pMg$ for ATP at 298.15 K and 0.25 M ionic strength as a function of pH and ionic strength.

The calculation of $\partial \Delta N_{MgH} / \partial pH$ yields Figure 2.11 again. This shows that

$$\frac{\partial \Delta N_{HMg}}{\partial pMg} = \frac{\partial \Delta N_{MgH}}{\partial pH} \tag{2.8-1}$$

Thus the changes in binding of hydrogen ions and magnesium ions are reciprocal effects.

2.9 Discussion

The use of binding polynomials provides a way to obtain $\Delta_f G°$ and $\Delta_f H°$ of a species if K' and the enthalpy of reaction have been measured at specified pHs and ionic strengths and the $\Delta_f G°$ and $\Delta_f H°$ values all the other species in the reaction are known but one. This calculation can also be made when K' has been measured as a function of temperature. If the pKs and perhaps pK_{Mg}s are known for all the reactants, $\Delta_f G°$ can be calculated for the species of a single reactant. The calculation of $\Delta_f H°$ for the species of a reactant requires information on the heats of dissociation of hydrogen ions. The calculations are much more complicated when magnesium ions are present. In general pMg effects are ignored and K' at the lowest concentration of total magnesium ions is used. If $\Delta_f G°$ is unknown for two species, one can be arbitrarily assigned $\Delta_f G° = 0$. This was done by Alberty and Goldberg (7) in order to calculate $\Delta_f G°$ for species in the ATP series. Later, Boeiro-Goates and coworkers (9) used the third law of thermodynamics to determine the standard entropy of adenosine crystals at 298.15 K With additional thermodynamic measurements they calculated $\Delta_f G°$ and $\Delta_f H°$ of adenosine in dilute aqueous solution. That made it possible to calculate $\Delta_f G°$ and $\Delta_f H°$ of all the species in the ATP series with respect to the elements. These values can be used to calculate other thermodynamic properties of reactants for which the dissociation constants of magnesium complex ions are known (10).

It is important to note that in this chapter symbols have not been given for the Gibbs energy, enthalpy, or entropy. Prior to 1992, the symbols $\Delta_f G°'$ and $\Delta_f H°'$ were used (4). The question of what symbols should be used is discussed in the next chapter.

References

1. J. Wyman, Heme Proteins, Adv. Protein Chem. 4, 407-531 (1948).
2. J. Shellman, Macromolecular binding, Biopolymers 14, 999-1018 (1975).
3. J. Wyman and S. J. Gill, Binding and Linkage, Wiley, Hoboken, NJ (1990).
4. I. Wadsö, H. Gutfreund, P. Privalov, J. T. Edsall, W. P. Jencks, G. T. Armstrong, and R. L. Biltonen, Recommendations for Measurement and Presentation of Biochemical Equilibrium Data, J. Biol. Chem. 251, 6879-6885; Q. Rev. of Biophys. 9, 439-456 (1976).
5. R. A. Alberty, A. Cornish-Bowden, Q. H. Gibson, R. N. Goldberg, G. G. Hammes, W. Jencks, K. F. Tipton, R. Veech, H. V. Westerhoff, and E. C. Webb, Recommendations for nomenclature and tables in biochemical thermodynamics, Pure Appl. Chem. 66, 1641-1666 (1994). Reprinted in Europ. J. Biochem. 240, 1-14 (1996).

http://www.chem.qmw.ac.uk/iubmb/thermod/

6. R. N. Goldberg and Y. Tewari, Thermodynamic and transport properties of carbohydrates and their monophosphates: The pentoses and hexoses, J. Phys. Chem. Ref. Data 18, 809-880 (1989).
7. R. A. Alberty and R. N. Goldberg, Calculation of thermodynamic formation properties for the ATP series at specified pH and pMg, Biochem. 31, 10610-10615 (1992).
8. R. A. Alberty, R. H. Smith, and R. M. Bock, Apparent ionization constants of the adenosine phosphates and related compounds, J. Biol. Chem. 193, 425 (1951).
9. J. Boerio-Goates, M. R. Francis, R. N. Goldberg, M. A. V. Ribeiro da Silva, M. D. M. C. Ribeiro da Silva, and Y. Tewari, Thermochemistry of adenosine, J. Chem. Thermo. 33, 929-947 (2001)
10. R. A. Alberty, Thermodynamics of the hydrolysis of adenosine triphosphate as a function of temperature, pH pMg, and ionic strength, J. Phys. Chem. 107 B, 12324-12330 (2003).

Chapter 3 Biochemical Reactions at Specified Temperature and Various pHs

3.1 The Fundamental Equation of Chemical Thermodynamics

The fundamental equation for the Gibbs energy G of a chemical reaction system at specified T and P is given by

$$dG = -SdT + VdP + \sum_{j=1}^{N} \mu_j \, dn_j \qquad (3.1\text{-}1)$$

where S is the entropy of the system, μ_j is the chemical potential of species j, n_j is the amount of species j, and N is the number of different species (1,2,3). This fundamental equation shows that T, P, and $\{n_j\}$ are the independent variables for the Gibbs energy of a chemical reaction system. The thermodynamic properties of a chemical reaction system are interrelated in several ways. The first type of interrelations are the partial derivatives of G:

$$-S = \frac{\partial G}{\partial T} \qquad (3.1\text{-}2)$$

$$V = \frac{\partial G}{\partial P} \qquad (3.1\text{-}3)$$

$$\mu_j = \frac{\partial G}{\partial n_j} \qquad (3.1\text{-}4)$$

These relations are often called **equations of state** because they relate different state properties. Since the variables T, P, and $\{n_j\}$ play this special role of yielding the other thermodynamic properties, they are referred to as the **natural variables** of G. Further information on natural variables is given in the Appendix of this chapter. In writing partial derivatives, subscripts are omitted to simplify the notation. The second type of interrelations are Maxwell equations (mixed partial derivatives). Ignoring the VdP term, equation 3.1-1 has two types of Maxwell relations:

$$-\frac{\partial S}{\partial n_j} = \frac{\partial \mu_j}{\partial T} \qquad (3.1\text{-}5)$$

$$\frac{\partial \mu_j}{\partial n_{j+1}} = \frac{\partial \mu_{j+1}}{\partial n_j} \qquad (3.1\text{-}6)$$

However, the two sides of Maxwell relation 3.1-6 will always be equal to zero for the systems treated in this book because it is assumed that solutions are ideal in the sense that

$$\mu_j = \mu_j{}^{\circ} + RT\ln[j] \qquad (3.1\text{-}7)$$

Equation 3.1-1 can be integrated at constant S, V, and $\{n_j\}$ to obtain $G = -TS + PV + \sum \mu_j n_j$. If we write out the total differential of G using this equation and subtract equation 3.1-1, we obtain the Gibbs-Duhem equation.

$$0 = -SdT + VdP - \sum n_j \, d\mu_j \qquad (3.1\text{-}8)$$

This shows that the intensive variables for this system (T, P, $\{\mu_j\}$) are not independent at equilibrium; all but one are independent. The derivation of the Gibbs-Duhem equation by use of a complete Legendre transform is shown in the Appendix.

When there are a number of chemical reactions between the species in a system, we can shift from the independent variables T, P, and $\{n_j\}$ to the smaller number of independent variables T, P, and $\{\xi_k\}$, where the ξ_k are the extents of the k independent reactions. The extent of a reaction is defined by $n_j = n_{j0} + \nu_{jk} \xi_k$, where n_{j0} is the amount of the jth species when

$\xi_k = 0$ and v_{jk} is the stoichiometric number of the jth species in the kth chemical reaction. Note that the stoichiometric numbers are positive for products and are negative for reactants. Thus, $dn_j = \Sigma v_{jk} \, d\xi_k$ can be used to write the summation in equation 3.1-1. The fundamental equation for a system involving R independent chemical reactions is

$$dG = -SdT + VdP + \sum_{k}^{R}\sum_{j}^{N} v_{jk} \, \mu_j \, d\xi_k \tag{3.1-9}$$

The reactions in a set are independent if none of the reactions can be obtained by adding and subtracting other reactions in the set. (In Chapter 7 we will see that there is a more operational method for determining the number of independent reactions by use of linear algebra.) Experimental data at a specified T and P can be interpreted by use of

$$dG = \sum_{k}^{R}\sum_{j}^{N} v_{jk} \, \mu_j \, d\xi_k \tag{3.1-10}$$

Thus for each reaction

$$\frac{\partial G}{\partial \xi_k} = \sum v_{jk} \, \mu_j \tag{3.1-11}$$

The left side of this equation is equal to the change in Gibbs energy $\Delta_r G_k$ per mole of reaction k. In derivations with the fundamental equation, chemical potentials μ_j are used, but in working with experimental data, the Gibbs energy $\Delta_f G_j$ of formation of species j from the elements is used instead. The right side of equation 3.1-11 can be written in terms of $\Delta_f G_j$ to obtain

$$\Delta_r G_k = \sum_{j=1}^{N} v_{jk} \Delta_f G_j \tag{3.1-12}$$

The change in Gibbs energy in a chemical reaction is related to the change in the standard Gibbs energy of reaction by

$$\Delta_r G_k = \Delta_r G_k° + RT\ln Q_k \tag{3.1-13}$$

where Q_k is the reaction quotient for the kth chemical reaction. For $ATP^{4-} + H_2O = ADP^{3-} + HPO_4^{2-} + H^+$, $Q_k = [ADP^{3-}][HPO_4^{2-}][H^+] / [ATP^{4-}]$. At chemical equilibrium, $\Delta_r G_k$ is equal to zero and Q_k is equal to the equilibrium constant K_k for the kth reaction.

$$\Delta_r G_k° = -RT\ln K_k = \sum_{j=1}^{N} v_{jk} \, \Delta_f G_j° \tag{3.1-14}$$

If the standard Gibbs energies of formation of all the species in a reaction but one are known, say from the National Bureau of Standards Tables (4), $\Delta_f G°$ for the species with unknown $\Delta_f G_j°$ can be calculated from an experimentally determined equilibrium constant by use of equation 3.1-14. Suffice it to say here that the $\Delta_f G°$ values for many species of biochemical interest have been determined (5).

3.2 Use of a Legendre Transform to Define a Transformed Gibbs Energy at a Specified pH

When enzyme-catalyzed reactions produce or consume hydrogen ions, their apparent equilibrium constants depend on the pH. The term apparent equilibrium constant and the symbol K' are used to indicate that a biochemical reaction and the expression for its apparent equilibrium constant are written in terms of **sums of species**. Hydrogen ions are omitted in writing a biochemical reaction, and $[H^+]$ is omitted in writing the expression for the apparent equilibrium constant of a biochemical reaction because $[H^+]$ is specified.

The Gibbs energy G does not provide the criterion for spontaneous change and equilibrium when the pH is specified because it does this when the only independent variables are T and P. When additional independent variables are specified, it is necessary to use a Legendre transform (see Appendix) to define a new thermodynamic potential. Callen (6) has emphasized the importance of the choice of independent variables to be used in the solution of a thermodynamic problem. The new thermodynamic potential, in this case the **transformed Gibbs energy** G', is defined by subtracting a product of conjugate variables from the Gibbs energy. Chemical potentials are used in writing the fundamental equation for G, and so the intensive property involved in making this Legendre transform is $\mu(H^+)$. The corresponding extensive property is the amount of the hydrogen component (7) in the system $n_c(H)$, which is the total amount of hydrogen atoms in the system. Components are the things that are conserved in a reaction system. The amount of the hydrogen component in a system is given by

$$n_c(H) = \sum N_H(j)\, n_j \tag{3.2-1}$$

where $N_H(j)$ is the number of hydrogen atoms in species j. The Legendre transform that defines G' is (8-10)

$$G' = G - n_c(H)\mu(H^+) \tag{3.2-2}$$

This transformed Gibbs energy provides the criterion for spontaneous change and equilibrium at constant T, P, pH and amounts of components other than hydrogen atoms: $dG' \leq 0$. The change in the transformed Gibbs energy is negative in a spontaneous process and is equal to zero at equilibrium. This use of a Legendre transform to introduce a concentration variable as an independent variable was preceded by the use of the partial pressure of ethylene in calculating the distribution of alkyl benzenes in the alklation of benzene (11).

3.3 Fundamental Equation of Thermodynamics for a System that Contains a Single Biochemical Reactant

The differential of the transformed Gibbs energy is obtained from equation 3.2-2.

$$dG' = dG - n_c(H)d\mu(H^+) - \mu(H^+)\, dn_c(H) \tag{3.3-1}$$

Substituting equation 3.1-1 for dG into equation 3.3-1 yields a form of the fundamental equation for the transformed Gibbs energy.

$$dG' = -SdT + VdP + \sum \mu_j\, d\, n_j - n_c(H)d\mu(H^+) - \mu(H^+)\, dn_c(H) \tag{3.3-2}$$

The differential of the amount of the hydrogen component can be obtained from equation 3.2-1: $dn_c(H) = \sum N_H(j)\, dn_j$. Substituting this relation in equation 3.3-2 yields

$$dG' = -SdT + VdP + \sum \{\mu_j - N_H(j)\,\mu(H^+)\}dn_j - n_c(H)d\mu(H^+) \tag{3.3-3}$$

$$= -SdT + VdP + \sum \mu_j' \, d\,n_j - n_c(H)d\mu(H^+)$$

where the **transformed chemical potential** of the *j*th species is given by

$$\mu_j' = \mu_j - N_H(j)\,\mu(H^+) \tag{3.3-4}$$

Note that the term for H^+ in the summation in equation 3.3-3 is equal to zero because $\mu_j - N_H(j)\,\mu(H^+) = 0$. Thus the effect of the Legendre transform is to replace the term for the hydrogen ion in the summation in equation 3.3-2 with a new term that is proportional to $d\mu(H^+)$. Specifically, the Legendre transform has replaced the term $\mu(H^+)\,d\,n(H^+)$ with the term $n_c(H)d\mu(H^+)$. As a consequence, G' has natural variables T, P, $\mu(H^+)$, and $\{n_j\}$, rather than just T, P, and $\{n_j\}$ for G.

However, $\mu(H^+)$ is not a very convenient independent variable because it is a function of both the temperature and the concentration of hydrogen ions. The differential of the chemical potential of hydrogen ions is given by

$$d\mu(H^+) = \frac{\partial \mu(H^+)}{\partial T}\,dT + \frac{\partial \mu(H^+)}{\partial [H^+]}\,d[H^+] = -\overline{S}(H^+)dT - RT\ln(10)dpH \tag{3.3-5}$$

where $\overline{S}(H^+)$ is the molar entropy of hydrogen ions and pH = $-\log[H^+]$. Substituting equation 3.3-5 in equation 3.3-3 yields

$$dG' = -S'dT + VdP + \sum_{j=1}^{N-1} \mu_j' \, d\,n_j + RT\ln(10)n_c(H)dpH \tag{3.3-6}$$

where the summation is over species other than H^+. The **transformed entropy** of the system is given by

$$S' = S - n_c(H)\overline{S}(H^+) \tag{3.3-7}$$

Since $G = H - TS$, the **transformed enthalpy** of the system is given by

$$H' = H - n_c(H)\overline{H}(H^+) \tag{3.3-8}$$

where the molar enthalpy of hydrogen ions is given by

$$\overline{H}(H^+) = \mu(H^+) + T\overline{S}(H^+) \tag{3.3-9}$$

Note the similarity of equations 3.2-2, 3.3-7, and 3.3-8. Thus the definition of G' by use of a Legendre transform automatically brings in a transformed entropy S' and a transformed enthalpy H'.

Equation 3.3-6 can be written in another way that makes it useful for treating biochemical reactions that are written in terms of reactants rather than species. At equilibrium at a specified pH, the transformed chemical potentials of the species ATP^{4-}, $HATP^{3-}$, and H_2ATP^{2-} are equal. This is demonstrated later in Section 3.12 by calculating the transformed Gibbs energies of formation of these species at equilibrium. Since the transformed chemical potentials of the various species in a reactant are equal at equilibrium at a specified pH, this transformed chemical potential can be associated with reactant *i* and the sum of the differentials of the amounts of the species of the reactant can be added and represented by dn_i', where n_i' is the **amount of reactant** *i*. Thus equation 3.3-6 can be written as

$$dG' = -S'dT + VdP + \sum_{i=1}^{N'} \mu_i' dn_i' + RT\ln(10)n_c(H)dpH \tag{3.3-10}$$

where N' is the number of reactants. Now the summation has fewer terms, N' rather than N. The natural variables for G' of a mixture of biochemical reactants are T, P, $\{n_i'\}$, and pH. The thermodynamic properties in these equations are taken to be functions of the ionic strength so that we do not have to explicitly show activity coefficients, as explained in Section 1.3.

Integration of equation 3.3-10 at constant T, P, and pH yields

$$G' = \sum \mu_i' n_i' \tag{3.3-11}$$

which is similar to the expression $G = \sum \mu_j n_j$ for a chemical reaction Thus the transformed Gibbs energy is additive in the transformed chemical potentials of reactants. At specified T and P, G is at a minimum at equilibrium for a system described in terms of species, and at specified T, P, and pH, G' is at a minimum at equilibrium for a system described in terms of reactants (sums of species).

3.4 Equations of State and Maxwell Relations for the Transformed Gibbs Energy

The coefficients of the differential terms in equation 3.3-10 are each equal to a partial derivative of G'.

$$-S' = \frac{\partial G'}{\partial T} \tag{3.4-1}$$

$$V = \frac{\partial G'}{\partial P} \tag{3.4-2}$$

$$\mu_i' = \frac{\partial G'}{\partial n_i'} \tag{3.4-3}$$

$$RT\ln(10)n_c(H) = \frac{\partial G'}{\partial pH} \tag{3.4-4}$$

These equations are often referred to as **equations of state** because they provide relations between state properties. If G' could be determined experimentally as a function of T, P, n_i', and pH, then S', V, μ_i', and $n_c(H)$ could be calculated by taking partial derivatives. This illustrates a very importnat concept: **when a thermodynamic potential can be determined as a function of its natural variables, all the other thermodynamic properties can be obtained by taking partial derivatives of this function.** However, since there is no direct method to determine G', we turn to the Maxwell relations of equation 3.3-10.

When the VdP term is ignored and only one reactant is present, equation 3.3-10 has the following Maxwell relations:

$$-\frac{\partial S'}{\partial n_i'} = \frac{\partial \mu_i'}{\partial T} \tag{3.4-5}$$

$$-\frac{\partial S'}{\partial pH} = R\ln(10)\frac{\partial (Tn_c(H))}{\partial T} \tag{3.4-6}$$

$$\frac{\partial \mu_i'}{\partial pH} = RT\ln(10)\frac{\partial n_c(H)}{\partial n_i'} \tag{3.4-7}$$

By substituting $S' = H'/T - G'/T$ into the first and second Maxwell relations we obtain two more Maxwell relations:

$$\overline{H}_i' = -T^2 \frac{\partial(\mu_i'/T)}{\partial T} \tag{3.4-8}$$

$$\frac{\partial H'}{\partial pH} = -RT^2 \ln(10) \frac{\partial n_c(H)}{\partial T} \tag{3.4-9}$$

The first of these is referred to as a **Gibbs-Helmholtz equation**.

In dealing with fundamental equations it is customary to use chemical potentials and molar properties (designated by overbars), but in making thermodynamic calculations it is customary to use transformed Gibbs energies of formation, transformed enthalpies of formation, and transformed entropies of formation. Since we are assuming ideal solutions, the transformed Gibbs energy of a species is given by

$$\Delta_f G_j' = \Delta_f G_j'^\circ + RT\ln[j] \tag{3.4-10}$$

and the transformed Gibbs energy of a reactant is given by

$$\Delta_f G_i' = \Delta_f G_i'^\circ + RT\ln[i] \tag{3.4-11}$$

Some of the five Maxwell relations can be written in terms of standard transformed properties. Also note that $n_c(H)/n_i'$ can be written as $\overline{N}_H(i)$, which is the **average number of hydrogen atoms** in reactant i. Thus the five Maxwell relations can be used in the following forms:

$$-\Delta_f S_i'^\circ = \frac{\partial \Delta_f G_i'^\circ}{\partial T} \tag{3.4-12}$$

$$-\frac{\partial \Delta_f S_i'^\circ}{\partial pH} = R\ln(10)\frac{\partial(T\overline{N}_H(i))}{\partial T} \tag{3.4-13}$$

$$\frac{\partial \Delta_f G_i'^\circ}{\partial pH} = RT\ln(10)\overline{N}_H(i) \tag{3.4-14}$$

$$\Delta_f H_i'^\circ = -T^2 \frac{\partial(\Delta_f G_i'^\circ/T)}{\partial T} \tag{3.4-15}$$

$$\frac{\partial \Delta_f H_i'^\circ}{\partial pH} = -RT^2 \ln(10)\frac{\partial \overline{N}_H(i)}{\partial T} \tag{3.4-16}$$

The standard formation properties of species are set by convention at zero for the elements in their reference forms at each temperature. The standard formation properties of H^+ in aqueous solution at zero ionic strength are also set at zero at each temperature. For other species the properties are determined by measuring equilibrium constants and heats of reaction. Standard transformed Gibbs energies of formation can be calculated from measurements of K', and so it is really these Maxwell relations that make it possible to calculate five transformed thermodynamic properties of a reactant.

3.5 Equations for the Standard Transformed Formation Properties of a Reactant

The equations in the preceding sections are general, but now we will concentrate on the interpretation of measurements of apparent equilibrium constants at a single temperature because this is the situation for most studies of biochemical reactions. The effect of temperature will be treated in the next chapter. Since we will be considering experimental data, formation properties will be used. Equation 3.3-4 for the transformed chemical potential of a species can be written as

$$\Delta_f G_j' = \Delta_f G_j - N_H(j)\,\Delta_f G(H^+)$$ (3.5-1)

Use of equation 3.4-10 leads to

$$\Delta_f G_j'^\circ + RT\ln[j] = \Delta_f G_j^\circ + RT\ln[j] - N_H(j)(\Delta_f G^\circ(H^+) + RT\ln[H^+])$$ (3.5-2)

Since the terms in $[j]$ on the two sides of the equation cancel, this equation can be written as

$$\Delta_f G_j'^\circ = \Delta_f G_j^\circ - N_H(j)\{\Delta_f G^\circ(H^+) - RT\ln(10)\,pH\}$$ (3.5-3)

The corresponding equation for the standard transformed enthalpy for a species can be obtained by applying the Gibbs-Helmholtz equation.

$$\Delta_f H_j'^\circ = \Delta_f H_j^\circ - N_H(j)\,\Delta_f H^\circ(H^+)$$ (3.5-4)

Note that the standard transformed Gibbs energy of formation of a species depends on the pH, but the standard transformed enthalpy of formation does not. When species have electric charges, their standard thermodynamic properties need to be adjusted for the ionic strength according to the extended Debye-Hückel theory (see Section 1.3) At zero ionic strength, $\Delta_f G^\circ(H^+) = 0$ and $\Delta_f H^\circ(H^+) = 0$ at each temperature.

The fundamental equation for G' involves the amount n_i' of a reactant, which is a sum of species, and so now we must consider the relation between the standard transformed Gibbs energies of formation of the species and the **standard transformed Gibbs energy of formation of the reactant**. Because of the entropy of mixing, the standard transformed Gibbs energy of formation is more negative than any of the species. (See Isomer Group Thermodynamics in the Appendix of this chapter.) The standard transformed Gibbs energy of formation of a reactant $\Delta_f G_i'^\circ$ is given in terms of the standard transformed Gibbs energies of formation $\Delta_f G_j'^\circ$ of the species it contains (pseudoisomers) by (12,13)

$$\Delta_f G_i'^\circ = -RT\ln\sum_{j=1}^{N}\exp(-\Delta_f G_j'^\circ/RT)$$ (3.5-5)

where N is the number of different species in the pseudoisomer group. The same result can be obtained by taking the mole-fraction-weighted average and adding the transformed Gibbs energy of mixing. The equilibrium mole fractions r_j of the species in the reactant at a specified pH are given by

$$r_j = \exp[(\Delta_f G_i'^\circ - \Delta_f G_j'^\circ)/RT]$$ (3.5-6)

The **standard transformed enthalpy of formation** of reactant i is given by

$$\Delta_f H_i'^\circ = \sum r_i \Delta_f H_j'^\circ$$ (3.5-7)

The **standard transformed entropy of formation** of reactant i is given by

$$\Delta_f G_i'^\circ = \Delta_f H_i'^\circ - T\Delta_f S_i'^\circ$$ (3.5-8)

Thus measurements of $\Delta_f G_i'^\circ$ and $\Delta_f H_i'^\circ$ at a single temperature yield $\Delta_f S_i'^\circ$ at that temperature. In the next chapter we will see that if $\Delta_f H_i'^\circ$ is known, the standard transformed Gibbs energy of formation can be expressed as a function of temperature, and then all the other thermodynamic properties can be calculated by taking partial derivatives of this function. Note that in equations 3.4-5 to 3.4-9, the only Maxwell relation that does not involve a partial derivative with respect to the temperature is the one that yields $\overline{N}_H(i)$.

3.6 Derivations of Functions of pH and Ionic Strength that Yield Transformed Thermodynamic Properties at 298.15 K

Now we are in a position to calculate the standard transformed thermodynamic properties of reactants from the standard properties of the species that make them up. In this chapter the transformed thermodynamic properties are calculated only at 298.15 K. Calculations at other temperatures are presented in the next chapter. The first step is to adjust the properties at zero ionic strength to the desired ionic strength in the range 0-0.35 M. Equations 1.3-5 and 1.3-6 Chapter 1 show how these calculations can be made using the extended Debye-Hückel equation. Substituting equation 1.3-5 in equation 3.5-3 in two places yields

$$\Delta_f G_j'^\circ(I) = \Delta_f G_j^\circ(I{=}0) + N_H(j) RT\ln(10)\, pH - RT\alpha(z_i^2 - N_H(j))\, I^{1/2} / (1 + 1.6\, I^{1/2})$$ (3.6-1)

Substituting equation 1.3-6 in equation 3.5-4 in two places yields

$$\Delta_f H_j'^\circ(I) = \Delta_f H_j^\circ(I{=}0) + RT^2(\partial\alpha/\partial T)(z_i^2 - N_H(j))\, I^{1/2} / (1 + 1.6\, I^{1/2})$$ (3.6-2)

As an illustration of the use of these equations we first calculate the standard transformed Gibbs energies and enthalpies of inorganic phosphate in kJ mol^{-1} at 0.25 M ionic strength and a series of pH values. The values of $\Delta_f G_j^\circ(I{=}0)$ and $\Delta_f H_j^\circ(I{=}0)$ of species of inorganic phosphate that are important in the pH range 5 to 9 are given in BasicBiochemData3 (5). First we will express equations 3.6-1 and 3.6-2 in *Mathematica* and use these "one-liners" to calculate functions of pH and ionic strength for standard transformed Gibbs energies and standard transformed enthalpies of inorganic phosphate at 0.25 M ionic strength and pHs 5, 6, 7, 8, and 9 at 298.15 K.

The basic data on inorganic phosphate at 298.15 K and zero ionic strength are given by

```
In[2]:=  Off[General::"spell"];
         Off[General::"spell1"];

In[4]:=  pisp = {{-1096.1, -1299., -2, 1}, {-1137.3, -1302.6, -1, 2}};
```

The rows in a species matrix are $\{\Delta_f G_j^\circ, \Delta_f H_j^\circ, z_j, N_H\ (j)\}$ at 298.15 K, where the energies are in kJ mol^{-1}.

(a) Calculation of $\Delta_f G_j'^\circ$ of the two species of inorganic phosphate at five pHs
   Equation 3.6-1 is applied to the base form (HPO_4^{2-}) and acid form ($H_2 PO_4^-$) as follows:

```
In[5]:=  gprimebaseform =
           -1096.1 + 8.31451 * .29815 * Log[10] * pH - 2.91482 * (2^2 - 1) * is^.5 / (1 + 1.6 * is^.5)
```

$$Out[5]= -1096.1 - \frac{8.74446\, is^{0.5}}{1 + 1.6\, is^{0.5}} + 5.70804\, pH$$

```
In[6]:=  gprimebaseform /. is → .25 /. pH → {5, 6, 7, 8, 9}
```

$$Out[6]= \{-1069.99, -1064.28, -1058.57, -1052.86, -1047.16\}$$

```
In[7]:=  gprimeacidform =
           -1137.3 + 2 * 8.31451 * .29815 * Log[10] * pH - 2.91482 * (1 - 2) * is^.5 / (1 + 1.6 * is^.5)
```

$$Out[7]= -1137.3 + \frac{2.91482\, is^{0.5}}{1 + 1.6\, is^{0.5}} + 11.4161\, pH$$

```
In[8]:=  gprimeacidform /. is → .25 /. pH → {5, 6, 7, 8, 9}
```

$$Out[8]= \{-1079.41, -1067.99, -1056.58, -1045.16, -1033.75\}$$

(b) Calculation of $\Delta_f H_j'°$ of the two species of inorganic phosphate

Equation 3.6-2 is applied to the base form (HPO_4^{2-}) and acid form ($H_2PO_4^-$) as follows:

In[9]:= **hprimebaseform = -1299. + 1.4775 * (2^2 - 1) * is^.5 / (1 + 1.6 * is^.5)**

Out[9]= $-1299. + \dfrac{4.4325 \; is^{0.5}}{1 + 1.6 \; is^{0.5}}$

In[10]:= **hprimebaseformpH = hprimebaseform /. is → .25 /. pH → {5, 6, 7, 8, 9}**

Out[10]= -1297.77

In[11]:= **hprimeacidform = -1302.6 + 1.4775 * (1 - 2) * is^.5 / (1 + 1.6 * is^.5)**

Out[11]= $-1302.6 - \dfrac{1.4775 \; is^{0.5}}{1 + 1.6 \; is^{0.5}}$

In[12]:= **hprimeacidformpH = hprimeacidform /. is → .25 /. pH → {5, 6, 7, 8, 9}**

Out[12]= -1303.01

Note that the standard transformed enthalpies of the two species of inorganic phosphate are independent of pH.

Now we are in position to calculate the standard transformed Gibbs energies of inorganic phosphate (the pseudoisomer group) for these conditions.

(c) Calculation of $\Delta_f G'°(P_i)$ using equation 3.5-5 at five pHs

In[13]:= **gprimephos = -8.31451 * .29815 ***
      Log[Exp[-gprimebaseform / (8.31451 * .29815)] + Exp[-gprimeacidform / (8.31451 * .29815)]]

Out[13]= $-2.47897 \, \mathrm{Log}\left[e^{0.403393 \left(1137.3 - \frac{2.91482 \, is^{0.5}}{1+1.6 \, is^{0.5}} - 11.4161 \, pH\right)} + e^{0.403393 \left(1096.1 + \frac{8.74446 \, is^{0.5}}{1+1.6 \, is^{0.5}} - 5.70804 \, pH\right)} \right]$

In[14]:= **gprimephospH = gprimephos /. is → .25 /. pH → {5, 6, 7, 8, 9}**

Out[14]= $\{-1079.46, -1068.49, -1059.49, -1052.97, -1047.17\}$

(d) Calculation of $\Delta_f H'°(P_i)$ using equation 3.5-7 at five pHs

In order to calculate the standard transformed enthalpy of formation of inorganic phosphate under these conditions we need to calculate the equilibrium mole fractions of the two species using equation 3.5-6.

In[15]:= **rbase = Exp[(gprimephos - gprimebaseform) / (8.31451 * .29815)]**

Out[15]= $e^{0.403393 \left(1096.1 + \frac{8.74446 \, is^{0.5}}{1+1.6 \, is^{0.5}} - 5.70804 \, pH - 2.47897 \, \mathrm{Log}\left[e^{0.403393 \left(1137.3 - \frac{2.91482 \, is^{0.5}}{1+1.6 \, is^{0.5}} - 11.4161 \, pH\right)} + e^{0.403393 \left(1096.1 + \frac{8.74446 \, is^{0.5}}{1+1.6 \, is^{0.5}} - 5.70804 \, pH\right)} \right] \right)}$

In[16]:= **rbasepH = rbase /. is → .25 /. pH → {5, 6, 7, 8, 9}**

Out[16]= $\{0.0218725, 0.182751, 0.690992, 0.957195, 0.995548\}$

In[17]:= **racid = Exp[(gprimephos - gprimeacidform) / (8.31451 * .29815)]**

Out[17]= $e^{0.403393 \left(1137.3 - \frac{2.91482 \, is^{0.5}}{1+1.6 \, is^{0.5}} - 11.4161 \, pH - 2.47897 \, \mathrm{Log}\left[e^{0.403393 \left(1137.3 - \frac{2.91482 \, is^{0.5}}{1+1.6 \, is^{0.5}} - 11.4161 \, pH\right)} + e^{0.403393 \left(1096.1 + \frac{8.74446 \, is^{0.5}}{1+1.6 \, is^{0.5}} - 5.70804 \, pH\right)} \right] \right)}$

In[18]:= **racidpH = racid /. is → .25 /. pH → {5, 6, 7, 8, 9}**

Out[18]= {0.978127, 0.817249, 0.309008, 0.0428052, 0.00445203}

In[19]:= **hprimephospH = rbasepH * hprimebaseformpH + racidpH * hprimeacidformpH**

Out[19]= {-1302.9, -1302.05, -1299.39, -1297.99, -1297.79}

Note that the standard transformed enthalpy of formation of inorganic phosphate changes a little with pH, although the standard transformed enthalpies of formation of the two species do not. This is because the composition of the mixture of species changes with the pH.

(e) Calculation of $\Delta_f S'°(P_i)$ in J K^{-1} mol^{-1} using equation 3.5-8 at five pHs

$$\Delta_f S_i'° = (\Delta_f H_i'° - \Delta_f G_i'°)/T \tag{3.6-3}$$

In[20]:= **((hprimephospH - gprimephospH) / .29815) /. is → .25 /. pH → {5, 6, 7, 8, 9}**

Out[20]= {-749.391, -783.359, -804.627, -821.801, -840.598}

(f) Program to derive the function of pH and ionic strength for $\Delta_f G'°$ for a reactant

The above steps show one way to use *Mathematica* to obtain various properties of reactants (sum of species), but it is much more efficient to use the symbolic capabilities of *Mathematica* to derive the function of pH and ionic strength that yields the desired property. The following program (14) derives the function for the standard transformed Gibbs energy of formation of a reactant as a function of pH and ionic strength at 298.15 K.

In[21]:= **calcdGmat[speciesmat_] :=**
　　　　Module[{dGzero,dHzero, zi, nH, pHterm, isterm,gpfnsp},(*This program derives the
　　　　function of pH and ionic strength (is) that gives the standard transformed Gibbs
　　　　energy of formation of a reactant (sum of species) at 298.15 K. The input
　　　　speciesmat is a matrix that gives the standard Gibbs energy of formation, the
　　　　standard enthalpy of formation, the electric charge, and the number of hydrogen
　　　　atoms in each species. There is a row in the matrix for each species of the
　　　　reactant. gpfnsp is a list of the functions for the species. Energies are expressed
　　　　in kJ mol^-1.*)
　　　　{dGzero,dHzero,zi,nH}=Transpose[speciesmat];
　　　　pHterm = nH*8.31451*.29815*Log[10^-pH];
　　　　isterm = 2.91482*((zi^2) - nH)*(is^.5)/(1 + 1.6*is^.5);
　　　　gpfnsp=dGzero - pHterm - isterm;
　　　　-8.31451*.29815*Log[Apply[Plus,Exp[-1*gpfnsp/(8.31451*.29815)]]]]]

The function that expresses the standard transformed Gibbs energy of formation of inorganic phosphate in the range pH 5 to 9 is calculated as follows:

In[22]:= **piG = calcdGmat[pisp]**

Out[22]= $-2.47897 \, Log\left[e^{-0.403393\left(-1137.3+\frac{2.91482 \, is^{0.5}}{1+1.6 \, is^{0.5}}-4.95794 \, Log[10^{-pH}]\right)} + e^{-0.403393\left(-1096.1-\frac{8.74446 \, is^{0.5}}{1+1.6 \, is^{0.5}}-2.47897 \, Log[10^{-pH}]\right)}\right]$

It is impractical to write out such functions by hand, but it is convenient to derive this type of function using *Mathematica*. The values of $\Delta_f G'°$ in kJ mol^{-1} at ionic strength 0.25 M and pHs 5, 6, 7, 8, and 9 are calculated as follows:

In[23]:= **piGpH = piG /. is → .25 /. pH → {5, 6, 7, 8, 9}**

Out[23]= {-1079.46, -1068.49, -1059.49, -1052.97, -1047.17}

These values agree with the values calculated earlier in this section.

(g) Program to derive the function of pH and ionic strength for $\Delta_f H'°$ for a reactant

The function of pH and ionic strength that gives the standard transformed enthalpy of formation in kJ mol^{-1} at 298.15 K is derived by the following program (3):

```
In[24]:= calcdHmat[speciesmat_] :=
           Module[{dGzero, dHzero, zi, nH, dhfnsp, pHterm, isenth, dgfnsp, dGreactant, ri},
             (*This program derives the function of ionic strength (is) that gives the standard
               transformed enthalpy of formation of a reactant (sum of species) at 298.15
               K.  The input is a matrix that gives the standard Gibbs energy of formation,
              the standard enthalpy of formation, the electric charge,
              and the number of hydrogen atoms in the species in the reactant.  There is
              a row in the matrix for each species of the reactant.  dhfnsp is a list
              of the functions for the species.  Energies are expressed in kJ mol^-1.*)
             {dGzero, dHzero, zi, nH} = Transpose[speciesmat];
             isenth = 1.4775 * ((zi^2) - nH) * (is^.5) / (1 + 1.6 * is^.5);
             dhfnsp = dHzero + isenth;
           (*Calculate the functions for the
             standard Gibbs energies of formation of the species.*)
           pHterm = nH * 8.31451 * .29815 * Log[10^-pH];
           gpfnsp = dGzero - pHterm - isenth * 2.91482 / 1.4775;
           (*Calculate the standard
             transformed Gibbs energy of formation for the reactant.*)
           dGreactant = -8.31451 * .29815 * Log[Apply[Plus, Exp[-1 * gpfnsp / (8.31451 * .29815)]]];
           (*Calculate the equilibrium mole fractions of the species
               in the reactant and the mole fraction-weighted average of the
               functions for the standard transformed enthalpies of the species.*)
           ri = Exp[(dGreactant - gpfnsp) / (8.31451 * .29815)];
           ri.dhfnsp]
```

The function of pH and ionic strength for the standard transformed enthalpy of formation of inorganic phosphate at 298.15 K is obtained by use of

```
In[25]:= piH = calcdHmat[pisp]
```

$$Out[25]= e^{0.403393\left(1137.3 - \frac{2.91482\,is^{0.5}}{1+1.6\,is^{0.5}} + 4.95794\,Log[10^{-pH}] - 2.47897\,Log\left[e^{-0.403393\left(-1137.3 + \frac{2.91482\,is^{0.5}}{1+1.6\,is^{0.5}} - 4.95794\,Log[10^{-pH}]\right)} + e^{-0.403393\left(-1096.1 - \frac{8.74446}{1+1.6\,i}\right)}\right]\right)}$$

$$\left(-1302.6 - \frac{1.4775\,is^{0.5}}{1+1.6\,is^{0.5}}\right) +$$

$$e^{0.403393\left(1096.1 + \frac{8.74446\,is^{0.5}}{1+1.6\,is^{0.5}} + 2.47897\,Log[10^{-pH}] - 2.47897\,Log\left[e^{-0.403393\left(-1137.3 + \frac{2.91482\,is^{0.5}}{1+1.6\,is^{0.5}} - 4.95794\,Log[10^{-pH}]\right)} + e^{-0.403393\left(-1096.1 - \frac{8.744}{1+1.}\right)}\right]\right)}$$

$$\left(-1299. + \frac{4.4325\,is^{0.5}}{1+1.6\,is^{0.5}}\right)$$

The standard transformed enthalpies of formation at ionic strength 0.25 M and pHs 5, 6, 7, 8, and 9 are

```
In[26]:= piHpH = piH /. is → .25 /. pH → {5, 6, 7, 8, 9}
```

```
Out[26]= {-1302.9, -1302.05, -1299.39, -1297.99, -1297.79}
```

These values agree with those calculated earlier in this section.

(h) Program to derive the function of pH and ionic strength for $\Delta_f S'°$ for a reactant

　　The function of pH and ionic strength that yields the standard transformed entropy of formation of a reactants is obtained by the following program:

```
In[27]:= derivetrS[speciesmat_] :=
            Module[{dG, dH}, (*This program derives the function of pH and ionic strength (is)
                that gives the standard transformed entropy of formation of a reactant
                (sum of species) at 298.15 K.  The entropy is given in J K^-1 mol^-1.*)
              dG = calcdGmat[speciesmat];
              dH = calcdHmat[speciesmat];
              (dH - dG) / .29815]
```

Note that this program calls on two previous programs.

```
In[28]:= piS = derivetrS[pisp]
```

```
Out[28]= 3.35402
```

$$\left(e^{0.403393\left(1137.3-\frac{2.91482\,\text{is}^{0.5}}{1+1.6\,\text{is}^{0.5}}+4.95794\,\text{Log}[10^{-\text{pH}}]-2.47897\,\text{Log}\left[e^{-0.403393\left(-1137.3+\frac{2.91482\,\text{is}^{0.5}}{1+1.6\,\text{is}^{0.5}}-4.95794\,\text{Log}[10^{-\text{pH}}]\right)}+e^{-0.403393\left(-1096.1-\frac{8.7\cdots}{1+\cdots}\right.}\right.\right)}\right.$$

$$\left(-1302.6-\frac{1.4775\,\text{is}^{0.5}}{1+1.6\,\text{is}^{0.5}}\right)+$$

$$e^{0.403393\left(1096.1+\frac{8.74446\,\text{is}^{0.5}}{1+1.6\,\text{is}^{0.5}}+2.47897\,\text{Log}[10^{-\text{pH}}]-2.47897\,\text{Log}\left[e^{-0.403393\left(-1137.3+\frac{2.91482\,\text{is}^{0.5}}{1+1.6\,\text{is}^{0.5}}-4.95794\,\text{Log}[10^{-\text{pH}}]\right)}+e^{-0.403393\left(-1096.1-\frac{8.\cdots}{\cdots}\right.}\right.\right)}$$

$$\left(-1299.+\frac{4.4325\,\text{is}^{0.5}}{1+1.6\,\text{is}^{0.5}}\right)+2.47897$$

$$\text{Log}\left[e^{-0.403393\left(-1137.3+\frac{2.91482\,\text{is}^{0.5}}{1+1.6\,\text{is}^{0.5}}-4.95794\,\text{Log}[10^{-\text{pH}}]\right)}+e^{-0.403393\left(-1096.1-\frac{8.74446\,\text{is}^{0.5}}{1+1.6\,\text{is}^{0.5}}-2.47897\,\text{Log}[10^{-\text{pH}}]\right)}\right]\right)$$

```
In[29]:= piS /. is → 0 /. pH → {5, 6, 7, 8, 9}
```

```
Out[29]= {-745.745, -782.979, -813.957, -834.129, -852.894}
```

These standard transformed entropies of formation are in J K^{-1} mol^{-1}.

(i) Program to derive the function of pH and ionic strength for $\overline{N}_H$ (i) for a reactant

　　One of the Maxwell relations in equation 3.4-14 shows that the average binding of hydrogen ions by a reactant can be calculated by taking the partial derivative of the standard transformed Gibbs energy of formation of a reactant with respect to pH.

```
In[30]:= calcavHbound[speciesmat_] := Module[{gfn},
            (*This program derives the function of pH and ionic strength that gives the average
                number of hydrogen atoms bound by a biochemical reactant at 298.15 K.*)
              gfn = calcdGmat[speciesmat];
              (1 / (8.31451 * .29815 * Log[10])) * D[gfn, pH]]
```

```
In[31]:= pinH = calcavHbound[pisp]
```

$Out[31]= -\left(0.434294 \left(-4.60517 \, e^{-0.403393 \left(-1137.3 + \frac{2.91482 \, is^{0.5}}{1+1.6 \, is^{0.5}} - 4.95794 \, \text{Log}[10^{-pH}]\right)} - \right.\right.$

$\qquad 2.30259 \, e^{-0.403393 \left(-1096.1 - \frac{8.74446 \, is^{0.5}}{1+1.6 \, is^{0.5}} - 2.47897 \, \text{Log}[10^{-pH}]\right)}\left.\left.\left.\right)\right)\right) \Bigg/$

$\qquad \left(e^{-0.403393 \left(-1137.3 + \frac{2.91482 \, is^{0.5}}{1+1.6 \, is^{0.5}} - 4.95794 \, \text{Log}[10^{-pH}]\right)} + e^{-0.403393 \left(-1096.1 - \frac{8.74446 \, is^{0.5}}{1+1.6 \, is^{0.5}} - 2.47897 \, \text{Log}[10^{-pH}]\right)}\right)\right)$

The average numbers of hydrogen ions bound by inorganic phosphate at ionic strength 0.25 M and pHs 5, 6, 7, 8, and 9 is given by

$In[32]:=$ **pinH /. is → .25 /. pH → {5, 6, 7, 8, 9}**

$Out[32]=$ {1.97813, 1.81725, 1.30901, 1.04281, 1.00445}

At pH 5, inorganic phosphate is primarily in the form $H_2 PO_4^-$, and at pH 9 it is primarily in the form HPO_4^{-2}.

3.7 Construction of Tables and Plots of Transformed Thermodynamic Properties of Inorganic Phosphate at 298.15 K

The functions derived in the preceding section can also be used to construct tables and plots.

Table 3.1 Standard transformed Gibbs energies of formation in kJ mol^{-1} of inorganic phosphate at 298.15 K.

$In[33]:=$ **TableForm[piG /. is → {0, .1, .25} /. pH → {5, 6, 7, 8, 9}, TableHeadings →**
$\qquad$ **{{"I=0", "I=0.10", "I=0.25"}, {"pH 5", "pH 6", "pH 7", "pH 8", "pH 9"}}]**

$Out[33]//TableForm=$

	pH 5	pH 6	pH 7	pH 8	pH 9
I=0	-1080.23	-1068.95	-1058.56	-1050.81	-1044.77
I=0.10	-1079.65	-1068.56	-1059.17	-1052.42	-1046.58
I=0.25	-1079.46	-1068.49	-1059.49	-1052.97	-1047.17

$In[34]:=$ **Plot[Evaluate[piG /. is → {0, .1, .25}], {pH, 5, 9}, AxesLabel → {"pH", "$\Delta_f G_i$'°"}];**

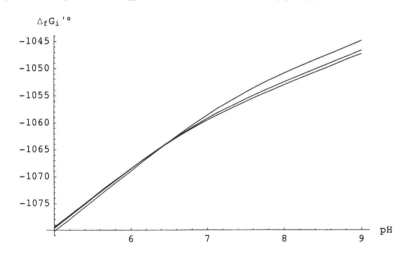

Figure 3.1 $\Delta_f G'° (P_i)$ at ionic strengths of 0, 0.10, and 0.25 M at 298.15 K.

Table 3.2 Standard transformed enthalpies of formation in kJ mol^{-1} of inorganic phosphate at 298.15 K.

In[35]:= **TableForm[piH /. is → {0, .1, .25} /. pH → {5, 6, 7, 8, 9}, TableHeadings →**
        {{"I=0", "I=0.10", "I=0.25"}, {"pH 5", "pH 6", "pH 7", "pH 8", "pH 9"}}]

Out[35]//TableForm=

	pH 5	pH 6	pH 7	pH 8	pH 9
I=0	-1302.58	-1302.39	-1301.24	-1299.51	-1299.06
I=0.10	-1302.83	-1302.23	-1299.91	-1298.35	-1298.1
I=0.25	-1302.9	-1302.05	-1299.39	-1297.99	-1297.79

In[36]:= **Plot[Evaluate[piH /. is → {0, .1, .25}], {pH, 5, 9}, AxesLabel → {"pH", "$\Delta_f H_i$'°"}];**

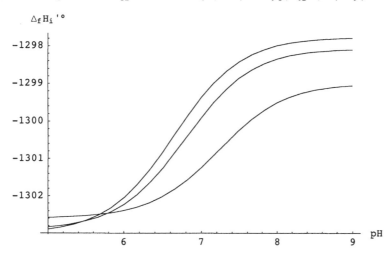

Figure 3.2 $\Delta_f H'°$ (P$_i$) at ionic strengths of 0, 0.10, and 0.25 M at 298.15 K.

Table 3.3 Standard transformed entropies of formation in J K^{-1} mol^{-1} of inorganic phosphate at 298.15 K.

In[37]:= **PaddedForm[TableForm[piS /. is → {0, .1, .25} /. pH → {5, 6, 7, 8, 9}, TableHeadings →**
{{"I=0", "I=0.10", "I=0.25"}, {"pH 5", "pH 6", "pH 7", "pH 8", "pH 9"}}], {5, 2}]

Out[37]//PaddedForm=

	pH 5	pH 6	pH 7	pH 8	pH 9
I=0	-745.74	-782.98	-813.96	-834.13	-852.89
I=0.10	-748.57	-783.73	-807.46	-824.85	-843.60
I=0.25	-749.39	-783.36	-804.63	-821.80	-840.60

In[38]:= **Plot[Evaluate[piS /. is → {0, .1, .25}], {pH, 5, 9},**
AxesLabel → {"pH", "$\Delta_f S_i'^\circ$"}, PlotRange → {-860, -740}];

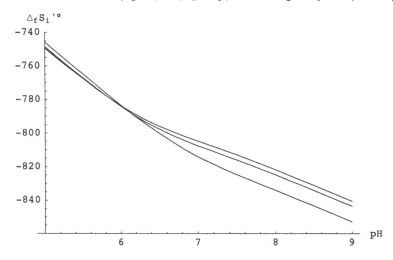

Figure 3.3 $\Delta_f S'^\circ (P_i)$ at ionic strengths of 0, 0.10, and 0.25 M at 298.15 K.

Table 3.4 Average number of hydrogen ions bound by inorganic phosphate at 298.15 K.

In[39]:= **TableForm[pinH /. is → {0, .1, .25} /. pH → {5, 6, 7, 8, 9}, TableHeadings →**
{{"I=0", "I=0.10", "I=0.25"}, {"pH 5", "pH 6", "pH 7", "pH 8", "pH 9"}}]

Out[39]//TableForm=

	pH 5	pH 6	pH 7	pH 8	pH 9
I=0	1.99398	1.94291	1.62286	1.14174	1.01625
I=0.10	1.984	1.86017	1.38086	1.05795	1.00611
I=0.25	1.97813	1.81725	1.30901	1.04281	1.00445

In[40]:= **Plot[Evaluate[pinH /. is → {0, .1, .25}], {pH, 5, 9}, AxesLabel → {"pH", "$\Delta_r N_H$"}];**

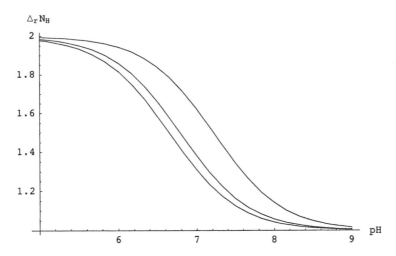

Figure 3.4 $\Delta_r N_H$ (P$_i$)at ionic strengths of 0, 0.10, and 0.25 M at 298.15 K.

The pK of inorganic phosphate is highest at zero ionic strength. Now we go on to consider a biochemical reaction at constant temperature.

3.8 Fundamental Equation for a Single Biochemical Reaction at 298.15 K and the Calculation of Reaction Properties

When a number of reactants are present, the equation for dG ' (3.3-10) can be written as

$$dG' = -S'dT + VdP + \sum \Delta_f G_i' \, dn_i' + RT\ln(10)n_c(\text{H})d\text{pH} \tag{3.8-1}$$

When the reactants are involved in a single biochemical reaction, $dn_{i'}' = \nu_i' d\xi'$, where ν_i' is the stoichiometric number of the ith reactant and ξ' is the extent of the biochemical reaction. The change in the transformed Gibbs energy in the reaction is given by

$$\Delta_r G' = \sum \nu_i' \Delta_f G_i' \tag{3.8-2}$$

Thus equation 3.8-1 can be written as

$$dG' = -S'dT + VdP + \Delta_r G' \, d\xi' + RT\ln(10)n_c(\text{H})d\text{pH} \tag{3.8-3}$$

This equation leads to five Maxwell relations like equations 3.4-12 to 3.4-16.

$$-\Delta_r S'^\circ = \frac{\partial \Delta_r G'^\circ}{\partial T} \tag{3.8-4}$$

$$-\frac{\partial \Delta_r S'^\circ}{\partial \text{pH}} = R\ln(10) \frac{\partial(T\Delta_r N_H)}{\partial T} \tag{3.8-5}$$

$$\frac{\partial \Delta_r G'^\circ}{\partial \text{pH}} = RT\ln(10)\Delta_r N_H \tag{3.8-6}$$

$$\Delta_r H'^\circ = -T^2 \frac{\partial(\Delta_r G'^\circ/T)}{\partial T} \tag{3.8-7}$$

$$\frac{\partial \Delta_r H'^\circ}{\partial \text{pH}} = -RT^2 \ln(10) \frac{\Delta_r N_H}{\partial T} \tag{3.8-8}$$

$\Delta_r N_H$ is the change in the number of hydrogen ions bound in the reaction. It is important to notice that if $\Delta_r G'^\circ$ can be determined as a function of temperature, pH, and ionic srength, all of the other standard transformed thermodynamic properties can be obtained by taking partial derivatives. If K' is determined as a function of pH at a single temperature and $\Delta_r H_i'^\circ$ is not known, $\Delta_r G'^\circ = -RT\ln K'$ and $\Delta_r N_H$ can only be calculated as functions of pH and ionic strength at that temperature.

Now we consider the hydrolysis of ATP to ADP at 298.15 K. The standard transformed properties of inorganic phosphate have been calculated in the preceding section, and the standard transformed properties for ATP, H_2O, and ADP are calculated from the species properties, which are as follows (5):

```
In[41]:=  atpsp = {{-2768.1, -3619.21, -4, 12},
             {-2811.48, -3612.91, -3, 13}, {-2838.18, -3627.91, -2, 14}};
```

```
In[42]:=  adpsp = {{-1906.13, -2626.54, -3, 12},
             {-1947.1, -2620.94, -2, 13}, {-1971.98, -2638.54, -1, 14}};
```

```
In[43]:=  h2osp = {{-237.19, -285.83, 0, 2}};
```

```
In[44]:=  atpG = calcdGmat[atpsp]
```

$$Out[44]= -2.47897 \, \text{Log}\Big[e^{-0.403393\left(-2838.18 + \frac{29.1482\, is^{0.5}}{1+1.6\, is^{0.5}} - 34.7056\, \text{Log}[10^{-pH}]\right)} +$$
$$e^{-0.403393\left(-2811.48 + \frac{11.6593\, is^{0.5}}{1+1.6\, is^{0.5}} - 32.2266\, \text{Log}[10^{-pH}]\right)} + e^{-0.403393\left(-2768.1 - \frac{11.6593\, is^{0.5}}{1+1.6\, is^{0.5}} - 29.7477\, \text{Log}[10^{-pH}]\right)} \Big]$$

```
In[45]:=  adpG = calcdGmat[adpsp]
```

$$Out[45]= -2.47897 \, \text{Log}\Big[e^{-0.403393\left(-1971.98 + \frac{37.8927\, is^{0.5}}{1+1.6\, is^{0.5}} - 34.7056\, \text{Log}[10^{-pH}]\right)} +$$
$$e^{-0.403393\left(-1947.1 + \frac{26.2334\, is^{0.5}}{1+1.6\, is^{0.5}} - 32.2266\, \text{Log}[10^{-pH}]\right)} + e^{-0.403393\left(-1906.13 + \frac{8.74446\, is^{0.5}}{1+1.6\, is^{0.5}} - 29.7477\, \text{Log}[10^{-pH}]\right)} \Big]$$

```
In[46]:=  h2oG = calcdGmat[h2osp]
```

$$Out[46]= -2.47897 \, \text{Log}\Big[e^{-0.403393\left(-237.19 + \frac{5.82964\, is^{0.5}}{1+1.6\, is^{0.5}} - 4.95794\, \text{Log}[10^{-pH}]\right)} \Big]$$

Now we need a program to calculate the function of pH and ionic strength that yields the change in a standard thermodynamic property for a biochemical reaction (3).

```
In[47]:=  deriverxfn[eq_] := Module[{function},
              (*Derives the function of pH and ionic strength that gives the thermodynamic
                 properties of a biochemical reaction typed in the form atpG+h2oG+de==
                 adpG+piG. Other suffixes can be used for H, S, and NH.*)
              function = Solve[eq, de]; function[[1, 1, 2]]]
```

This program can be used to calculate standard transformed Gibbs energies of reaction, standard transformed enthalpies of reaction, standard transformed entropies of reaction and the change in the binding of hydrogen ions in a reaction.

In[48]:= **atphydfnG = deriverxfn[atpG + h2oG + de == adpG + piG]**

Out[48]= $2.47897 \, \text{Log}\left[e^{-0.403393 \left(-237.19 + \frac{5.82964 \, is^{0.5}}{1.+1.6 \, is^{0.5}} - 4.95794 \, \text{Log}[10.^{-1.\,pH}] \right)} \right] + 2.47897$

$\text{Log}\left[e^{-0.403393 \left(-2838.18 + \frac{29.1482 \, is^{0.5}}{1.+1.6 \, is^{0.5}} - 34.7056 \, \text{Log}[10.^{-1.\,pH}] \right)} + e^{-0.403393 \left(-2811.48 + \frac{11.6593 \, is^{0.5}}{1.+1.6 \, is^{0.5}} - 32.2266 \, \text{Log}[10.^{-1.\,pH}] \right)} + \right.$

$\left. e^{-0.403393 \left(-2768.1 - \frac{11.6593 \, is^{0.5}}{1.+1.6 \, is^{0.5}} - 29.7477 \, \text{Log}[10.^{-1.\,pH}] \right)} \right] - 2.47897$

$\text{Log}\left[e^{-0.403393 \left(-1971.98 + \frac{37.8927 \, is^{0.5}}{1.+1.6 \, is^{0.5}} - 34.7056 \, \text{Log}[10.^{-1.\,pH}] \right)} + e^{-0.403393 \left(-1947.1 + \frac{26.2334 \, is^{0.5}}{1.+1.6 \, is^{0.5}} - 32.2266 \, \text{Log}[10.^{-1.\,pH}] \right)} + \right.$

$\left. e^{-0.403393 \left(-1906.13 + \frac{8.74446 \, is^{0.5}}{1.+1.6 \, is^{0.5}} - 29.7477 \, \text{Log}[10.^{-1.\,pH}] \right)} \right] - 2.47897$

$\text{Log}\left[e^{-0.403393 \left(-1137.3 + \frac{2.91482 \, is^{0.5}}{1.+1.6 \, is^{0.5}} - 4.95794 \, \text{Log}[10.^{-1.\,pH}] \right)} + e^{-0.403393 \left(-1096.1 - \frac{8.74446 \, is^{0.5}}{1.+1.6 \, is^{0.5}} - 2.47897 \, \text{Log}[10.^{-1.\,pH}] \right)} \right]$

Table 3.5 Standard transformed Gibbs energies of reaction in kJ mol^{-1} for ATP + H$_2$O = ADP + P$_i$ at 298.15 K

In[49]:= **PaddedForm[TableForm[atphydfnG /. is → {0, .1, .25} /. pH → {5, 6, 7, 8, 9}, TableHeadings →**
{{"I=0", "I=0.10", "I=0.25"}, {"pH 5", "pH 6", "pH 7", "pH 8", "pH 9"}}], {4, 2}]

Out[49]//PaddedForm=

	pH 5	pH 6	pH 7	pH 8	pH 9
I=0	-35.30	-35.91	-37.60	-42.50	-48.29
I=0.10	-33.30	-33.87	-36.50	-41.48	-47.10
I=0.25	-32.56	-33.22	-36.04	-41.07	-46.70

In[50]:= **Plot[Evaluate[atphydfnG /. is → {0, .1, .25}],**
{pH, 5, 9}, AxesLabel → {"pH", "Δ_rG'°"}, PlotRange → {-46, -32}];

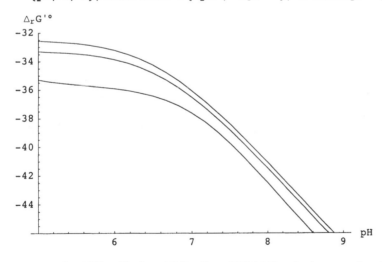

Figure 3.5 $\Delta_r G'°$ for ATP + H$_2$O = ADP + P$_i$ at 298.15 K at ionic strengths of 0, 0.10, and 0.25 M at 298.15 K.

3.9 Calculation of Standard Transformed Enthalpies of Reaction for ATP + H$_2$O = ADP + P$_i$ at 298.15 K

In these calculations we will not print out the functions because they are long, but they can be printed out by removing the semicolons. For the changes in standard transformed enthalpy, an H is appended to the name of the reactant.

In[51]:= **atpH = calcdHmat[atpsp];**

In[52]:= **adpH = calcdHmat[adpsp];**

In[53]:= **h2oH = calcdHmat[h2osp];**

In[54]:= **atphydfnH = deriverxfn[atpH + h2oH + de == adpH + piH];**

Table 3.6 Standard transformed enthalpies of reaction in kJ mol^{-1} for ATP + H$_2$O = ADP + P$_i$ at 298.15 K

In[55]:= **PaddedForm[TableForm[atphydfnH /. is → {0, .1, .25} /. pH → {5, 6, 7, 8, 9}, TableHeadings → {{"I=0", "I=0.10", "I=0.25"}, {"pH 5", "pH 6", "pH 7", "pH 8", "pH 9"}}], {4, 2}]**

Out[55]//PaddedForm=

	pH 5	pH 6	pH 7	pH 8	pH 9
I=0	-23.22	-24.49	-24.38	-22.07	-20.72
I=0.10	-25.82	-25.59	-23.37	-21.48	-21.16
I=0.25	-26.38	-25.71	-23.07	-21.57	-21.35

In[56]:= **Plot[Evaluate[atphydfnH /. is → {0, .1, .25}], {pH, 5, 9}, AxesLabel → {"pH", "Δ$_r$H$_i$'°"}];**

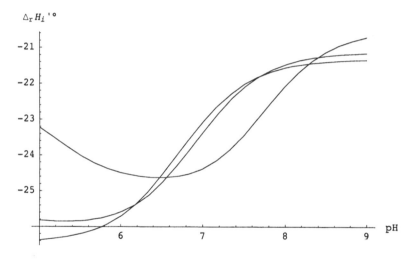

Figure 3.6 $\Delta_r H'°$ kJ mol^{-1} for ATP + H$_2$O = ADP + P$_i$ at 298.15 K at ionic strengths of 0, 0.10, and 0.25 M at 298.15 K.

3.10 Calculation of Standard Transformed Entropies of Reaction for ATP + H$_2$O = ADP + P$_i$ at 298.15 K

The functions of pH and ionic strength that give the standard transformed Gibbs energies of formation for the reactants are given the suffix S.

In[57]:= **atpS = derivetrS[atpsp];**

In[58]:= **adpS = derivetrS[adpsp];**

In[59]:= **h2oS = derivetrS[h2osp];**

In[60]:= **atphydfnS = deriverxfn[atpS + h2oS + de == adpS + piS];**

Table 3.7 Standard transformed entropies of reaction in J K^{-1} mol^{-1} for ATP + H$_2$O = ADP + P$_i$ at 298.15 K

In[61]:= **PaddedForm[TableForm[atphydfnS /. is → {0, .1, .25} /. pH → {5, 6, 7, 8, 9}, TableHeadings → {{"I=0", "I=0.10", "I=0.25"}, {"pH 5", "pH 6", "pH 7", "pH 8", "pH 9"}}], {4, 2}]**

Out[61]//PaddedForm=

	pH 5	pH 6	pH 7	pH 8	pH 9
I=0	40.50	38.31	44.35	68.52	92.49
I=0.10	25.08	27.76	44.05	67.08	87.00
I=0.25	20.75	25.17	43.48	65.43	85.04

In[62]:= **Plot[Evaluate[atphydfnS /. is → {0, .1, .25}], {pH, 5, 9}, AxesLabel → {"pH", "$\Delta_f S_i$'°"}];**

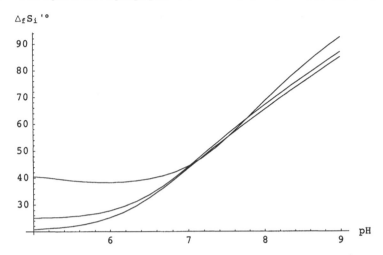

Figure 3.7 $\Delta_r S'°$ in J K^{-1} mol^{-1} for ATP + H_2 O = ADP + P_i at ionic strengths of 0, 0.10, and 0.25 M at 298.15 K.

3.11 Calculation of Changes in the Binding of Hydrogen Ions in the Reaction ATP + H_2 O = ADP + P_i at 298.15 K

In[63]:= **atpnH = calcavHbound[atpsp];**

In[64]:= **adpnH = calcavHbound[adpsp];**

In[65]:= **h2onH = calcavHbound[h2osp];**

In[66]:= **atphydfnnH = deriverxfn[atpnH + h2onH + de == adpnH + pinH];**

Table 3.8 Changes in the binding of hydrogen ions in the reaction ATP + H_2 O = ADP + P_i at 298.15 K

In[67]:= **PaddedForm[**
    TableForm[atphydfnnH /. is → {0, .1, .25} /. pH → {5, 6, 7, 8, 9}, TableHeadings →
      {{"I=0", "I=0.10", "I=0.25"}, {"pH 5", "pH 6", "pH 7", "pH 8", "pH 9"}}], {3, 4}]

Out[67]//PaddedForm=

	pH 5	pH 6	pH 7	pH 8	pH 9
I=0	-0.1470	-0.1180	-0.5780	-1.0100	-1.0100
I=0.10	-0.0459	-0.2150	-0.7200	-0.9610	-0.9960
I=0.25	-0.0384	-0.2490	-0.7420	-0.9650	-0.9960

In[68]:= **Plot[Evaluate[atphydfnnH /. is → {0, .1, .25}],**
    {pH, 5, 9}, AxesLabel → {"pH", "$\Delta_r N_H$"}, AxesOrigin → {5, -1}];

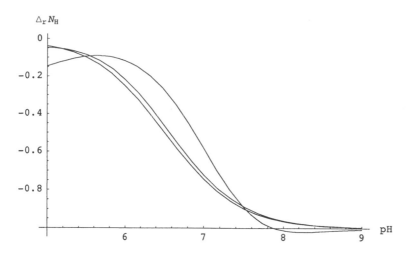

Figure 3.8 $\Delta_r N_H$ for ATP + H$_2$O = ADP + P$_i$ at 298.15 K at ionic strengths of 0, 0.10, and 0.25 M at 298.15 K.

3.12 Transformed Gibbs Energies of Formation of the Species of ATP at Equilibrium

It is important to understand that at equilibrium at a specified pH the $\Delta_f G$ ' values for the species of a reactant are equal. This is what makes it possible to aggregate the species and take this $\Delta_f G$ ' to be the $\Delta_f G$ ' for the sum of species. As an example, consider ATP at 298.15 K, pH 5.9, and ionic strength 0.25 M.

The two pKs for ATP at 298.15 K and 0.25 M strength can be calculated using **calcpK** (see Section 1.6).

```
In[69]:= calcpK[speciesmat_, no_, is_] :=
            Module[{lnkzero, sigmanuzsq, lnK}, (*Calculates pKs for a weak acid at 298.15
                K at specified ionic strengths (is) when the number no of the pK is
                specified. pKs are numbered 1,2,3,... from the highest pK to the lowest pK,
                but the highest pK for a weak acid may be omitted if it is outside
                of the range 5 t0 9. For h3PO4,pK1=calc[pisp,1,{0}]=7.22.*)
            lnkzero = (speciesmat[[no + 1, 1]] - speciesmat[[no, 1]]) / (8.31451 * 0.29815);
            sigmanuzsq = speciesmat[[no, 3]]^2 - speciesmat[[no + 1, 3]]^2 + 1;
            lnK = lnkzero + (1.17582 * is^0.5 * sigmanuzsq) / (1 + 1.6 * is^0.5);
            N[-(lnK / Log[10])]]
```

```
In[70]:= pK1ATP = calcpK[atpsp, 1, .25]
```

```
Out[70]= 6.46502
```

```
In[71]:= pK2ATP = calcpK[atpsp, 2, .25]
```

```
Out[71]= 3.82652
```

The equilibrium mole fractions r_j of the three species of ATP were calculated in Section 2.1 using

```
In[72]:= patp = 1 + 10^(pK1ATP-pH) + 10^(pK1ATP+pK2ATP-2 pH) ;
```

```
In[73]:= r1 = 1 / patp;
```

```
In[74]:= r2 = (10^(pK1ATP-pH)) / patp;
```

```
In[75]:= r3 = 10^(-2 pH+pK1ATP+pK2ATP) / patp;
```

These equations can be used to calculate equilibrium mole fractions of the three species of ATP.

In[76]:= `{r1, r2, r3} /. pH → 5.9`

Out[76]= `{0.212584, 0.780823, 0.0065929}`

    The standard transformed Gibbs energy of formation of a species is given as a function of pH and ionic strength by equation 3.6-1. This property can be calculated by use of the following program, which is a part of **derivetrG** (see Section 3.6f):

In[77]:= `derivetrGspecies[speciesmat_] :=`
`Module[{dGzero,dHzero, zi, nH, pHterm, isterm},(*This program derives the functions`
`of pH and ionic strength (is) that gives the standard transformed Gibbs energies of`
`formation of the species of a reactant at 298.15 K.  The input speciesmat is a`
`matrix that gives the standard Gibbs energy of formation, the standard enthalpy of`
`formation, the electric charge, and the number of hydrogen atoms in each species.`
`There is a row in the matrix for each species of the reactant. The output is a list`
`of the functions for the species of the reactant.  Energies are expressed in kJ`
`mol^-1.*)`
`{dGzero,dHzero,zi,nH}=Transpose[speciesmat];`
`pHterm = nH*8.31451*.29815*Log[10^-pH];`
`isterm = 2.91482*((zi^2) - nH)*(is^.5)/(1 + 1.6*is^.5);`
`dGzero - pHterm - isterm]`

In[78]:= `atpspeciesfns = derivetrGspecies[atpsp];`

In[79]:= `atpspeciesfns /. pH -> 5.9 /. is → .25`

Out[79]= `{-2367.21, -2370.43, -2358.6}`

These are the standard transformed Gibbs energies of formation of the three species of ATP, but now we need to calculate the transformed Gibbs energies of the species at their equilibrium concentrations when [ATP] = 0.01 M, T = 298.15 K, pH = 5.9, and I = 0.25 M. Equation 3.4-10 is used to calculate $\Delta_f G$ ' for the three species:

In[80]:= `(atpspeciesfns /. pH -> 5.9 /. is → .25) + 8.31451*.29815*Log[.01*{r1, r2, r3} /. pH → 5.9]`

Out[80]= `{-2382.46, -2382.46, -2382.46}`

    Since the three species have the same $\Delta_f G'$ at equilibrium, this can be taken as $\Delta_f G$ ' (ATP,0.01 M,298.15 K,pH 5.9, I = 0.25 M) in the equation $\Delta_f G' = \Delta_f G'^\circ + RT\ln[\text{ATP}]$. Therefore, $\Delta_f G$ '°(ATP,0.01 M,298.15 K,pH 5.9, I = 0.25 M) is given by

In[81]:= `-2382.46 - 8.31451*.29815*Log[.01]`

Out[81]= `-2371.04`

The usual way to calculate this property is

In[82]:= `calcdGmat[atpsp] /. pH → 5.9 /. is → .25`

Out[82]= `-2371.05`

3.13 Discussion

    Chapters 1 and 2 dealt with species properties. But in this chapter we have found that when the pH is specified, it is necessary to use a Legendre transform to define a transformed Gibbs energy G ' of a system and that this automatically brings in a transformed enthalpy H ' and transformed entropy S '. In fact we have entered a whole new world of thermodynamics where attention is focussed on reactants, which are sums of species, rather than on species. This world of biochemical thermodynamics is very similar to the world of chemical thermodynamics in that there are fundamental equations and Maxwell equations, but there is a new type of term in the fundamental equation for G' that is proportional to dpH. This new term brings in the average binding of hydrogen ions $\overline{N}_H$ (i) by a reactant and the change in binding of hydrogen ions $\Delta_r N_H$ in a biochemical reaction. When enzyme-catalyzed reactions are studied at a single temperature and there is no calorimetric data, the measurement of apparent equilibrium constants leads to $\Delta_r G'^\circ$ and $\Delta_r N_H$ that are functions of pH, provided that pKs are known.

When enzyme-catalyzed reactions are studied at a series of temperatures or there are calorimetric data, it is possible to calculate in addition $\Delta_r H'^\circ$ and $\Delta_r S'^\circ$, provided that the temperature dependencies of the pKs have been determined. In this chapter we have emphasized calculations at 298.15 K, including $\Delta_r H'^\circ$ and $\Delta_r S'^\circ$, but we have not fully utilized the enthalpy information. In Chapter 4, we will use the enthalpy information to calculate transformed thermodynamic properties at other temperatures. This will make it possible to utilize more Maxwell relations that show how various transformed thermodynamic properties are necessarily interrelated.

Appendix

Natural Variables

The choice of independent variables is a very important decision in thermodynamics (6). In chemical thermodynamics, the independent variables are usually T, P, and amounts of species, and the criterion of spontaneous change and equilibrium is provided by the Gibbs energy G. When G can be expressed as a function of T, P, and $\{n_j\}$, the total differential of G can be expressed by a fundamental equation made up of additive terms proportional to dT, dP, and $\{dn_j\}$. For example, if g is a function of x and y, the total differential of g is given by

$$dg = \frac{\partial g}{\partial x}\,dx + \frac{\partial g}{\partial y}\,dy = Mdx + Ndy \tag{A3-1}$$

In this book it is always to be understood that in taking the partial derivative of g with respect to x, then y is held constant, and in taking the partial derivative of g with respect to y, then x is held constant. Since we will be dealing later with functions of many variables, it is impractical to indicate the variables held constant as subscripts. If we take the differential of the coefficient of the first term of equation A3-1 with respect to y we obtain $\frac{\partial^2 g}{\partial x\,\partial y}$, and if we take the differential of the coefficient of the second term of equation A3-1 with respect to x we obtain $\frac{\partial^2 g}{\partial y\,\partial x}$. These two second derivatives are equal, as indicated by

$$\frac{\partial^2 g}{\partial x\,\partial y} = \frac{\partial^2 g}{\partial y\,\partial x} \quad \text{or} \quad \frac{\partial M}{\partial y} = \frac{\partial N}{\partial x} \tag{A3-2}$$

These relations are very important in thermodynamics, where they are referred to as **Maxwell relations**.

The fundamental equation for a thermodynamic potential like G is very important because it shows that the coefficients of the differential terms can be obtained by taking partial derivatives of G with respect to T, P, and $\{n_j\}$. The Gibbs energy of a chemical reaction system can also be expressed as a function of T, V, and $\{n_j\}$, but, when this is done, the other thermodynamic properties of the system **cannot** be calculated from this function. Because of the special importance of T, P, and $\{n_j\}$, these particular variables are referred to as the **natural variables** of G. In this chapter, pH has been introduced as a natural variable by making a Legendre transform. When pH is an independent variable, G no longer provides the criterion for spontaneous change and equilibrium. Therefore, it is necessary to define a transformed Gibbs energy G' that does.

Legendre Transforms

The choice of independent variables in studying a system is very important decision in thermodynamics and mechanics, and doubtlessly elsewhere. There are two ways to introduce a new independent variable: (1) Simply substitute the expression for one of the variables for the system in terms of a new variable. (2) Make a Legendre transform to define a new property that has the desired independent variables. An example of the first way is changing the equation for a property in terms of the Celsius temperature t to an equation in terms of the thermodynamic temperature T. This is done by simply substituting $t = T - 273.15$. The advantage of a Legendre transform is that it makes it possible to introduce the derivative of a property to be a new variable. This involves defining a new property of a system by subtracting a product of conjuugate variables from a property of the system (6,15-17). We have already seen examples of this in the definitions in Section 1.1 of the enthalpy ($H = U + PV$) and Gibbs energy ($G = H - TS$). For a system of chemical reactions the internal energy U provides a criterion for spontaneous change and equilibrium; $dU \leq 0$ at constant S, V, and $\{n_j\}$. The enthalpy H provides a criterion for spontaneous change and equilibrium; $dH \leq 0$ at constant S, P, and $\{n_j\}$. The Gibbs energy G provides a criterion for spontaneous change and equilibrium; $dG \leq 0$ at constant T, P, and $\{n_j\}$. This shows why the Gibbs energy is so useful in chemistry. However, when the species are involved in chemical reactions that are at equilibrium, the $\{n_j\}$ are not independent variables, but the amounts of components $\{n_c\}$ are. In contrast with amounts of species, amounts of components are conserved in a reaction system. Since amounts of elements are conserved, we can think of components as the elements, but we will see later (Chapter 6) that for enzyme-catalyzed reactions groups of atoms of elements can be conserved in addition. For each component there is a conservation equation, but only independent conservation equations are useful as constraints (7).

The first application of a Legendre transform in biochemical thermodynamics introduces the pH as an independent variable. Specifying the pH really amounts to specifying the chemical potential of hydrogen ions, and the conjugate variable is the total amount of hydrogen atoms in the system. When the pH is specified, it is necessary to define a transformed Gibbs energy G' with the Legendre transform

$$G' = G - n_c(\text{H})\mu(\text{H}^+) \tag{A3-3}$$

The transformed Gibbs energy provides the criterion for spontaneous change and equilibrium in systems of enzyme-catalyzed reactions when the independent variables for the system are T, P, pH, and $\{n_c\}$. Notice that making this Legendre transform has introduced $\mu(\text{H}^+)$ as a natural variable, but it has not changed the number of natural variables because there is now one less component that is conserved, the hydrogen atom component.

In considering systems of enzyme-catalyzed reactions at steady state concentrations of coenzymes, a Legendre transform can be used to define a further transformed Gibbs energy G'' (17).

$$G'' = G' - \sum n_c(\text{coenzyme})\Delta_f G_i'(\text{coenzyme}) \tag{A3-4}$$

This is discussed in Chapter 7.

There is a limit to the number of Legendre transforms that can be made on a given system. If all the extensive variables are eliminated as natural variables by making Legendre transforms, the new thermodynamic potential that is defined is equal to zero. This is one way to derive the Gibbs-Duhem equation for a system; this equation provides a relation between the intensive variables for the system. Thus one intensive variable for a system can always be expressed in terms of all the others. The rule for making Legendre transforms that yield criteria for spontaneous change and equilibrium is that one component must remain.

More complete descriptions of Legendre transforms are provided by Callen (6) and Alberty (15). There is an IUPAC Technical Report on Legendre transforms (18). It is interesting to note that a Legendre transform is used in defining the Hamiltonian for a mechanical system on the basis of the Lagrangian (19).

Isomer Group Thermodyanmics

When isomers are in equilibrium in a chemical reaction system, they can be considered to be a single reactant because the mole fractions of the various isomers are independent of the reactions they are involved in. But the isomers do have different standard Gibbs energies of formation. The question is "What is the standard Gibbs energy of formation of the isomer group?" The answer is not a mole fraction-weighted average because there is a free energy of mixing of the isomers. It is readily shown that the standard Gibbs energy of formation of an isomer group $\Delta_f G_i°$ is given in terms of the standard Gibbs energies of formation $\Delta_f G_j°$ of the species it contains by (12)

$$\Delta_f G_i° = -RT\ln\sum_{j=1}^{N}\exp(-\Delta_f G_j°/RT) \tag{A3-5}$$

where N is the number of different species in the pseudoisomer group. The same result can be obtained by taking the mole-fraction-weighted average and adding the Gibbs energy of mixing. The equilibrium mole fractions r_j of the species in the isomer group are given by

$$r_j = \exp[(\Delta_f G_i° - \Delta_f G_j°)/RT] \tag{A3-6}$$

The standard enthalpy of formation of reactant i is given by the mole fraction weighted average of the species:

$$\Delta_f H_i° = \sum r_j \Delta_f H_j° \tag{A3-7}$$

In a biochemical reaction system at a specified pH, the various protonated forms of a reactant are pseudoisomers, as discussed in Section 3.12. They are pseudoisomers in the sense that their equilibrium mole fractions in the pseudoisomer group are dependent only on the pH, which is specified. Thus at specified pH, equations A3-5, A3-6, and A3-7 are replaced by equations 3.5-5, 3.5-6, and 3.5-7.

Gibbs-Duhem Equation

There are two ways to derive the Gibbs-Duhem equation for a system: (1) Subtract the fundamental equation from the total differential of the thermodynamic potential. (2) Use a complete Legendre transform. As an example of the second method, consider fundamental equation 3.3-10 for a single reactant:

$$dG' = -S'dT + VdP + \Delta_f G_i'd n_i' + RT\ln(10)n_c(H)dpH \tag{A3-8}$$

The natural variables of G' are T, P, n_i', and pH of which only n_i' is an extensive variable. This extensive variable can be eliminated as a natural variable by using the following Legendre transform:

$$G^* = G' - n_i' \Delta_f G_i'$$
(A3-9)

This leads to

$$dG^* = -S'dT + VdP - n_i'd\Delta_f G_i' + RT\ln(10)n_c(H)dpH$$
(A3-10)

Substituying equation 3.8-1 and the differential of equation 3.3-11 in this equation shows that $dG^* = 0$. With dG^* set equal to zero, equation A3-10 is the Gibbs-Duhem equation. All the differentials in equation A3-10 are intensive variables, and so the intensive variables for the system are not all independent. Equation A3-10 shows, for example, that $\Delta_f G_i'^\circ$ is a function of T, P, and pH.

References

1. J. W. Gibbs, The Scientific Papers of J. Willard Gibbs,Vol. 1, Thermodynamics, Dover, New York, 1961.
2. R. J. Silbey, R. A. Alberty, and M. Bawendi, Physical Chemistry, Wiley, Hoboken, NJ (2005).
3. R. A. Alberty, Thermodynamics of Biochemical Reactions, Wiley, Hoboken, NJ (2003).
4. D. D. Wagman, W. H. Evans, V. B. Parker, R. H. Schumm, I. Halow, S. M. Bailey, K. L. Churney, and R. L. Nuttall, The NBS tables of chemical thermodynamic properties, J. Phys. Chem. Ref. Data, 11, Supplement 2 (1982).
5. R. A. Alberty, BasicBiochemData3 (2005).

In[83]:= **http : // library.wolfram.com / infocenter / MathSource / 5704**

6. H. B. Callen, Thermodynamics and an Introduction to Thermostatistics, Wiley, Hoboken, NJ (1985).
7. J. A. Beattie and I. Oppenheim, Principles of Thermodynamics, Elsevier, Amsterdam, 1979.
8. R. A. Alberty, Equilibrium calculations on systems of biochemical reactions, Biophys. Chem. 42, 117-131 (1992).
9. R. A. Alberty, Calculation of transformed thermodynamic properties of biochemical reactants at specified dpH and pMg, Biophys. Chem. 43, 239-254 (1992).
10. R. A. Alberty and R. N. Goldberg, Calculation of thermodynamic formation properties for the ATP series at specified pH and pMg. Biochem. 31, 10610-10615 (1992).
11. R. A. Alberty and I. Oppenheim, Use of semigrand ensembles in chemical equilibrium calculations on complex systems, J. Chem. Phys. 91, 1824-1828 (1989).
12. W. R. Smith and R. W. Missen, Chemical Equilibrium Reaction Analysis: Theory and Algorithms, Wiley, Hoboken, NJ (1982).
13. R. A. Alberty, Chemical thermodynamic properties of isomer groups, I&EC Fund. 22, 218-321 (1983).
14. R. A. Alberty, Inverse Legendre transform in biochemical thermodynamics: Applied to the last five reactions of glycolysis, J. Phys. Chem. 106, 6594-6599 (2002).
15. R. A. Alberty, Legendre transforms in chemical thermodynamics, Chem. Rev. 94, 1457-1482 (1994).
16. R. A. Alberty, A. Cornish-Bowden, Q. H. Gibson, R. N. Goldberg, G. G. Hammes, W. Jencks, K. F. Tipton, R. Veech, H. V. Westerhof, and E. C. Webb, Recommendations for nomenclature and tables for biochemical thermodynamics, Pure Appl. Chem. 66, 1641-1666 (1994).

http://www.chem.qmw.ac.uk/iubmb/thermod/

17. R. A. Alberty, Calculation of equilibrium compositions of large systems of biochemical reactions, J. Phys. Chem. 104 B 4807-4814 (2000).
18. R. A. Alberty J. M. G. Barthel, E. R. Cohen, M. B. Ewing, R. N. Goldberg, and E. Wilhelm, Use of Legendre transforms

in chemical thermodynamics (an IUPAC technical report), Pure. Appl. Chem. 73, No. 8 (2001).

19. H. Goldstein, Classical Mechanics, Addison-Wesley, Reading, MA (1980).

Chapter 4 Biochemical Reactions at Various pHs and Various Temperatures

4.1 Dependence of the Standard Transformed Gibbs Energy of a Reactant on T, pH, and Ionic Strength

This chapter is concerned with reactants for which $\Delta_f G°$ and $\Delta_f H°$ are known for all the species that have significant concentrations in the pH range 5 to 9. The small data matrices for these reactants provide the information to derive the expression for $\Delta_f G'°$ for reactants as functions of T, pH, and ionic strength. This mathematical function contains all the thermodynamic information about the reactant in dilute aqueous solutions in the temperature range 273.15 to about 313.15 K, pH 5 to 9, and ionic strengths zero to 0.35 M. The other standard transformed thermodynamic properties are calculated from the functions for $\Delta_f G'°$ using the Maxwell relations derived in Chapter 3. These relations make it possible to obtain the mathematical functions that yield $\Delta_f H'°$, $\Delta_f S'°$, and $\overline{N}_H$ for a reactant by simply taking partial derivatives. These functions are complicated, but *Mathematica* is very useful because of its symbolic capabilities and convenience in making tables and plots. As mentioned in the *Mathematica* book, partial derivatives of almost any function can be taken because the chain rule can be used over and over again. .

When $\Delta_f G°$ and $\Delta_f H°$ are known for all species in an enzyme-catalyzed reaction, the standard transformed Gibbs energies of the reactants can be added and subtracted to obtain $\Delta_r G'°$. This very complicated function contains all of the information needed to derive functions of temperature, pH and ionic strength for the reaction properties $\Delta_r G'°$, $\Delta_r H'°$, $\Delta_r S'°$, $\Delta_r N_H$, and more. These very complicated functions can be used to construct tables and plots of standard transformed properties.

In 2001 a *Mathematica* program was written (1) to calculate $\Delta_f G_i'^\circ$ and $\Delta_f H_i'^\circ$ as functions of T, pH, and ionic strength for a reactant. In 2003 the *Mathematica* program **calcthprops** was written (2) to calculate $\Delta_f G_i'^\circ$ and other properties for a biochemical reactant. In this chapter we will use a program **derivetrGibbsT** to calculate $\Delta_r G'^\circ$ as a function of T, pH, and ionic strength for a reactant. This chapter requires the data of BasicBiochemData3 (2).

```
In[2]:=  Off[General::spell1];
         Off[General::spell];

In[4]:=  << BiochemThermo`BasicBiochemData3`

In[5]:=  derivetrGibbsT[speciesmat_]:=Module[{dGzero,dGzeroT,dHzero,zi,nH,gibbscoeff,pHterm,
         isterm,gpfnsp},(*This program derives the function of T (in Kelvin), pH, and ionic
         strength (is) that gives the standard transformed Gibbs energy of formation of a
         reactant (sum of species).  The input speciesmat is a matrix that gives the standard
         Gibbs energy of formation in kJ mol^-1 at 298.15 K and zero ionic strength, the
         standard enthalpy of formation in kJ mol^-1 at 298.15 K and zero ionic strength, the
         electric charge, and the number of hydrogen atoms in each species.  There is a row in
         the matrix for each species of the reactant.  gpfnsp is a list of the functions for
         the transformed Gibbs energies of the species.  The corresponding functions for other
         transformed properties can be obtained by taking partial derivatives.  The standard
         transformed Gibbs energy of formation of a reactant in kJ mol^-1 can be calculated at
         any temperature in the range 273.15 K to 313.15 K, any pH in the range 5 to 9, and
         any ionic strength in the range 0 to 0.35 M by use of the assignment operator (/.).*)
         {dGzero,dHzero,zi,nH}=Transpose[speciesmat];
         gibbscoeff=(9.20483*t)/10^3-(1.284668*t^2)/10^5+(4.95199*t^3)/10^8;
         dGzeroT=(dGzero*t)/298.15+dHzero*(1-t/298.15);
         pHterm=(nH*8.31451*t*Log[10^(-pH)])/1000;
         istermG=(gibbscoeff*(zi^2-nH)*is^0.5)/(1+1.6*is^0.5);
         gpfnsp=dGzeroT-pHterm-istermG;
         -((8.31451*t*Log[Plus@@(E^(-(gpfnsp/((8.31451*t)/1000))))])/1000)]
```

This program is especially important because when $\Delta_f G^\circ$ and $\Delta_f H^\circ$ are known for all the species of a reactant that have significant concentrations in the pH range 5 to 9, all the other standard transformed properties can be calculated by taking partial derivatives. The first three sections of this chapter will be concerned with making these calculations for 94 reactants. These properties of reactants can be added and subtracted to obtain the corresponding reaction properties and the apparent equilibrium constant K'. The thermodynamic properties for an enzyme-catalyzed reaction can be calculated using the program **derivefnGHSNHrx** (3), which is given later.

When K' has been determined over a range of temperatures or the heat of reaction has been determined calorimetrically, it is possible to learn more about the thermodynamics of an enzyme-catalyzed reaction or a macromolecule-ligand binding reaction. Since it is generally not possible to determine K' over a very wide range of temperature, it is usually found that the plot of $\log K'$ versus $1/T$ is linear so that the standard transformed enthalpy of reaction is independent of temperature. If the plot of $\log K'$ versus $1/T$ is curved, the change in transformed heat capacity in the reaction can be calculated. The transformed heat capacities of reactants can be determined using special calorimeters designed for that purpose. To a first approximation, the standard enthalpy of formation of a species at zero ionic strength is given as a function of temperature by

$$\Delta_f H_j^\circ(T,I=0) = \Delta_f H_j^\circ(298.15,I=0) + C_{Pm}^\circ(j, I=0)(T - 298.15) \qquad (4.1\text{-}1)$$

It is necessary to specify zero ionic strength here because Debye-Hückel adjustments for ionic strength depend on the temperature. Heat capacities and transformed heat capacities are discussed in an Appendix to this chapter. However, since there is not very much information in the literature on heat capacities of species or transformed heat capacities of reactants, the treatments described here are based on the assumption that heat capacities of species are equal to zero. When molar heat capacities of species can be taken as zero, both standard enthalpies of formation and standard entropies of formation of species are independent of temperature. When $\Delta_f H^\circ$ and $\Delta_f S^\circ$ are independent of temperature, standard Gibbs energies of formation of species at zero ionic strength can be calculated using

$$\Delta_f G_j^\circ(T,I=0) = \Delta_f H_j^\circ(298.15,I=0) - T\Delta_f S_j^\circ(298.15,I=0) \qquad (4.1\text{-}2)$$

This equation can be written in terms of $\Delta_f G_j°(298.15, I=0)$ and $\Delta_f H_j°(298.15, I=0)$ by substituting $\Delta_f H_j°(298.15, I=0) - \Delta_f G_j°(298.15, I=0)$ for $T\Delta_f S_j°(298.15, I=0)$ to obtain

$$\Delta_f G_j°(T, I=0) = (T/298.15)\Delta_f G_j°(298.15, I=0) + (1 - T/298.15)\Delta_f H_j°(298.15, I=0) \qquad (4.1-3)$$

This equation was introduced in Chapter 1 (see equation 1.7-1), and it is used in the calculations discussed in this chapter. Substituting this equation for a species into equation 3.6-1 of the previous chapter yields a mathematical function of T, pH, and ionic strength for the standard transformed Gibbs energy of a species.

$$\Delta_f G_j'°(T, \text{pH}, I) = (T/298.15)\Delta_f G_j°(298.15, I=0) + (1 - T/298.15)\Delta_f H_j°(298.15, I=0) +$$
$$N_H(j)RT\ln(10)\,\text{pH} - RT\alpha(z_i^2 - N_H(j))\,I^{1/2}/(1 + 1.6\,I^{1/2}) \qquad (4.1-4)$$

The temperature dependence of α is given by an empirical equation in Section 1.4. When this empirical function is substituted in equation 4.1-4, it possible to express $\Delta_f G_j'°$ for a species as a function of temperature, pH and ionic strength. Equation 3.5-5 makes it possible to express $\Delta_f G'°$ for a reactant as a function of temperature, pH, and ionic strength. Although it is impractical to write out such an equation for a reactant, it can be derived by using **derivetrGibbsT**. As an example, the function of T, pH, and ionic strength for inorganic phosphate is derived from the properties of its species that are given by

In[6]:= **phossp = {{-1096.1, -1299., -2, 1}, {-1137.3, -1302.6, -1, 2}};**

We will use this name for the species matrix, rather than the pisp in BasicBiochemData3 to avoid shadowing.

In[7]:= **phosGT = derivetrGibbsT[phossp]**

$$Out[7]= -0.00831451\, t\, \text{Log}\left[e^{-\frac{120.272\left(-1302.6\,(1-0.00335402\,t)-3.81452\,t+\frac{is^{0.5}\,(0.00920483\,t-0.0000128467\,t^2+4.95199\times10^{-8}\,t^3)}{1+1.6\,is^{0.5}}-0.016629\,t\,\text{Log}[10^{-\text{pH}}]\right)}{t}}\right] +$$

$$e^{-\frac{120.272\left(-1299.\,(1-0.00335402\,t)-3.67634\,t-\frac{3\,is^{0.5}\,(0.00920483\,t-0.0000128467\,t^2+4.95199\times10^{-8}\,t^3)}{1+1.6\,is^{0.5}}-0.00831451\,t\,\text{Log}[10^{-\text{pH}}]\right)}{t}}\,\Big]$$

The function phosGT can be used to construct the following table of the standard transformed Gibbs energies of formation of inorganic phosphate.

Table 4.1 Standard transformed Gibbs energies of formation of inorganic phosphate in kJ mol^{-1} at three temperatures, five pHs, and three ionic strengths

In[8]:= **TableForm[phosGT /. t → {273.15, 298.15, 313.15} /. pH → {5, 6, 7, 8, 9} /. is → {0, .1, .25},**
      TableHeadings → {{"273.15 K", "298.15 K", "313.15 K"},
         {"   pH 5", "pH 6", "pH 7", "pH 8", "pH 9"}, {"I=0", "I=0.10", "I=0.25"}}]

Out[8]//TableForm=

		pH 5	pH 6	pH 7	pH 8	pH 9
	I=0	-1098.88	-1088.52	-1078.91	-1071.67	-1066.09
273.15 K	I=0.10	-1098.36	-1088.16	-1079.37	-1073.05	-1067.68
	I=0.25	-1098.2	-1088.08	-1079.62	-1073.53	-1068.2
	I=0	-1080.23	-1068.95	-1058.56	-1050.81	-1044.77
298.15 K	I=0.10	-1079.65	-1068.56	-1059.17	-1052.42	-1046.58
	I=0.25	-1079.46	-1068.49	-1059.49	-1052.97	-1047.17
	I=0	-1069.05	-1057.2	-1046.35	-1038.3	-1031.97
313.15 K	I=0.10	-1068.42	-1056.81	-1047.06	-1040.05	-1033.93
	I=0.25	-1068.22	-1056.75	-1047.43	-1040.65	-1034.57

4.2 Calculation of Other Standard Transformed Thermodynamic Properties of a Reactant by Taking Partial Derivatives

As shown by the Maxwell relations in equations 3.4-12 to 3.4-16, all the other thermodynamic properties of a biochemical reactant can be calculated by taking partial derivatives of the function of T, pH, and ionic strength for $\Delta_f G_i'^\circ$. This is illustrated by calculating the other standard transformed thermodynamic properties of inorganic phosphate.

Table 4.2 Standard transformed enthalpies of formation of inorganic phosphate in kJ mol^{-1} at three temperatures, five pHs, and three ionic strengths

```
In[9]:= TableForm[-t^2*D[phosGT/t, t] /. t → {273.15, 298.15, 313.15} /. pH → {5, 6, 7, 8, 9} /.
           is → {0, .1, .25}, TableHeadings → {{"273.15 K", "298.15 K", "313.15 K"},
              {"      pH 5", "pH 6", "pH 7", "pH 8", "pH 9"}, {"I=0", "I=0.10", "I=0.25"}}]
```

```
Out[9]//TableForm=
```

		pH 5	pH 6	pH 7	pH 8	pH 9
	I=0	-1302.58	-1302.42	-1301.35	-1299.57	-1299.07
273.15 K	I=0.10	-1302.76	-1302.28	-1300.23	-1298.64	-1298.36
	I=0.25	-1302.81	-1302.15	-1299.79	-1298.36	-1298.14
	I=0	-1302.58	-1302.39	-1301.24	-1299.51	-1299.06
298.15 K	I=0.10	-1302.83	-1302.23	-1299.91	-1298.35	-1298.1
	I=0.25	-1302.9	-1302.05	-1299.39	-1297.99	-1297.79
	I=0	-1302.58	-1302.38	-1301.18	-1299.48	-1299.05
313.15 K	I=0.10	-1302.88	-1302.2	-1299.7	-1298.15	-1297.91
	I=0.25	-1302.96	-1301.99	-1299.12	-1297.73	-1297.54

Since these calculations are based on the assumption that $C_{Pm}^\circ(j,I=0) = 0$, it might have been expected that these $\Delta_f H_i'^\circ$ should be independent of temperature and ionic strength. There are small changes for two reasons: The first is that the Debye-Hückel adjustments for ionic strength are functions of temperature. The second is that the composition of the phosphate pseudoisomer group changes with the pH. According to Le Chatelier's principle, as the temperature is raised the equilibrium shifts in the direction to absorb heat. These effects are discussed in an Appendix of this chapter.

Table 4.3 Standard transformed entropies of formation of inorganic phosphate in J K^{-1} mol^{-1} at three temperatures, five pHs, and three ionic strengths

```
In[10]:= PaddedForm[
            TableForm[-D[phosGT, t] * 1000 /. t → {273.15, 298.15, 313.15} /. pH → {5, 6, 7, 8, 9} /.
               is → {0, .1, .25}, TableHeadings → {{"273.15 K", "298.15 K", "313.15 K"},
                  {"      pH 5", "pH 6", "pH 7", "pH 8", "pH 9"}, {"I=0", "I=0.10", "I=0.25"}}], 5]
```

```
Out[10]//PaddedForm=
```

		pH 5	pH 6	pH 7	pH 8	pH 9
	I=0	-745.75	-783.06	-814.34	-834.34	-852.92
273.15 K	I=0.10	-748.32	-783.9	-808.56	-825.86	-844.53
	I=0.25	-749.08	-783.68	-806.02	-823.09	-841.82
	I=0	-745.74	-782.98	-813.96	-834.13	-852.89
298.15 K	I=0.10	-748.57	-783.73	-807.46	-824.84	-843.59
	I=0.25	-749.4	-783.36	-804.62	-821.79	-840.58
	I=0	-745.74	-782.93	-813.76	-834.03	-852.88
313.15 K	I=0.10	-748.74	-783.62	-806.78	-824.19	-842.97
	I=0.25	-749.6	-783.14	-803.74	-820.94	-839.77

Table 4.4 Average numbers of hydrogen ions bound by inorganic phosphate at three temperatures, five pHs, and three ionic strengths

```
In[11]:= PaddedForm[TableForm[
         (1 / (8.31451 * (t / 1000) * Log[10])) * D[phosGT, pH] /. t → {273.15, 298.15, 313.15} /.
           pH → {5, 6, 7, 8, 9} /. is → {0, .1, .25},
         TableHeadings → {{"273.15 K", "298.15 K", "313.15 K"},
            {"        pH 5", "pH 6", "pH 7", "pH 8", "pH 9"}, {"I=0", "I=0.10", "I=0.25"}}], 3]
```

Out[11]//PaddedForm=

		pH 5	pH 6	pH 7	pH 8	pH 9
	I=0	1.99	1.95	1.65	1.16	1.02
273.15 K	I=0.10	1.99	1.88	1.42	1.07	1.01
	I=0.25	1.98	1.84	1.35	1.05	1.01
	I=0	1.99	1.94	1.62	1.14	1.02
298.15 K	I=0.10	1.98	1.86	1.38	1.06	1.01
	I=0.25	1.98	1.82	1.31	1.04	1.
	I=0	1.99	1.94	1.61	1.13	1.02
313.15 K	I=0.10	1.98	1.85	1.36	1.05	1.01
	I=0.25	1.98	1.8	1.29	1.04	1.

The rate of change of the standard transformed enthalpy of formation with pH can be calculated in two ways: (1) using equation 3.4-15 and (2) from the rate of change of the average number of hydrogen ions bound with temperature, as shown in equation 3.4-16.

Table 4.5 Rate of change of standard transformed enthalpy of formation of inorganic phosphate with pH at three temperatures, five pHs, and three ionic strengths calculated directly using equation 3.4-15

```
In[12]:= PaddedForm[TableForm[
         D[-t^2 * D[phosGT / t, t], pH] /. t → {273.15, 298.15, 313.15} /. pH → {5, 6, 7, 8, 9} /.
           is → {0, .1, .25},
         TableHeadings → {{"273.15 K", "298.15 K", "313.15 K"}, {"        pH 5", "pH 6",
            "pH 7", "pH 8", "pH 9"}, {"I=0", "I=0.10", "I=0.25"}}], {5, 4}]
```

Out[12]//PaddedForm=

		pH 5	pH 6	pH 7	pH 8	pH 9
	I=0	0.0435	0.3963	1.8769	1.1067	0.1506
273.15 K	I=0.10	0.1377	1.0950	2.5221	0.6560	0.0744
	I=0.25	0.1972	1.4544	2.5018	0.5328	0.0585
	I=0	0.0496	0.4462	1.9472	1.0084	0.1325
298.15 K	I=0.10	0.1756	1.3420	2.6309	0.6091	0.0678
	I=0.25	0.2585	1.8047	2.5800	0.4951	0.0536
	I=0	0.0531	0.4745	1.9785	0.9588	0.1239
313.15 K	I=0.10	0.2027	1.5108	2.6988	0.5882	0.0649
	I=0.25	0.3037	2.0469	2.6298	0.4782	0.0513

The same table can be calculated from the average number of hydrogen ions bound by inorganic phosphate by using

```
In[13]:= -8.31451 * 10^-3 * t^2 * Log[10] * D[(1 / (8.31451 * (t / 1000) * Log[10])) * D[phosGT, pH], t];
```

The rate of change of the standard transformed entropy of formation with pH can be calculated in two ways: (1) directly and (2) from the rate of change of the temperature times the average number of hydrogen ions bound with temperature, as shown in equation 3.4-13.

Table 4.6 Rate of change of standard transformed entropy of formation in J K^{-1} mol^{-1} of inorganic phosphate with pH at three temperatures, five pHs, and three ionic strengths calculated directly

```
In[14]:= PaddedForm[TableForm[
            -D[-D[phosGT, t], pH] * 1000 /. t → {273.15, 298.15, 313.15} /. pH → {5, 6, 7, 8, 9} /.
            is → {0, .1, .25}, TableHeadings → {{"273.15 K", "298.15 K", "313.15 K"},
            {"        pH 5", "pH 6", "pH 7", "pH 8", "pH 9"}, {"I=0", "I=0.10", "I=0.25"}}], 4]
```

```
Out[14]//PaddedForm=
                         pH 5        pH 6        pH 7        pH 8        pH 9
                I=0      38.03       35.87       24.79       18.13       18.95
     273.15 K   I=0.10   37.53       31.98       17.99       18.05       19.01
                I=0.25   37.22       29.96       16.68       18.17       19.03

                I=0      38.01       35.7        24.54       18.48       19.01
     298.15 K   I=0.10   37.39       31.11       17.61       18.21       19.03
                I=0.25   37.         28.74       16.41       18.3        19.05

                I=0      38.         35.61       24.44       18.64       19.04
     313.15 K   I=0.10   37.31       30.56       17.39       18.28       19.04
                I=0.25   36.86       27.95       16.24       18.36       19.06
```

The same table of standard transformed entropy of formation of inorganic phosphate can be calculated from the average number of hydrogen ions bound using

```
In[15]:= 8.31451 * 1000 * Log[10] *
            D[(t / 1000) * (1 / (8.31451 * (t / 1000) * Log[10])) * D[phosGT, pH], t];
```

These calculations can be checked in additional ways by use of $\Delta_f G_j'° = \Delta_f H_j'° - T\Delta_f S_j'°$ and $\partial\Delta_f G_j'°/\partial pH = \partial\Delta_f H_j'°/\partial pH - T\partial\Delta_f S_j'°/\partial pH$. More partial derivatives can be taken, but taking a second derivative with respect to the same variable is not likely to be very accurate. An example of a second derivative is the standard transformed heat capacity since $\Delta_f C_P'° = -T\partial^2 \Delta_f G_i'°/\partial T^2$. Another example is the binding capacity, defined by di Cera, Gill, and Wyman (4).

```
In[16]:= Clear[phosGT];
```

4.3 Derivations of Functions of Temperature, pH, and Ionic Strength for Standard Transformed Properties of Reactants

The package BasicBiochemData3 (see Appendix 1) has been written to (1) provide functions of pH and ionic strength at 298.15 K that yield $\Delta_f G'°$ and $\overline{N}_H$ for 199 reactants and (2) provide functions of temperature, pH and ionic strength that yield $\Delta_f G'°$, $\Delta_f H'°$, $\Delta_f S'°$ and $\overline{N}_H$ for 94 reactants. The program **derivetrGibbsT** has been applied to the 94 reactants by use of **Map** to produce functions that are given names ending with GT. These functions are very complicated, but Replace-All (/.x->) makes it possible to calculate these properties from 273.15 K to 313.25 K, pHs 5 to 9, and ionic strengths zero to about 0.35 M. For example, consider atp:

```
In[17]:= atpGT /. t → 298.15 /. pH → 7 /. is → .25
```

```
Out[17]= -2292.5
```

```
In[18]:= atpHT /. t → 298.15 /. pH → 7 /. is → .25
```

```
Out[18]= -3616.886007691
```

```
In[19]:= atpST /. t → 298.15 /. pH → 7 /. is → .25
```

```
Out[19]= -4.44202
```

```
In[20]:= ((atpHT - atpGT) / .29815) /. t → 298.15 /. pH → 7 /. is → .25
```

Out[20]= -4442.02

In[21]:= **atpNHT /. t → 298.15 /. pH → 7 /. is → .25**

Out[21]= 12.22611909409359

The names of the 94 reactants for which $\Delta_f H$ ° is known are given by

In[22]:= **nameswithH = {"acetaldehyde", "acetate", "acetone", "adenine", "adenosine",**
    "adp", "alanine", "ammonia", "amp", "arabinose", "asparagineL", "aspartate",
    "atp", "citrate", "co2g", "co2tot", "coaq", "cog", "ethaneaq", "ethanol",
    "ethylacetate", "ferric", "ferrous", "formate", "fructose", "fructose16phos",
    "fructose6phos", "fumarate", "galactose", "glucose", "glucose6phos", "glutamate",
    "glutamine", "glycerol", "glycerol3phos", "glycine", "glycylglycine", "h2aq",
    "h2g", "h2o", "h2o2aq", "h2saq", "i2cr", "idp", "imp", "indole", "inosine",
    "iodideion", "isomaltose", "itp", "ketoglutarate", "lactate", "lactose",
    "leucineL", "malate", "maltose", "mannose", "mannose6phos", "methaneaq",
    "methaneg", "methanol", "n2aq", "n2g", "n2oaq", "nadox", "nadpox", "nadpred",
    "nadred", "nitrate", "nitrite", "noaq", "o2aq", "o2g", "oxaloacetate",
    "pep", "pi", "ppi", "propanol2", "pyruvate", "ribose", "ribose1phos",
    "ribose5phos", "ribulose", "sorbose", "succinate", "sucrose", "sulfate",
    "sulfite", "sulfurcr", "tryptophanL", "urea", "valineL", "xylose", "xylulose"};

In[23]:= **Dimensions[nameswithH]**

Out[23]= {94}

The functions of temperature, pH, and ionic strength for $\Delta_f G$ ' ° are available because the BasicBiochemData3 was loaded at the beginning of the chapter. They can be used to calculate $\Delta_f G$ ' ° for any reactant with known $\Delta_f H$ ° at any temperature in the range 273.25 K to 313.5 K, pHs in the range 5 to 9, and ionic strengths in the range zero to about 0.35 M. Since the tables in Appendix 2 are for 298.15 K, the tables here are calculated at 273.15 K and 313.15 K.

Table 4.7 Standard Transformed Gibbs Energies of Formation of Reactants in kJ mol^{-1} at 273.15 K, Ionic Strength 0.25 M, and pHs 5, 6, 7, 8, and 9

In[24]:= **table7 = PaddedForm[**
    TableForm[{acetaldehydeGT, acetateGT, acetoneGT, adenineGT, adenosineGT, adpGT,
        alanineGT, ammoniaGT, ampGT, arabinoseGT, asparagineLGT, aspartateGT, atpGT,
        citrateGT, co2gGT, co2totGT, coaqGT, cogGT, ethaneaqGT, ethanolGT, ethylacetateGT,
        ferricGT, ferrousGT, formateGT, fructoseGT, fructose16phosGT, fructose6phosGT,
        fumarateGT, galactoseGT, glucoseGT, glucose6phosGT, glutamateGT, glutamineGT,
        glycerolGT, glycerol3phosGT, glycineGT, glycylglycineGT, h2aqGT, h2gGT, h2oGT,
        h2o2aqGT, h2saqGT, i2crGT, idpGT, impGT, indoleGT, inosineGT, iodideionGT,
        isomaltoseGT, itpGT, ketoglutarateGT, lactateGT, lactoseGT, leucineLGT, malateGT,
        maltoseGT, mannoseGT, mannose6phosGT, methaneaqGT, methanegGT, methanolGT,
        n2aqGT, n2gGT, n2oaqGT, nadoxGT, nadpoxGT, nadpredGT, nadredGT, nitrateGT,
        nitriteGT, noaqGT, o2aqGT, o2gGT, oxaloacetateGT, pepGT, piGT, ppiGT, propanol2GT,
        pyruvateGT, riboseGT, ribose1phosGT, ribose5phosGT, ribuloseGT, sorboseGT,
        succinateGT, sucroseGT, sulfateGT, sulfiteGT, sulfurcrGT, tryptophanLGT, ureaGT,
        valineLGT, xyloseGT, xyluloseGT} /. t → 273.15 /. is → .25 /. pH → {5, 6, 7, 8, 9},
      TableHeadings → {nameswithH, {" pH 5", " pH 6", " pH 7", " pH 8", " pH 9"}},
      TableSpacing → {1, 1}], {6, 2}]

Out[24]//PaddedForm=

	pH 5	pH 6	pH 7	pH 8	pH 9
acetaldehyde	-37.70	-16.78	4.13	25.05	45.97
acetate	-299.83	-283.61	-267.86	-252.16	-236.48

acetone	-3.74	27.63	59.01	90.39	121.76
adenine	431.63	458.35	484.56	510.72	536.87
adenosine	118.77	186.86	254.85	322.83	390.82
adp	-1657.65	-1590.29	-1525.52	-1462.42	-1399.63
alanine	-198.39	-161.79	-125.18	-88.58	-51.97
ammonia	22.96	43.87	64.79	85.69	106.45
amp	-776.99	-709.97	-645.63	-582.64	-519.86
arabinose	-498.92	-446.63	-394.33	-342.04	-289.74
asparagineL	-331.19	-289.36	-247.52	-205.68	-163.85
aspartate	-556.19	-524.81	-493.44	-462.06	-430.68
atp	-2536.26	-2468.70	-2403.56	-2340.35	-2277.54
citrate	-1068.43	-1039.03	-1012.16	-985.92	-959.77
co2g	-394.29				
co2tot	-575.94	-566.43	-559.33	-553.82	-548.74
coaq	-119.99				
cog	-134.94				
ethaneaq	137.01	168.39	199.77	231.14	262.52
ethanol	-29.43	1.95	33.33	64.70	96.08
ethylacetate	-134.88	-93.04	-51.21	-9.37	32.46
ferric	-14.79				
ferrous	-82.61				
formate	-331.10	-325.87	-320.65	-315.42	-310.19
fructose	-622.03	-559.28	-496.52	-433.77	-371.02
fructose16phos	-2411.60	-2354.87	-2301.90	-2249.53	-2197.23
fructose6phos	-1514.55	-1453.88	-1395.55	-1337.93	-1280.39
fumarate	-565.94	-555.28	-544.80	-534.34	-523.88
galactose	-615.65	-552.90	-490.14	-427.39	-364.64
glucose	-622.62	-559.87	-497.12	-434.36	-371.61
glucose6phos	-1519.04	-1457.99	-1399.41	-1341.75	-1284.21
glutamate	-506.99	-465.15	-423.32	-381.48	-339.65
glutamine	-282.65	-230.35	-178.06	-125.77	-73.47
glycerol	-297.62	-255.78	-213.95	-172.11	-130.28
glycerol3phos	-1210.36	-1169.69	-1131.51	-1094.66	-1058.03
glycine	-257.61	-231.46	-205.32	-179.17	-153.02
glycylglycine	-323.27	-281.44	-239.60	-197.77	-155.93
h2aq	69.49	79.95	90.41	100.87	111.33
h2g	53.72	64.18	74.64	85.10	95.55
h2o	-187.55	-177.09	-166.63	-156.17	-145.71
h2o2aq	-85.10	-74.64	-64.18	-53.73	-43.27
h2saq	24.87	35.17	44.42	51.17	56.63
i2cr	0.00				
idp	-1897.84	-1835.94	-1776.44	-1718.79	-1662.74
imp	-1017.31	-955.71	-896.62	-838.94	-782.09
indole	401.23	437.83	474.44	511.04	547.65
inosine	-121.27	-58.52	4.22	66.82	128.46
iodideion	-52.59				
isomaltose	-1051.87	-936.82	-821.77	-706.73	-591.68
itp	-2776.37	-2714.28	-2654.40	-2596.57	-2540.13
ketoglutarate	-709.84	-688.92	-668.00	-647.09	-626.17
lactate	-397.38	-371.24	-345.09	-318.94	-292.79
lactose	-1032.24	-917.19	-802.15	-687.10	-572.05
leucineL	-27.49	40.50	108.48	176.46	244.44
malate	-758.91	-737.15	-716.13	-695.20	-674.28
maltose	-1039.40	-924.36	-809.31	-694.26	-579.21

mannose	-616.92	-554.17	-491.41	-428.66	-365.91
mannose6phos	-1515.17	-1454.08	-1395.46	-1337.80	-1280.26
methaneaq	68.52	89.44	110.36	131.27	152.19
methaneg	54.70	75.62	96.53	117.45	138.37
methanol	-73.79	-52.88	-31.96	-11.04	9.88
n2aq	16.25				
n2g	0.00				
n2oaq	102.34				
nadox	697.64	833.60	969.57	1105.53	1241.50
nadpox	-184.55	-53.81	76.92	207.66	338.39
nadpred	-141.79	-5.83	130.14	266.10	402.07
nadred	740.43	881.63	1022.82	1164.01	1305.21
nitrate	-117.52	-117.52	-117.52	-117.52	-117.52
nitrite	-39.02	-38.99	-38.98	-38.98	-38.98
noaq	86.86				
o2aq	14.04				
o2g	0.00				
oxaloacetate	-756.39	-745.93	-735.47	-725.01	-714.56
pep	-1252.53	-1237.89	-1225.72	-1214.99	-1204.51
pi	-1098.20	-1088.08	-1079.62	-1073.53	-1068.20
ppi	-1985.35	-1976.45	-1970.13	-1965.43	-1963.09
propanol2	17.44	59.27	101.11	142.94	184.78
pyruvate	-402.80	-387.11	-371.42	-355.73	-340.04
ribose	-507.05	-454.76	-402.46	-350.17	-297.87
ribose1phos	-1392.13	-1341.28	-1292.93	-1245.69	-1198.61
ribose5phos	-1399.53	-1348.68	-1300.34	-1253.09	-1206.01
ribulose	-491.42	-439.12	-386.83	-334.53	-282.24
sorbose	-619.10	-556.34	-493.59	-430.84	-368.08
succinate	-606.11	-583.50	-562.34	-541.40	-520.48
sucrose	-1027.05	-912.00	-796.95	-681.91	-566.86
sulfate	-761.19	-761.19	-761.19	-761.19	-761.19
sulfite	-509.95	-505.24	-502.52	-501.92	-501.85
sulfurcr	0.00				
tryptophanL	183.26	246.01	308.76	371.52	434.27
urea	-104.99	-84.07	-63.16	-42.24	-21.32
valineL	-84.44	-26.91	30.61	88.13	145.66
xylose	-506.67	-454.37	-402.08	-349.78	-297.49
xylulose	-501.33	-449.03	-396.74	-344.44	-292.15

Notice that when $\Delta_f G'^\circ$ is independent of pH, *Mathematica* only prints it once.

Table 4.8 Standard Transformed Gibbs Energies of Formation of Reactants in kJ mol^{-1} at 313.15 K, Ionic Strength 0.25 M, and pHs 5, 6, 7, 8, and 9

```
In[25]:= table8 = PaddedForm[
           TableForm[{acetaldehydeGT, acetateGT, acetoneGT, adenineGT, adenosineGT, adpGT,
               alanineGT, ammoniaGT, ampGT, arabinoseGT, asparagineLGT, aspartateGT, atpGT,
               citrateGT, co2gGT, co2totGT, coaqGT, cogGT, ethaneaqGT, ethanolGT, ethylacetateGT,
               ferricGT, ferrousGT, formateGT, fructoseGT, fructose16phosGT, fructose6phosGT,
               fumarateGT, galactoseGT, glucoseGT, glucose6phosGT, glutamateGT, glutamineGT,
               glycerolGT, glycerol3phosGT, glycineGT, glycylglycineGT, h2aqGT, h2gGT, h2oGT,
               h2o2aqGT, h2saqGT, i2crGT, idpGT, impGT, indoleGT, inosineGT, iodideionGT,
               isomaltoseGT, itpGT, ketoglutarateGT, lactateGT, lactoseGT, leucineLGT, malateGT,
               maltoseGT, mannoseGT, mannose6phosGT, methaneaqGT, methanegGT, methanolGT,
               n2aqGT, n2gGT, n2oaqGT, nadoxGT, nadpoxGT, nadpredGT, nadredGT, nitrateGT,
               nitriteGT, noaqGT, o2aqGT, o2gGT, oxaloacetateGT, pepGT, piGT, ppiGT, propanol2GT,
               pyruvateGT, riboseGT, ribose1phosGT, ribose5phosGT, ribuloseGT, sorboseGT,
               succinateGT, sucroseGT, sulfateGT, sulfiteGT, sulfurcrGT, tryptophanLGT, ureaGT,
               valineLGT, xyloseGT, xyluloseGT} /. t → 313.15 /. is → .25 /. pH → {5, 6, 7, 8, 9},
           TableHeadings → {nameswithH, {"   pH 5", "   pH 6", "   pH 7", "   pH 8", "   pH 9"}},
           TableSpacing → {1, 1}], {6, 2}]
```

```
Out[25]//PaddedForm=
```

	pH 5	pH 6	pH 7	pH 8	pH 9
acetaldehyde	-11.92	12.06	36.04	60.02	84.00
acetate	-272.43	-253.85	-235.80	-217.81	-199.82
acetone	28.52	64.49	100.46	136.43	172.40
adenine	476.63	506.85	536.85	566.83	596.80
adenosine	227.96	305.94	383.89	461.83	539.76
adp	-1515.87	-1438.68	-1364.20	-1291.78	-1219.79
alanine	-145.81	-103.84	-61.88	-19.91	22.06
ammonia	45.89	69.87	93.81	117.46	139.39
amp	-651.21	-574.33	-500.31	-428.03	-356.05
arabinose	-418.57	-358.61	-298.66	-238.71	-178.76
asparagineL	-267.05	-219.09	-171.13	-123.17	-75.21
aspartate	-499.20	-463.23	-427.26	-391.29	-355.32
atp	-2378.18	-2300.78	-2225.87	-2153.31	-2081.30
citrate	-1002.51	-969.31	-938.70	-908.65	-878.67
co2g	-394.40				
co2tot	-557.80	-547.34	-539.77	-533.60	-528.08
coaq	-119.85				
cog	-138.51				
ethaneaq	172.37	208.34	244.31	280.28	316.25
ethanol	8.82	44.79	80.76	116.74	152.71
ethylacetate	-83.59	-35.63	12.33	60.29	108.25
ferric	-10.35				
ferrous	-81.88				
formate	-317.27	-311.28	-305.28	-299.29	-293.29
fructose	-528.02	-456.08	-384.13	-312.19	-240.25
fructose16phos	-2275.30	-2210.41	-2149.73	-2089.71	-2029.75
fructose6phos	-1404.27	-1334.70	-1267.81	-1201.75	-1135.79
fumarate	-535.11	-522.87	-510.86	-498.86	-486.87
galactose	-521.32	-449.38	-377.43	-305.49	-233.55
glucose	-528.29	-456.34	-384.40	-312.46	-240.52
glucose6phos	-1407.76	-1337.76	-1270.59	-1204.48	-1138.52
glutamate	-437.34	-389.38	-341.42	-293.46	-245.49

glutamine	-205.59	-145.64	-85.69	-25.74	34.22
glycerol	-241.68	-193.72	-145.75	-97.79	-49.83
glycerol3phos	-1134.93	-1088.28	-1044.49	-1002.25	-960.25
glycine	-218.46	-188.49	-158.51	-128.54	-98.56
glycylglycine	-262.64	-214.68	-166.71	-118.75	-70.79
h2aq	80.40	92.39	104.38	116.37	128.36
h2g	61.70	73.69	85.68	97.67	109.66
h2o	-173.04	-161.05	-149.06	-137.07	-125.08
h2o2aq	-69.46	-57.47	-45.48	-33.49	-21.50
h2saq	34.39	45.76	54.80	61.45	67.52
i2cr	0.00				
idp	-1762.29	-1691.19	-1622.84	-1557.64	-1495.69
imp	-897.74	-826.93	-758.99	-693.38	-630.51
indole	446.10	488.07	530.03	572.00	613.96
inosine	-18.30	53.64	125.49	196.72	265.48
iodideion	-52.26				
isomaltose	-875.98	-744.09	-612.20	-480.30	-348.41
itp	-2624.49	-2553.20	-2484.36	-2418.52	-2355.73
ketoglutarate	-660.90	-636.91	-612.93	-588.95	-564.97
lactate	-354.80	-324.82	-294.85	-264.87	-234.89
lactose	-855.15	-723.26	-591.36	-459.47	-327.57
leucineL	63.44	141.37	219.31	297.25	375.19
malate	-711.83	-686.97	-662.88	-638.88	-614.90
maltose	-862.63	-730.74	-598.84	-466.95	-335.05
mannose	-522.27	-450.33	-378.38	-306.44	-234.50
mannose6phos	-1403.73	-1333.67	-1266.46	-1200.35	-1134.38
methaneaq	91.82	115.80	139.78	163.76	187.74
methaneg	73.89	97.87	121.85	145.83	169.81
methanol	-48.36	-24.38	-0.40	23.58	47.56
n2aq	20.17				
n2g	0.00				
n2oaq	105.31				
nadox	801.21	957.08	1112.96	1268.83	1424.71
nadpox	-63.14	86.74	236.62	386.50	536.38
nadpred	-10.18	145.69	301.57	457.44	613.32
nadred	854.83	1016.70	1178.57	1340.45	1502.32
nitrate	-104.77	-104.77	-104.77	-104.77	-104.77
nitrite	-29.45	-29.43	-29.43	-29.43	-29.43
noaq	86.36				
o2aq	17.81				
o2g	0.00				
oxaloacetate	-726.70	-714.71	-702.72	-690.73	-678.74
pep	-1198.84	-1182.09	-1168.17	-1155.88	-1143.85
pi	-1068.22	-1056.75	-1047.43	-1040.65	-1034.57
ppi	-1940.11	-1930.10	-1923.03	-1917.84	-1915.49
propanol2	68.89	116.85	164.81	212.77	260.74
pyruvate	-374.36	-356.37	-338.39	-320.40	-302.42
ribose	-429.32	-369.37	-309.42	-249.46	-189.51
ribose1phos	-1298.08	-1239.24	-1183.19	-1128.89	-1074.90
ribose5phos	-1306.57	-1247.73	-1191.68	-1137.38	-1083.38
ribulose	-413.00	-353.05	-293.10	-233.15	-173.20
sorbose	-524.08	-452.14	-380.20	-308.25	-236.31
succinate	-561.63	-535.86	-511.62	-487.62	-463.63
sucrose	-854.06	-722.17	-590.27	-458.38	-326.48

sulfate	-739.74	-739.73	-739.73	-739.73	-739.73
sulfite	-492.90	-487.30	-483.61	-482.63	-482.51
sulfurcr	0.00				
tryptophanL	270.11	342.05	413.99	485.93	557.88
urea	-73.62	-49.64	-25.66	-1.68	22.30
valineL	-6.56	59.38	125.33	191.28	257.23
xylose	-427.13	-367.18	-307.23	-247.28	-187.32
xylulose	-423.39	-363.44	-303.49	-243.54	-183.59

The standard transformed enthalpies of formation of these reactants are now calculated at 273.15 K.

Table 4.9 Standard Transformed Enthalpies of Formation of Reactants in kJ mol^{-1} at 273.15 K, Ionic Strength 0.25 M, and pHs 5, 6, 7, 8, and 9

In[26]:= **table9 =**
PaddedForm[TableForm[{acetaldehydeHT, acetateHT, acetoneHT, adenineHT, adenosineHT,
adpHT, alanineHT, ammoniaHT, ampHT, arabinoseHT, asparagineLHT, aspartateHT,
atpHT, citrateHT, co2gHT, co2totHT, coaqHT, cogHT, ethaneaqHT, ethanolHT,
ethylacetateHT, ferricHT, ferrousHT, formateHT, fructoseHT, fructose16phosHT,
fructose6phosHT, fumarateHT, galactoseHT, glucoseHT, glucose6phosHT,
glutamateHT, glutamineHT, glycerolHT, glycerol3phosHT, glycineHT,
glycylglycineHT, h2aqHT, h2gHT, h2oHT, h2o2aqHT, h2saqHT, i2crHT, idpHT,
impHT, indoleHT, inosineHT, iodideionHT, isomaltoseHT, itpHT, ketoglutarateHT,
lactateHT, lactoseHT, leucineLHT, malateHT, maltoseHT, mannoseHT,
mannose6phosHT, methaneaqHT, methanegHT, methanolHT, n2aqHT, n2gHT, n2oaqHT,
nadoxHT, nadpoxHT, nadpredHT, nadredHT, nitrateHT, nitriteHT, noaqHT,
o2aqHT, o2gHT, oxaloacetateHT, pepHT, piHT, ppiHT, propanol2HT, pyruvateHT,
riboseHT, ribose1phosHT, ribose5phosHT, ribuloseHT, sorboseHT, succinateHT,
sucroseHT, sulfateHT, sulfiteHT, sulfurcrHT, tryptophanLHT, ureaHT, valineLHT,
xyloseHT, xyluloseHT} /. t → 273.15 /. is → .25 /. pH → {5, 6, 7, 8, 9} // N,
TableHeadings → {nameswithH, {" pH 5", " pH 6", " pH 7", " pH 8", " pH 9"}},
TableSpacing → {1, 1}], {7,
2}]

Out[26]//PaddedForm=

	pH 5	pH 6	pH 7	pH 8	pH 9
acetaldehyde	-213.41	-213.41	-213.41	-213.41	-213.41
acetate	-486.68	-486.61	-486.60	-486.60	-486.60
acetone	-223.48	-223.48	-223.48	-223.48	-223.48
adenine	123.01	127.38	127.96	128.02	128.03
adenosine	-625.96	-625.21	-625.14	-625.13	-625.13
adp	-2625.75	-2625.07	-2626.82	-2627.35	-2627.42
alanine	-556.86	-556.86	-556.86	-556.86	-556.86
ammonia	-133.39	-133.39	-133.35	-132.97	-129.46
amp	-1635.40	-1635.47	-1637.26	-1637.67	-1637.72
arabinose	-1046.73	-1046.73	-1046.73	-1046.73	-1046.73
asparagineL	-768.45	-768.45	-768.45	-768.45	-768.45
aspartate	-944.88	-944.88	-944.88	-944.88	-944.88
atp	-3616.03	-3615.34	-3617.23	-3617.93	-3618.02
citrate	-1518.65	-1515.50	-1514.14	-1513.95	-1513.93
co2g	-393.50				
co2tot	-699.75	-697.11	-693.14	-691.97	-690.61
coaq	-120.96				
cog	-110.53				
ethaneaq	-103.86	-103.86	-103.86	-103.86	-103.86
ethanol	-290.07	-290.07	-290.07	-290.07	-290.07

ethylacetate	-484.36	-484.36	-484.36	-484.36	-484.36
ferric	-45.85				
ferrous	-87.92				
formate	-425.55	-425.55	-425.55	-425.55	-425.55
fructose	-1262.91	-1262.91	-1262.91	-1262.91	-1262.91
fructose16phos	-3341.96	-3341.64	-3341.50	-3341.49	-3341.48
fructose6phos	-2266.71	-2267.02	-2267.20	-2267.23	-2267.23
fumarate	-776.64	-776.78	-776.80	-776.80	-776.80
galactose	-1258.73	-1258.73	-1258.73	-1258.73	-1258.73
glucose	-1265.72	-1265.72	-1265.72	-1265.72	-1265.72
glucose6phos	-2277.96	-2278.24	-2278.46	-2278.50	-2278.50
glutamate	-981.95	-981.95	-981.95	-981.95	-981.95
glutamine	-807.94	-807.94	-807.94	-807.94	-807.94
glycerol	-678.91	-678.91	-678.91	-678.91	-678.91
glycerol3phos	-1724.90	-1725.13	-1725.40	-1725.47	-1725.47
glycine	-524.47	-524.47	-524.47	-524.47	-524.47
glycylglycine	-736.61	-736.61	-736.61	-736.61	-736.61
h2aq	-4.79	-4.79	-4.79	-4.79	-4.79
h2g	-0.59	-0.59	-0.59	-0.59	-0.59
h2o	-286.42	-286.42	-286.42	-286.42	-286.42
h2o2aq	-191.76	-191.76	-191.76	-191.76	-191.76
h2saq	-40.10	-38.52	-29.91	-20.00	-17.86
i2cr	0.00				
idp	-2822.83	-2823.96	-2825.57	-2823.20	-2809.71
imp	-1832.86	-1834.41	-1836.16	-1835.26	-1825.91
indole	95.44	95.44	95.44	95.44	95.44
inosine	-823.37	-823.35	-823.16	-821.42	-811.42
iodideion	-54.90				
isomaltose	-2250.96	-2250.96	-2250.96	-2250.96	-2250.96
itp	-3813.28	-3814.25	-3816.15	-3815.15	-3805.40
ketoglutarate	-1044.06	-1044.06	-1044.06	-1044.06	-1044.06
lactate	-687.82	-687.82	-687.82	-687.82	-687.82
lactose	-2239.56	-2239.56	-2239.56	-2239.56	-2239.56
leucineL	-647.20	-647.20	-647.20	-647.20	-647.20
malate	-1080.25	-1079.86	-1079.80	-1079.79	-1079.79
maltose	-2244.54	-2244.54	-2244.54	-2244.54	-2244.54
mannose	-1262.19	-1262.19	-1262.19	-1262.19	-1262.19
mannose6phos	-2275.21	-2275.50	-2275.73	-2275.77	-2275.77
methaneaq	-90.22	-90.22	-90.22	-90.22	-90.22
methaneg	-75.99	-75.99	-75.99	-75.99	-75.99
methanol	-247.11	-247.11	-247.11	-247.11	-247.11
n2aq	-10.54				
n2g	0.00				
n2oaq	82.05				
nadox	-7.36	-7.36	-7.36	-7.36	-7.36
nadpox	-1012.19	-1012.19	-1012.19	-1012.19	-1012.19
nadpred	-1039.60	-1039.60	-1039.60	-1039.60	-1039.60
nadred	-38.71	-38.71	-38.71	-38.71	-38.71
nitrate	-204.71	-204.71	-204.71	-204.71	-204.71
nitrite	-104.53	-104.33	-104.31	-104.31	-104.31
noaq	90.25				
o2aq	-11.70				
o2g	0.00				
oxaloacetate	-959.31	-959.31	-959.31	-959.31	-959.31

pep	-1619.29	-1619.30	-1619.31	-1619.32	-1619.32
pi	-1302.81	-1302.15	-1299.79	-1298.36	-1298.14
ppi	-2294.53	-2293.52	-2292.54	-2291.45	-2289.50
propanol2	-333.19	-333.19	-333.19	-333.19	-333.19
pyruvate	-596.81	-596.81	-596.81	-596.81	-596.81
ribose	-1036.94	-1036.94	-1036.94	-1036.94	-1036.94
ribose1phos	-2033.82	-2038.11	-2042.10	-2042.86	-2042.94
ribose5phos	-2033.82	-2038.11	-2042.10	-2042.86	-2042.94
ribulose	-1025.96	-1025.96	-1025.96	-1025.96	-1025.96
sorbose	-1266.83	-1266.83	-1266.83	-1266.83	-1266.83
succinate	-909.66	-908.83	-908.70	-908.68	-908.68
sucrose	-2206.35	-2206.35	-2206.35	-2206.35	-2206.35
sulfate	-908.09	-908.09	-908.09	-908.09	-908.09
sulfite	-626.45	-628.04	-632.24	-634.05	-634.29
sulfurcr	0.00				
tryptophanL	-408.73	-408.73	-408.73	-408.73	-408.73
urea	-318.83	-318.83	-318.83	-318.83	-318.83
valineL	-615.23	-615.23	-615.23	-615.23	-615.23
xylose	-1048.88	-1048.88	-1048.88	-1048.88	-1048.88
xylulose	-1032.59	-1032.59	-1032.59	-1032.59	-1032.59

Table 4.10 Standard Transformed Enthalpies of Formation of Reactants in kJ mol^{-1} at 313.15 K, Ionic Strength 0.25 M, and pHs 5, 6, 7, 8, and 9

```
In[27]:= table10 =
    PaddedForm[TableForm[{acetaldehydeHT, acetateHT, acetoneHT, adenineHT, adenosineHT,
        adpHT, alanineHT, ammoniaHT, ampHT, arabinoseHT, asparagineLHT, aspartateHT,
        atpHT, citrateHT, co2gHT, co2totHT, coaqHT, cogHT, ethaneaqHT, ethanolHT,
        ethylacetateHT, ferricHT, ferrousHT, formateHT, fructoseHT, fructose16phosHT,
        fructose6phosHT, fumarateHT, galactoseHT, glucoseHT, glucose6phosHT,
        glutamateHT, glutamineHT, glycerolHT, glycerol3phosHT, glycineHT,
        glycylglycineHT, h2aqHT, h2gHT, h2oHT, h2o2aqHT, h2saqHT, i2crHT, idpHT,
        impHT, indoleHT, inosineHT, iodideionHT, isomaltoseHT, itpHT, ketoglutarateHT,
        lactateHT, lactoseHT, leucineLHT, malateHT, maltoseHT, mannoseHT,
        mannose6phosHT, methaneaqHT, methanegHT, methanolHT, n2aqHT, n2gHT, n2oaqHT,
        nadoxHT, nadpoxHT, nadpredHT, nadredHT, nitrateHT, nitriteHT, noaqHT,
        o2aqHT, o2gHT, oxaloacetateHT, pepHT, piHT, ppiHT, propanol2HT, pyruvateHT,
        riboseHT, ribose1phosHT, ribose5phosHT, ribuloseHT, sorboseHT, succinateHT,
        sucroseHT, sulfateHT, sulfiteHT, sulfurcrHT, tryptophanLHT, ureaHT, valineLHT,
        xyloseHT, xyluloseHT} /. t → 313.15 /. is → .25 /. pH → {5, 6, 7, 8, 9} // N,
      TableHeadings → {nameswithH, {"  pH 5", "  pH 6", "  pH 7", "  pH 8", "  pH 9"}},
      TableSpacing → {1, 1}], {7,
      2}]
```

Out[27]//PaddedForm=

	pH 5	pH 6	pH 7	pH 8	pH 9
acetaldehyde	-214.21	-214.21	-214.21	-214.21	-214.21
acetate	-487.17	-487.02	-487.00	-487.00	-487.00
acetone	-224.68	-224.68	-224.68	-224.68	-224.68
adenine	125.08	126.81	127.00	127.02	127.03
adenosine	-628.07	-627.77	-627.74	-627.73	-627.73
adp	-2626.27	-2626.26	-2627.54	-2627.97	-2628.02
alanine	-558.26	-558.26	-558.26	-558.26	-558.26
ammonia	-133.99	-133.91	-133.20	-127.04	-102.37
amp	-1636.72	-1637.28	-1638.87	-1639.28	-1639.32
arabinose	-1048.74	-1048.74	-1048.74	-1048.74	-1048.74

asparagineL	-770.05	-770.05	-770.05	-770.05	-770.05
aspartate	-945.88	-945.88	-945.88	-945.88	-945.88
atp	-3615.75	-3615.53	-3616.68	-3617.16	-3617.22
citrate	-1518.48	-1514.66	-1513.32	-1513.15	-1513.13
co2g	-393.50				
co2tot	-699.85	-696.36	-692.75	-691.69	-688.76
coaq	-120.96				
cog	-110.53				
ethaneaq	-105.06	-105.06	-105.06	-105.06	-105.06
ethanol	-291.27	-291.27	-291.27	-291.27	-291.27
ethylacetate	-485.96	-485.96	-485.96	-485.96	-485.96
ferric	-44.05				
ferrous	-87.12				
formate	-425.55	-425.55	-425.55	-425.55	-425.55
fructose	-1265.32	-1265.32	-1265.32	-1265.32	-1265.32
fructose16phos	-3342.83	-3340.90	-3340.35	-3340.29	-3340.28
fructose6phos	-2268.78	-2268.69	-2268.64	-2268.64	-2268.63
fumarate	-776.31	-776.39	-776.40	-776.40	-776.40
galactose	-1261.14	-1261.14	-1261.14	-1261.14	-1261.14
glucose	-1268.13	-1268.13	-1268.13	-1268.13	-1268.13
glucose6phos	-2280.06	-2279.98	-2279.92	-2279.91	-2279.90
glutamate	-983.35	-983.35	-983.35	-983.35	-983.35
glutamine	-809.95	-809.95	-809.95	-809.95	-809.95
glycerol	-680.51	-680.51	-680.51	-680.51	-680.51
glycerol3phos	-1726.24	-1726.18	-1726.09	-1726.08	-1726.07
glycine	-525.47	-525.47	-525.47	-525.47	-525.47
glycylglycine	-738.21	-738.21	-738.21	-738.21	-738.21
h2aq	-5.19	-5.19	-5.19	-5.19	-5.19
h2g	-0.99	-0.99	-0.99	-0.99	-0.99
h2o	-286.82	-286.82	-286.82	-286.82	-286.82
h2o2aq	-192.16	-192.16	-192.16	-192.16	-192.16
h2saq	-40.01	-35.29	-23.30	-18.33	-17.61
i2cr	0.00				
idp	-2824.35	-2824.98	-2824.64	-2813.55	-2798.29
imp	-1834.96	-1836.04	-1836.63	-1828.53	-1808.72
indole	94.04	94.04	94.04	94.04	94.04
inosine	-825.77	-825.68	-824.79	-818.28	-803.74
iodideion	-54.70				
isomaltose	-2255.37	-2255.37	-2255.37	-2255.37	-2255.37
itp	-3813.81	-3814.30	-3814.78	-3808.44	-3793.39
ketoglutarate	-1044.06	-1044.06	-1044.06	-1044.06	-1044.06
lactate	-688.62	-688.62	-688.62	-688.62	-688.62
lactose	-2243.97	-2243.97	-2243.97	-2243.97	-2243.97
leucineL	-649.80	-649.80	-649.80	-649.80	-649.80
malate	-1080.48	-1079.89	-1079.80	-1079.79	-1079.79
maltose	-2248.95	-2248.95	-2248.95	-2248.95	-2248.95
mannose	-1264.60	-1264.60	-1264.60	-1264.60	-1264.60
mannose6phos	-2277.32	-2277.25	-2277.19	-2277.18	-2277.17
methaneaq	-91.02	-91.02	-91.02	-91.02	-91.02
methaneg	-76.79	-76.79	-76.79	-76.79	-76.79
methanol	-247.91	-247.91	-247.91	-247.91	-247.91
n2aq	-10.54				
n2g	0.00				
n2oaq	82.05				

nadox	-12.37	-12.37	-12.37	-12.37	-12.37
nadpox	-1015.40	-1015.40	-1015.40	-1015.40	-1015.40
nadpred	-1041.61	-1041.61	-1041.61	-1041.61	-1041.61
nadred	-43.32	-43.32	-43.32	-43.32	-43.32
nitrate	-204.51	-204.51	-204.51	-204.51	-204.51
nitrite	-104.21	-104.12	-104.11	-104.11	-104.11
noaq	90.25				
o2aq	-11.70				
o2g	0.00				
oxaloacetate	-958.91	-958.91	-958.91	-958.91	-958.91
pep	-1619.01	-1618.60	-1618.06	-1617.93	-1617.92
pi	-1302.96	-1301.99	-1299.12	-1297.73	-1297.54
ppi	-2293.94	-2292.31	-2290.88	-2289.10	-2286.42
propanol2	-334.79	-334.79	-334.79	-334.79	-334.79
pyruvate	-597.21	-597.21	-597.21	-597.21	-597.21
ribose	-1038.95	-1038.95	-1038.95	-1038.95	-1038.95
ribose1phos	-2035.19	-2038.23	-2042.68	-2043.81	-2043.94
ribose5phos	-2035.19	-2038.23	-2042.68	-2043.81	-2043.94
ribulose	-1027.97	-1027.97	-1027.97	-1027.97	-1027.97
sorbose	-1269.24	-1269.24	-1269.24	-1269.24	-1269.24
succinate	-910.03	-908.90	-908.70	-908.68	-908.68
sucrose	-2210.76	-2210.76	-2210.76	-2210.76	-2210.76
sulfate	-907.28	-907.29	-907.29	-907.29	-907.29
sulfite	-626.35	-627.37	-630.98	-633.15	-633.48
sulfurcr	0.00				
tryptophanL	-411.14	-411.14	-411.14	-411.14	-411.14
urea	-319.63	-319.63	-319.63	-319.63	-319.63
valineL	-617.43	-617.43	-617.43	-617.43	-617.43
xylose	-1050.89	-1050.89	-1050.89	-1050.89	-1050.89
xylulose	-1034.60	-1034.60	-1034.60	-1034.60	-1034.60

The standard transformed entropies of formation of these reactants are now calculated at 273.15 K.

Table 4.11 Standard Transformed Entropies of Formation of Reactants in kJ K^{-1} mol^{-1} at 273.15 K, Ionic Strength 0.25 M, and pHs 5, 6, 7, 8, and 9

```
In[28]:= table11 =
        PaddedForm[TableForm[{acetaldehydeST, acetateST, acetoneST, adenineST, adenosineST,
            adpST, alanineST, ammoniaST, ampST, arabinoseST, asparagineLST, aspartateST,
            atpST, citrateST, co2gST, co2totST, coaqST, cogST, ethaneaqST, ethanolST,
            ethylacetateST, ferricST, ferrousST, formateST, fructoseST, fructose16phosST,
            fructose6phosST, fumarateST, galactoseST, glucoseST, glucose6phosST, glutamateST,
            glutamineST, glycerolST, glycerol3phosST, glycineST, glycylglycineST, h2aqST,
            h2gST, h2oST, h2o2aqST, h2saqST, i2crST, idpST, impST, indoleST, inosineST,
            iodideionST, isomaltoseST, itpST, ketoglutarateST, lactateST, lactoseST,
            leucineLST, malateST, maltoseST, mannoseST, mannose6phosST, methaneaqST,
            methanegST, methanolST, n2aqST, n2gST, n2oaqST, nadoxST, nadpoxST, nadpredST,
            nadredST, nitrateST, nitriteST, noaqST, o2aqST, o2gST, oxaloacetateST, pepST,
            piST, ppiST, propanol2ST, pyruvateST, riboseST, ribose1phosST, ribose5phosST,
            ribuloseST, sorboseST, succinateST, sucroseST, sulfateST, sulfiteST,
            sulfurcrST, tryptophanLST, ureaST, valineLST, xyloseST, xyluloseST} /.
            t → 273.15 /. is → .25 /. pH → {5, 6, 7, 8, 9}, TableHeadings →
          {nameswithH, {"    pH 5", "    pH 6", "    pH 7", "    pH 8", "    pH 9"}},
        TableSpacing → {1, 1}], {6, 4}]
```

Out[28]//PaddedForm=

	pH 5	pH 6	pH 7	pH 8	pH 9
acetaldehyde	-0.6433	-0.7198	-0.7964	-0.8730	-0.9496
acetate	-0.6841	-0.7432	-0.8008	-0.8583	-0.9157
acetone	-0.8044	-0.9193	-1.0342	-1.1491	-1.2639
adenine	-1.1298	-1.2117	-1.3055	-1.4011	-1.4968
adenosine	-2.7264	-2.9730	-3.2216	-3.4705	-3.7194
adp	-3.5442	-3.7883	-4.0319	-4.2648	-4.4949
alanine	-1.3123	-1.4464	-1.5804	-1.7144	-1.8484
ammonia	-0.5724	-0.6490	-0.7254	-0.8005	-0.8637
amp	-3.1426	-3.3882	-3.6303	-3.8625	-4.0925
arabinose	-2.0055	-2.1970	-2.3884	-2.5799	-2.7713
asparagineL	-1.6008	-1.7539	-1.9071	-2.0603	-2.2134
aspartate	-1.4230	-1.5379	-1.6527	-1.7676	-1.8825
atp	-3.9530	-4.1978	-4.4432	-4.6772	-4.9075
citrate	-1.6483	-1.7444	-1.8378	-1.9331	-2.0288
co2g	0.0029				
co2tot	-0.4533	-0.4784	-0.4899	-0.5058	-0.5194
coaq	-0.0036				
cog	0.0894				
ethaneaq	-0.8818	-0.9967	-1.1116	-1.2264	-1.3413
ethanol	-0.9542	-1.0691	-1.1839	-1.2988	-1.4137
ethylacetate	-1.2794	-1.4326	-1.5858	-1.7389	-1.8921
ferric	-0.1137				
ferrous	-0.0195				
formate	-0.3458	-0.3649	-0.3841	-0.4032	-0.4223
fructose	-2.3463	-2.5760	-2.8058	-3.0355	-3.2652
fructose16phos	-3.4061	-3.6126	-3.8060	-3.9976	-4.1891
fructose6phos	-2.7537	-2.9769	-3.1911	-3.4022	-3.6128
fumarate	-0.7714	-0.8109	-0.8493	-0.8876	-0.9259
galactose	-2.3543	-2.5841	-2.8138	-3.0435	-3.2733
glucose	-2.3544	-2.5841	-2.8139	-3.0436	-3.2733
glucose6phos	-2.7784	-3.0029	-3.2182	-3.4294	-3.6401
glutamate	-1.7388	-1.8920	-2.0452	-2.1983	-2.3515
glutamine	-1.9231	-2.1146	-2.3060	-2.4974	-2.6889
glycerol	-1.3959	-1.5490	-1.7022	-1.8554	-2.0085
glycerol3phos	-1.8837	-2.0335	-2.1743	-2.3094	-2.4435
glycine	-0.9770	-1.0727	-1.1684	-1.2642	-1.3599
glycylglycine	-1.5132	-1.6664	-1.8195	-1.9727	-2.1258
h2aq	-0.2719	-0.3102	-0.3485	-0.3868	-0.4251
h2g	-0.1988	-0.2371	-0.2754	-0.3137	-0.3520
h2o	-0.3620	-0.4003	-0.4385	-0.4768	-0.5151
h2o2aq	-0.3905	-0.4288	-0.4670	-0.5053	-0.5436
h2saq	-0.2379	-0.2698	-0.2721	-0.2606	-0.2727
i2cr	0.0000				
idp	-3.3864	-3.6171	-3.8409	-4.0432	-4.1990
imp	-2.9857	-3.2169	-3.4397	-3.6475	-3.8214
indole	-1.1195	-1.2535	-1.3875	-1.5215	-1.6555
inosine	-2.5704	-2.8000	-3.0290	-3.2518	-3.4409
iodideion	-0.0085				
isomaltose	-4.3899	-4.8110	-5.2322	-5.6534	-6.0746
itp	-3.7961	-4.0270	-4.2532	-4.4612	-4.6321
ketoglutarate	-1.2236	-1.3002	-1.3767	-1.4533	-1.5299
lactate	-1.0633	-1.1590	-1.2547	-1.3505	-1.4462

lactose	-4.4200	-4.8412	-5.2623	-5.6835	-6.1047
leucineL	-2.2688	-2.5176	-2.7665	-3.0154	-3.2643
malate	-1.1764	-1.2546	-1.3314	-1.4080	-1.4846
maltose	-4.4120	-4.8332	-5.2544	-5.6756	-6.0967
mannose	-2.3623	-2.5921	-2.8218	-3.0516	-3.2813
mannose6phos	-2.7825	-3.0072	-3.2226	-3.4339	-3.6446
methaneaq	-0.5811	-0.6577	-0.7343	-0.8109	-0.8875
methaneg	-0.4784	-0.5550	-0.6316	-0.7082	-0.7848
methanol	-0.6345	-0.7111	-0.7877	-0.8642	-0.9408
n2aq	-0.0981				
n2g	0.0000				
n2oaq	-0.0743				
nadox	-2.5810	-3.0788	-3.5765	-4.0743	-4.5721
nadpox	-3.0300	-3.5086	-3.9872	-4.4659	-4.9445
nadpred	-3.2869	-3.7847	-4.2824	-4.7802	-5.2780
nadred	-2.8524	-3.3694	-3.8863	-4.4032	-4.9201
nitrate	-0.3192	-0.3192	-0.3192	-0.3192	-0.3192
nitrite	-0.2399	-0.2392	-0.2392	-0.2391	-0.2391
noaq	0.0124				
o2aq	-0.0942				
o2g	0.0000				
oxaloacetate	-0.7429	-0.7812	-0.8195	-0.8578	-0.8960
pep	-1.3427	-1.3963	-1.4409	-1.4802	-1.5186
pi	-0.7491	-0.7837	-0.8060	-0.8231	-0.8418
ppi	-1.1319	-1.1608	-1.1803	-1.1936	-1.1950
propanol2	-1.2836	-1.4368	-1.5899	-1.7431	-1.8963
pyruvate	-0.7103	-0.7677	-0.8251	-0.8826	-0.9400
ribose	-1.9399	-2.1314	-2.3228	-2.5143	-2.7057
ribose1phos	-2.3492	-2.5511	-2.7427	-2.9184	-3.0911
ribose5phos	-2.3221	-2.5240	-2.7156	-2.8913	-3.0640
ribulose	-1.9570	-2.1484	-2.3399	-2.5313	-2.7228
sorbose	-2.3714	-2.6011	-2.8308	-3.0606	-3.2903
succinate	-1.1113	-1.1910	-1.2680	-1.3446	-1.4212
sucrose	-4.3174	-4.7386	-5.1598	-5.5810	-6.0022
sulfate	-0.5378	-0.5378	-0.5378	-0.5378	-0.5378
sulfite	-0.4265	-0.4496	-0.4749	-0.4837	-0.4849
sulfurcr	0.0000				
tryptophanL	-2.1673	-2.3970	-2.6267	-2.8565	-3.0862
urea	-0.7829	-0.8594	-0.9360	-1.0126	-1.0892
valineL	-1.9432	-2.1538	-2.3644	-2.5750	-2.7856
xylose	-1.9851	-2.1765	-2.3679	-2.5594	-2.7508
xylulose	-1.9450	-2.1364	-2.3279	-2.5193	-2.7108

Table 4.12 Standard Transformed Entropies of Formation of Reactants in kJ K^{-1} mol^{-1} at 273.15 K, Ionic Strength 0.25 M, and pHs 5, 6, 7, 8, and 9

```
In[29]:=  table12 =
            PaddedForm[TableForm[{acetaldehydeST, acetateST, acetoneST, adenineST, adenosineST,
                adpST, alanineST, ammoniaST, ampST, arabinoseST, asparagineLST, aspartateST,
                atpST, citrateST, co2gST, co2totST, coaqST, cogST, ethaneaqST, ethanolST,
                ethylacetateST, ferricST, ferrousST, formateST, fructoseST, fructose16phosST,
                fructose6phosST, fumarateST, galactoseST, glucoseST, glucose6phosST, glutamateST,
                glutamineST, glycerolST, glycerol3phosST, glycineST, glycylglycineST, h2aqST,
                h2gST, h2oST, h2o2aqST, h2saqST, i2crST, idpST, impST, indoleST, inosineST,
                iodideionST, isomaltoseST, itpST, ketoglutarateST, lactateST, lactoseST,
                leucineLST, malateST, maltoseST, mannoseST, mannose6phosST, methaneaqST,
                methanegST, methanolST, n2aqST, n2gST, n2oaqST, nadoxST, nadpoxST, nadpredST,
                nadredST, nitrateST, nitriteST, noaqST, o2aqST, o2gST, oxaloacetateST, pepST,
                piST, ppiST, propanol2ST, pyruvateST, riboseST, ribose1phosST, ribose5phosST,
                ribuloseST, sorboseST, succinateST, sucroseST, sulfateST, sulfiteST,
                sulfurcrST, tryptophanLST, ureaST, valineLST, xyloseST, xyluloseST} /.
              t → 313.15 /. is → .25 /. pH → {5, 6, 7, 8, 9}, TableHeadings →
                {nameswithH, {"    pH 5", "    pH 6", "    pH 7", "    pH 8", "    pH 9"}},
            TableSpacing → {1, 1}], {6, 4}]
```

Out[29]//PaddedForm=

	pH 5	pH 6	pH 7	pH 8	pH 9
acetaldehyde	-0.6460	-0.7226	-0.7991	-0.8757	-0.9523
acetate	-0.6857	-0.7446	-0.8022	-0.8596	-0.9171
acetone	-0.8085	-0.9234	-1.0383	-1.1531	-1.2680
adenine	-1.1226	-1.2136	-1.3088	-1.4044	-1.5002
adenosine	-2.7336	-2.9817	-3.2305	-3.4794	-3.7282
adp	-3.5459	-3.7924	-4.0343	-4.2669	-4.4970
alanine	-1.3171	-1.4511	-1.5851	-1.7192	-1.8532
ammonia	-0.5744	-0.6507	-0.7249	-0.7808	-0.7720
amp	-3.1471	-3.3944	-3.6358	-3.8679	-4.0979
arabinose	-2.0124	-2.2038	-2.3953	-2.5867	-2.7782
asparagineL	-1.6062	-1.7594	-1.9126	-2.0657	-2.2189
aspartate	-1.4264	-1.5413	-1.6561	-1.7710	-1.8859
atp	-3.9520	-4.1984	-4.4414	-4.6746	-4.9048
citrate	-1.6477	-1.7415	-1.8350	-1.9304	-2.0261
co2g	0.0029				
co2tot	-0.4536	-0.4759	-0.4885	-0.5048	-0.5131
coaq	-0.0036				
cog	0.0894				
ethaneaq	-0.8859	-1.0008	-1.1157	-1.2305	-1.3454
ethanol	-0.9583	-1.0732	-1.1880	-1.3029	-1.4178
ethylacetate	-1.2849	-1.4381	-1.5912	-1.7444	-1.8975
ferric	-0.1076				
ferrous	-0.0167				
formate	-0.3458	-0.3649	-0.3841	-0.4032	-0.4223
fructose	-2.3545	-2.5842	-2.8139	-3.0437	-3.2734
fructose16phos	-3.4090	-3.6100	-3.8021	-3.9936	-4.1850
fructose6phos	-2.7607	-2.9826	-3.1960	-3.4070	-3.6176
fumarate	-0.7702	-0.8096	-0.8480	-0.8863	-0.9246
galactose	-2.3625	-2.5922	-2.8220	-3.0517	-3.2815
glucose	-2.3626	-2.5923	-2.8221	-3.0518	-3.2815
glucose6phos	-2.7856	-3.0088	-3.2231	-3.4342	-3.6449
glutamate	-1.7436	-1.8968	-2.0499	-2.2031	-2.3563
glutamine	-1.9299	-2.1214	-2.3128	-2.5043	-2.6957
glycerol	-1.4013	-1.5545	-1.7077	-1.8608	-2.0140

glycerol3phos	-1.8883	-2.0370	-2.1766	-2.3114	-2.4456
glycine	-0.9804	-1.0761	-1.1718	-1.2676	-1.3633
glycylglycine	-1.5187	-1.6718	-1.8250	-1.9781	-2.1313
h2aq	-0.2733	-0.3116	-0.3499	-0.3882	-0.4265
h2g	-0.2002	-0.2385	-0.2768	-0.3151	-0.3533
h2o	-0.3633	-0.4016	-0.4399	-0.4782	-0.5165
h2o2aq	-0.3918	-0.4301	-0.4684	-0.5067	-0.5450
h2saq	-0.2376	-0.2588	-0.2494	-0.2548	-0.2718
i2cr	0.0000				
idp	-3.3915	-3.6206	-3.8378	-4.0106	-4.1597
imp	-2.9929	-3.2224	-3.4413	-3.6249	-3.7624
indole	-1.1243	-1.2583	-1.3923	-1.5263	-1.6603
inosine	-2.5785	-2.8080	-3.0346	-3.2413	-3.4144
iodideion	-0.0078				
isomaltose	-4.4049	-4.8261	-5.2472	-5.6684	-6.0896
itp	-3.7979	-4.0271	-4.2485	-4.4385	-4.5910
ketoglutarate	-1.2236	-1.3002	-1.3767	-1.4533	-1.5299
lactate	-1.0660	-1.1617	-1.2575	-1.3532	-1.4489
lactose	-4.4350	-4.8562	-5.2774	-5.6985	-6.1197
leucineL	-2.2776	-2.5265	-2.7754	-3.0243	-3.2732
malate	-1.1772	-1.2547	-1.3314	-1.4080	-1.4846
maltose	-4.4270	-4.8482	-5.2694	-5.6906	-6.1117
mannose	-2.3705	-2.6003	-2.8300	-3.0597	-3.2895
mannose6phos	-2.7897	-3.0132	-3.2276	-3.4387	-3.6493
methaneaq	-0.5839	-0.6604	-0.7370	-0.8136	-0.8902
methaneg	-0.4812	-0.5577	-0.6343	-0.7109	-0.7875
methanol	-0.6372	-0.7138	-0.7904	-0.8670	-0.9436
n2aq	-0.0981				
n2g	0.0000				
n2oaq	-0.0743				
nadox	-2.5980	-3.0958	-3.5936	-4.0913	-4.5891
nadpox	-3.0409	-3.5195	-3.9981	-4.4768	-4.9554
nadpred	-3.2937	-3.7915	-4.2892	-4.7870	-5.2848
nadred	-2.8681	-3.3850	-3.9020	-4.4189	-4.9358
nitrate	-0.3185	-0.3185	-0.3185	-0.3185	-0.3185
nitrite	-0.2387	-0.2385	-0.2385	-0.2385	-0.2385
noaq	0.0124				
o2aq	-0.0942				
o2g	0.0000				
oxaloacetate	-0.7415	-0.7798	-0.8181	-0.8564	-0.8947
pep	-1.3417	-1.3939	-1.4367	-1.4755	-1.5139
pi	-0.7496	-0.7831	-0.8037	-0.8209	-0.8398
ppi	-1.1299	-1.1567	-1.1747	-1.1856	-1.1845
propanol2	-1.2891	-1.4422	-1.5954	-1.7486	-1.9017
pyruvate	-0.7116	-0.7691	-0.8265	-0.8839	-0.9414
ribose	-1.9468	-2.1382	-2.3297	-2.5211	-2.7126
ribose1phos	-2.3538	-2.5515	-2.7446	-2.9217	-3.0945
ribose5phos	-2.3267	-2.5244	-2.7175	-2.8946	-3.0674
ribulose	-1.9638	-2.1552	-2.3467	-2.5381	-2.7296
sorbose	-2.3795	-2.6093	-2.8390	-3.0688	-3.2985
succinate	-1.1126	-1.1913	-1.2680	-1.3446	-1.4212
sucrose	-4.3324	-4.7536	-5.1748	-5.5960	-6.0172
sulfate	-0.5350	-0.5351	-0.5351	-0.5351	-0.5351
sulfite	-0.4262	-0.4473	-0.4706	-0.4807	-0.4821

```
sulfurcr          0.0000
tryptophanL      -2.1755    -2.4052    -2.6349    -2.8647    -3.0944
urea             -0.7856    -0.8622    -0.9387    -1.0153    -1.0919
valineL          -1.9507    -2.1613    -2.3719    -2.5825    -2.7931
xylose           -1.9919    -2.1833    -2.3748    -2.5662    -2.7577
xylulose         -1.9518    -2.1432    -2.3347    -2.5261    -2.7176
```

These standard transformed entropies of formation determine the contributions these reactants make to the apparent equilibrium constant for an enzyme-catalyzed reaction, but the more fundamental property, the standard molar entropies of species, are discussed in Chapter 15.

The average numbers of hydrogen atoms in these reactants can also be calculated at 273.15 K and 313.15 K.

4.4 Calculations of Standard Transformed Thermodynamic Properties of Enzyme-catalyzed Reactions

The functions derived in the preceding section can be added and subtracted to obtain standard transformed thermodynamic properties for enzyme-catalyzed reactions. The program **derivefnGHSNHrx** is used to produce a list of functions for the reaction properties $\Delta_r G'°$, $\Delta_r H'°$, $\Delta_r S'°$, and $\Delta_r N_H$ for a typed-in reaction that can be used to calculate tables or make plots.

```
In[30]:= derivefnGHSNHrx[eq_] := Module[{function, functionG, functionH, functionS, functionNH},
            (*Derives the functions of temperature, pH,
             and ionic strength that give the standard transformed reaction Gibbs energy,
             standard transformed reaction enthalpy, standard transformed reaction entropy,
             and the change in binding of hydrogen ions of a biochemical reaction
               or half reaction typed in the form atpGT+h2oGT+de==adpGT+piGT.*)
            function = Solve[eq, de];
            functionG = function[[1, 1, 2]];
            functionH = -t^2 * D[function[[1, 1, 2]] / t, t];
            functionS = -1000 * D[function[[1, 1, 2]], t];
            functionNH = (1 / (8.31451 * (t / 1000) * Log[10])) * D[function[[1, 1, 2]], pH];
            {functionG, functionH, functionS, functionNH}]
```

This program is applied to the following reactions for which standard enthalpies of formation are known for all species. It is convenient to keep track of biochemical reactions by using their EC numbers (5).

EC 1.1.1.1   ethanol+nadox=acetaldehyde+nadred

EC 1.1.1.27b lactate+nadox=pyruvate+nadred

Ec 1.2.1.2 formate+h2o+nadox=co2tot+nadred

EC 1.1.1.70 propanol2+nadox=acetone+nadred

EC 1.4.1.1 alanine+h2o+nadox=pyruvate+ammonia+nadred

EC 2.7.1.1 atp+glucose=adp+glucose6phos

EC 2.7.1.23 atp+nadox=adp+nadpox

EC 2.7.4.3 2adp=atp+amp

EC 3.1.3.1 amp+h2o=adenosine+pi

EC 3.1.3.1c glucose6phos+h2o=glucose+pi

EC 3.1.3.1e ppi + h2o = 2pi

EC 3.2.1.3 maltose+h2o=2glucose

EC 3.2.1.23 lactose+h2o=glucose+galactose

EC 3.5.1.2 glutamine+h2o=glutamate+ammonia

EC 4.1.99.1 tryptophane+h2o=indole+pyruvate+ammonia

EC 4.3.1.1 aspartate=fumarate+ammonia

EC 5.3.1.5 glucose=fructose

EC 5.3.1.5b xylose=xylulose

EC 5.3.1.7 manose=fructose

EC 6.1.1.2 atp+glutamate+ammonia=adp+pi+glutamine

```
In[31]:= rx1x1x1x1 = derivefnGHSNHrx[ethanolGT + nadoxGT + de == acetaldehydeGT + nadredGT];
```

```
In[32]:= PaddedForm[
            TableForm[rx1x1x1x1 /. pH → {5, 6, 7, 8, 9} /. t → {273.15, 298.15, 313.15} /. is → .25,
              TableHeadings → {{"Δr G'°", "Δr H'°", "Δr S'°", "Δr NH"},
                {"      pH 5", "pH 6", "pH 7", "pH 8", "pH 9"}, {273.15, 298.15, 313.15}}]], {4, 2}]
```

```
Out[32]//PaddedForm=
```

		pH 5	pH 6	pH 7	pH 8	pH 9
	273.20	34.52	29.29	24.06	18.83	13.60
Δr G'°	298.20	33.51	27.80	22.09	16.39	10.68
	313.20	32.89	26.89	20.90	14.90	8.90
	273.20	45.31	45.31	45.31	45.31	45.31
Δr H'°	298.20	45.78	45.78	45.78	45.78	45.78
	313.20	46.11	46.11	46.11	46.11	46.11
	273.20	39.50	58.64	77.79	96.93	116.10
Δr S'°	298.20	41.14	60.29	79.43	98.58	117.70
	313.20	42.23	61.37	80.52	99.66	118.80
Δr NH		-1.00				

These properties for the alcohol dehydrogenase reaction can be plotted as functions of temperature and pH.

```
In[33]:= plot1 = Plot3D[Evaluate[rx1x1x1x1[[1]] /. is → 0.25], {pH, 5, 9}, {t, 273.15, 313.15},
            AxesLabel → {"pH", "   T/K", "Δr G'°  "}, DisplayFunction → Identity];
```

```
In[34]:= plot2 = Plot3D[Evaluate[rx1x1x1x1[[2]] /. is → 0.25], {pH, 5, 9}, {t, 273.15, 313.15},
            AxesLabel → {"pH", "   T/K", "Δr H'°   "}, DisplayFunction → Identity];
```

```
In[35]:= plot3 = Plot3D[Evaluate[rx1x1x1x1[[3]] /. is → 0.25], {pH, 5, 9}, {t, 273.15, 313.15},
            AxesLabel → {"pH", "   T/K", "Δr S'°   "}, DisplayFunction → Identity];
```

```
In[36]:=  plot4 = Plot3D[Evaluate[rx1x1x1x1[[4]] /. is → 0.25], {pH, 5, 9}, {t, 273.15, 313.15},
             AxesLabel → {"pH", "    T/K", "Δr NH    "}, DisplayFunction → Identity];

In[37]:=  Show[GraphicsArray[{{plot1, plot2}, {plot3, plot4}}]];
```

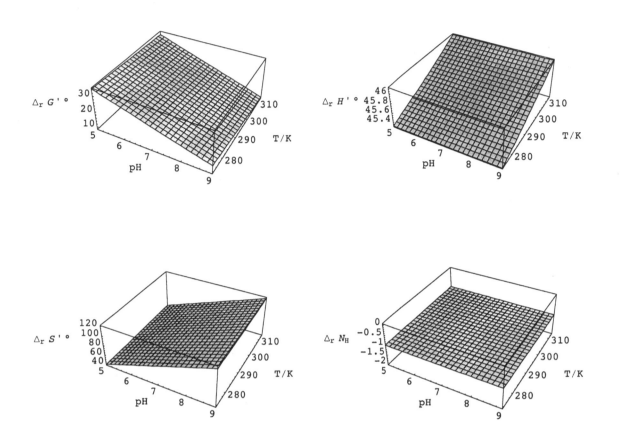

Figure 4.1 Plots of $\Delta_r G'^\circ$, $\Delta_r H'^\circ$, $\Delta_r S'^\circ$, and $\Delta_r N_H$ for ethanol + NAD_{ox} = acetaldehyde + NAD_{red}.

```
In[38]:=  rx1x1x1x27b = derivefnGHSNHrx[lactateGT + nadoxGT + de == pyruvateGT + nadredGT];

In[39]:=  PaddedForm[
            TableForm[rx1x1x1x27b /. pH → {5, 6, 7, 8, 9} /. t → {273.15, 298.15, 313.15} /. is → .25,
            TableHeadings → {{"Δr G'°", "Δr H'°", "Δr S'°", "Δr NH"},
               {"       pH 5", "pH 6", "pH 7", "pH 8", "pH 9"}, {273.15, 298.15, 313.15}}]], {5, 2}]
```

Out[39]//PaddedForm=

		pH 5	pH 6	pH 7	pH 8	pH 9
	273.15	37.38	32.15	26.92	21.69	16.46
$\Delta_r G'^\circ$	298.15	35.32	29.61	23.90	18.20	12.49
	313.15	34.06	28.07	22.07	16.08	10.08
	273.15	59.66	59.66	59.66	59.66	59.66
$\Delta_r H'^\circ$	298.15	60.13	60.13	60.13	60.13	60.13
	313.15	60.46	60.46	60.46	60.46	60.46
	273.15	81.56	100.70	119.85	138.99	158.14
$\Delta_r S'^\circ$	298.15	83.20	102.35	121.49	140.64	159.78
	313.15	84.29	103.43	122.58	141.72	160.87
$\Delta_r N_H$		-1.00				

This reaction is nonspontaneous because of the change in the transformed enthalpy. These properties for the formate dehydrogenase reaction can also be plotted.

In[40]:= **rx1x2x1x2 = derivefnGHSNHrx[formateGT + h2oGT + nadoxGT + de == co2totGT + nadredGT];**

In[41]:= **PaddedForm[**
TableForm[rx1x2x1x2 /. pH → {5, 6, 7, 8, 9} /. t → {273.15, 298.15, 313.15} /. is → .25,
TableHeadings → {{"Δ_r *G* ' °", "Δ_r *H* ' °", "Δ_r *S* ' °", "Δ_r *N*$_H$"},
{"   pH 5", "pH 6", "pH 7", "pH 8", "pH 9"}, {273.15, 298.15, 313.15}}], {5, 2}]

Out[41]//PaddedForm=

		pH 5	pH 6	pH 7	pH 8	pH 9
	273.15	−14.49	−15.44	−18.80	−23.74	−29.12
Δ_r *G* ' °	298.15	−14.09	−15.38	−19.41	−24.90	−30.94
	313.15	−13.86	−15.39	−19.81	−25.63	−32.10
	273.15	−19.13	−16.49	−12.52	−11.36	−10.00
Δ_r *H* ' °	298.15	−18.72	−15.53	−11.79	−10.72	−8.48
	313.15	−18.43	−14.94	−11.34	−10.27	−7.34
	273.15	−16.98	−3.86	23.00	45.35	70.03
Δ_r *S* ' °	298.15	−15.54	−0.51	25.57	47.56	75.33
	313.15	−14.61	1.44	27.05	49.06	79.07
	273.15	−0.06	−0.38	−0.86	−0.99	−1.09
Δ_r *N*$_H$	298.15	−0.08	−0.45	−0.89	−1.01	−1.15
	313.15	−0.09	−0.49	−0.91	−1.01	−1.19

This reaction is spontaneous because of the change in the transformed enthalpy.

In[42]:= **plota = Plot3D[Evaluate[rx1x2x1x2[[1]] /. is → 0.25], {pH, 5, 9}, {t, 273.15, 313.15},**
AxesLabel → {"pH", " T/K", "Δ_r *G* ' °   "}, DisplayFunction → Identity];

In[43]:= **plotb = Plot3D[Evaluate[rx1x2x1x2[[2]] /. is → 0.25], {pH, 5, 9}, {t, 273.15, 313.15},**
AxesLabel → {"pH", " T/K", "Δ_r *H* ' °   "}, DisplayFunction → Identity];

In[44]:= **plotc = Plot3D[Evaluate[rx1x2x1x2[[3]] /. is → 0.25], {pH, 5, 9}, {t, 273.15, 313.15},**
AxesLabel → {"pH", " T/K", "Δ_r *S* ' °   "}, DisplayFunction → Identity];

In[45]:= **plotd = Plot3D[Evaluate[rx1x2x1x2[[4]] /. is → 0.25], {pH, 5, 9}, {t, 273.15, 313.15},**
AxesLabel → {"pH", " T/K", "Δ_r*N*$_H$ "}, DisplayFunction → Identity];

In[46]:= **Show[GraphicsArray[{{plota, plotb}, {plotc, plotd}}]];**

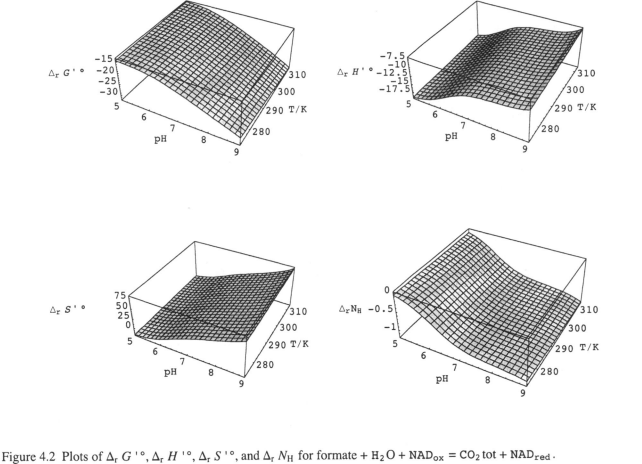

Figure 4.2 Plots of $\Delta_r G'^\circ$, $\Delta_r H'^\circ$, $\Delta_r S'^\circ$, and $\Delta_r N_H$ for formate + H_2O + NAD_{ox} = CO_2 tot + NAD_{red}.

In[47]:= **rx1x1x1x80 = derivefnGHSNHrx[propanol2GT + nadoxGT + de == acetoneGT + nadredGT];**

In[48]:= **PaddedForm[**
    TableForm[rx1x1x1x80 /. pH → {5, 6, 7, 8, 9} /. t → {273.15, 298.15, 313.15} /. is → .25,
      TableHeadings → {{"$\Delta_r G'^\circ$", "$\Delta_r H'^\circ$", "$\Delta_r S'^\circ$", "$\Delta_r N_H$"},
        {"   **pH 5", "pH 6", "pH 7", "pH 8", "pH 9"}, {273.15, 298.15, 313.15}}}], {5, 2}]**

Out[48]//PaddedForm=

		pH 5	pH 6	pH 7	pH 8	pH 9
$\Delta_r G'^\circ$	273.15	21.61	16.39	11.16	5.93	0.70
	298.15	16.40	10.69	4.98	-0.72	-6.43
	313.15	13.25	7.26	1.26	-4.73	-10.73
$\Delta_r H'^\circ$	273.15	78.36	78.36	78.36	78.36	78.36
	298.15	78.83	78.83	78.83	78.83	78.83
	313.15	79.16	79.16	79.16	79.16	79.16
$\Delta_r S'^\circ$	273.15	207.74	226.88	246.03	265.17	284.32
	298.15	209.38	228.52	247.67	266.81	285.96
	313.15	210.47	229.61	248.76	267.90	287.04
$\Delta_r N_H$		-1.00				

This reaction is nonspontaneous because of the change in the transformed enthalpy.

In[49]:= **rx1x4x1x1 =**
    derivefnGHSNHrx[alanineGT + h2oGT + nadoxGT + de == pyruvateGT + ammoniaGT + nadredGT];

In[50]:= **PaddedForm[**
TableForm[rx1x4x1x1 /. pH → {5, 6, 7, 8, 9} /. t → {273.15, 298.15, 313.15} /. is → .25,
TableHeadings → {{"$\Delta_r G'°$", "$\Delta_r H'°$", "$\Delta_r S'°$", "$\Delta_r N_H$"},
{"   pH 5", "pH 6", "pH 7", "pH 8", "pH 9"}, {273.15, 298.15, 313.15}}], {5, 2}]

Out[50]//PaddedForm=

		pH 5	pH 6	pH 7	pH 8	pH 9
	273.15	48.90	43.67	38.44	33.19	27.80
$\Delta_r G'°$	298.15	45.86	40.15	34.43	28.60	21.93
	313.15	44.01	38.01	31.98	25.65	17.61
	273.15	81.73	81.73	81.77	82.15	85.66
$\Delta_r H'°$	298.15	82.43	82.46	82.72	85.20	101.18
	313.15	82.94	83.01	83.72	89.88	114.56
	273.15	120.19	139.35	158.64	179.24	211.82
$\Delta_r S'°$	298.15	122.67	141.90	161.97	189.83	265.80
	313.15	124.31	143.70	165.22	205.11	309.60
	273.15	-1.00	-1.00	-1.00	-1.01	-1.08
$\Delta_r N_H$	298.15	-1.00	-1.00	-1.01	-1.05	-1.36
	313.15	-1.00	-1.00	-1.02	-1.13	-1.61

This reaction is nonspontaneous because of the change in the transformed enthalpy.

In[51]:= **rx2x7x1x1 = derivefnGHSNHrx[atpGT + glucoseGT + de == adpGT + glucose6phosGT];**

In[52]:= **PaddedForm[**
TableForm[rx2x7x1x1 /. pH → {5, 6, 7, 8, 9} /. t → {273.15, 298.15, 313.15} /. is → .25,
TableHeadings → {{"$\Delta_r G'°$", "$\Delta_r H'°$", "$\Delta_r S'°$", "$\Delta_r N_H$"},
{"   pH 5", "pH 6", "pH 7", "pH 8", "pH 9"}, {273.15, 298.15, 313.15}}], {5, 2}]

Out[52]//PaddedForm=

		pH 5	pH 6	pH 7	pH 8	pH 9
	273.15	-17.80	-19.71	-24.25	-29.46	-34.69
$\Delta_r G'°$	298.15	-17.41	-19.47	-24.42	-30.11	-35.82
	313.15	-17.17	-19.32	-24.51	-30.50	-36.49
	273.15	-21.95	-22.25	-22.33	-22.19	-22.17
$\Delta_r H'°$	298.15	-22.20	-22.45	-22.52	-22.42	-22.41
	313.15	-22.45	-22.58	-22.65	-22.58	22.57
	273.15	-15.19	-9.32	7.04	26.60	45.82
$\Delta_r S'°$	298.15	-16.04	-9.99	6.36	25.79	45.00
	313.15	-16.86	-10.44	5.95	25.27	44.46
	273.15	-0.14	-0.66	-0.98	-1.00	-1.00
$\Delta_r N_H$	298.15	-0.14	-0.65	-0.98	-1.00	-1.00
	313.15	-0.14	-0.65	-0.99	-1.00	-1.00

This reaction goes because of the change in the transformed enthalpy.

In[53]:= **rx2x7x1x23 = derivefnGHSNHrx[atpGT + nadoxGT + de == adpGT + nadpoxGT];**

In[54]:= **PaddedForm[**
TableForm[rx2x7x1x23 /. pH → {5, 6, 7, 8, 9} /. t → {273.15, 298.15, 313.15} /. is → .25,
TableHeadings → {{"$\Delta_r G'°$", "$\Delta_r H'°$", "$\Delta_r S'°$", "$\Delta_r N_H$"},
{"   pH 5", "pH 6", "pH 7", "pH 8", "pH 9"}, {273.15, 298.15, 313.15}}], {5, 2}]

Out[54]//PaddedForm=

		pH 5	pH 6	pH 7	pH 8	pH 9
	273.15	-3.57	-9.00	-14.60	-19.94	-25.19
$\Delta_r G'°$	298.15	-2.60	-8.51	-14.63	-20.47	-26.20
	313.15	-2.04	-8.24	-14.66	-20.80	-26.82
	273.15	-14.55	-14.56	-14.42	-14.25	-14.22
$\Delta_r H'°$	298.15	-13.91	-14.09	-14.12	-14.01	-13.99
	313.15	-13.54	-13.76	-13.89	-13.83	-13.82
	273.15	-40.18	-20.37	0.64	20.84	40.13
$\Delta_r S'°$	298.15	-37.95	-18.71	1.71	21.69	40.96
	313.15	-36.74	-17.63	2.46	22.26	41.50
	273.15	-1.01	-1.07	-1.05	-1.01	-1.00
$\Delta_r N_H$	298.15	-1.02	-1.07	-1.05	-1.01	-1.00
	313.15	-1.02	-1.06	-1.05	-1.01	-1.00

This reaction goes because of the change in transformed enthalpy.

In[55]:= `rx2x7x4x3 = derivefnGHSNHrx[2 * adpGT + de == atpGT + ampGT];`

In[56]:= `PaddedForm[`
`    TableForm[rx2x7x4x3 /. pH → {5, 6, 7, 8, 9} /. t → {273.15, 298.15, 313.15} /. is → .25,`
`        TableHeadings → {{"Δ`$_r$` G'°", "Δ`$_r$` H'°", "Δ`$_r$` S'°", "Δ`$_r$` N`$_H$`"}, {"          pH 5", "pH 6",`
`            "pH 7", "pH 8", "pH 9"}, {273.15, 298.15, 313.15}}, TableSpacing → {1, 2}], {5, 2}]`

Out[56]//PaddedForm=

		pH 5	pH 6	pH 7	pH 8	pH 9
	273.15	2.05	1.90	1.84	1.85	1.85
$\Delta_r G'°$	298.15	2.23	2.13	2.07	2.09	2.09
	313.15	2.35	2.25	2.21	2.22	2.23
	273.15	0.07	-0.67	-0.84	-0.90	-0.91
$\Delta_r H'°$	298.15	-0.01	-0.45	-0.62	-0.67	-0.68
	313.15	0.06	-0.29	-0.46	-0.50	-0.51
	273.15	-7.22	-9.41	-9.80	-10.06	-10.10
$\Delta_r S'°$	298.15	-7.52	-8.65	-9.02	-9.24	-9.28
	313.15	-7.29	-8.12	-8.52	-8.70	-8.73
	273.15	-0.03	-0.03	0.00	0.00	0.00
$\Delta_r N_H$	298.15	-0.01	-0.02	0.00	0.00	0.00
	313.15	-0.01	-0.02	0.00	0.00	0.00

This reaction is nonspontaneous because of the transformed entropy.

In[57]:= `rx3x1x3x1 = derivefnGHSNHrx[ampGT + h2oGT + de == adenosineGT + piGT];`

In[58]:= `PaddedForm[`
`    TableForm[rx3x1x3x1 /. pH → {5, 6, 7, 8, 9} /. t → {273.15, 298.15, 313.15} /. is → .25,`
`        TableHeadings → {{"Δ`$_r$` G'°", "Δ`$_r$` H'°", "Δ`$_r$` S'°", "Δ`$_r$` N`$_H$`"},`
`            {"          pH 5", "pH 6", "pH 7", "pH 8", "pH 9"}, {273.15, 298.15, 313.15}}], {5, 2}]`

Out[58]//PaddedForm=

		pH 5	pH 6	pH 7	pH 8	pH 9
	273.15	14.89	-14.16	-12.51	-11.89	-11.81
$\Delta_r G'°$	298.15	-15.60	-14.95	-13.54	-13.03	-12.97
	313.15	-16.01	-15.42	-14.16	-13.72	-13.66
	273.15	-6.95	-5.47	-1.25	0.60	0.87
$\Delta_r H'°$	298.15	-7.36	-5.61	-1.20	0.62	0.87
	313.15	-7.50	-5.65	-1.17	0.63	0.87
	273.15	29.06	31.83	41.23	45.73	46.41
$\Delta_r S'°$	298.15	27.62	31.34	41.39	45.80	46.41
	313.15	27.18	31.20	41.51	45.83	46.42
	273.15	0.01	0.28	0.24	0.04	0.00
$\Delta_r N_H$	298.15	0.02	0.22	0.18	0.03	0.00
	313.15	0.02	0.19	0.15	0.02	0.00

The change in the transformed entropy determines the direction of tis reaction.

```
In[59]:=  rx3x1x3x1c = derivefnGHSNHrx[glucose6phosGT + h2oGT + de == glucoseGT + piGT];
```

```
In[60]:=  PaddedForm[
             TableForm[rx3x1x3x1c /. pH → {5, 6, 7, 8, 9} /. t → {273.15, 298.15, 313.15} /. is → .25,
                TableHeadings → {{"Δ_r G ' °", "Δ_r H ' °", "Δ_r S ' °", "Δ_r N_H"},
                   {"        pH 5", "pH 6", "pH 7", "pH 8", "pH 9"}, {273.15, 298.15, 313.15}}], {5, 2}]
```

Out[60]//PaddedForm=

		pH 5	pH 6	pH 7	pH 8	pH 9
	273.15	-14.23	-12.87	-10.69	-9.97	-9.88
$\Delta_r G$ ' °	298.15	-15.15	-13.75	-11.62	-10.96	-10.88
	313.15	-15.70	-14.27	-12.17	-11.56	-11.48
	273.15	-4.16	-3.21	-0.63	0.83	1.05
$\Delta_r H$ ' °	298.15	-4.19	-3.27	-0.55	0.85	1.06
	313.15	-4.21	-3.32	-0.51	0.87	1.06
	273.15	36.88	35.37	36.83	39.55	40.02
$\Delta_r S$ ' °	298.15	36.77	35.14	37.11	39.63	40.03
	313.15	36.70	35.00	37.24	39.66	40.04
	273.15	0.11	0.43	0.28	0.04	0.00
$\Delta_r N_H$	298.15	0.10	0.40	0.24	0.04	0.00
	313.15	0.10	0.38	0.22	0.03	0.00

This reaction is driven by the change in the transformed entropy.

```
In[61]:=  rx3x1x3x1e = derivefnGHSNHrx[ppiGT + h2oGT + de == 2 * piGT];
```

```
In[62]:=  PaddedForm[
             TableForm[rx3x1x3x1e /. pH → {5, 6, 7, 8, 9} /. t → {273.15, 298.15, 313.15} /. is → .25,
                TableHeadings → {{"Δ_r G ' °", "Δ_r H ' °", "Δ_r S ' °", "Δ_r N_H"},
                   {"        pH 5", "pH 6", "pH 7", "pH 8", "pH 9"}, {273.15, 298.15, 313.15}}], {5, 2}]
```

Out[62]//PaddedForm=

		pH 5	pH 6	pH 7	pH 8	pH 9
	273.15	-23.49	-22.62	-22.48	-25.47	-27.59
$\Delta_r G$ ' °	298.15	-23.37	-22.45	-22.66	-26.06	-28.22
	313.15	-23.29	-22.33	-22.76	-26.40	-28.56
	273.15	-24.66	-24.35	-20.61	-18.85	-20.36
$\Delta_r H$ ' °	298.15	-24.95	-24.63	-20.55	-19.23	-21.22
	313.15	-25.16	-24.84	-20.54	-19.54	-21.84
	273.15	-4.29	-6.34	6.84	24.23	26.47
$\Delta_r S$ ' °	298.15	-5.29	-7.32	7.08	22.89	23.47
	313.15	-5.97	-8.00	7.10	21.88	21.45
	273.15	0.07	0.24	-0.34	-0.62	-0.19
$\Delta_r N_H$	298.15	0.08	0.21	-0.41	-0.60	-0.17
	313.15	0.08	0.20	-0.44	-0.59	-0.15

This reaction is driven by the change in the transformed enthalpy.

```
In[63]:=  rx3x2x1x3 = derivefnGHSNHrx[maltoseGT + h2oGT + de == 2 * glucoseGT];
```

```
In[64]:=  PaddedForm[
             TableForm[rx3x2x1x3 /. pH → {5, 6, 7, 8, 9} /. t → {273.15, 298.15, 313.15} /. is → .25,
                TableHeadings → {{"Δ_r G ' °", "Δ_r H ' °", "Δ_r S ' °", "Δ_r N_H"},
                   {"        pH 5", "pH 6", "pH 7", "pH 8", "pH 9"}, {273.15, 298.15, 313.15}}], {5, 2}]
```

Out[64]//PaddedForm=

		pH 5	pH 6	pH 7	pH 8	pH 9
$\Delta_r G'°$	273.15	-18.29	-18.29	-18.29	-18.29	-18.29
	298.15	-19.92	-19.92	-19.92	-19.92	-19.92
	313.15	-20.90	-20.90	-20.90	-20.90	-20.90
$\Delta_r H'°$	273.15	-0.49	-0.49	-0.49	-0.49	-0.49
	298.15	-0.49	-0.49	-0.49	-0.49	-0.49
	313.15	-0.49	-0.49	-0.49	-0.49	-0.49
$\Delta_r S'°$	273.15	65.17	65.17	65.17	65.17	65.17
	298.15	65.17	65.17	65.17	65.17	65.17
	313.15	65.17	65.17	65.17	65.17	65.17
$\Delta_r N_H$		2.90×10^{-15}				

This reaction is driven by the change in the transformed entropy.

In[65]:= **rx3x2x1x23 = derivefnGHSNHrx[lactoseGT + h2oGT + de == glucoseGT + galactoseGT];**

In[66]:= **PaddedForm[**
  TableForm[rx3x2x1x23 /. pH → {5, 6, 7, 8, 9} /. t → {273.15, 298.15, 313.15} /. is → .25,
    TableHeadings → {{"$\Delta_r G'°$", "$\Delta_r H'°$", "$\Delta_r S'°$", "$\Delta_r N_H$"},
      {"   pH 5", "pH 6", "pH 7", "pH 8", "pH 9"}, {273.15, 298.15, 313.15}}], {5, 2}]

Out[66]//PaddedForm=

		pH 5	pH 6	pH 7	pH 8	pH 9
$\Delta_r G'°$	273.15	-18.48	-18.48	-18.48	-18.48	-18.48
	298.15	-20.31	-20.31	-20.31	-20.31	-20.31
	313.15	-21.41	-21.41	-21.41	-21.41	-21.41
$\Delta_r H'°$	273.15	1.52	1.52	1.52	1.52	1.52
	298.15	1.52	1.52	1.52	1.52	1.52
	313.15	1.52	1.52	1.52	1.52	1.52
$\Delta_r S'°$	273.15	73.22	73.22	73.22	73.22	73.22
	298.15	73.22	73.22	73.22	73.22	73.22
	313.15	73.22	73.22	73.22	73.22	73.22
$\Delta_r N_H$		2.90×10^{-15}				

This reaction is driven by the change in the transformed entropy.

In[67]:= **rx3x5x1x2 = derivefnGHSNHrx[glutamineGT + h2oGT + de == glutamateGT + ammoniaGT];**

In[68]:= **PaddedForm[**
  TableForm[rx3x5x1x2 /. pH → {5, 6, 7, 8, 9} /. t → {273.15, 298.15, 313.15} /. is → .25,
    TableHeadings → {{"$\Delta_r G'°$", "$\Delta_r H'°$", "$\Delta_r S'°$", "$\Delta_r N_H$"}, {"   pH 5", "pH 6",
      "pH 7", "pH 8", "pH 9"}, {273.15, 298.15, 313.15}}, TableSpacing → {1, 1}], {5, 2}]

Out[68]//PaddedForm=

		pH 5	pH 6	pH 7	pH 8	pH 9
$\Delta_r G'°$	273.15	-13.83	-13.83	-13.84	-13.85	-14.01
	298.15	-13.19	-13.19	-13.20	-13.32	-14.29
	313.15	-12.81	-12.82	-12.85	-13.19	-15.24
$\Delta_r H'°$	273.15	-20.98	-20.98	-20.94	-20.56	-17.05
	298.15	-20.74	-20.72	-20.46	-17.98	-2.00
	313.15	-20.57	-20.50	-19.79	-13.63	11.05
$\Delta_r S'°$	273.15	-26.17	-26.15	-26.01	-24.56	-11.12
	298.15	-25.34	-25.24	-24.32	-15.61	41.22
	313.15	-24.78	-24.53	-22.15	-1.41	83.93
$\Delta_r N_H$	273.15	-8.14×10^{-6}	-0.00	-0.00	-0.01	-0.08
	298.15	-0.00	-0.00	-0.01	-0.05	-0.36
	313.15	-0.00	-0.00	-0.02	-0.13	-0.61

This reaction is driven by the change in the transformed enthalpy.

In[69]:= `rx4x1x99x1 = derivefnGHSNHrx[tryptophanLGT + h2oGT + de == indoleGT + pyruvateGT + ammoniaGT]`

In[70]:= `PaddedForm[`
`        TableForm[rx4x1x99x1 /. pH → {5, 6, 7, 8, 9} /. t → {273.15, 298.15, 313.15} /. is → .25,`
`          TableHeadings → {{"Δ`$_r$` G ' °", "Δ`$_r$` H ' °", "Δ`$_r$` S ' °", "Δ`$_r$` N`$_H$`"}, {"         pH 5", "pH 6",`
`            "pH 7", "pH 8", "pH 9"}, {273.15, 298.15, 313.15}}, TableSpacing → {1, 1}], {5, 2}]`

Out[70]//PaddedForm=

		pH 5	pH 6	pH 7	pH 8	pH 9
$\Delta_r G' °$	273.15	25.68	25.68	25.68	25.66	25.50
	298.15	22.49	22.49	22.48	22.36	21.39
	313.15	20.57	20.56	20.53	20.20	18.15
$\Delta_r H' °$	273.15	60.39	60.39	60.43	60.81	64.32
	298.15	60.63	60.65	60.91	63.39	79.37
	313.15	60.80	60.87	61.58	67.74	92.42
$\Delta_r S' °$	273.15	127.08	127.09	127.24	128.69	142.13
	298.15	127.91	128.00	128.92	137.64	194.47
	313.15	128.47	128.71	131.09	151.83	237.18
$\Delta_r N_H$	273.15	-8.14×10^{-6}	-0.00	-0.00	-0.01	-0.08
	298.15	-0.00	-0.00	-0.01	-0.05	-0.36
	313.15	-0.00	-0.00	-0.02	-0.13	-0.61

This reaction is driven by the change in the transformed enthalpy.

In[71]:= `rx4x3x1x1 = derivefnGHSNHrx[aspartateGT + de == fumarateGT + ammoniaGT];`

In[72]:= `PaddedForm[`
`        TableForm[rx4x3x1x1 /. pH → {5, 6, 7, 8, 9} /. t → {273.15, 298.15, 313.15} /. is → .25,`
`          TableHeadings → {{"Δ`$_r$` G ' °", "Δ`$_r$` H ' °", "Δ`$_r$` S ' °", "Δ`$_r$` N`$_H$`"},`
`            {"         pH 5", "pH 6", "pH 7", "pH 8", "pH 9"}, {273.15, 298.15, 313.15}}], {5, 2}]`

Out[72]//PaddedForm=

		pH 5	pH 6	pH 7	pH 8	pH 9
$\Delta_r G' °$	273.15	13.20	13.40	13.42	13.41	13.25
	298.15	11.20	11.43	11.44	11.33	10.36
	313.15	9.99	10.23	10.22	9.89	7.84
$\Delta_r H' °$	273.15	34.85	34.71	34.73	35.11	38.62
	298.15	35.29	35.20	35.45	37.93	53.90
	313.15	35.59	35.58	36.28	42.44	67.12
$\Delta_r S' °$	273.15	79.25	78.00	78.01	79.45	92.88
	298.15	80.77	79.72	80.51	89.22	146.04
	313.15	81.77	80.96	83.23	103.95	189.30
$\Delta_r N_H$	273.15	0.09	0.01	0.00	-0.01	-0.08
	298.15	0.10	0.01	-0.00	-0.05	-0.36
	313.15	0.10	0.01	-0.01	-0.13	-0.61

This reaction is driven by the change in the transformed enthalpy.

In[73]:= `rx5x3x1x5 = derivefnGHSNHrx[glucoseGT + de == fructoseGT];`

In[74]:= `PaddedForm[`
`        TableForm[rx5x3x1x5 /. pH → {5, 6, 7, 8, 9} /. t → {273.15, 298.15, 313.15} /. is → .25,`
`          TableHeadings → {{"Δ`$_r$` G ' °", "Δ`$_r$` H ' °", "Δ`$_r$` S ' °", "Δ`$_r$` N`$_H$`"},`
`            {"         pH 5", "    pH 6", "    pH 7", "    pH 8", "    pH 9"},`
`            {273.15, 298.15, 313.15}}], {5, 2}]`

Out[74]//PaddedForm=

		pH 5	pH 6	pH 7	pH 8	pH 9
	273.15	0.59	0.59	0.59	0.59	0.59
$\Delta_r G'°$	298.15	0.39	0.39	0.39	0.39	0.39
	313.15	0.27	0.27	0.27	0.27	0.27
	273.15	2.81	2.81	2.81	2.81	2.81
$\Delta_r H'°$	298.15	2.81	2.81	2.81	2.81	2.81
	313.15	2.81	2.81	2.81	2.81	2.81
	273.15	8.12	8.12	8.12	8.12	8.12
$\Delta_r S'°$	298.15	8.12	8.12	8.12	8.12	8.12
	313.15	8.12	8.12	8.12	8.12	8.12
$\Delta_r N_H$		0.00				

This reaction is not spontaneousbecause of the change in the transformed enthalpy.

In[75]:= **rx5x3x1x5b = derivefnGHSNHrx[xyloseGT + de == xyluloseGT];**

In[76]:= **PaddedForm[**
 TableForm[rx5x3x1x5b /. pH → {5, 6, 7, 8, 9} /. t → {273.15, 298.15, 313.15} /. is → .25,
 TableHeadings → {{"$\Delta_r G'°$", "$\Delta_r H'°$", "$\Delta_r S'°$", "$\Delta_r N_H$"}, {"   pH 5",
 "   pH 6", "pH 7", "   pH 8", "   pH 9"}, {273.15, 298.15, 313.15}}], {5, 2}]

Out[76]//PaddedForm=

		pH 5	pH 6	pH 7	pH 8	pH 9
	273.15	5.34	5.34	5.34	5.34	5.34
$\Delta_r G'°$	298.15	4.34	4.34	4.34	4.34	4.34
	313.15	3.74	3.74	3.74	3.74	3.74
	273.15	16.29	16.29	16.29	16.29	16.29
$\Delta_r H'°$	298.15	16.29	16.29	16.29	16.29	16.29
	313.15	16.29	16.29	16.29	16.29	16.29
	273.15	40.08	40.08	40.08	40.08	40.08
$\Delta_r S'°$	298.15	40.08	40.08	40.08	40.08	40.08
	313.15	40.08	40.08	40.08	40.08	40.08
$\Delta_r N_H$		0.00				

This reaction is not spontaneous because of the change in the transformed enthalpy.

In[77]:= **rx5x3x1x7 = derivefnGHSNHrx[mannoseGT + de == fructoseGT];**

In[78]:= **PaddedForm[**
 TableForm[rx5x3x1x7 /. pH → {5, 6, 7, 8, 9} /. t → {273.15, 298.15, 313.15} /. is → .25,
 TableHeadings → {{"$\Delta_r G'°$", "$\Delta_r H'°$", "$\Delta_r S'°$", "$\Delta_r N_H$"},
 {"   pH 5", "   pH 6", "pH 7", "   pH 8", "   pH 9"},
 {273.15, 298.15, 313.15}}], {5, 2}]

Out[78]//PaddedForm=

		pH 5	pH 6	pH 7	pH 8	pH 9
	273.15	−5.11	−5.11	−5.11	−5.11	−5.11
$\Delta_r G'°$	298.15	−5.51	−5.51	−5.51	−5.51	−5.51
	313.15	−5.75	−5.75	−5.75	−5.75	−5.75
	273.15	−0.72	−0.72	−0.72	−0.72	−0.72
$\Delta_r H'°$	298.15	−0.72	−0.72	−0.72	−0.72	−0.72
	313.15	−0.72	−0.72	−0.72	−0.72	−0.72
	273.15	16.07	16.07	16.07	16.07	16.07
$\Delta_r S'°$	298.15	16.07	16.07	16.07	16.07	16.07
	313.15	16.07	16.07	16.07	16.07	16.07
$\Delta_r N_H$		0.00				

This reaction is spontaneous mainly because of the change in the transformed entropy.

In[79]:= **rx5x3x1x20 = derivefnGHSNHrx[riboseGT + de == ribuloseGT];**

In[80]:= **PaddedForm[**
    TableForm[rx5x3x1x20 /. pH → {5, 6, 7, 8, 9} /. t → {273.15, 298.15, 313.15} /. is → .25,
      TableHeadings → {{"Δ_r G'°", "Δ_r H'°", "Δ_r S'°", "Δ_r N_H"},
        {"   pH 5", "   pH 6", "   pH 7", "   pH 8", "   pH 9"},
      {273.15, 298.15, 313.15}}], {5, 2}]

Out[80]//PaddedForm=

		pH 5	pH 6	pH 7	pH 8	pH 9
	273.15	15.63	15.63	15.63	15.63	15.63
$\Delta_r G'°$	298.15	16.06	16.06	16.06	16.06	16.06
	313.15	16.32	16.32	16.32	16.32	16.32
	273.15	10.98	10.98	10.98	10.98	10.98
$\Delta_r H'°$	298.15	10.98	10.98	10.98	10.98	10.98
	313.15	10.98	10.98	10.98	10.98	10.98
	273.15	-17.04	-17.04	-17.04	-17.04	-17.04
$\Delta_r S'°$	298.15	-17.04	-17.04	-17.04	-17.04	-17.04
	313.15	-17.04	-17.04	-17.04	-17.04	-17.04
$\Delta_r N_H$		0.00				

This reaction is nonspontaneous because of enthalpy.

In[81]:= **rx6x1x1x2 = derivefnGHSNHrx[atpGT + glutamateGT + ammoniaGT + de == adpGT + piGT + glutamineGT]**

In[82]:= **PaddedForm[**
    TableForm[rx6x1x1x2 /. pH → {5, 6, 7, 8, 9} /. t → {273.15, 298.15, 313.15} /. is → .25,
      TableHeadings → {{"Δ_r G'°", "Δ_r H'°", "Δ_r S'°", "Δ_r N_H"},
        {"   pH 5", "   pH 6", "   pH 7", "   pH 8", "   pH 9"},
      {273.15, 298.15, 313.15}}], {5, 2}]

Out[82]//PaddedForm=

		pH 5	pH 6	pH 7	pH 8	pH 9
	273.15	-18.20	-18.74	-21.11	-25.58	-30.55
$\Delta_r G'°$	298.15	-19.37	-20.03	-22.83	-27.75	-32.41
	313.15	-20.05	-20.77	-23.83	-28.87	-32.74
	273.15	-5.13	-4.48	-2.02	-0.80	-4.07
$\Delta_r H'°$	298.15	-5.64	-5.00	-2.62	-3.59	-19.35
	313.15	-6.08	-5.40	-3.37	-8.09	-32.56
	273.15	47.86	52.20	69.87	90.70	96.96
$\Delta_r S'°$	298.15	46.07	50.39	67.79	81.03	43.81
	313.15	44.62	49.09	65.35	66.35	0.56
	273.15	-0.03	-0.23	-0.70	-0.95	-0.92
$\Delta_r N_H$	298.15	-0.04	-0.25	-0.74	-0.91	-0.64
	313.15	-0.04	-0.26	-0.75	-0.84	-0.39

This reaction is spontaneous mainly because of the change in the transformed entropy. Note the entropy increases because H$^+$ is produced. For most of these reactions, the enthalpy change at pH 7 determines whether the reaction goes to the right or the left, but sometimes the entropy change does, especially when hydrogen ions are produced by the reaction.

Standard transformed thermodynamic properties are given for more enzyme-catalyzed reactions in Chapter 13.

4.5 Two-dimensional Plots of Transformed Thermodynamic Properties of Biochemical Reactions

Since **derivefnGHSNHrx** produces a list of functions for $\Delta_r G'°$, $\Delta_r H'°$, $\Delta_r S'°$, and $\Delta_r N_H$, any one of them can be selected and plotted. The first function in the list, that for the standard transformed Gibbs energy, is selected by use of

In[83]:= **rx1x1x1x1[[1]];**

This is used to make plots for the reaction ethanol + nadox = acetaldehyde + nadred.

```
In[84]:= ploteth1 =
           Plot[Evaluate[rx1x1x1x1[[1]] /. is → 0.25 /. t → {273.15, 298.15, 313.15}], {pH, 5, 9},
             AxesLabel → {"pH", "Δr G' °"}, AxesOrigin → {5, 10}, DisplayFunction → Identity];
```

```
In[85]:= ploteth2 = Plot[Evaluate[rx1x1x1x1[[2]] /. is → 0.25 /. t → {273.15, 298.15, 313.15}],
             {pH, 5, 9}, AxesLabel → {"pH", "Δr H' °"}, AxesOrigin → {5, 45.2},
             PlotRange → {45.2, 46.3}, DisplayFunction → Identity];
```

```
In[86]:= ploteth3 =
           Plot[Evaluate[rx1x1x1x1[[3]] /. is → 0.25 /. t → {273.15, 298.15, 313.15}], {pH, 5, 9},
             AxesLabel → {"pH", "Δr S' °"}, AxesOrigin → {5, 38}, DisplayFunction → Identity];
```

```
In[87]:= ploteth4 =
           Plot[Evaluate[rx1x1x1x1[[4]] /. is → 0.25 /. t → {273.15, 298.15, 313.15}], {pH, 5, 9},
             AxesLabel → {"pH", "Δr N_H"}, AxesOrigin → {5, -2}, DisplayFunction → Identity];
```

```
In[88]:= Show[GraphicsArray[{{ploteth1, ploteth2}, {ploteth3, ploteth4}}]];
```

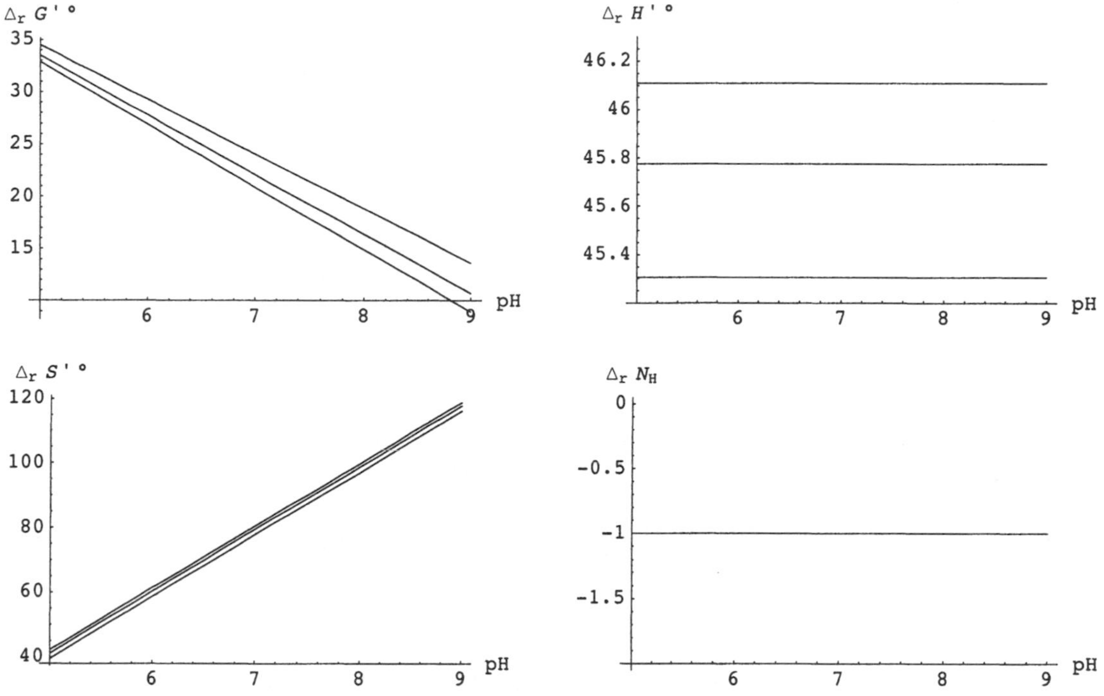

Figure 4.3 Plots of transformed thermodynamic properties for the reaction ethanol + nadox = acetaldehyde + nadred as a function of pH at 0.25 M ionic strength and temperatures of 273.15, 298.15, and 313.15 K.

Plots for the reaction formate+h2o + nadox = co2tot + nadred are made as follows:

```
In[89]:= plotfor1 =
           Plot[Evaluate[rx1x2x1x2[[1]] /. is → 0.25 /. t → {273.15, 298.15, 313.15}], {pH, 5, 9},
             AxesLabel → {"pH", "Δr G' °"}, AxesOrigin → {5, -32.5}, DisplayFunction → Identity];
```

```
In[90]:= plotfor2 =
           Plot[Evaluate[rx1x2x1x2[[2]] /. is → 0.25 /. t → {273.15, 298.15, 313.15}], {pH, 5, 9},
             AxesLabel → {"pH", "Δr H' °"}, AxesOrigin → {5, -19}, DisplayFunction → Identity];
```

```
In[91]:= plotfor3 =
            Plot[Evaluate[rx1x2x1x2[[3]] /. is → 0.25 /. t → {273.15, 298.15, 313.15}], {pH, 5, 9},
              AxesLabel → {"pH", "Δr S ' °"}, AxesOrigin → {5, -20}, DisplayFunction → Identity];

In[92]:= plotfor4 =
            Plot[Evaluate[rx1x2x1x2[[4]] /. is → 0.25 /. t → {273.15, 298.15, 313.15}], {pH, 5, 9},
              AxesLabel → {"pH", "Δr NH"}, AxesOrigin → {5, -1.2}, DisplayFunction → Identity];

In[93]:= Show[GraphicsArray[{{plotfor1, plotfor2}, {plotfor3, plotfor4}}]];
```

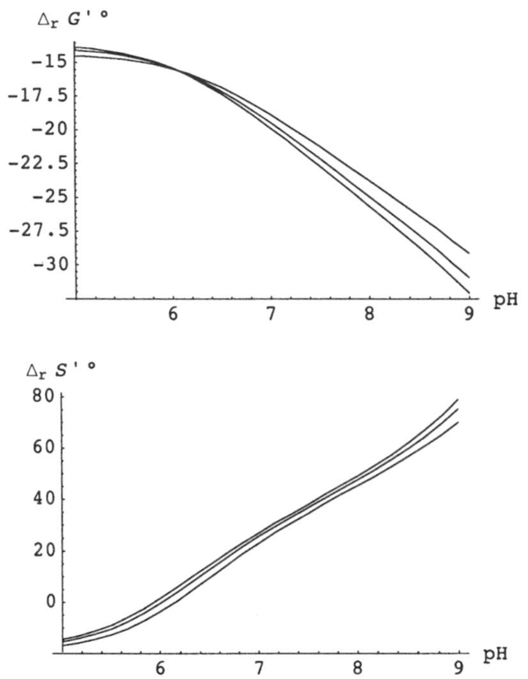

Figure 4.4 Plots of transformed thermodynamic properties for the reaction formate+h2o + nadox = co2tot + nadred as a function of pH at 0.25 M ionic strength and temperatures of 273.15, 298.15, and 313.15 K.

4.6 Changes of Standard Transformed Enthalpy of Reaction and Standard Transformed Entropy of Reaction with pH

For reactions for which standard enthalpies of formation of species are known, the preceding chapter shows that $\partial \Delta_r H'°/\partial pH$ and $\partial \Delta_r S'°/\partial pH$ can be calculated in two ways; directly and from the change in binding of hydrogen ions with temperature. This provides a check on the calculations and it provides an explanation of the magnitudes of these derivatives.

The following calculations show the two ways to calculate $\partial \Delta_r H'°/\partial pH$:

```
In[94]:= Plot[Evaluate[D[rx1x2x1x2[[2]], pH] /. is → 0.25 /. t → {273.15, 298.15, 313.15}],
            {pH, 5, 9}, AxesOrigin → {5, 0}, AxesLabel → {"pH", "∂Δr H'°/∂pH"}];
```

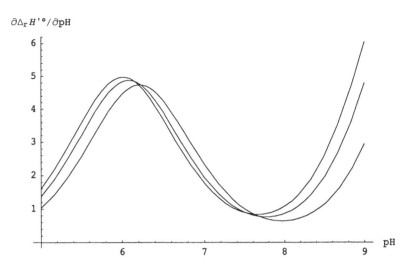

Figure 4.5 Rate of change of $\Delta_r H'^\circ$ for EC 1.2.1.2 in kJ mol^{-1} with pH at temperatures of 273.15, 298.15, and 313.15 K at 0.25 M ionic strength calculated using the functiion for the enthalpy.

The same plot is obtained from the function for $\Delta_r N_H$ by use of

```
In[95]:= Plot[Evaluate[-8.31451 * (t^2) * Log[10] * D[rx1x2x1x2[[4]], t] /. is → 0.25 /.
            t → {273.15, 298.15, 313.15}], {pH, 5, 9}, AxesOrigin → {5, 0},
        AxesLabel → {"pH", "∂Δ_r H'°/∂pH"}, DisplayFunction → Identity];
```

as can be seen by deleting the DisplayFunction->Identity.

The following calculations show the two ways to calculate $\partial\Delta_r S'^\circ/\partial pH$:

```
In[96]:= Plot[Evaluate[-D[rx1x2x1x2[[3]] * 1000, pH] /. is → 0.25 /. t → {273.15, 298.15, 313.15}],
        {pH, 5, 9}, AxesOrigin → {5, -40000}, AxesLabel → {"pH", "∂Δ_r S'°/∂pH"}];
```

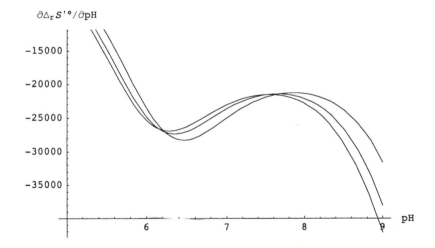

Figure 4.6 Rate of change of $\Delta_r S'^\circ$ for EC 1.2.1.2 in J K^{-1} mol^{-1} with pH at temperatures of 273.15, 298.15, and 313.15 K at 0.25 M ionic strength calculated using the function for the entropy.

The same plot is obtained from the function for $\Delta_r N_H$ by use of

```
In[97]:=  Plot[Evaluate[8.31451*Log[10]*D[t*rx1x2x1x2[[4]], t] /. is → 0.25 /.
          t → {273.15, 298.15, 313.15}], {pH, 5, 9}, AxesOrigin → {5, -40000},
          AxesLabel → {"pH", "∂ΔᵣS'°/∂pH"}, DisplayFunction → Identity];
```

4.7 Discussion

This chapter shows that more thermodynamic information can be obtained for enzyme-catalyzed reactions for which $\Delta_f H°$ is known for all the species in addition to $\Delta_f G°$. When $\Delta_f H°$ is independent of temperature, $\Delta_f G°$ of the species can be expressed as a function of temperature. Since the temperature dependence of the parameter in the Debye-Hückel equation is known, the $\Delta_f G'°$ for a reactant can be derived as a function of temperature, pH, and ionic strength by use of **derivetrGibbsT**. As shown in Chapter 3, this means that $\Delta_f G'°$, $\Delta_f H'°$, $\Delta_f S'°$, $\overline{N}_H$, and derivatives of these properties of reactants with respect to temperature and pH can be calculated. *Mathematica* makes it possible to derive these functions and use them to produce tables and plots. At the present time this can be done for 94 reactants. This chapter emphasizes the usefulness of calorimetric data in biochemical thermodynamics.

These functions for reactants can be added and subtracted to obtain $\Delta_r G'°$, $\Delta_r H'°$, $\Delta_r S'°$, and $\Delta_r N_H$ for reactions. Maxwell relations show how $\partial\Delta_r H'°/\partial pH$ and $\partial\Delta_r S'°/\partial pH$ are determined by the temperature coefficient of $\Delta_r N_H$. The functions for $\Delta_r G'°$, $\Delta_r H'°$, $\Delta_r S'°$, and $\Delta_r N_H$ are calculated using the *Mathematica* program **derivefnGHSNHrx**. This program can be used to calculate tables of the four properties for all the reactants in BasicBiochemData3 for which $\Delta_f H°$ are known for all species.

It is of interest to see the extents to which the standard transformed enthalpy and standard transformed entropy of reaction determine whether the reaction goes spontaneously to the right or the left under specified conditions.

Appendix

Heat Capacities

There is some literature data on heat capacities and transformed heat capacities, but not enough to justify including them in the general treatments here, but they are of interest and may not be negligible. The adjustment of the heat capacity of a species for the ionic strength depends on both the first and second derivatives of the coefficient alpha in the Debye-Hückel equation.

$$C_{Pm}°(I) = C_{Pm}°(I=0) + RT[2\frac{\partial\alpha}{\partial T} + T\frac{\partial^2\alpha}{\partial T^2}](z_j^2 - N_H(j))\frac{I^{1/2}}{1+BI^{1/2}} \tag{A4-1}$$

Thus the transformed heat capacity of a species can be positive or negative. The transformed heat capacity of a reactant involving two or more species is not simply a weighted average, but is given by the following equation.

$$C_{Pm}'°(i) = \Sigma r_j C_{Pm}'°(j) + \frac{1}{RT}\{\Sigma r_j(\Delta_f H_i'°)^2 - (\Delta_f H_i'°)^2\} \tag{A4-2}$$

The second term in this equation is always positive because the weighted average of the squares of the individual standard transforme enthalpies of formation of the species is always greater that the square of the weighted average enthalpy of formation (6). This is the quantitative expression of Le Chatelier's principle for the heat capacity.

Le Chatelier's Principle

According to this principle, when a system that is in equilibrium is perturbed, the equilibrium will always be displaced in such a way as to oppose the applied change. Thus when the temperature is raised on a system of chemical reactions at equilibrium, the reactions will shift in such a way as to absorb heat. This is illustrated by equation A4-2. When the pH is lowered on a biochemical reaction system, the reaction will shift in such a way that $\Delta_r N_H$ will increase, that is the binding of hydrogen ions in the reaction will become more positive.

References

1. R. A. Alberty, Effect of temperature on standard transformed Gibbs energies of formation of reactants at specified pH and ionic strength and apparent equilibrium constants of biochemical reactions, J. Phys. Chem. 105 B, 7865-7870 (2001). (Supporting Information is available.)
2. R. A. Alberty, Effect of temperature on the standard transformed thermodynamic properties of biochemical reactions with emphasis on the Maxwell equations, J. Phys. Chem. 107 B, 3631-3635 (2003). (Supporting Information is available.)
3. R. A. Alberty, BasicBiochemData3 (2005).

In[98]:= **http : // library.wolfram.com / infocenter / MathSource / 5704**

4. E. Di Cera, S. J. Gill, and J. Wyman, Binding capacity: cooperativity and buffering in biopolymers, Proc. Natl. Acad, Sci. USA 85, 449-452 (1988).
5. Webb, Enzyme Nomenclature 1992, Academic Press (1992).

   http : // www.chem.qmw.ac.uk / iubmb / enzyme /

6. R. A. Alberty. Chemical thermodynamic properties of isomer groups, I&EC Fund. 22, 218-221 (1983).

Chapter 5 Biochemical Reactions at Specified pHs, pMgs, and Various Temperatures

5.1 Fundamental Equation and Maxwell Relations for the Transformed Gibbs Energy of a Reactant at Specified T, pH, pMg, and Ionic Strength

5.2 Calculation of Standard Transformed Thermodynamic Properties of Inorganic Phosphate at Specified T, pH, pMg and Ionic Strength by Use of Maxwell Relations

5.3 Calculation of All the Standard Transformed Thermodynamic Properties of a Reactant by Use of a Single Program

5.4 Calculation of Standard Transformed Thermodynamic Properties of ATP + H_2O = ADP + P_i at Specified T, pH, pMg, and Ionic Strength

5.5 Calculation of Standard Transformed Thermodynamic Properties of ATP + $2H_2O$ = AMP + $2P_i$ at Specified T, pH, pMg, and Ionic Strength

5.6 Discussion

References

5.1 Fundamental Equation and Maxwell Relations for the Transformed Gibbs Energy of a Reactant at Specified T, pH, pMg, and Ionic Strength

When a solution contains metal ions that are bound by species of a reactant, for example, $MgATP^{2-}$, $MgHATP^-$, and Mg_2ATP, the apparent equilibrium constant for any reaction involving that reactant will depend on pMg = $-\log[Mg^{2+}]$. When pMg is specified as well as pH, the transformed Gibbs energy G' that provides the criterion for spontaneous change and equilibrium is defined by the following Legendre transform (1-3):

$$G' = G - n_c(H)\mu(H^+) - n_c(Mg)\mu(Mg^{2+}) \tag{5.1-1}$$

where $n_c(Mg)$ is the amount of the magnesium component that is given by

$$n_c(Mg) = \sum N_{Mg}(j) n_j \tag{5.1-2}$$

where N_{Mg} is the number of magnesium atoms in species j. Carrying out the mathematical operations described in Section 3.3 leads to a new fundamental equation for G' that replaces equation 3.3-10:

$$dG' = -S'dT + VdP + \sum \mu_i' d n_i' + RT\ln(10)n_c(H)dpH + RT\ln(10)n_c(Mg)dpMg \tag{5.1-3}$$

The transformed entropy of the system is given by

$$S' = S - n_c(H)\overline{S}(H^+) - n_c(Mg)\overline{S}(Mg^{2+}) \tag{5.1-4}$$

and the transformed enthalpy of the system is given by

$$H' = H - n_c(\text{H})\overline{H}(\text{H}^+) - n_c(\text{Mg})\overline{H}(\text{Mg}^{2+}) \tag{5.1-5}$$

The natural variables of G' are now T, pH, pMg, and $\{n_i'\}$. The equation $G' = \Sigma\mu_j' n_j'$ applies, but since there is an additional term in the fundamental equation 5.1-3, there are more Maxwell relations. When there is a single reactant, the five Maxwell relations in equations 3.4-12 to 3.4-16 apply, but now there are four more Maxwell relations that involve magnesium:

$$\frac{\partial\Delta_f G_i'^\circ}{\partial\text{pMg}} = RT\ln(10)\overline{N}_{\text{Mg}}(i) \tag{5.1-6}$$

$$\frac{\partial\Delta_f H_i'^\circ}{\partial\text{pMg}} = -RT^2\ln(10)\frac{\partial\overline{N}_{\text{Mg}}(i)}{\partial T} \tag{5.1-7}$$

$$-\frac{\partial\Delta_f S_i'^\circ}{\partial\text{pMg}} = R\ln(10)\frac{\partial(T\,\overline{N}_{\text{Mg}}(i))}{\partial T} \tag{5.1-8}$$

$$\frac{\partial\overline{N}_{\text{H}}}{\partial\text{pMg}} = \frac{\partial\overline{N}_{\text{Mg}}}{\partial\text{pH}} \tag{5.1-9}$$

The last of these nine Maxwell relations expresses the reciprocal effect between the binding of hydrogen ions and magnesium ions. Some higher partial derivatives can also be calculated.

5.2 Calculation of Standard Transformed Thermodynamic Properties of Inorganic Phosphate at Specified T, pH, pMg and Ionic Strength by Use of Maxwell Relations

A number of investigators have added to our knowledge of dissociation contants of magnesium complex ions of biochemical interest. Goldberg and Tewari (4) and Larson, Tewari, and Goldberg (5) have contributed and have also evaluated the existing data. The strategy here is to derive $\Delta_f G_i'^\circ$ of a reactant as a function of T, pH, pMg, and ionic strength, and then derive the functions for all the other properties using the Maxwell relations in equations 3.4-12 to 3.4-16 and 5.1-6 to 5.1-9. The standard transformed Gibbs energy of formation of a species is given by the following equation (6):

$$\Delta_f G_j'^\circ = \Delta_f G_j^\circ - N_{\text{H}}(j)[\Delta_f G^\circ(\text{H}^+) - RT\ln(10)\,\text{pH}] - N_{\text{Mg}}(j)[\Delta_f G^\circ(\text{Mg}^{2+}) - RT\ln(10)\,\text{pMg}] \tag{5.2-1}$$

Since $\Delta_f G_j^\circ$, $\Delta_f G^\circ(\text{H}^+)$ and $\Delta_f G^\circ(\text{Mg}^{2+})$ depend on temperature and ionic strength, these dependencies must be substituted into this equation. The first term on the right side of equation 5.2-1 is replaced with equation 4.1-3. The $\Delta_f G^\circ(\text{H}^+)$ in the second term of equation 5.2-1 is replaced with equation 1.3-5, which is

$$\Delta_f G_j^\circ(\text{H}^+, I) = -RT\alpha I^{1/2}/(1 + BI^{1/2}) \tag{5.2-2}$$

for H^+ because $\Delta_f G_j^\circ(\text{H}^+, I=0) = 0$ at each temperature. The $\Delta_f G^\circ(\text{Mg}^{2+})$ in the next term is replaced with

$$\Delta_f G^\circ(\text{Mg}^{2+}) = (T/298.15)\Delta_f G_{\text{Mg}}^\circ(298.15, I=0) + (1 - T/298.15)\Delta_f H_{\text{Mg}}^\circ(298.15, I=0) - 4RT\alpha I^{1/2}/(1 + BI^{1/2}) \tag{5.2-3}$$

This yields the following expression for the standard transformed Gibbs energy of formation of species j:

$$\Delta_f G_j{}'^{\circ}(T,\text{pH},\text{pMg},I) = (T/298.15)\Delta_f G_j{}^{\circ}(298.15,I{=}0) + (1 - T/298.15)\Delta_f H_j{}^{\circ}(298.15,I{=}0)$$

$$- N_{\text{Mg}}(j)\{ (T/298.15) \Delta_f G_{\text{Mg}}{}^{\circ}(298.15, I = 0) + (1 - T/298.15) \Delta_f H_{\text{Mg}}{}^{\circ}(298.15, I = 0) \}$$

$$+ N_{\text{H}}(j) RT\ln(10)\,\text{pH} + N_{\text{Mg}}(j) RT\ln(10)\,\text{pMg} - RT\alpha(z_i{}^2 - N_{\text{H}}(j) - 4 N_{\text{Mg}}(j))I^{1/2}/(1 + 1.6 I^{1/2}) \tag{5.2-4}$$

The temperature dependence of $RT\alpha$ that is given in equation 1.7-2 still has to be substituted in this equation.

The next step is to use equation 3.5-5 to derive the expression for the standard transformed Gibbs energy of the reactant (pseudoisomer group). The function of T, pH, pMg, and ionic strength is too long to write out. Fortunately, *Mathematica* can be used to derive this function, given the species matrix for a reactant. These calculations are discussed in greater detail in reference 6. The following program derives the function of temperature, pH, pMg and ionic strength for $\Delta_f G_j{}'^{\circ}(T,\text{pH},\text{pMg},I)$:

```
In[2]:=  Off[General::"spell"];
         Off[General::"spell1"];

In[4]:=  ClearAll["Global`*"]

In[5]:=  deriveGMgT[speciesmat_] :=
         Module[{dGzero,dHzero,dGzeroT,zi,nH,nMg,pHterm,stdGMg,
         pMgterm,coeffis,isterm,gpfnsp},(*This program derives the function of T, pH, pMg and
         ionic strength (is) that gives the standard transformed Gibbs energy of formation of
         a reactant (sum of species).  The input speciesmat is a matrix that gives the
         standard Gibbs energy of formation, the standard enthalpy of formation, the electric
         charge, the number of hydrogen atoms, and the number of magnesium atoms in each
         species. There is a row in the matrix for each species of the reactant. gpfnsp is a
         list of the functions for the species.  Energies in the output are in kJ mol^-1.*)
         {dGzero,dHzero,zi,nH,nMg} = Transpose[speciesmat];
         pHterm = nH*8.31451*t*Log[10]*pH/1000;
         stdGMg=(t/298.15)*(-455.3)+(1-t/298.15)*(-467.00);
         pMgterm = nMg*(-stdGMg+8.31451*(t/1000)*Log[10]*pMg);
         coeffis=(9.20483*t)/10^3-(1.284668*t^2)/10^5+(4.95199*t^3)/10^8;
         dGzeroT=(t/298.15)*dGzero+(1-t/298.15)*dHzero;
         isterm = coeffis*((zi^2)-nH-4*nMg)*(is^.5)/(1 + 1.6*is^.5);
         gpfnsp=dGzeroT+pHterm+pMgterm-isterm;
         -8.31451*(t/1000)*Log[Apply[Plus,Exp[-1*gpfnsp/((8.31451*(t/1000)))]]]]];
```

The use of this program is illustrated with the calculation of the transformed thermodynamic properties of inorganic phosphate, for which the basic data including the complex ion are given by

```
In[6]:=  piMgsp = {{-1096.1, -1299., -2, 1, 0},
            {-1137.3, -1302.6, -1, 2, 0}, {-1566.87, -1753.8, 0, 1, 1}};
```

The last entry in each row is the number of magnesium atoms in a species. The third species is $MgHPO_4$. Now we are in position to calculate all the transformed thermodynamic properties of inorganic phosphate. The function of temperature, pH, pMg, and ionic strength for the standard transformed Gibbs energy of formation of inorganic phosphate is given by

```
In[7]:=  piGMgT = deriveGMgT[piMgsp]
```

$$Out[7]= -0.00831451\, t \,\text{Log}\Big[e^{-\frac{120.272\left(-1299.(1-0.00335402\,t)-3.67634\,t+0.0191449\,\text{pH}\,t-\frac{3\,is^{0.5}\left(0.00920483\,t-0.0000128467\,t^2+4.95199\times10^{-8}\,t^3\right)}{1+1.6\,is^{0.5}}\right)}{t}}$$
$$+ e^{-\frac{120.272\left(-1302.6(1-0.00335402\,t)-3.81452\,t+0.0382897\,\text{pH}\,t+\frac{is^{0.5}\left(0.00920483\,t-0.0000128467\,t^2+4.95199\times10^{-8}\,t^3\right)}{1+1.6\,is^{0.5}}\right)}{t}}$$
$$+ e^{-\frac{120.272\left(-1286.8(1-0.00335402\,t)-3.72822\,t+0.0191449\,\text{pH}\,t+0.0191449\,\text{pMg}\,t+\frac{5\,is^{0.5}\left(0.00920483\,t-0.0000128467\,t^2+4.95199\times10^{-8}\,t^3\right)}{1+1.6\,is^{0.5}}\right)}{t}} \Big]$$

At 298.15 K, pH 7, pMg 3.0, and ionic strength 0.25 M the standard transformed Gibbs energy of formation of inorganic phosphate in kJ mol^{-1} is given by

In[8]:= **piGMgT /. t -> 298.15 /. pH → 7.0 /. pMg → 3.0 /. is → .25**

Out[8]= -1059.55

The function for the standard transformed enthalpy of formation of inorganic phosphate can be obtained by use of the piGMgT and the Gibbs-Helmholtz equation (3.4-15).

In[9]:= **piHMgT = -t^2 * D[piGMgT / t, t];**

This is an even more complicated function that can be seen by removing the semicolon, but it can readily be evaluated at specified temperature, pH, pMg, and ionic strength. The standard transformed enthalpy of formation of inorganic phosphate in kJ mol^{-1} for specified conditions is gven by

In[10]:= **piHMgT /. t -> 298.15 /. pH → 7.0 /. pMg → 3.0 /. is → .25**

Out[10]= -1299.12

The function for the standard transformed entropy of formation of inorganic phosphate in J K^{-1} mol^{-1} is given by (see equation 3.4-12)

In[11]:= **piSMgT = -1000 * D[piGMgT, t];**

The standard transformed entropy of formation of inorganic phosphate in J K^{-1} mol^{-1} for specified conditions is given by

In[12]:= **piSMgT /. t -> 298.15 /. pH → 7.0 /. pMg → 3.0 /. is → .25**

Out[12]= -803.51

The function for the average number of hydrogen atoms in inorganic phosphate is given by (see equation 3.4-14)

In[13]:= **pinHMgT = (1 / (8.31451 * (t / 1000) * Log[10])) * D[piGMgT, pH];**

The average number of hydrogen ions bound at specified conditions is given by

In[14]:= **pinHMgT /. t -> 298.15 /. pH → 7.0 /. pMg → 3.0 /. is → .25**

Out[14]= 1.30117

The average number of magnesium atoms in inorganic phosphate is given by (see equation 5.1-6)

In[15]:= **pinMgMgT = (1 / (8.31451 * (t / 1000) * Log[10])) * D[piGMgT, pH];**

The average number of magnesium atoms in inorganic phosphate under specified conditions is given by

In[16]:= **pinMgMgT /. t -> 298.15 /. pH → 7.0 /. pMg → 3.0 /. is → .25**

Out[16]= 1.30117

5.3 Calculation of All the Standard Transformed Thermodynamic Properties of a Reactant by Use of a Single Program

All the standard transformed properties can be calculated using the following program:

```
In[17]:= deriveGHSNHNMg[speciesmat_] :=
         Module[{dGzero,dHzero,dGzeroT,zi,nH,nMg,pHterm,stdGMg,
         pMgterm,coeffis,isterm,gpfnsp,gibbsfn,enthalfn,entfn,nHfn,nMgfn},(*This program
         derives the function of T, pH, pMg and ionic strength (is) that gives the standard
         transformed Gibbs energy of formation of a reactant (sum of species).  Then partial
         differentiation is used to derive the functions that yield the standard transformed
         enthalp of formation, standard transformed entropy of formation, average numbe of
         hydrogen ions bound and average number of magnesium ions bound.  The input
         speciesmat is a matrix that gives the standard Gibbs energy of formation, the
         standard enthalpy of formation, the electric charge, the number of hydrogen atoms,
         and the number of magnesium atoms in each species. There is a row in the matrix for
         each species of the reactant. gpfnsp is a list of the functions for the species.
         Energies are expressed in kJ mol^-1 in the program until gibbsfn is needed, but the
         output is changed to J mol^-1 so that differentiations can be made with respect to t
         in K. The output is a list of five functions.  Energies in the output are in kJ
         mol^-1 and entropies in J K^-1 mol^-1.*)
         {dGzero,dHzero,zi,nH,nMg} = Transpose[speciesmat];
         pHterm = nH*8.31451*t*Log[10]*pH/1000;
         stdGMg=(t/298.15)*(-455.3)+(1-t/298.15)*(-467.00);
         pMgterm = nMg*(-stdGMg+8.31451*(t/1000)*Log[10]*pMg);
         coeffis=(9.20483*t)/10^3-(1.284668*t^2)/10^5+(4.95199*t^3)/10^8;
         dGzeroT=(t/298.15)*dGzero+(1-t/298.15)*dHzero;
         isterm = coeffis*((zi^2)-nH-4*nMg)*(is^.5)/(1 + 1.6*is^.5);
         gpfnsp=dGzeroT+pHterm+pMgterm-isterm;
         gibbsfn=-8.31451*(t/1000)*Log[Apply[Plus,Exp[-1*gpfnsp/((8.31451*(t/1000)))]]];
         enthalfn=-t^2*D[gibbsfn/t,t];
         entfn=-1000*D[gibbsfn,t];
         nHfn=(1/(8.31451*(t/1000)*Log[10]))*D[gibbsfn,pH];
         nMgfn=(1/(8.31451*(t/1000)*Log[10]))*D[gibbsfn,pMg];
         {gibbsfn,enthalfn,entfn,nHfn,nMgfn}]
```

A vector of the mathematical functions for the properties $\Delta_f G'°$, $\Delta_f H'°$, $\Delta_f S'°$, $\overline{N}_H$, and $\overline{N}_{Mg}$ is produced by this program. This vector is given the name pi.

```
In[18]:= Clear[pi];
```

```
In[19]:= pi = deriveGHSNHNMg[piMgsp];
```

The five functions in the list for inorganic phosphate are given names by use of

```
In[20]:= {piG, piH, piS, piNH, piNMg} = pi;
```

These five functions are used to calculate values of the various transformed thermodynamic properties at 298.15 K, pH 7, pMg 3, and ionic strength 0.25 M. Since the specified concentration of magnesium ions is so low, these values are not very different from the values calculated in the preceding chapter for the absence of magnesium ions.

```
In[21]:= piG /. t → 298.15 /. pH → 7 /. pMg → 3 /. is → .25
```

```
Out[21]= -1059.55
```

```
In[22]:= piH /. t → 298.15 /. pH → 7 /. pMg → 3 /. is → .25
```

```
Out[22]= -1299.12
```

```
In[23]:= piS /. t → 298.15 /. pH → 7 /. pMg → 3 /. is → .25
```

```
Out[23]= -803.51
```

```
In[24]:=  piNH /. t → 298.15 /. pH → 7 /. pMg → 3 /. is → .25

Out[24]=  1.30117

In[25]:=  piNMg /. t → 298.15 /. pH → 7 /. pMg → 3 /. is → .25

Out[25]=  0.0253359
```

Since the transformed thermodynamic properties of inorganic phosphate are each functions of four variables and we want to see how five transformed properties are affected, it is difficult to describe these dependencies, but *Mathematica* provides two ways. The first is to make a table, and the second is to make three-dimensional plots at a specified ionic strength.

Table 5.1 Transformed thermodynamic properties of inorganic phosphate at temperatures 298.15 K, 313.15 K, pHs 5, 7, and 9, pMgs at 2, 4, and 6, and ionic strengths of zero and 0.25 M. Deeply nested lists like this are by default printed with successive dimensions alternating between rows and columns.

```
In[26]:= PaddedForm[
        TableForm[{piG, piH, piS, piNH, piNMg} /. t → {298.15, 313.15} /. pMg → {2, 3, 6} /.
          is → {0, .25} /. pH → {5, 7, 9},
        TableHeadings → {{"piG", "piH", "piS", "piNH", "piNMg"}, {"298.15 K", "313.15 K"},
          {"pMg 2", "pMg 3", "pMg 6"}, {"I=0", "I=0.25"}, {"pH 5", "pH 7", "pH 9"}}], {5, 2}]
```

Out[26]//PaddedForm=

			298.15 K		313.15 K	
			I=0	I=0.25	I=0	I=0.25
piG	pMg 2	pH 5	-1080.30	-1079.50	-1069.20	-1068.30
		pH 7	-1061.20	-1060.10	-1049.70	-1048.10
		pH 9	-1049.20	-1048.00	-1037.20	-1035.50
	pMg 3	pH 5	-1080.20	-1079.50	-1069.10	-1068.20
		pH 7	-1059.00	-1059.60	-1046.90	-1047.50
		pH 9	-1045.80	-1047.30	-1033.30	-1034.70
	pMg 6	pH 5	-1080.20	-1079.50	-1069.00	-1068.20
		pH 7	-1058.60	-1059.50	-1046.40	-1047.40
		pH 9	-1044.80	-1047.20	-1032.00	-1034.60
piH	pMg 2		-1302.10	-1302.80	-1301.90	-1302.80
			-1291.70	-1297.20	-1290.80	-1296.80
			-1288.80	-1295.40	-1288.50	-1295.00
	pMg 3	pH 5	-1302.50	-1302.90	-1302.50	-1302.90
		pH 7	-1298.90	-1299.10	-1298.30	-1298.80
		pH 9	-1294.90	-1297.50	-1294.30	-1297.20
	pMg 6	pH 5	-1302.60	-1302.90	-1302.60	-1303.00
		pH 7	-1301.20	-1299.40	-1301.20	-1299.10
		pH 9	-1299.10	-1297.80	-1299.00	-1297.50
piS	pMg 2		-743.91	-748.94	-743.37	-749.05
			-773.07	-795.41	-770.20	-793.90
			-803.61	-829.78	-802.40	-828.63
	pMg 3	pH 5	-745.56	-749.35	-745.49	-749.54
		pH 7	-804.63	-803.51	-802.52	-802.52
		pH 9	-835.70	-839.20	-833.50	-838.29
	pMg 6	pH 5	-745.74	-749.40	-745.74	-749.60
		pH 7	-813.95	-804.62	-813.75	-803.74
		pH 9	-852.87	-840.58	-852.85	-839.76
piNH	pMg 2		1.96	1.97	1.95	1.97
			1.21	1.25	1.17	1.22
			1.00	1.00	1.00	1.00
	pMg 3	pH 5	1.99	1.98	1.99	1.97
		pH 7	1.52	1.30	1.48	1.28
		pH 9	1.01	1.00	1.01	1.00
	pMg 6	pH 5	1.99	1.98	1.99	1.98
		pH 7	1.62	1.31	1.61	1.29
		pH 9	1.02	1.00	1.02	1.00
piNMg	pMg 2		0.03	0.01	0.04	0.01
			0.66	0.21	0.72	0.24
			0.83	0.27	0.86	0.31
	pMg 3	pH 5	0.00	0.00	0.00	0.00
		pH 7	0.16	0.03	0.20	0.03
		pH 9	0.34	0.04	0.39	0.04
	pMg 6	pH 5	3.09×10^{-6}	8.23×10^{-7}	4.19×10^{-6}	1.08×10^{-6}
		pH 7	0.00	0.00	0.00	0.00
		pH 9	0.00	0.00	0.00	0.00

The lack of "pH 5, pH 7, and pH 9" for four of the five properties is a known bug in *Mathematica* that is being worked on.

Three dimensional plots of these five transformed properties and the reciprocal relation at 298.15 K and 0.25 M ionic strength are obtained as follows:

```
In[27]:=  plot1 = Plot3D[Evaluate[piG /. t -> 298.15 /. is → .25],
             {pH, 5, 9}, {pMg, 2, 6}, AxesLabel → {"pH", "pMg", " "},
             PlotLabel -> "ΔfG'°/kJ mol⁻¹", DisplayFunction → Identity];

In[28]:=  plot2 = Plot3D[Evaluate[piH /. t -> 298.15 /. is → .25],
             {pH, 5, 9}, {pMg, 2, 6}, AxesLabel → {"pH", "pMg", " "},
             PlotLabel -> "ΔfH'°/kJ mol⁻¹", DisplayFunction → Identity];

In[29]:=  plot3 = Plot3D[Evaluate[piS /. t -> 298.15 /. is → .25],
             {pH, 5, 9}, {pMg, 2, 6}, AxesLabel → {"pH", "pMg", " "},
             PlotLabel -> "ΔfS'°/J K⁻¹ mol⁻¹", DisplayFunction → Identity];

In[30]:=  plot4 = Plot3D[Evaluate[piNH /. t -> 298.15 /. is → .25], {pH, 5, 9}, {pMg, 2, 6},
             AxesLabel → {"pH", "pMg", " "}, PlotLabel -> "N̄_H", DisplayFunction → Identity];

In[31]:=  plot5 = Plot3D[Evaluate[piNMg /. t -> 298.15 /. is → .25],
             {pH, 5, 9}, {pMg, 2, 6}, AxesLabel → {"pH", "pMg", " "},
             PlotLabel -> "N̄_Mg", PlotRange → {0, .4}, DisplayFunction → Identity];

In[32]:=  plot6 = Plot3D[Evaluate[D[piNH, pMg] /. t -> 298.15 /. is → .25],
             {pH, 5, 9}, {pMg, 2, 6}, AxesLabel → {"pH", "pMg", " "},
             PlotLabel -> "(dN̄_H/dpMg),(dN̄_Mg/dpH)", PlotRange → {0, .2}, DisplayFunction → Identity];
```

These plots can be viewed individually by deleting the DisplayFunction->Identity.

```
In[33]:=  Show[GraphicsArray[{{plot1, plot2}, {plot3, plot4}, {plot5, plot6}}]];
```

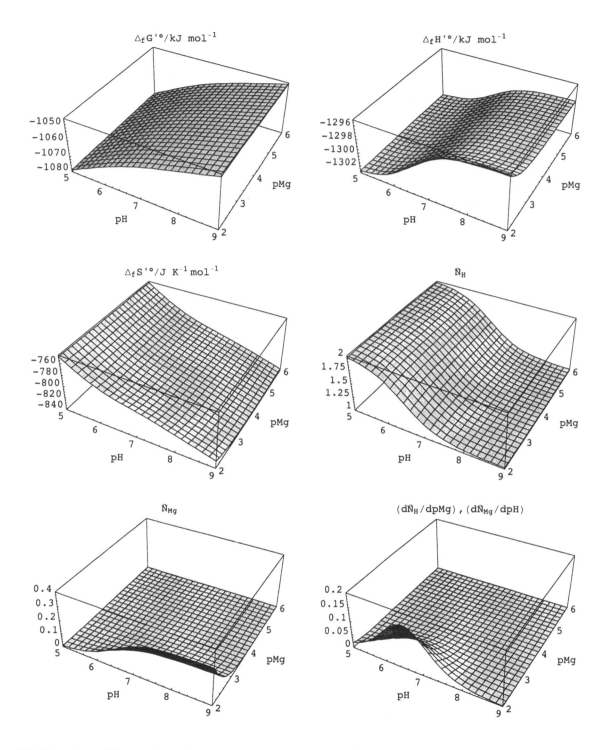

Figure 5.1 Transformed thermodynamic properties of inorganic phosphate at 298.15 K and 0.25 M ionic strength.

5.4 Calculation of Standard Transformed Thermodynamic Properties of ATP + H_2O = ADP + P_i at Specified *T*, pH, pMg, and Ionic Strength

In order to calculate standard transformed thermodynamic properties for ATP + H_2O = ADP + P_i at specified *T*, pH, pMg, and ionic strength, we need to carry out the same process for ATP, ADP, and H_2O that we have used for inorganic phosphate. The basic data on species at 298.15 K and zero ionic strength are as follows (4,5,6):

In[34]:= **atpMgsp = {{-2768.1, -3619.21, -4, 12, 0}, {-2811.48, -3612.91, -3, 13, 0},**
           {-2838.18, -3627.91, -2, 14, 0}, {-3258.68, -4063.31, -2, 12, 1},
           {-3287.5, -4063.02, -1, 13, 1}, {-3729.33, -4519.51, 0, 12, 2}};

In[35]:= **adpMgsp = {{-1906.13, -2626.54, -3, 12, 0},**
           {-1947.1, -2620.94, -2, 13, 0}, {-1971.98, -2638.54, -1, 14, 0},
           {-2387.97, -3074.54, -1, 12, 1}, {-2416.67, -3075.44, 0, 13, 1}};

In[36]:= **ampMgsp = {{-1040.45, -1635.37, -2, 12, 0}, {-1078.86, -1629.97, -1, 13, 0},**
           {-1101.63, -1648.07, 0, 14, 0}, {-1511.68, -2091.07, 0, 12, 1}};

In[37]:= **h2oMgsp = {{-237.19, -285.83, 0, 2, 0}};**

In[38]:= **hionMgsp = {{0, 0, 1, 1, 0}};**

In[39]:= **mgionMgsp = {{-455.30, -467.00, 2, 0, 1}};**

ATP properties

In[40]:= **atp = deriveGHSNHNMg[atpMgsp];**

In[41]:= **{atpG, atpH, atpS, atpNH, atpNMg} = atp;**

ADP properties

In[42]:= **adp = deriveGHSNHNMg[adpMgsp];**

In[43]:= **{adpG, adpH, adpS, adpNH, adpNMg} = adp;**

H_2O properties

In[44]:= **h2o = deriveGHSNHNMg[h2oMgsp];**

In[45]:= **{h2oG, h2oH, h2oS, h2oNH, h2oNMg} = h2o;**

Plots like Figure 5.1 can be made for ATP, ADP, and H_2O, but in this section we are interested in an enzyme-catalyzed reaction. Properties of the reaction ATP + H_2O = ADP + P_i at 298.15 K, pH 7, pMg 6, and 0.25 M ionic strength

In[46]:= **trGATPhyd = adpG + piG - atpG - h2oG;**

In[47]:= **trGATPhyd /. t → 298.15 /. pH → 7 /. pMg → 3 /. is → .25**

Out[47]= **-32.4522**

In[48]:= **trHATPhyd = adpH + piH - atpH - h2oH;**

In[49]:= **trHATPhyd /. t → 298.15 /. pH → 7 /. pMg → 3 /. is → .25**

Out[49]= **-30.8748**

In[50]:= **trSATPhyd = adpS + piS - atpS - h2oS;**

In[51]:= **trSATPhyd /. t → 298.15 /. pH → 7 /. pMg → 3 /. is → .25**

Out[51]= 5.29064

In[52]:= **nHATPhyd = adpNH + piNH - atpNH - h2oNH;**

In[53]:= **nHATPhyd /. t → 298.15 /. pH → 7 /. pMg → 3 /. is → .25**

Out[53]= -0.627989

In[54]:= **nMgATPhyd = adpNMg + piNMg - atpNMg - h2oNMg;**

In[55]:= **nMgATPhyd /. t → 298.15 /. pH → 7 /. pMg → 3 /. is → .25**

Out[55]= -0.448624

In[56]:= **logKATPhyd = -trGATPhyd / (8.31451 * .29815 * Log[10]);**

In[57]:= **logKATPhyd /. t → 298.15 /. pH → 7 /. pMg → 3 /. is → .25**

Out[57]= 5.68534

Table 5.2 Standard transformed thermodynamic properties for the reaction ATP + H_2O = ADP + P_i at temperatures 298.15 K, 313.15 K, pH 5, 7, and 9, pMg 2, 3, and 6, and ionic strengths of zero and 0.25 M.

```
In[58]:= PaddedForm[
            TableForm[{trGATPhyd, trHATPhyd, trSATPhyd, nHATPhyd, nMgATPhyd, logKATPhyd} /.
                t → {298.15, 313.15} /. pMg → {2, 3, 6} /. is → {0, .25} /. pH → {5, 7, 9},
             TableHeadings → {{"rxG", "rxH", "rxS", "rxNH", "rxNMg", "logK'"},
                {"298.15 K", "313.15 K"}, {"pMg 2", "pMg 3", "pMg 6"},
                {"I=0", "I=0.25"}, {"pH 5", "pH 7", "pH 9"}}], {5, 2}]
```

Out[58]//PaddedForm=

			298.15 K		313.15 K	
			I=0	I=0.25	I=0	I=0.25
rxG	pMg 2	pH 5	-26.90	-29.63	-26.35	-29.38
		pH 7	-28.88	-30.80	-29.03	-31.04
		pH 9	-39.68	-41.48	-40.51	-42.35
	pMg 3	pH 5	-31.24	-31.98	-30.89	-32.10
		pH 7	-30.16	-32.45	-30.25	-32.55
		pH 9	-39.69	-42.80	-40.48	-43.51
	pMg 6	pH 5	-35.29	-32.56	-35.89	-32.87
		pH 7	-36.99	-36.02	-37.40	-36.67
		pH 9	-46.17	-46.68	-46.76	-47.95
rxH	pMg 2		-38.11	-34.81	-37.73	-34.67
			-26.16	-26.42	-25.56	-25.71
			-23.24	-24.41	-23.15	-23.83
	pMg 3	pH 5	-38.33	-29.37	-38.09	-30.02
		pH 7	-28.61	-30.87	-28.28	-29.95
		pH 9	-24.30	-29.24	-24.02	-28.20
	pMg 6	pH 5	-23.33	-26.39	-23.47	-26.66
		pH 7	-28.31	-23.16	-29.31	-23.27
		pH 9	-33.36	-21.47	-35.29	-21.66
rxS	pMg 2		-37.59	-17.36	-36.33	-16.90
			9.12	14.69	11.08	17.02
			55.17	57.24	55.46	59.14
	pMg 3	pH 5	-23.79	8.75	-23.01	6.62
		pH 7	5.21	5.29	6.29	8.32
		pH 9	51.63	45.51	52.55	48.89
	pMg 6	pH 5	40.09	20.72	39.65	19.82
		pH 7	29.09	43.12	25.82	42.78
		pH 9	42.96	84.58	36.64	83.94
rxNH	pMg 2		0.43	0.40	0.40	0.39
			-0.78	-0.73	-0.82	-0.76
			-1.00	-1.00	-1.00	-1.00
	pMg 3	pH 5	0.41	0.13	0.39	0.16
		pH 7	-0.45	-0.63	-0.49	-0.65
		pH 9	-0.99	-0.99	-0.99	-1.00
	pMg 6	pH 5	-0.14	-0.04	-0.13	-0.04
		pH 7	-0.40	-0.74	-0.36	-0.76
		pH 9	-0.99	-1.00	-0.98	-1.00
rxNMg	pMg 2		-0.83	-0.61	-0.82	-0.62
			-0.17	-0.16	-0.14	-0.14
			0.00	-0.08	0.01	-0.06
	pMg 3	pH 5	-0.70	-0.20	-0.71	-0.25
		pH 7	-0.21	-0.45	-0.21	-0.42
		pH 9	-0.01	-0.42	-0.00	-0.38
	pMg 6	pH 5	-0.00	-0.00	-0.01	-0.00
		pH 7	-0.22	-0.01	-0.28	-0.01
		pH 9	-0.55	-0.01	-0.63	-0.01
logK'	pMg 2		4.71	5.19	4.62	5.15
			5.06	5.40	5.09	5.44
			6.95	7.27	7.10	7.42
	pMg 3	pH 5	5.47	5.60	5.41	5.62
		pH 7	5.28	5.69	5.30	5.70
		pH 9	6.95	7.50	7.09	7.62
	pMg 6	pH 5	6.18	5.70	6.29	5.76
		pH 7	6.48	6.31	6.55	6.42
		pH 9	8.09	8.18	8.19	8.40

Three dimensional plots of these six transformed properties at 298.15 K and 0.25 M ionic strength are obtained as follows:

These plots can be viewed individually by deleting the DisplayFunction->Identity.

In[59]:= **plot1 = Plot3D[Evaluate[trGATPhyd /. t -> 298.15 /. is → .25],**
     {pH, 5, 9}, {pMg, 2, 6}, AxesLabel → {"pH", "pMg", " "},
     PlotLabel -> "$\Delta_r G'^\circ/kJ\ mol^{-1}$", DisplayFunction → Identity];

```
In[60]:= plot2 = Plot3D[Evaluate[trHATPhyd /. t -> 298.15 /. is → .25],
            {pH, 5, 9}, {pMg, 2, 6}, AxesLabel → {"pH", "pMg", " "},
            PlotLabel -> "Δ_rH'°/kJ mol⁻¹", DisplayFunction → Identity];

In[61]:= plot3 = Plot3D[Evaluate[trSATPhyd /. t -> 298.15 /. is → .25],
            {pH, 5, 9}, {pMg, 2, 6}, AxesLabel → {"pH", "pMg", " "},
            PlotLabel -> "Δ_rS'°/J K⁻¹ mol⁻¹", DisplayFunction → Identity];

In[62]:= plot4 = Plot3D[Evaluate[nHATPhyd /. t -> 298.15 /. is → .25], {pH, 5, 9},
            {pMg, 2, 6}, PlotRange → {-1, .5}, AxesLabel → {"pH", "pMg", " "},
            PlotLabel -> "Δ_rN_H", DisplayFunction → Identity];

In[63]:= plot5 = Plot3D[Evaluate[nMgATPhyd /. t -> 298.15 /. is → .25],
            {pH, 5, 9}, {pMg, 2, 6}, AxesLabel → {"pH", "pMg", " "},
            PlotLabel -> "Δ_r N_Mg", PlotRange → {-1, 0}, DisplayFunction → Identity];

In[64]:= plot6 = Plot3D[Evaluate[logKATPhyd /. t -> 298.15 /. is → .25],
            {pH, 5, 9}, {pMg, 2, 6}, AxesLabel → {"pH", "pMg", " "},
            PlotLabel -> "logK'", PlotRange → {4, 8}, DisplayFunction → Identity];

In[65]:= Show[GraphicsArray[{{plot1, plot2}, {plot3, plot4}, {plot5, plot6}}]];
```

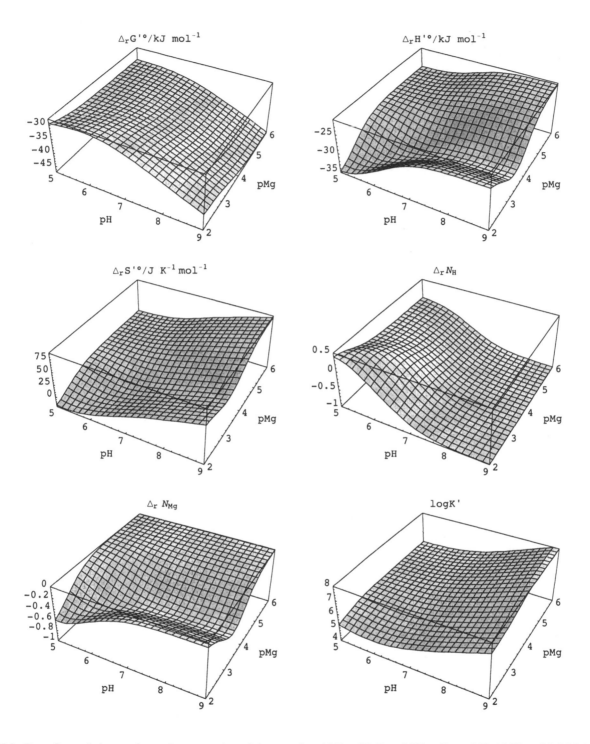

Figure 5.2 Transformed thermodynamic properties of the reaction ATP + H_2O = ADP + P_i at 298.15 K and 0.25 M ionic strength.

5.5 Calculation of Standard Transformed Thermodynamic Properties of ATP + 2H$_2$O = AMP + 2P$_i$ at Specified T, pH, pMg, and Ionic Strength

These calculations are now repeated for the reaction ATP + 2 H$_2$ O = AMP + 2 P$_i$.

The functions for AMP properties are obtained as follows:

In[66]:= **amp = deriveGHSNHNMg[ampMgsp];**

In[67]:= **{ampG, ampH, ampS, ampNH, ampNMg} = amp;**

Properties of the reaction ATP + 2H$_2$O = AMP + 2P$_i$ at 298.15 K, pH 7, pMg 3, and 0.25 M ionic strength

In[68]:= **trGATPhyd2 = ampG + 2 * piG - atpG - 2 * h2oG;**

In[69]:= **trGATPhyd2 /. t → 298.15 /. pH → 7 /. pMg → 3 /. is → .25**

Out[69]= -65.2078

In[70]:= **trHATPhyd2 = ampH + 2 * piH - atpH - 2 * h2oH;**

In[71]:= **trHATPhyd2 /. t → 298.15 /. pH → 7 /. pMg → 3 /. is → .25**

Out[71]= -59.7333

In[72]:= **trSATPhyd2 = ampS + 2 * piS - atpS - 2 * h2oS;**

In[73]:= **trSATPhyd2 /. t → 298.15 /. pH → 7 /. pMg → 3 /. is → .25**

Out[73]= 18.3614

In[74]:= **nHATPhyd2 = ampNH + 2 * piNH - atpNH - 2 * h2oNH;**

In[75]:= **nHATPhyd2 /. t → 298.15 /. pH → 7 /. pMg → 3 /. is → .25**

Out[75]= -1.30811

In[76]:= **nMgATPhyd2 = ampNMg + 2 * piNMg - atpNMg - 2 * h2oNMg;**

In[77]:= **nMgATPhyd2 /. t → 298.15 /. pH → 7 /. pMg → 3 /. is → .25**

Out[77]= -0.808825

In[78]:= **logKATPhyd2 = -trGATPhyd2 / (8.31451 * .29815 * Log[10]);**

In[79]:= **logKATPhyd2 /. t → 298.15 /. pH → 7 /. pMg → 3 /. is → .25**

Out[79]= 11.4238

Table 5.3 Transformed thermodynamic properties for the reaction $ATP + 2H_2O = AMP + 2P_i$ at temperatures 298.15 K, 313.15 K, pH 5, 7, and 9, pMg 2, 3, and 6, and ionic strengths of zero and 0.25 M.

```
In[80]:=  PaddedForm[
            TableForm[{trGATPhyd2, trHATPhyd2, trSATPhyd2, nHATPhyd2, nMgATPhyd2, logKATPhyd2} /.
                t → {298.15, 313.15} /. pMg → {2, 3, 6} /. is → {0, .25} /. pH → {5, 7, 9},
              TableHeadings → {{"rxG", "rxH", "rxS", "rxNH", "rxNMg", "logK'"},
                {"298.15 K", "313.15 K"}, {"pMg 2", "pMg 3", "pMg 6"},
                {"I=0", "I=0.25"}, {"pH 5", "pH 7", "pH 9"}}], {5, 2}]
```

`Out[80]//PaddedForm=`

			298.15 K		313.15 K	
			I=0	I=0.25	I=0	I=0.25
	pMg 2	pH 5	-54.43	-58.91	-53.43	-58.58
		pH 7	-57.06	-60.90	-57.44	-61.10
		pH 9	-78.55	-82.13	-80.29	-83.57
rxG	pMg 3	pH 5	-61.91	-62.18	-61.59	-62.45
		pH 7	-58.53	-65.21	-58.28	-65.50
		pH 9	-77.08	-86.05	-78.15	-87.51
	pMg 6	pH 5	-66.92	-62.89	-67.88	-63.39
		pH 7	-71.33	-69.98	-72.16	-71.14
		pH 9	-90.69	-91.29	-92.36	-93.69
			-74.59	-65.20	-74.29	-65.45
	pMg 2		-50.37	-57.69	-48.63	-56.18
			-44.30	-54.13	-43.58	-52.81
rxH	pMg 3	pH 5	-67.75	-56.37	-68.68	-57.30
		pH 7	-64.10	-59.73	-62.88	-59.08
		pH 9	-56.51	-57.18	-55.09	-56.49
	pMg 6	pH 5	-47.51	-52.78	-48.01	-53.25
		pH 7	-54.29	-46.87	-55.34	-46.90
		pH 9	-56.26	-43.50	-58.52	-43.70
			-67.62	-21.09	-66.62	-21.93
	pMg 2		22.44	10.76	28.12	15.70
			114.88	93.91	117.23	98.25
rxS	pMg 3	pH 5	-19.59	19.47	-22.63	16.42
		pH 7	-18.67	18.36	-14.69	20.50
		pH 9	68.99	96.80	73.64	99.07
	pMg 6	pH 5	65.11	33.92	63.47	32.37
		pH 7	57.16	77.53	53.73	77.41
		pH 9	115.47	160.29	108.07	159.65
			0.73	0.55	0.70	0.58
	pMg 2		-1.51	-1.42	-1.60	-1.48
			-1.99	-1.99	-2.00	-1.99
rxNH	pMg 3	pH 5	0.50	0.11	0.53	0.15
		pH 7	-0.72	-1.31	-0.79	-1.34
		pH 9	-1.98	-1.99	-1.98	-1.99
	pMg 6	pH 5	-0.26	-0.09	-0.22	-0.09
		pH 7	-1.02	-1.48	-0.99	-1.52
		pH 9	-1.98	-1.99	-1.98	-1.99
			-1.53	-0.94	-1.55	-1.00
	pMg 2		0.29	-0.55	0.41	-0.49
			0.70	-0.40	0.76	-0.33
rxNMg	pMg 3	pH 5	-1.02	-0.26	-1.09	-0.31
		pH 7	-0.71	-0.81	-0.64	-0.83
		pH 9	-0.28	-0.81	-0.16	-0.82
	pMg 6	pH 5	-0.01	-0.00	-0.01	-0.00
		pH 7	-0.23	-0.01	-0.30	-0.01
		pH 9	-0.59	-0.01	-0.69	-0.01
			9.54	10.32	9.36	10.26
	pMg 2		10.00	10.67	10.06	10.70
			13.76	14.39	14.07	14.64
logK'	pMg 3	pH 5	10.85	10.89	10.79	10.94
		pH 7	10.25	11.42	10.21	11.47
		pH 9	13.50	15.07	13.69	15.33
	pMg 6	pH 5	11.72	11.02	11.89	11.11
		pH 7	12.50	12.26	12.64	12.46
		pH 9	15.89	15.99	16.18	16.41

Three dimensional plots of these six transformed properties at 298.15 K and 0.25 M ionic strength are obtained as follows:

These plots can be viewed individually by deleting the DisplayFunction->Identity.

```
In[81]:= plot1 = Plot3D[Evaluate[trGATPhyd2 /. t -> 298.15 /. is -> .25],
            {pH, 5, 9}, {pMg, 2, 6}, AxesLabel → {"pH", "pMg", " "},
            PlotLabel -> "ΔrG'°/kJ mol⁻¹", DisplayFunction → Identity];
```

```
In[82]:= plot2 = Plot3D[Evaluate[trHATPhyd2 /. t -> 298.15 /. is → .25],
            {pH, 5, 9}, {pMg, 2, 6}, AxesLabel → {"pH", "pMg", " "},
            PlotLabel -> "Δ_rH'°/kJ mol⁻¹", DisplayFunction → Identity];

In[83]:= plot3 = Plot3D[Evaluate[trSATPhyd2 /. t -> 298.15 /. is → .25],
            {pH, 5, 9}, {pMg, 2, 6}, AxesLabel → {"pH", "pMg", " "},
            PlotLabel -> "Δ_rS'°/J K⁻¹ mol⁻¹", DisplayFunction → Identity];

In[84]:= plot4 = Plot3D[Evaluate[nHATPhyd2 /. t -> 298.15 /. is → .25],
            {pH, 5, 9}, {pMg, 2, 6}, PlotRange → {-2, 1}, AxesLabel → {"pH", "pMg", " "},
            PlotLabel -> "Δ_r N_H", DisplayFunction → Identity];

In[85]:= plot5 = Plot3D[Evaluate[nMgATPhyd2 /. t -> 298.15 /. is → .25],
            {pH, 5, 9}, {pMg, 2, 6}, AxesLabel → {"pH", "pMg", " "},
            PlotLabel -> "Δ_r N_Mg", PlotRange → {-2, 0}, DisplayFunction → Identity];

In[86]:= plot6 = Plot3D[Evaluate[logKATPhyd2 /. t -> 298.15 /. is → .25],
            {pH, 5, 9}, {pMg, 2, 6}, AxesLabel → {"pH", "pMg", " "},
            PlotLabel -> "logK'", PlotRange → {8, 16}, DisplayFunction → Identity];

In[87]:= Show[GraphicsArray[{{plot1, plot2}, {plot3, plot4}, {plot5, plot6}}]];
```

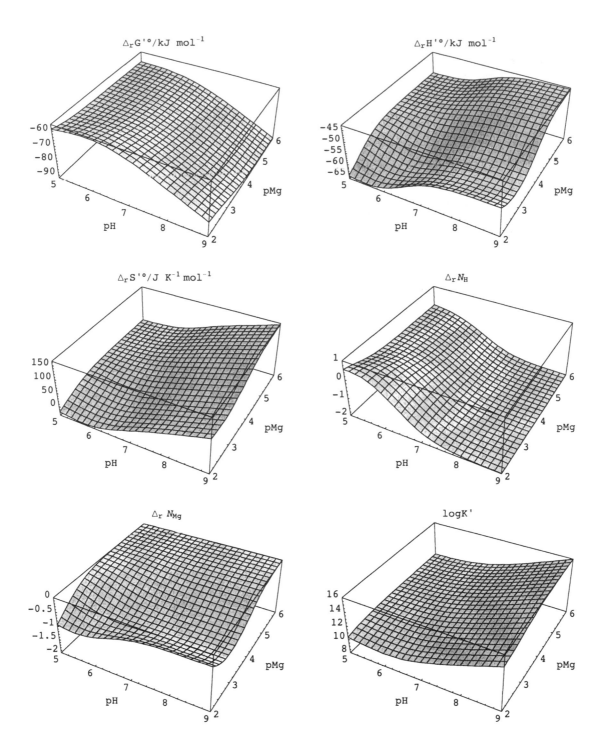

Figure 5.3 Transformed thermodynamic properties of the reaction $ATP + 2\,H_2O = AMP + 2\,P_i$ at 298.15 K and 0.25 M ionic strength.

5.6 Discussion

In this chapter we have seen that when metal ions are bound by a reactant in addition to hydrogen ions, the Legendre transform defining the transformed Gibbs energy of the system can be extended by subtracting a product of conjugate variables like $n_c(Mg) \mu(Mg^{2+})$. This adds a term to the fundamental equation of the form $RT\ln(10) n_c(Mg) dpMg$, increases the number of Maxwell relations to nine, and adds a new measurable thermodynamic property $\overline{N}_{Mg}$ for a reactant. The change in binding of magnesium ions $\Delta_r N_{Mg}$ in a reaction can in principle be measured experimentally. At the present time there may not be sufficiently sensitive magnesium ion electrodes to do this, but if $[Mg^{2+}]$ can be calculated from the composition of the equilibrium solution and known pK and pK_{Mg}, $\Delta_r G'^\circ = -RT\ln K'$ can be determined as a function of T, pH, pMg, and ionic strength. The hydrolysis of ATP is one of the few reactions for which all the standard transformed Gibbs energies of formation and standard transformed enthalpies of formation can be calculated at specified pMg. The *Mathematica* program **deriveGMgT** has been written to derive the function of T, pH, pMg, and ionic strength that yields $\Delta_f G'^\circ$ of a reactant. It is shown that the mathematical functions for all the other transformed thermodynamic properties of a reactant can be obtained by use of Maxwell relations. This makes it possible to write a single program to derive the mathematical functions for a reactant. This program is called **deriveGHSNHNMg**. When $\Delta_f G^\circ$ and $\Delta_f H^\circ$ are known for all the species in an enzyme-catalyzed reaction, it is possible to derive the mathematical function for $\Delta_r G'^\circ$ for an enzyme-catalyzed reaction. Tables and plots of standard transformed thermodynamic properties $\Delta_r G'^\circ$, $\Delta_r H'^\circ$, $\Delta_r S'^\circ$, $\Delta_r N_H$, $\Delta_r N_{Mg}$ and K' can be constructed for such a reaction. These six functions of T, pH, pMg and ionic strength for the reactions ATP + H_2O = ADP + P_i and ATP + $2H_2O$ = AMP + $2P_i$ have been used to produce tables and plots. It is difficult to visualize functions of four variables, but *Mathematica* provides the facility to make multi-dimensional tables and three-dimensional plots, which are utilized in this chapter.

If K' for an enzyme-catalyzed reaction can be determined experimentally as a function of T, pH, pMg, and ionic strength, the function for $\Delta_r G'^\circ$ can be obtained, and species properties of all the reactants involved can be calculated from this experimental data if all the pK and pK_{Mg} are known. The process for obtaining species properties from experimental data is complicated, but it is discussed in the next chapter.

References

1. R. A. Alberty, Equilibrium calculations on systems of biochemical reactions, Biophys. Chem. 42, 117-131 (1992).

2. R. A. Alberty, Calculation of transformed thermodynamic properties of biochemical reactants at specified dpH and pMg, Biophys. Chem. 43, 239-254 (1992).

3. R. A. Alberty and R. N. Goldberg, Calculation of thermodynamic formation properties for the ATP series at specified pH and pMg, Biochem. 31, 10610-10615 (1992).

4. R. N. Goldberg and Y. Tewari, Thermodynamic and transport properties of carbohydrates and their monophosphates: The pentoses and hexoses, J. Phys. Chem. Ref. Data 18, 809-880 (1989).

5. J. W. Larson, Y. B. Tewari, and R. N. Goldberg, Thermodynamics of the reactions between adenosine, adenosine 5' monophosphate, inosine, and inosine 5'-monophosphate. The conversion of L-histidine to (urocanic acid and ammonia), J. Chem. Thermodyn. 25, 73-90 (1993).

6. R. A. Alberty, Thermodynamics of the hydrolysis of adenosine triphosphate as a function of temperature, pH pMg, and ionic strength, J. Phys. Chem. 107 B, 12324-12330 (2003).

Chapter 6 Development of a Database on Species

6.1 Interpretation of Experimental Data on Biochemical Reactions in Terms of Chemical Reactions

Chapters 3-5 have described the calculation of various transformed thermodynamic properties of biochemical reactants and reactions from standard thermodynamic properties of species, but they have not discussed how these species properties were determined. Of course, some species properties came directly out of the National Bureau of Standard Tables (1) and CODATA Tables (2). One way to calculate standard thermodynamic properties of species not in the tables of chemical thermodynamic properties is to express the apparent equilibrium constant K' in terms of the equilibrium constant K of a reference chemical reaction, that is a reference reaction written in terms of species, and binding polynomials of reactants, as described in Chapter 2. In order to do this the pKs of the reactants in the pH range of interest must be known, and if metal ions are bound, the dissociation constants of the metal ion complexes must also be known. For the hydrolysis of adenosine triphosphate to adenosine diphosphate, the apparent equilibrium constant is given by

$$K' = \frac{K}{[H^+]} \frac{P(ADP)\, P(P_i)}{P(ATP)} \qquad (6.1\text{-}1)$$

where

$$K = \frac{[ADP^{3-}][HPO_4{}^{2-}][H^+]}{[ATP^{4-}]} \qquad (6.1\text{-}2)$$

is the equilibrium constant K for the chemical reference reaction

$$ATP^{4-} + H_2O = ADP^{3-} + HPO_4{}^{2-} + H^+ \qquad (6.1\text{-}3)$$

The binding polynomial for ATP is

$$P(ATP) = 1 + 10^{pK_1 - pH} + 10^{pK_1 + pK_2 - 2\,pH} \qquad (6.1\text{-}4)$$

where pK_1 is the pK of $HATP^{3-}$ and pK_2 is the pK of H_2ATP^{2-}.

The equilibrium constant for chemical reaction 6.1-3 is related to the standard Gibbs energies of formation of the five species by

$$\Delta_r G° = -RT\ln K = \Delta_f G° \, (ADP^{3-}) + \Delta_f G° \, (HPO_4{}^{2-}) + \Delta_f G° \, (H^+) - \Delta_f G° \, (ATP^{4-}) - \Delta_f G° \, (H_2O) \qquad (6.1\text{-}5)$$

When standard Gibbs energies of formation for ATP species were first calculated (3), $\Delta_f G° \, (H^+)$, $\Delta_f G° \, (HPO_4{}^{2+})$, and $\Delta_f G° \, (H_2O)$ were known, but it was not possible to calculate $\Delta_f G°$ or $\Delta_f H°$ of any species in the ATP series with respect to the reference forms of the elements involved. Therefore, the convention was adopted that $\Delta_f G°$ (adenosine) = 0 and $\Delta_f H°$ (adenosine) = 0 at 298.15 K and zero ionic strength. Subsequently, Boeiro-Goates and coworkers (4) determined the heat capacity of crystalline adenosine down to about 10 K. The third law made it possible for them to calculate the molar entropy of adenosine at 298.15 K. With additional calorimetric information they were able to calculate $\Delta_f G°$ (adenosine) and $\Delta_f H°$ (adenosine) in dilute aqueous solutions at 298.15 K. Boyer and coworkers (5) carried out similar studies for adenine. This changed the tabulated standard properties of species throughout the ATP series, but did not change the calculated apparent equilibrium constants and transformed enthalpies of reaction calculated earlier. The use of these properties to calculate standard transformed thermodynamic properties of enzyme-catalyzed reactions has been discussed (6).

Calorimetric measurements on a reaction like the hydrolysis of ATP yields $\Delta_r H'°$ at the experimental T, pH, and ionic strength. The calorimetric heat of reaction $\Delta_r H_c$ must be corrected for the heat effect of the hydrogen ions produced by the enzyme-catalyzed reaction on the acid dissociation of the buffer, as described in Chapter 15. If K' is measured at several temperatures and the acid dissociation constants of all the reactants are known at these temperature, the equilibrium constant K for the reference reaction can be calculated at each temperature. Plotting $\ln K$ versus $1/T$ yields $\Delta_r H°$, which is given by

$$\Delta_r H° = \Delta_f H° \, (ADP^{3-}) + \Delta_f H° \, (HPO_4{}^{2-}) + \Delta_f H° \, (H^+) - \Delta_f H° \, (ATP^{3-}) - \Delta_f H° \, (H_2O) \qquad (6.1\text{-}6)$$

Since $\Delta_f H°$ (adenosine) is known, equations like this and other experimental measurements can be used to calculate the standard formation properties of all the species in the ATP series. Effects of ionic strength on $\Delta_f G°$ and $\Delta_f H°$ of species have been discussed in Chapter 1

6.2 Use of an Inverse Legendre Transform

The calculations in Chapters 3 to 5 have been based on the use of Legendre transforms to introduce pH and pMg as independent intensive variables. But now we need to discuss the reverse process - that is the transformation of $\Delta_f G'°$ values calculated from measured apparent equilibrium constants in the literature to $\Delta_f G°$ values of species and the transformation of $\Delta_f H'°$ values calculated from calorimetric measurements in the literature to $\Delta_f H°$ of species. This is accomplished by use of the inverse Legendre transform defined by (7):

$$G = G' + n_c(H)\,\mu(H^+) \qquad (6.2\text{-}1)$$

Callen (8) pointed out that this type of transform is "symmetrical with its inverse except for a change in sign in the equation for the Legendre transform." In calculating species properties from experimental measurements of K' and $\Delta_r H'°$ it is necessary to go from systems described by the transformed Gibbs energy and transformed enthalpy to systems described by the Gibbs energy and the enthalpy. Reference 7 shows the steps by which equation 6.2-1 can be used to convert G' back to G, but here we are primarily concerned with the use the use of $\Delta_r G_i'°$ and $\Delta_r H_i'°$ to calculate $\Delta_f G°$ and $\Delta_f H°$ values. The differential of G is obtained using equation 6.2-1.

$$dG = dG' + n_c(H)d\mu(H^+) + \mu(H^+)\,dn_c(H) \qquad (6.2\text{-}2)$$

Substituting the expression for dG' in equation 3.3-3 yields

$$dG = -SdT + VdP + \sum \mu_j'\,dn_j + \mu(H^+)\,dn_c(H) \qquad (6.2\text{-}3)$$

Substituting the expression for $n_c(H)$ in equation 6.2-3 yields

$$dG = -SdT + VdP + \sum \mu_j\,dn_j \qquad (6.2\text{-}4)$$

This shows that the fundamental equation for G can be obtained from the fundmental equation for G' by use of the inverse Legendre transform.

6.3 Calculation of Species Properties at 298.15 K When the Reactant Consists of One, Two, or Three Species

If the species properties are known for all the reactants in a biochemical reaction but one, the standard transformed Gibbs energy of formation of that one reactant is readily calculated from the value of K'. The type of program mentioned earlier in Chapter 2 for calculating $\Delta_r G'^{\circ}$ for a biochemical reaction using **Solve** can be used with x replacing the one reactant with unknown properties. This type of program yields the function of pH and ionic strength that gives $\Delta_r G'^{\circ}$ for the one reactant. Now the problem is to interpret this function in terms of species properties. The key to this calculation is to write the expression for $\Delta_f G_j'^{\circ}$ in terms of the binding polynomial. If the reactant has one acidic group, there are two species and the expression for $\Delta_f G_j'^{\circ}$ (see equation 3.5-5) can be rearranged as follows:

$$\Delta_f G'^{\circ}(\text{reactant}) = -RT\ln\{\exp(-\Delta_f G_1'^{\circ}/RT) + \exp(-\Delta_f G_2'^{\circ}/RT)\} \qquad (6.3\text{-}1)$$

$$= \Delta_f G_1'^{\circ} - RT\ln\{1 + 10^{pK_1 - pH}\} = \Delta_f G_1'^{\circ} - RT\ln P(\text{reactant})$$

where $\Delta_f G_1'^{\circ}$ is for the more basic form and $P(\text{reactant})$ is the binding polynomial. This equation is important because when pK_1 and $\Delta_f G'^{\circ}(\text{reactant})$ are known at 298.15 K and a specified ionic strength, $\Delta_f G_1'^{\circ}$ for species one can be calculated at the experimental temperature, pH, and ionic strength. Equation 6.3-1 can be used to calculate $\Delta_f G_1'^{\circ}$ from experimental data, and the following equation (see equation 3.6-1) can be used to calculate $\Delta_f G_1^{\circ}$ at zero ionic strength:

$$\Delta_f G_1^{\circ}(I=0) = \Delta_f G_1'^{\circ}(I) - N_H(1)RT\ln(10)\,pH + RT\alpha(z_1^2 - N_H(1))\,I^{1/2}/(1 + 1.6\,I^{1/2}) \qquad (6.3\text{-}2)$$

where pH and I are the experimental values and $RT\alpha = 2.91482$ kJ mol$^{-3/2}$ kg$^{1/2}$ at 298.15 K. Once $\Delta_f G_1^{\circ}(I=0)$ has been calculated with this equation, $\Delta_f G_2^{\circ}(I=0)$ for the third species can be calculated using the value of $pK_1(I=0)$. The program **calcGef2sp** (7) given below yields the species data matrix for a reactant consisting of two species. The programs for reactants with one and three species have a similar structure. These three programs are based on the concept of the inverse Legendre transform (8).

When the standard thermodynamic properties of species are unknown for two reactants in a biochemical equation, the $\Delta_f G_j^{\circ}(I=0)$ and $\Delta_f H_j^{\circ}(I=0)$ of the more basic species of this reactant can be assigned values of zero, so $\Delta_f G_1'^{\circ}$ for that reactant can be calculated under the experimental conditions. These assigned values become conventions of the thermodynamic table, like $\Delta_f G^{\circ}(H^+) = 0$ and $\Delta_f H^{\circ}(H^+) = 0$ at each temperature. As described in the preceding section, this was done for adenosine in dilute aqueous solution (3) in 1992, but the determination of the thermodynamic properties of adenosine in dilute aqueous (4) made it possible to drop this convention for the ATP series.

The calculation of $\Delta_f H_j^{\circ}(I=0)$ of species is a two-step process in the sense that $\Delta_f G_j^{\circ}(I=0)$ of the species of the reactant has to be obtained first. This calculation is more complicted because the pK and the standard enthalpy of dissociation

$\Delta_{diss} H°(I=0)$ of the weak acid are needed. If a reactant consists of a single species, the standard enthalpy of formation of that species at zero ionic strength can be calculated using the following equation:

$$\Delta_f H_i°(I=0) = \Delta_f H_i{}'°(I) - N_H(i) RT\ln(10) \, pH + RT^2(\partial\alpha/\partial T)(z_i{}^2 - N_H(i)) I^{1/2}/(1 + 1.6\,I^{1/2}) \tag{6.3-3}$$

The program calcH1sp carries out this calculation at 298.15 K. The programs given in this section are based on the inverse Legendre transform, and they have been used to calculate 32 new species matrices providing $\Delta_f G°$ and 8 new species matrices providing $\Delta_f H°$ values (7).

```
In[2]:=  Off[General::"spell"];
         Off[General::"spell1"];
```

```
In[4]:=  calcGef1sp[equat_, pHc_, ionstr_, z1_, nH1_] :=
           Module[{energy, trGereactant},(*This program uses ∑viΔfGi'°=-RTlnK' to calculate
         the standard Gibbs energy of formation of the species of a reactant that does not
         have a pK in the range 4 to 10. The equation is of the form
         pyruvate+atp-x-adp==-8.31451*.29815*Log[K'], where K' is the apparent equilibrium
         constant at 298.15 K, pHc, and ionic strength is.  The reactant has charge number z1
         and hydrogen atom number nH1.  The output is the species vector without the standard
         enthalpy of formation.*)
               energy = Solve[equat, x] /. pH -> pHc /. is -> ionstr;
               trGereactant = energy[[1,1,2]];
               gef1 = trGereactant - nH1*8.31451*0.29815*Log[10]*pHc +
                  (2.91482*(z1^2 - nH1)*ionstr^0.5)/
                  (1 + 1.6*ionstr^0.5);
               {{gef1, _, z1, nH1}}]
```

```
In[5]:=  calcGef2sp[equat_, pHc_, ionstr_, z1_, nH1_, pK0_] :=
           Module[{energy, trGereactant, pKe, trgefpHis,gef1, gef2},(*This program uses ∑
         viΔfGi'°=-RTlnK' to calculate the standard Gibbs energies of formation of the two
         species of a reactant for which the pK at zero ionic strength is pK0. The equation is
         of the form pyruvate+atp-x-adp==-8.31451*.29815*Log[K'], where K' is the apparent
         equilibrium constant at 298.15 K, pHc, and ionic strength is.  The more basic form of
         the reactant has charge number z1 and hydrogen atom number nH1.  The output is the
         species matrix without the standard enthalpies of formation.*)
             energy = Solve[equat, x] /. pH -> pHc /. is -> ionstr;
               trGereactant = energy[[1,1,2]];
               pKe = pK0 + (0.510651*ionstr^0.5*2*z1)/
                  (1 + 1.6*ionstr^0.5); trgefpHis =
                trGereactant + 8.31451*0.29815*Log[1 + 10^(pKe - pHc)];
               gef1 = trgefpHis - nH1*8.31451*0.29815*Log[10]*pHc +
                  (2.91482*(z1^2 - nH1)*ionstr^0.5)/
                  (1 + 1.6*ionstr^0.5);
               gef2 = gef1 + 8.31451*0.29815*Log[10^(-pK0)];
               {{gef1, _, z1, nH1}, {gef2, _, z1 + 1, nH1 + 1}}]
```

```
In[6]:= calcGef3sp[equat_, pHc_, ionstr_, z1_, nH1_, pK10_,
          pK20_] := Module[{energy, trGereactant, pKe, trgefpHis,
          gef1, gef2, gef3, pK1e, pK2e},(*This program uses ∑viΔfGi'°=-RTlnK' to calculate
   the standard Gibbs energies of formation of the three species of a reactant for which
   the pKs at zero ionic strength is pK10 and pK20. The equation is of the form
   pyruvate+atp-x-adp==-8.31451*.29815*Log[K'], where K' is the apparent equilibrium
   constant at 298.15 K, pHc, and ionic strength is.  The more basic form of the
   reactant has charge number z1 and hydrogen atom number nH1.  The output is the
   species matrix without the standard enthalpies of formation of the three species.*)
        energy = Solve[equat, x] /. pH -> pHc /. is -> ionstr;
        trGereactant = energy[[1,1,2]];
        pK1e = pK10 + (0.510651*ionstr^0.5*2*z1)/
          (1 + 1.6*ionstr^0.5);
        pK2e = pK20 + (0.510651*ionstr^0.5*(2*z1 + 2))/
          (1 + 1.6*ionstr^0.5); trgefpHis =
         trGereactant + 8.31451*0.29815*
           Log[1 + 10^(pK1e - pHc) + 10^(pK1e + pK2e - 2*pHc)];
        gef1 = trgefpHis - nH1*8.31451*0.29815*Log[10]*pHc +
          (2.91482*(z1^2 - nH1)*ionstr^0.5)/
          (1 + 1.6*ionstr^0.5);
        gef2 = gef1 + 8.31451*0.29815*Log[10^(-pK10)];
        gef3 = gef2 + 8.31451*0.29815*Log[10^(-pK20)];
        {{gef1, _, z1, nH1}, {gef2, _, z1 + 1, nH1 + 1},
         {gef3, _, z1 + 2, nH1 + 2}}]

In[7]:= calcHf1sp[equat_, spmat_, pHc_, ionstr_] :=
        Module[{energy, trHreactant, enthf1, gef1, dHzero1, z1, nH1},
          (*This program uses ∑viΔfHi'°=ΔrH'° (298.15 K) to calculate the standard
            enthalpy of formation (I=0) of the single species of a reactant for which
            the species matrix (spmat) contains ΔfG° at zero ionic strength.  The
            reaction equation (equat) is of the form x+nadredh-malateh-nadoxh==89.5,
          where 89.5 kJ mol^-1 is the heat of reaction and x is oxaloacetate.  The species
           matrix (spmat) is that for oxaloacetate.  The calorimetric experiment is at pHc
           and ionic strength ionstr.  The reactant x has charge number z1 and hydrogen
           atom number nH1.  The output is the complete species matrix for x. 11-21-04*)
          {gef1, dHzero1, z1, nH1} = Transpose[spmat];
          energy = Solve[equat, x] /. pH -> pHc /. is → ionstr;
          trHreactant = energy[[1, 1, 2]];
          enthf1 = trHreactant - 1.4775 * (z1^2 - nH1) * ionstr^0.5 / (1 + 1.6 * ionstr^0.5);
          Flatten[{gef1, enthf1, z1, nH1}]]
```

```
In[8]:= calcHf2sp[equat_, spmat_, pHc_, ionstr_, dHdisszero_] :=
        Module[{dGzero, dHzero, zi, nHi, pHterm, isterm, gpfnsp, energy,
          trHreactant, stdtrGereactant, r1, r2, solution, dH1zero, dH2zero, dH1, dH2},
          (*This program uses ∑viΔfH'°=ΔrH'° (298.15 K) to calculate the standard enthalpy
              of formation (I=0) of the two species of a reactant for which the species
              matrix (spmat) contains ΔfG° at zero ionic strength for the two species of the
              reactant.  The reaction equation (equat) is of the form mannoseh+pih-x-h2oh==
              1.7, where 1.7 kJ mol^-1 is the heat of reaction and x is mannnose6phosh.  The
              species matrix (spmat) is that for mannose6phos.  The calorimetric experiment
              is at pH_c and ionic strength ionstr.  The first step in the calculation
              is to use the information on the standard Gibbs energies of formation of
              the species of the reactant of interest to calculate the equilibrium mole
              fractions r1 (base form) and r2 (acidform) of the two species of the reactant
              of interest.  The final output is the complete species matrix for x.*)
          {dGzero, dHzero, zi, nHi} = Transpose[spmat];
          pHterm = nHi * 8.31451 * .29815 * Log[10^-pH] /. pH -> pHc;
          isterm = 2.91482 * ((zi^2) - nHi) * (is^.5) / (1 + 1.6 * is^.5) /. is -> ionstr;
          gpfnsp = dGzero - pHterm - isterm;
          stdtrGereactant =
            -8.31451 * .29815 * Log[Apply[Plus, Exp[-1 * gpfnsp / (8.31451 * .29815)]]];
          r1 = Exp[(stdtrGereactant - gpfnsp[[1]]) / (8.31451 * .29815)];
          r2 = Exp[(stdtrGereactant - gpfnsp[[2]]) / (8.31451 * .29815)];
          (*Now calculate dfH'° (reactant) from ΔrH'° (298.15 K) for the reaction.*)
          energy = Solve[equat, x] /. pH -> pHc /. is -> ionstr;
          trHreactant = energy[[1, 1, 2]];
          (*dH1xero is given by the following equation. dH2zero is
            calculated from the equation for the enthalpy of dissociation.*)
          solution = Solve[trHreactant ==
              r1 * (dH1zero + 1.4775 * (zi[[1]]^2 - nHi[[1]]) * ionstr^0.5 / (1 + 1.6 * ionstr^0.5)) +
              r2 * (dH1zero - dHdisszero + 1.4775 * (zi[[2]]^2 - nHi[[2]]) *
                  ionstr^0.5 / (1 + 1.6 * ionstr^0.5)), dH1zero];
          dH1 = solution[[1, 1, 2]];
          dH2 = dH1 - dHdisszero;
          Transpose[{dGzero, {dH1, dH2}, zi, nHi}]]
```

```
In[9]:= calcHf3sp[equat_, spmat_, pHc_, ionstr_, dHdisszero1_, dHdisszero2_] :=
        Module[{dGzero, dHzero, zi, nHi, pHterm, isterm, gpfnsp, energy, trHreactant,
          stdtrGereactant, r1, r2, r3, solution, dH1zero, dH2zero, dH3zero,
          dH1expt, dH2expt, dH3expt, dH1, dH2, dH3}, (*This program uses ∑viΔfHi'°=
            ΔrH'° (298.15 K) to calculate the standard enthalpy of formation (I=0) of the
              three species of a reactant for which the species matrix (spmat) contains
              ΔfG° at zero ionic strength for the three species of the reactant.  The
              reaction equation (equat) is of the form adph+x-fructose6phos-atph=-84.2,
          where 84.2 kJ mol^-1 is the heat of reaction and x is fructose16phosh.  The
            species matrix (spmat) is that for fructose16phos.  The calorimetric experiment
            is at pH_c and ionic strength ionstr.  The first step in the calculation
            is to use the information on the standard Gibbs energies of formation of
            the species of the reactant of interest to calculate the equilibrium mole
            fractions r1 (base form) and r2 (acidform) of the two species of the reactant
            of interest.  The final output is the complete species matrix for x.*)
          {dGzero, dHzero, zi, nHi} = Transpose[spmat];
          pHterm = nHi * 8.31451 * .29815 * Log[10^-pH] /. pH -> pHc;
        isterm = 2.91482 * ((zi^2) - nHi) * (is^.5) / (1 + 1.6 * is^.5) /. is → ionstr;
        gpfnsp = dGzero - pHterm - isterm;
          stdtrGereactant =
            -8.31451 * .29815 * Log[Apply[Plus, Exp[-1 * gpfnsp / (8.31451 * .29815)]]];
          r1 = Exp[(stdtrGereactant - gpfnsp[[1]]) / (8.31451 * .29815)];
          r2 = Exp[(stdtrGereactant - gpfnsp[[2]]) / (8.31451 * .29815)];
          r3 = Exp[(stdtrGereactant - gpfnsp[[3]]) / (8.31451 * .29815)];
        (*Now calculate dfH'° (reactant) from ΔrH'° (298.15 K) for the reaction.*)
          energy = Solve[equat, x] /. pH -> pHc /. is → ionstr;
        trHreactant = energy[[1, 1, 2]];
          (*The standard transformed enthalpies of formation of the three
            species are given by the following six equations. dH1zero, dH2zero,
            and dH3zero are also related by the equations for the enthalpy of dissociation.*)
          solution = Solve[{dH1expt == dH1zero +
              1.4775 * (zi[[1]]^2 - nHi[[1]]) * ionstr^0.5 / (1 + 1.6 * ionstr^0.5),
            dH2expt == dH2zero + 1.4775 * (zi[[2]]^2 - nHi[[2]]) * ionstr^0.5 / (1 + 1.6 * ionstr^0.5),
            dH3expt == dH3zero + 1.4775 * (zi[[3]]^2 - nHi[[3]]) * ionstr^0.5 / (1 + 1.6 * ionstr^0.5),
            trHreactant == r1 * dH1expt + r2 * dH2expt + r3 * dH3expt,
            dH2zero == dH1zero - dHdisszero1, dH3zero == dH2zero - dHdisszero2},
            {dH1zero, dH2zero, dH3zero}, {dH1expt, dH2expt, dH3expt}];
          dHzerocalc = {solution[[1, 1, 2]], solution[[1, 2, 2]],
            solution[[1, 3, 2]]};
          Transpose[{dGzero, dHzerocalc, zi, nHi}]
          ]
```

The following example is concerned with biochemical reaction EC 1.1.1.37.

$$\text{malate} + \text{acetylcoA} + \text{nadred} + \text{h2o} = \text{citrate} + \text{coA} + \text{nadred} \tag{6.3-4}$$

Since the standard Gibbs energies of formation of the species of both coA and acetyl coA are unknown, the convention is adopted that the standard Gibbs energy of formation of RS^- (that is the basic form of coA) is zero. The standard Gibbs energy of formation of RSH can be calculated using the $pK = 8.38$ at 298.15 K and zero ionic strength.

```
In[10]:= coA2sp = {{0, _, -1, 0}, {-47.83, _, 0, 1}};
```

```
In[11]:= calcdGmat[speciesmat_] :=
         Module[{dGzero, dHzero,zi, nH, pHterm, isterm,gpfnsp},(*This program produces the
         function of pH and ionic strength (is) that gives the standard transformed Gibbs
         energy of formation of a reactant (sum of species) at 298.15 K.  The input
         speciesmat is a matrix that gives the standard Gibbs energy of formation, the
         standard enthalpy of formation, the electric charge, and the number of hydrogen
         atoms in each species. There is a row in the matrix for each species of the
         reactant. gpfnsp is a list of the functions for the species.  Energies are expressed
         in kJ mol^-1.*)
         {dGzero,dHzero,zi,nH} = Transpose[speciesmat];
         pHterm = nH*8.31451*.29815*Log[10^-pH];
         isterm = 2.91482*((zi^2) - nH)*(is^.5)/(1 + 1.6*is^.5);
         gpfnsp=dGzero - pHterm - isterm;
         -8.31451*.29815*Log[Apply[Plus,Exp[-1*gpfnsp/(8.31451*.29815)]]]]]
```

```
In[12]:= coA2 = calcdGmat[coA2sp];
```

The standard properties of the species in citrate, H_2O, malate, NAD_{ox}, and NAD_{red} are given by (9)

```
In[13]:= citratesp =
             {{-1162.69, -1515.11, -3, 5}, {-1199.18, -1518.46, -2, 6}, {-1226.33, -1520.88, -1, 7}}
```

```
In[14]:= h2osp = {{-237.19, -285.83, 0, 2}};
```

```
In[15]:= malatesp = {{-842.66, _, -2, 4}, {-872.68, _, -1, 5}};
```

```
In[16]:= nadoxsp = {{0, 0, -1, 26}};
```

```
In[17]:= nadredsp = {{22.65, -31.94, -2, 27}};
```

The functions of pH and ionic strength at 298.15 K for the standard transformed Gibbs energies of formation of these reactants are given by

```
In[18]:= citrate = calcdGmat[citratesp];
```

```
In[19]:= h2o = calcdGmat[h2osp];
```

```
In[20]:= malate = calcdGmat[malatesp];
```

```
In[21]:= nadox = calcdGmat[nadoxsp];
```

```
In[22]:= nadred = calcdGmat[nadredsp];
```

Now the data entry for acetylcoA can be calculated from the apparent equilibrium constant (10.8) of reaction 6.3-3 at pH 7.12 and ionic strength 0.05 M (7).

```
In[23]:= acetylcoAsp2 = calcGef1sp[
             citrate + coA2 + nadred - malate - x - nadox - h2o == -8.31451 0.29815 Log[10.8], 7.12, 0.05, (
```

```
Out[23]= {{-188.523, _, 0, 3}}
```

This is the data matrix for the single species of acetylcoA based on the convention that $\Delta_f G°\ (CoA^-) = 0$ at 298.15 K and zero ionic strength. The function that represents $\Delta_f G'°$(acetylcoA) can be calculated using calcdGmat.

```
In[24]:= acetylcoA2 = calcdGmat[acetylcoAsp2];
```

This calculation can be verified by using the data on coA and acetylcoA to calculate the apparent equilibrium constant for reaction 6.3-4 at the experimental conditions 298.15K, pH 7.12, and ionic strength 0.05 M. The apparent equilibrium constant under the experimental conditions can be calculated using **calckprime**.

```
In[25]:= calckprime[eq_, pHlist_, islist_] := Module[{energy, dG},(*Calculates the apparent
             equilibrium constant at specified pHs and ionic strengths for a biochemical reaction
             typed in the form atp+h2o+de==adp+pi.  The names of reactants call the appropriate
             functions of pH and ionic strength.  pHlist and is list can be lists.*)
                 energy = Solve[eq, de];
                 dG = energy[[1,1,2]] /. pH -> pHlist /. is -> islist;
                 E^(-(dG/(8.31451*0.29815)))]
```

```
In[26]:= calckprime[malate + acetylcoA2 + nadox + h2o + de == citrate + coA2 + nadred, 7.12, 0.05]
```

```
Out[26]= 10.8
```

Thus the data matrix for acetylcoA leads to the correct value for the apparent equilibriium constant under the experimental conditions. The dependence of log$K$ ' on pH is shown by the following plot.

```
In[27]:= Plot[
             Log[10, calckprime[malate + acetylcoA2 + nadox + h2o + de == citrate + coA2 + nadred, pH, 0.0!
             {pH, 5, 9}, AxesLabel -> {"pH", "logK'"}];
```

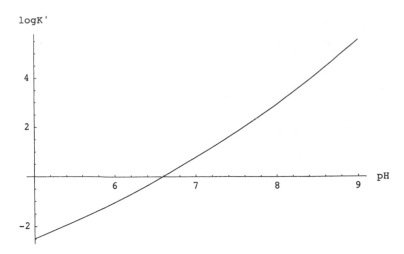

Figure 6.1 LogK' for reaction 6.3-4 at 298.15 K and 0.25 M ionic strength as a function of pH.

The change in binding of hydrogen ions can be calculated as follows:

```
In[28]:= Plot[Evaluate[-D[Log[10, calckprime[
                 malate + acetylcoA2 + nadox + h2o + de == citrate + coA2 + nadred, pH, 0.05]], pH]],
         {pH, 5, 9}, AxesLabel -> {"pH", "ArNH"}, AxesOrigin → {5, -3}, PlotRange → {-3, -1}];
```

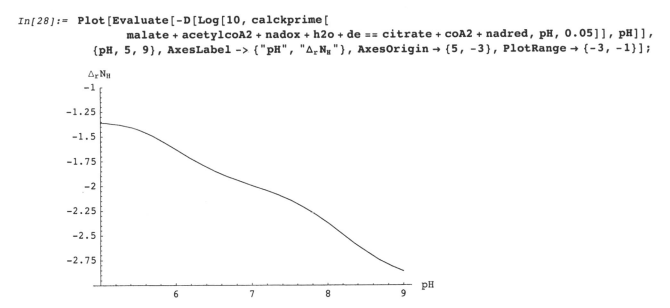

Figure 6.2 Change in the binding of hydrogen ions in reaction 6.4-3 at 298.15 K and 0.25 M ionic strength as a function of pH.

Over this whole range of pH there is a decrease in the binding of hydrogen ions in the reaction. Therefore, hydrogen ions are produced. Since hydrogen ions are produced, raising the pH pulls the reaction to the right and the apparent equilibrium constant increases.

6.4 Calculation of Standard Thermodynamic Properties of Species of a One-species Reactant from the Apparent Equilibrium Constant at 298.15 K and the Standard Transformed Enthalpy of Reaction at 313.15 K

If the apparent equilibrium constant and standard transformed enthalpy of a reaction are measured at a temperature different from 298.15 K and the species properties are known for all the reactants but one at 298.15 K, the question is how to calculate the species properties of that one reactant at 298.15 K for entry into a database. First we consider the simplest case where the reactant with unknown species properties consists of a single species. The reaction chosen as an example is EC 3.5.1.3.

glutamine + H_2O = glutamate + ammonia (6.4-1)

All the species properties for these four reactants are known at 298.15 K, but the objective of this calculation is to use these properties to calculate the apparent equilibrium constant and standard transformed reaction enthalpy at 313.15 K, and then assume these values to be experimental data. Assuming the species properties of glutamine are unknown, we want to write a program that can be used to calculate $\Delta_f G°(298.15, I = 0)$ and $\Delta_f H°(298.15, I = 0)$ for glutamine from the "experimental" data at 313.15 K. These calculations are based on the assumtion that the standard enthalpies of formation of the species at zero ionic strength are independent of temperature. The standard Gibbs energies of species j at temperture T can be calculated using equation 1.7-1, which is:

$$\Delta_f G_j°(T) = \frac{T}{298.15} \Delta_f G_j°(298.15 \text{ K}) + (1 - \frac{T}{298.15}) \Delta_f H_j°(298.15 \text{ K})$$ (6.4-2)

The properties of the species in the four reactants at 298.15 K and zero ionic strength are given by (9)

```
In[29]:= ammoniasp = {{-26.5, -80.29, 0, 3}, {-79.31, -132.51, 1, 4}};
```

```
In[30]:= glutamatesp = {{-697.47, -979.89, -1, 8}};
```

```
In[31]:= glutaminesp = {{-528.02, -805., 0, 10}};
```

```
In[32]:= h2osp = {{-237.19, -285.83, 0, 2}};
```

The standard Gibbs energies of formation at 313.15 K of these five species can be calculated by using equation 6.4-2.

```
In[33]:= -26.5 * 313.15 / 298.15 + (1 - 313.15 / 298.15) * -80.29

Out[33]= -23.7938
```

```
In[34]:= -79.31 * 313.15 / 298.15 + (1 - 313.15 / 298.15) * -132.51

Out[34]= -76.6335
```

```
In[35]:= -697.47 * 313.15 / 298.15 + (1 - 313.15 / 298.15) * -979.89

Out[35]= -683.261
```

```
In[36]:= -528.02 * 313.15 / 298.15 + (1 - 313.15 / 298.15) * -805.0

Out[36]= -514.085
```

```
In[37]:= -237.19 * 313.15 / 298.15 + (1 - 313.15 / 298.15) * -285.83
```

Out[37]= -234.743

Thus the species data matrices at 313.15 K are as follows:

In[38]:= **ammoniasp313 = {{-23.79, -80.29, 0, 3}, {-76.63, -132.51, 1, 4}};**

In[39]:= **glutamatesp313 = {{-683.26, -979.89, -1, 8}};**

In[40]:= **glutaminesp313 = {{-514.09, -805., 0, 10}};**

In[41]:= **h2osp313 = {{-234.74, -285.83, 0, 2}};**

The standard transformed Gibbs energies of formation of the four reactants at 313.15 K are calculated using

In[42]:= **calcdGmat313[speciesmat_] :=**
Module[{dGzero, dHzero,zi, nH, pHterm, isterm,gpfnsp},(*This program produces the
function of pH and ionic strength (is) that gives the standard transformed Gibbs
energy of formation of a reactant (sum of species) at 313.15 K. The input
speciesmat is a matrix that gives the standard Gibbs energy of formation, the
standard enthalpy of formation, the electric charge, and the number of hydrogen
atoms in each species. There is a row in the matrix for each species of the
reactant. gpfnsp is a list of the functions for the species. Energies are expressed
in kJ mol^-1.*)
{dGzero,dHzero,zi,nH}=Transpose[speciesmat];
pHterm = nH*8.31451*.31315*Log[10^-pH];
isterm = 3.14338*((zi^2) - nH)*(is^.5)/(1 + 1.6*is^.5);
gpfnsp=dGzero - pHterm - isterm;
-8.31451*.31315*Log[Apply[Plus,Exp[-1*gpfnsp/(8.31451*.31315)]]]]]

Note that the temperature has been changed from 0.29815 to 0.31315 in three places, and that the value of $RT\alpha$ has been changed. The value of $RT\alpha$ at 313.15 K is calculated as follows:

$$RT\alpha = 9.20483 \times 10^{-3} T - 1.28467 \times 10^{-5} T^2 + 4.95199 \times 10^{-8} T^3 \tag{6.4-3}$$

In[43]:= **rtalpha = 9.20483 * 10^-3 * t - 1.28467 * 10^-5 * t^2 + 4.95199 * 10^-8 * t^3;**

In[44]:= **rtalpha /. t → {298.15, 313.15}**

Out[44]= {2.91489, 3.14338}

The functions for the standard transformed Gibbs energies of formation at 313.15 K of the four reactants are derived as follows:

In[45]:= **ammonia313 = calcdGmat313[ammoniasp313]**

Out[45]= $-2.60369 \, \text{Log}\left[e^{-0.38407\left(-76.63+\frac{9.43014 \, is^{0.5}}{1+1.6 \, is^{0.5}}-10.4148 \, \text{Log}[10^{-pH}]\right)} + e^{-0.38407\left(-23.79+\frac{9.43014 \, is^{0.5}}{1+1.6 \, is^{0.5}}-7.81107 \, \text{Log}[10^{-pH}]\right)} \right]$

In[46]:= **glutamate313 = calcdGmat313[glutamatesp313]**

Out[46]= $-2.60369 \, \text{Log}\left[e^{-0.38407\left(-683.26+\frac{22.0037 \, is^{0.5}}{1+1.6 \, is^{0.5}}-20.8295 \, \text{Log}[10^{-pH}]\right)} \right]$

In[47]:= **glutamine313 = calcdGmat313[glutaminesp313]**

Out[47]= $-2.60369 \, \text{Log}\left[e^{-0.38407\left(-514.09+\frac{31.4338 \, is^{0.5}}{1+1.6 \, is^{0.5}}-26.0369 \, \text{Log}[10^{-pH}]\right)} \right]$

In[48]:= **h2o313 = calcdGmat313[h2osp313]**

$Out[48]=$ $-2.60369\,Log\left[e^{-0.38407\,\left(-234.74+\frac{6.28676\,is^{0.5}}{1+1.6\,is^{0.5}}-5.20738\,Log[10^{-pH}]\right)}\right]$

The hypothetical experiment is carried out at pH 5.9 and ionic strength, and so $\Delta_r G'^\circ(313.15, I = 0.10)$ is given in kJ mol^{-1} by

$In[49]:=$ **(glutamate313 + ammonia313 - glutamine313 - h2o313) /. pH → 5.9 /. is → .1**

$Out[49]=$ -12.3833

The apparent equilibrium constant at 313.15 K, pH 5.9, and ionic strength 0.10 M is given by

$In[50]:=$ **Exp[12.38 / (8.41451 * .31315)]**

$Out[50]=$ 109.759

The standard transformed enthalpies of formation of the reactants are calculated using calcdHmat313:

```
In[51]:= calcdHmat313[speciesmat_] :=
           Module[{dHzero, zi, nH, dhfnsp, dGzero, pHterm, isenth, dgfnsp, dGreactant, ri},
             (*This program produces the function of ionic strength (is) that gives the standard
                transformed enthalpy of formation of a reactant (sum of species) at 298.15
                K.  The input is a matrix that gives the standard Gibbs energy of formation,
             the standard enthalpy of formation, the electric charge,
             and the number of hydrogen atoms in the species in the reactant.  There is
             a row in the matrix for each species of the reactant.  dhfnsp is a list
             of the functions for the species.  Energies are expressed in kJ mol^-1.*)
             {dGzero, dHzero, zi, nH} = Transpose[speciesmat];
             isenth = 1.78158 * ((zi^2) - nH) * (is^.5) / (1 + 1.6 * is^.5);
             dhfnsp = dHzero + isenth;
           (*Now calculate the functions for
             the standard Gibbs energies of formation of the species.*)
             dGzero = speciesmat[[All, 1]];
             pHterm = nH * 8.31451 * .31315 * Log[10^-pH];
             gpfnsp = dGzero - pHterm - isenth * 3.14338 / 1.78158;
           (*Now calculate the standard
             transformed Gibbs energy of formation for the reactant.*)
             dGreactant = -8.31451 * .31315 * Log[Apply[Plus, Exp[-1 * gpfnsp / (8.31451 * .31315)]]];
           (*Now calculate the equilibrium mole fractions of the species
             in the reactant and the mole fraction-weighted average of the
             functions for the standard transformed enthalpies of the species.*)
             ri = Exp[ (dGreactant - gpfnsp) / (8.31451 * .31315)];
             ri.dhfnsp]
```

Note that 0.29815 has been changed to 0.31315 in four places. Also $RT^2(\partial\alpha/\partial T)$ has been changed to 1.78158 in two places. The value of 2.91482 has been changed to 3.14338 in one place. The value of this coefficient is obtained as follows:

$In[52]:=$ **rt2dalpha = -1.28466 * 10^-5 * t^2 + 9.90399 * 10^-8 * t^3**

$Out[52]=$ $-0.0000128466\,t^2 + 9.90399 \times 10^{-8}\,t^3$

$In[53]:=$ **rt2dalpha /. t → {298.15, 313.15}**

$Out[53]=$ $\{1.48293,\ 1.78158\}$

The functions for the standard transformed enthalpies of formation at 313.15 K of the four reactants are derived as follows:

$In[54]:=$ **hammonia313 = calcdHmat313[ammoniasp313]**

$Out[54]=$ $e^{0.38407\left(76.63-\frac{9.43014\ \text{is}^{0.5}}{1+1.6\ \text{is}^{0.5}}+10.4148\ \text{Log}[10^{-\text{pH}}]-2.60369\ \text{Log}\left[e^{-0.38407\left(-76.63+\frac{9.43014\ \text{is}^{0.5}}{1+1.6\ \text{is}^{0.5}}-10.4148\ \text{Log}[10^{-\text{pH}}]\right)}+e^{-0.38407\left(-23.79+\frac{9.43014\ \text{is}^{0.5}}{1+1.6\ \text{is}^{0.5}}\right)}\right)}$

$\left(-132.51-\frac{5.34474\ \text{is}^{0.5}}{1+1.6\ \text{is}^{0.5}}\right)+$

$e^{0.38407\left(23.79-\frac{9.43014\ \text{is}^{0.5}}{1+1.6\ \text{is}^{0.5}}+7.81107\ \text{Log}[10^{-\text{pH}}]-2.60369\ \text{Log}\left[e^{-0.38407\left(-76.63+\frac{9.43014\ \text{is}^{0.5}}{1+1.6\ \text{is}^{0.5}}-10.4148\ \text{Log}[10^{-\text{pH}}]\right)}+e^{-0.38407\left(-23.79+\frac{9.43014\ \text{is}^{0.}}{1+1.6\ \text{is}^{0.5}}\right)}\right)}$

$\left(-80.29-\frac{5.34474\ \text{is}^{0.5}}{1+1.6\ \text{is}^{0.5}}\right)$

$In[55]:=$ **hglutamate313 = calcdHmat313[glutamatesp313]**

$Out[55]=$ $e^{0.38407\left(683.26-\frac{22.0037\ \text{is}^{0.5}}{1+1.6\ \text{is}^{0.5}}+20.8295\ \text{Log}[10^{-\text{pH}}]-2.60369\ \text{Log}\left[e^{-0.38407\left(-683.26+\frac{22.0037\ \text{is}^{0.5}}{1+1.6\ \text{is}^{0.5}}-20.8295\ \text{Log}[10^{-\text{pH}}]\right)}\right]\right)}$

$\left(-979.89-\frac{12.4711\ \text{is}^{0.5}}{1+1.6\ \text{is}^{0.5}}\right)$

$In[56]:=$ **hglutamine313 = calcdHmat313[glutaminesp313]**

$Out[56]=$ $e^{0.38407\left(514.09-\frac{31.4338\ \text{is}^{0.5}}{1+1.6\ \text{is}^{0.5}}+26.0369\ \text{Log}[10^{-\text{pH}}]-2.60369\ \text{Log}\left[e^{-0.38407\left(-514.09+\frac{31.4338\ \text{is}^{0.5}}{1+1.6\ \text{is}^{0.5}}-26.0369\ \text{Log}[10^{-\text{pH}}]\right)}\right]\right)}$

$\left(-805.-\frac{17.8158\ \text{is}^{0.5}}{1+1.6\ \text{is}^{0.5}}\right)$

$In[57]:=$ **hglutamine313 /. is → .1 /. pH → {5, 6, 7, 8, 9}**

$Out[57]=$ {-808.741, -808.741, -808.741, -808.741, -808.741}

$In[58]:=$ **hh2o313 = calcdHmat313[h2osp313]**

$Out[58]=$ $e^{0.38407\left(234.74-\frac{6.28676\ \text{is}^{0.5}}{1+1.6\ \text{is}^{0.5}}+5.20738\ \text{Log}[10^{-\text{pH}}]-2.60369\ \text{Log}\left[e^{-0.38407\left(-234.74+\frac{6.28676\ \text{is}^{0.5}}{1+1.6\ \text{is}^{0.5}}-5.20738\ \text{Log}[10^{-\text{pH}}]\right)}\right]\right)}$

$\left(-285.83-\frac{3.56316\ \text{is}^{0.5}}{1+1.6\ \text{is}^{0.5}}\right)$

We assume the hypothetical experiment is carried out at pH 5.9 and ionic strength, and so $\Delta_r H'°(313.15,\ I = 0.10)$ is given in kJ mol^{-1} by

$In[59]:=$ **(hglutamate313 + hammonia313 - hglutamine313 - hh2o313) /. pH → 5.9 /. is → .1**

$Out[59]=$ -20.7582

The standard transformed enthalpy of reaction does not change much with pH in the region 5 to 7, but changes rapidly as the pK of ammonia is approached.

$In[60]:=$ **(hglutamate313 + hammonia313 - hglutamine313 - hh2o313) /. pH → {5, 5.9, 6, 7, 8, 9} /. is → .1**

$Out[60]=$ {-20.8138, -20.7582, -20.7417, -20.032, -13.87, 10.8042}

Now calculate the species properties of glutamine from $\Delta_r G'°(313.15,\ \text{pH}\ 5.9,\ I = 0.10) = -12.38$ kJ mol^{-1} and $\Delta_r H'°(313.15,\ \text{pH}\ 5.9,\ I = 0.10) = -20.76$ kJ mol^{-1}. Since we know $\Delta_f G_i'°$ values for glutamate, ammonia, and water, we can use $\Delta_r G'^\circ = -12.38 \text{ kJ mol}^{-1} = \sum \nu_i \Delta_f G_i'^\circ$ to calculate $\Delta_f G'^\circ(\text{glutamine}, 313.15 \text{ K}, \text{pH } 5.9, I = 0.1 \text{ M}) = -153.77 \text{ kJ mol}^{-1}$. Similarly, $\Delta_f H'^\circ(\text{glutamine}, 313.15 \text{ K}, \text{pH } 5.9, I = 0.1 \text{ M}) = -808.74 \text{ kJ mol}^{-1}$.

Equation 3.6-1 for $\Delta_f G'^\circ$ of a species at 313.15 K yields $\Delta_f G^\circ(313.15, I = 0)$:

```
In[61]:= -153.77 - 10 * 8.31451 * .31315 * Log[10] * 5.9 - 3.14338 * 10 * .1^.5 / (1 + 1.6 * .1^.5)
```

```
Out[61]= -514.088
```

$\Delta_f H^\circ(313.15, I = 0)$ is needed in order to adjust $\Delta_f G^\circ(313.15, I = 0)$ to 298.15 K. This is calculated using equation 3.6-2 for $\Delta_f H'^\circ$ of a species at 313.15 K.

```
In[62]:= -808.74 + 1.78158 * 10 * .1^.5 / (1 + 1.6 * .1^.5)
```

```
Out[62]= -804.999
```

Equation 6.4-2 can be rearranged to

$$\Delta_f G_j^\circ(298.15 \text{ K}) = \frac{298.15}{T} \Delta_f G_j^\circ(T) - \left(\frac{298.15}{T} - 1\right) \Delta_f H_j^\circ(298.15 \text{ K}) \tag{6.4-4}$$

so that we can calculate $\Delta_f G^\circ(\text{glutamine}, 298.15 \text{ K}, I=0)$

```
In[63]:= (298.15 / 313.15) * (-514.09) - ((298.15 / 313.15) - 1) * -805.00
```

```
Out[63]= -528.025
```

```
In[64]:= glutaminesp = {{-528.02, -805., 0, 10}};
```

This is the data entry calculated from K' and $\Delta_r H_i'^\circ$ at 313.15 K, and it agrees with what we expected.

This sequence of steps makes it possible to write a program that is more general in the sense that it an be applied to any reactant consisting of a single species and any temperature.

```
In[65]:= calconespeciesprops[trGT_, trHT_, nH_, z_, t_, pH_, is_] :=
           Module[{gcoeff, hcoeff, gadjust, hadjust},
             (*This program calculates the standard Gibbs energy of formation of a single
               species and its standard enthalpy of formation at 298.15 K and zero ionic
               strength. The input data are the standard transformed Gibbs energy of
               formation (trGT) and standard transformed enthalpy of formation (trHT)
               at a specified temperature (t in K), pH, and ionic strength (is). The
               energies are in kJ mol^-1. The output is the usual data matrix for 298.15 K,
              including the charge number and the number of hydrogen atoms in the species.*)
             gcoeff = 9.20483 * 10^-3 * t - 1.28467 * 10^-5 * t^2 + 4.95199 * 10^-8 * t^3;
             hcoeff = -1.28466 * 10^-5 * t^2 + 9.90399 * 10^-8 * t^3;
             gadjust = -nH * 8.31451 * (t / 1000) * Log[10] * pH - gcoeff * nH * is^.5 / (1 + 1.6 * is^.5);
             hadjust = hcoeff * nH * is^.5 / (1 + 1.6 * is^.5);
             htab = trHT + hadjust;
             gspT = trGT + gadjust;
             gtab = (298.15 / t) * gspT - ((298.15 / t) - 1) * htab;
             {{gtab, htab, z, nH}}]
```

```
In[66]:= calconespeciesprops[-153.77, -808.74, 10, 0, 313.15, 5.9, .1]
```

```
Out[66]= {{-528.023, -804.999, 0, 10}}
```

This program makes it possible to calculate $\Delta_f G°$ and $\Delta_f H°$ of the one species of glutamine at 298.15 K and zero ionic strength from the apparent equilibrium constant and the standard transformed enthalpy for reaction EC 3.5.1.3 at 313.15 K, pH 5.9, and 0.10 M ionic strength. Similar programs can be written to calculate species properties of reactants with two or three species, provided that the pKs and corresponding standard enthalpies of acid dissociation are known.

6.5 Discussion

We have seen that calculating species properties from experimental values of K' and $\Delta_r H'°$ is more complicated than calculating K' and $\Delta_r H'°$ from species values. Thermodynamic calculations can be made by alternate paths, and so there is more than one way to calculate species properties from experimental properties. This chapter emphasizes the concept of the inverse Legendre transform discussed by Callen (8). Biochemical reaction systems are described by transformed thermodynamic properties, and the inverse transform given in equation 6.2-1 provides the transformation from experimental reactant properties to calculated species properties. In this chapter we first considered calculations of species properties at 298.15 K from measurements of K' and $\Delta_r H'°$ at 298.15 K. Then we considered the more difficult problem of calculating $\Delta_f G°(298.15\ K)$ and $\Delta_f H°(298.15\ K)$ from $\Delta_r G'°(313.15\ K)$ and $\Delta_r H'°(313.15\ K)$. The programs developed here make it possible to go from $\Delta_r G'°(T,pH,I)$ and $\Delta_r H'°(T,pH,I)$ to $\Delta_f G°(298.15\ K,I=0)$ and $\Delta_f H°(298.15\ K,I=0)$ in one step. These calculations emphasize the importance of calorimetric measurements on enzyme-catalyzed reaction or measurements of K' at a series of temperatures.

This chapter has been about calculating species properties from apparent equilibrium constants and transformed enthalpies of reaction, but there is a prior question. Where is the experimental data? Fortunately, Goldberg, Tewari, and coworkers have searched the literature for these data, have evaluated it, and have published a series of review articles (10-15). These review articles provide thermodynamic data on about 500 enzyme-catalyzed reactions involving about 1000 reactants. In principle all these reactants can be put into thermodynamic tables. Goldberg, Tewari, and Bhat (16) have produced a web site to assist in the acquisition of data from the review articles.

References

1. D. D. Wagman, W. H. Evans, V. B. Parker, R. H. Schumm, I. Halow, S. M. Bailey, K. L. Churney, and R. L. Nuttall, The NBS tables of chemical thermodynamic properties, J. Phys. Chem. Ref. Data, 11, Supplement 2 (1982).

2. J. D. Cox, D.D. Wagman, and V. A. Medvedev, CODATA Key Values for Thermodynamics, Hemisphere, Washington, D. C. 1989.

3. R. A. Alberty and R. N. Goldberg, Calculation of thermodynamic formation properties for the ATP series at specified pH and pMg, Biochem. 31, 10610-10615 (1992).

4. J. Boerio-Goates, M. R. Francis, R. N. Goldberg, M. A. V. Ribeiro da Silva, M. D. M. C. Ribeiro da Silva, and Y. Tewari, Thermochemistry of adenosine, J. Chem. Thermo. 33, 929-947 (2001).

5. J. S. Boyer, M. R. Francis, and J. Boeiro-Goates, Heat-capacity measurements and thermodynamic funnctions of crystalline adenine: revised thermodynamic properties of aqueous adenine, J. Chem. Thermo. 35, 1917-1928 (2003).

6. R. A. Alberty, Use of standard Gibbs energies and standard enthalpies of adenosine (aq) and adenine(aq) in the thermodynamics of enzyme-catalyzed reactions, J. Chem. Thermo. 36, 593-601 (2004).

7. R. A. Alberty, Calculation of thermodynamic properties of species of biochemical reactants using the inverse Legendre transform, J. Phys. Chem. 109 B, 9132-9139 (2005).

8. H. B. Callen, Thermodynamics and an Introduction to Thermostatistics, Wiley, Hoboken, NJ (1985).

9. R. A. Alberty, BasicBiochemData3, 2005.

             http : // library.wolfram.com / infocenter / MathSource / 5704

10. R. N. Goldberg, Y. B. Tewari, D. Bell, and K. Fasio, Thermodynamics of enzyme-catalyzed reactions: Part I. Oxidoreductases, J. Phys. Chem. Ref. Data 22, 515 (1993).

11. R. N. Goldberg and Y. B. Tewari, Thermodynamics of Enzyme-catalyzed Reactions: Part 2 Transferases, J. Phys. Chem.

23, 547-617 (1994).

12. R. N. Goldberg and Y. B. Tewari, Thermodynamics of enzyme-catalyzed reactions: Part 3. Hydrolases, J. Phys. Chem. Ref. Data, 23, 1035-1103 (1994).

13. R. N. Goldberg and Y. B. Tewari, Thermodynamics of enzyme-catalyzed reactions: Part 4. Lyases, J. Phys. Chem. Ref, Data 24, 1669-1698 (1995).

14. R. N. Goldberg and Y. B. Tewari, Thermodynamics of enzyme-catalyzed reactions: Part 5. Isomerases and ligases, J. Phys. Chem. Ref, Data 24, 1765-1801 (1995).

15.   R. N. Goldberg, Thermodynamics of enzyme-catalyzed reactions: Part 6-1999 Update, J. Phys. Chem. Ref, Data 28, 931-965 (1999).

16. R. N. Goldberg, Y. B. Tewari, and T. N. Bhat, Thermodynamics of enzyme-catalyzed reactions, 2005.

```
In[67]:= http : // xpdb.nist.gov / enzyme_thermodynamics /
```

Chapter 7 Uses of Matrices in Biochemical Thermodynamics

7.1 Chemical Equations and Systems of Chemical Equations Can be Represented by Matrices

7.2 Biochemical Equations and Systems of Biochemical Equations Can be Represented by Matrices

7.3 A Problem with Conservation Matrices and Stoichiometric Number Matrices When a Reaction or a System of Reactions Involves H_2O

7.4 Coupling

7.5 Components in Reactions Involving Coenzymes

7.6 Use of Matrix Multiplication in Calculating Apparent Equilibrium Constants and of LinearSolve in Calculating Standard Transformed Gibbs Energies of Formation of Reactants

7.7 Uses of Matrices in Calculations of Equilibrium Compositions

7.8 Calculation of Equilibrium Concentrations in a Two-Reaction System

7.9 Discussion

Appendix

       Printing Enzyme-catalyzed Reactions from Stoichiometric Number Matrices

References

7.1 Chemical Equations and Systems of Chemical Equations Can be Represented by Matrices

Consider the oxidation of methane to carbon dioxide in the gas phase:

$$CH_4 + 2O_2 = CO_2 + 2H_2O \tag{7.1-1}$$

This equality can also be written as

$$- CH_4 - 2O_2 + CO_2 + 2H_2O = 0 \tag{7.1-2}$$

Each species in the reaction can be represented by a column vector giving the number of carbon atoms, hydrogen atoms, and oxygen atoms:

$$- \begin{pmatrix} 1 \\ 4 \\ 0 \end{pmatrix} - 2 \begin{pmatrix} 0 \\ 0 \\ 2 \end{pmatrix} + \begin{pmatrix} 1 \\ 0 \\ 2 \end{pmatrix} + 2 \begin{pmatrix} 0 \\ 2 \\ 1 \end{pmatrix} = \begin{pmatrix} 0 \\ 0 \\ 0 \end{pmatrix} \tag{7.1-3}$$

This equation can be written as a matrix multiplication (1,2,3).

$$\begin{pmatrix} 1 & 0 & 1 & 0 \\ 4 & 0 & 0 & 2 \\ 0 & 2 & 2 & 1 \end{pmatrix} \begin{pmatrix} -1 \\ -2 \\ 1 \\ 2 \end{pmatrix} = \begin{pmatrix} 0 \\ 0 \\ 0 \end{pmatrix} \tag{7.1-4}$$

The matrix on the left (a 3x4 matrix) is referred to as the **conservation matrix** because it gives the coefficients in the conservation equations for the reaction. The conservation equations for amounts n of C, H, and O atoms in reaction 7.1-1 are

$$n(C) = n(CH_4) + n(CO_2) \tag{7.1-5}$$

$$n(H) = 4\,n(CH_4) + 2\,n(H_2O) \tag{7.1-6}$$

$$n(O) = 2\,n(O_2) + 2\,n(CO_2) + n(H_2O) \tag{7.1-7}$$

The 4x1 column vector in equation 7.1-4 is referred to as the **stoichiometric number matrix** because it gives the stoichiometric coefficients ν in chemical equation 7.1-2.

Equation 7.1-4 can be generalized to

$$A\nu = 0 \tag{7.1-8}$$

where A is the conservation matrix, ν is the stoichiometric number matrix, and 0 is the corresponding zero matrix. The conservation matrix A has the dimensions CxN where C is the number of components and N is the number of different species in the chemical equation. Note that $C = 3$ and $N = 4$ in this case. The stoichiometric number matrix ν has the dimensions NxR where R is the number of independent chemical equations. Note that $N = 4$ and $R = 1$ in this case. The zero matrix therefore is

$(CxN)(NxR) = CxR$. Equation 7.1-8 is useful because it makes it possible to calculate a stoichiometric matrix from a conservation matrix. This operation is called taking the null space of A, and the *Mathematica* operation for doing this is called **NullSpace**. The use of **NullSpace** yields a basis for the stoichiometric number matrix. We will see what this means and how it is handled.

The order of multiplication in equation 7.1-8 can be changed by using the transposes A^T and ν^T:

$$\nu^T A^T = 0 \tag{7.1-9}$$

The transpose of a matrix is obtained by exchanging rows and columns. Equation 7.1-9 for reaction 7.1-1 is

$$(-1,-2,1,2) \begin{pmatrix} 1 & 4 & 0 \\ 0 & 0 & 2 \\ 1 & 0 & 2 \\ 0 & 2 & 1 \end{pmatrix} = (0,\,0,\,0) \tag{7.1-10}$$

In equation 7.1-10 the xero matrix is $(RxN)(NxC) = RxC = 1x3$.

The number N of different species is equal to the number C of components plus the number R of independent reactions.

$$N = C + R \tag{7.1-11}$$

One way to recognize the significance of this equation is to remember that the ultimate objective of chemical thermodynamics is to calculate the equilibrium composition of a system of reactions. A chemical reaction system has R independent equilibrium constant expressions and C conservation equations, and this is just enough information to calculate the equilibrium concentrations of N species. Equation 7.1-9 is useful because it makes it possible to calculate a conservation matrix from a stoichiometric number matrix. In doing this with the operation **NullSpace** we will see again that it yields a basis for the conservation matrix.

$\quad$ C can be referred to as the rank of the conservation matrix, and R can be referred to as the rank of the stoichiometric number matrix:

$$C = \text{rank } A \tag{7.1-12}$$

$$R = \text{rank } \nu \tag{7.1-13}$$

$\quad$ Consider that reaction 7.1-1 occurs in two steps:

$$CH_4 + O_2 = CH_2O + H_2O \tag{7.1-14}$$

$$CH_2O + O_2 = CO_2 + H_2O \tag{7.1-15}$$

Equation 7.1-8 for this system of two chemical reactions is

$$\begin{pmatrix} 1 & 0 & 1 & 1 & 0 \\ 4 & 0 & 2 & 0 & 2 \\ 0 & 2 & 1 & 2 & 1 \end{pmatrix} \begin{pmatrix} -1 & 0 \\ -1 & -1 \\ 1 & -1 \\ 0 & 1 \\ 1 & 1 \end{pmatrix} = \begin{pmatrix} 0 & 0 \\ 0 & 0 \\ 0 & 0 \end{pmatrix} \tag{7.1-16}$$

A is 3x5, ν is 5x2, and so $C = 3$, $R = 2$, and $N = 5$. The zero matrix is $(CxN)(NxR) = CxR$.

$\quad$ Now let us see how these equations look in *Mathematica*. The conservation matrix for reaction 7.1-1 is given by

```
In[2]:=  Off[General::spell1];
         Off[General::spell];
```

```
In[4]:=  conmat = {{1, 0, 1, 0}, {4, 0, 0, 2}, {0, 2, 2, 1}};
```

This conservation matrix can be labelled as follows:

```
In[5]:=  TableForm[conmat, TableHeadings → {{"C", "H", "O"}, {"CH₄", "O₂", "CO₂", "H₂O"}}]
```

```
Out[5]//TableForm=
```

	CH_4	O_2	CO_2	H_2O
C	1	0	1	0
H	4	0	0	2
O	0	2	2	1

The stoichiometric number matrix for reaction 7.1-1 is given by

```
In[6]:= TableForm[snmat = {{-1}, {-2}, {1}, {2}},
          TableHeadings → {{"CH₄", "O₂", "CO₂", "H₂O"}, {""}}]
```

Out[6]//TableForm=

CH₄	-1
O₂	-2
CO₂	1
H₂O	2

The matrix multiplication in reaction 7.1-8 is given by

```
In[7]:= conmat.snmat
```

Out[7]= {{0}, {0}, {0}}

Note that dot (.) is the matrix multiplication operator in *Mathematica*.
   Now we apply **NullSpace** to *A* according to equation 7.1-8.

```
In[8]:= NullSpace[conmat]
```

Out[8]= {{-1, -2, 1, 2}}

This stoichiometric matrix happens to be exactly what we might expect from equation 7.1-8, but it is presented in *Mathematica* as a row vector rather than a column vector. This agreement is not encountered very often because this operation yields a basis.
   Now we want to use equation 7.1-9 to obtain conservation matrix *A* by applying **NullSpace** to v^T.

```
In[9]:= NullSpace[Transpose[snmat]]
```

Out[9]= {{2, 0, 0, 1}, {1, 0, 1, 0}, {-2, 1, 0, 0}}

This looks different from the 3x4 matrix on the left side of equation 7.1-4. However, the fact that these two conservation matrices have the same information content can be demonstrated by looking at the row-reduced forms of the two conservation matrices:

```
In[10]:= RowReduce[conmat]
```

Out[10]= $\{\{1, 0, 0, \frac{1}{2}\}, \{0, 1, 0, 1\}, \{0, 0, 1, -\frac{1}{2}\}\}$

```
In[11]:= RowReduce[NullSpace[Transpose[snmat]]]
```

Out[11]= $\{\{1, 0, 0, \frac{1}{2}\}, \{0, 1, 0, 1\}, \{0, 0, 1, -\frac{1}{2}\}\}$

This is why calculating the null space yields a basis for the conservation matrix. The row reduced conservation matrix can be labelled as follows:

```
In[12]:= TableForm[RowReduce[conmat],
          TableHeadings → {{"CH₄", "O₂", "CO₂"}, {"CH₄", "O₂", "CO₂", "H₂O"}}]
```

```
Out[12]//TableForm=
                    CH₄        O₂        CO₂        H₂O
        CH₄          1          0          0          ½
        O₂           0          1          0          1
        CO₂          0          0          1         -½
```

It is important to notice that row reduction changes the components from atoms of elements to combinations of atoms. The first three species are selected as components if they contain all the different atoms. Thus a set of conservation equations can be written to conserve the first three species, rather than the atoms of C, H, and O. When row reduction yields a matrix of this form, the chemical reaction can be read from the last column. This shows how H_2O is made up from the three components: $H_2O = (1/2)CH_4 + O_2 - (1/2)CO_2$. This can be rearranged to give equation 7.1-1.

When the amounts of species in a system are known, the conservation matrix can be used to calculate the amounts of components in the system because

$$An = n_c \tag{7.1-17}$$

where n is the vector of amounts of species and n_c is the vector of amounts of components. If there is a mole of each of the species in reaction 7.1-1 in the system, the amounts of the components C, H, and O are given by

```
In[13]:= conmat.{{1}, {1}, {1}, {1}}

Out[13]= {{2}, {6}, {5}}
```

The amounts of the components CH_4, O_2, and CO_2 when there is a mole of each of the species in reaction 7.1-1 are given by

```
In[14]:= rowredconmat = RowReduce[conmat]
```

$$Out[14]= \left\{\left\{1, 0, 0, \frac{1}{2}\right\}, \{0, 1, 0, 1\}, \left\{0, 0, 1, -\frac{1}{2}\right\}\right\}$$

```
In[15]:= rowredconmat.{{1}, {1}, {1}, {1}}
```

$$Out[15]= \left\{\left\{\frac{3}{2}\right\}, \{2\}, \left\{\frac{1}{2}\right\}\right\}$$

The same operations can be carried out for the system with two reactions in 7.1-14 and 7.1-15.

```
In[16]:= conmat2 = {{1, 0, 1, 1, 0}, {4, 0, 2, 0, 2}, {0, 2, 1, 2, 1}};

In[17]:= snmat2 = {{-1, 0}, {-1, -1}, {1, -1}, {0, 1}, {1, 1}};

In[18]:= conmat2.snmat2

Out[18]= {{0, 0}, {0, 0}, {0, 0}}

In[19]:= NullSpace[conmat2]

Out[19]= {{-1, -1, 1, 0, 1}, {1, 0, -2, 1, 0}}

In[20]:= TableForm[RowReduce[NullSpace[conmat2]]]

Out[20]//TableForm=
        1        0       -2        1        0
        0        1        1       -1       -1
```

This stoichiometric number matrix is the same as the following:

In[21]:= **TableForm[RowReduce[Transpose[snmat2]]]**

Out[21]//TableForm=

1	0	-2	1	0
0	1	1	-1	-1

A basis for the conservation matrix for the two reactions in 7.1-14 and 7.1-15 can be obtained as follows:

In[22]:= **TableForm[RowReduce[NullSpace[Transpose[snmat2]]]]**

Out[22]//TableForm=

1	0	0	-1	1
0	1	0	0	1
0	0	1	2	-1

This agrees with

In[23]:= **TableForm[RowReduce[conmat2],**
　　　　TableHeadings → {{"CH₄", "O₂", "CH₂O"}, {"CH₄", "O₂", "CH₂O", "CO₂", "H₂O"}}]

Out[23]//TableForm=

	CH_4	O_2	CH_2O	CO_2	H_2O
CH_4	1	0	0	-1	1
O_2	0	1	0	0	1
CH_2O	0	0	1	2	-1

This shows that for the system of two reactions, the components can be taken to be atoms of the elements C, H, and O or molecules of CH_4, O_2, and CH_2O. The last two columns indicate that $CO_2 = -CH_4 + 2CH_2O$ and $H_2O = CH_4 + O_2 - CH_2O$. The second reaction is the same as reaction 7.1-15, but the first reaction is not the same as reaction 7.1-14. However, it does balance atoms. The use of **NullSpace** and **RowReduce** above provides a more organized way to compare conservation matrices with stoichiometric matrices for larger systems.

7.2 Biochemical Equations and Systems of Biochemical Equations Can be Represented by Matrices

The chemical reactions involved in the hydrolysis of glucose 6-phosphate can be taken to be

$$G6P^{2-} + H_2O = glucose + HPO_4{}^{2-} \tag{7.2-1}$$

$$HG6P^- = H^+ + G6P^{2-} \tag{7.2-2}$$

$$H_2PO_4{}^- = H^+ + HPO_4{}^{2-} \tag{7.2-3}$$

These reactions are arbitrary in the sense that $H_2PO_4{}^-$ could be used in the first reaction. The conservation matrix for the system of reactions is

In[24]:= **conmat3 = {{6, 0, 0, 0, 6, 0, 6},**
　　　　{11, 1, 2, 1, 12, 2, 12}, {15, 0, 1, 4, 12, 4, 15}, {1, 0, 0, 1, 0, 1, 1}};

In[25]:= **TableForm[conmat3, TableHeadings →**
　　　　{{"C", "H", "O", "P"}, {"G6P²⁻", "H⁺", "H₂O", "HPO₄²⁻", "glucose", "H₂PO₄⁻", "HG6P⁻"}}]

```
Out[25]//TableForm=
              G6P²⁻      H⁺      H₂O     HPO₄²⁻    glucose    H₂PO₄⁻    HG6P⁻
       C       6         0       0       0         6          0         6
       H      11         1       2       1        12          2        12
       O      15         0       1       4        12          4        15
       P       1         0       0       1         0          1         1
```

The conservation matrix A is useful in chemical thermodynamics, but biochemistry takes a more global view. When the pH is specified, hydrogen ions are not conserved, and so the second row and the second column in **conmat3** are deleted.

In[26]:= **conmat4 = {{6, 0, 0, 6, 0, 6}, {15, 1, 4, 12, 4, 15}, {1, 0, 1, 0, 1, 1}};**

In[27]:= **TableForm[conmat4, TableHeadings →**
       {{"C", "O", "P"}, {"G6P²⁻", "H₂O", "HPO₄²⁻", "glucose", "H₂PO₄⁻", "HG6P⁻"}}]

```
Out[27]//TableForm=
              G6P²⁻      H₂O     HPO₄²⁻    glucose    H₂PO₄⁻    HG6P⁻
       C       6         0       0         6          0         6
       O      15         1       4        12          4        15
       P       1         0       1         0          1         1
```

Notice that the first and sixth columns are redundant and so one has to be eliminated. Delete the sixth and label the first as glucose6phos, which represents the pseudoisomer group G6P²⁻ and HG6P⁻. Notice that the third and fifth columns are also redundant, and so we will delete the fifth. Label the third as pi. Now the columns are for glucose6phos, h2o, pi, and glucose. Notice that we have gone back to the names of reactants (sums of species).

In[28]:= **conmat5 = {{6, 0, 0, 6}, {15, 1, 4, 12}, {1, 0, 1, 0}};**

In[29]:= **TableForm[conmat5,**
       TableHeadings → {{"C", "O", "P"}, {"glucose6phos", "h2o", "pi", "glucose"}}]

```
Out[29]//TableForm=
              glucose6phos    h2o     pi     glucose
       C       6              0       0      6
       O      15              1       4      12
       P       1              0       1      0
```

This shows that when the pH is specified, the conservation equations can be written in terms of reactants (sums of species), rather than species. We can use linear algebra to show that this is the conservation matrix for a biochemical reaction system with the following single reaction.

glucose6phos + h2o = glucose + pi   (7.2-4)

The row reduced form of **conmat5** is

In[30]:= **TableForm[RowReduce[conmat5], TableHeadings →**
       {{"glucose6phos", "h2o", "pi"}, {"glucose6phos", "h2o", "pi", "glucose"}}]

```
Out[30]//TableForm=
                     glucose6phos    h2o     pi     glucose
       glucose6phos   1              0       0      1
       h2o            0              1       0      1
       pi             0              0       1      -1
```

Notice that the components are now glucose6phos, h2o, and pi.

To indicate that the pH is held constant, primes are added to the symbols in equations 7.1-8 and 7.1.9:

$$A'\nu' = 0 \qquad (7.2\text{-}5)$$

A' is the apparent conservation matrix, and ν' is the stoichiometric number matrix for the biochemical reaction system. This equation makes it possible to calculate a basis for the stoichiometric number matrix from the apparent conservation matrix by use of **NullSpace**. A' has the dimensions $C'\text{x}N'$ where C' is the apparent number of components $(C - 1)$ and N' is the number of reactants (sums of species). ν' has the dimensions $N'\text{x}R'$. Note that $N' = C' + R'$, where R' is the number of independent biochemical reactions. Equation 7.2-5 makes it possible to obtain a basis for the apparent stoichiometric number matrix by use of **NullSpace**.

The order of multiplication in equation 7.2-5 can be changed by use of the transposes $(A')^T$ and $(\nu')^T$

$$(\nu')^T (A')^T = 0 \qquad (7.2\text{-}6)$$

This makes it possible to obtain a basis for the apparent conservation matrix A' from the stoichiometric number matrix.

A basis for the stoichiometric number matrix for the biochemical reaction in the system being discussed can be obtained by applying equation 7.2-5 to **conmat5**.

```
In[31]:= TableForm[RowReduce[NullSpace[conmat5]],
             TableHeadings → {{""}, {"glucose6phos", "h2o", "pi", "glucose"}}]
```

```
Out[31]//TableForm=
              glucose6phos        h2o        pi        glucose
              1                   1          -1        -1
```

Since this stoichiometric number matrix is a basis, the signs can be changed to make it agree with equatuon 7.2-4.

The stoichiometric number matrix for equation 7.2-4 can be used to calculate a basis for the apparent conservation matrix A. The stoichiometric number matrix is given by

```
In[32]:= bionu = {{-1}, {-1}, {1}, {1}};
```

Many biochemical reactions, perhaps most, have exactly this stoichiometric number matrix. A basis for the apparent conservation matrix can be obtained by use of equation 7.2-6.

```
In[33]:= TableForm[RowReduce[NullSpace[Transpose[bionu]]],
             TableHeadings → {{"G6P", "H₂O", "Pᵢ"}, {"G6P", "H₂O", "Pᵢ", "glucose"}}]
```

```
Out[33]//TableForm=
                      G6P      H₂O      Pᵢ      glucose
              G6P     1        0        0       1
              H₂O     0        1        0       1
              Pᵢ      0        0        1       -1
```

This is the same row reduced result obtained from the apparent conservation matrix **conmat5**. This completes the demonstration that we can go from a conservation matrix for a biochemical reaction to a stoichiometric number matrix or from a stoichiometric number matrix to a conservation matrix.

The amounts of apparent components in a system of biochemical reactions can be calculated using

$$A'n' = n_c' \qquad (7.2\text{-}7)$$

where n' is the amount vector for reactants and n_c' is the amount vector for components.

7.3 A Problem with Conservation Matrices When a Reaction or a System of Reactions Involves H_2O

When an enzyme-catalyzed reaction does not involve h2o as a reactant, it is clear that oxygen atoms are conserved and $A'\nu' = 0$ and $(\nu')^T(A')^T = 0$ can be used to interchange conservation matrices and stoichiometric number matrices when the reaction is carried out at specified pH. But when an enzyme-catalyzed reaction involves h2o there is a problem (4,5). The convention is that for reactions in dilute solutions, the activity of h2o is taken as unity, which means that [h2o] does not appear in the expression for the apparent equilibrium constant. Since the activity of h2o does not depend on the extent of reaction, there is no conservation equation for oxygen atoms. Oxygen atoms in h2o can be taken from the solvent or be contributed to the solvent without significantly changing the amount of solvent, which is effectively infinite and of constant activity. Leaving out the row for oxygen atoms and the column for h2o in A' leads to conservation matrix A'' and a corresponding stoichiometric matrix ν'' that does not have a stoichiometric number for h2o, in agreement with the expression for the apparent equilibrium constant K'.

When the availability of oxygen atoms from h2o is specified by $\mu'^{\circ}(h2o)$, a Legendre transform can be used to define a **further transformed Gibbs energy** G'' by (4,5)

$$G'' = G' - n_c(O)\mu'^{\circ}(h2o) \tag{7.3-1}$$

where $n_c(O)$ is the amount of the oxygen component in the system and $\mu'^{\circ}(h2o)$ is the standard transformed chemical potential of h2o at the specified pH. The standard further transformed Gibbs energy of formation of a reactant can be calculated using

$$\Delta_f G_i''^{\circ} = \Delta_f G_i'^{\circ} - N_o(i)\Delta_f G'^{\circ}(h2o) \tag{7.3-2}$$

where $N_o(i)$ is the number of oxygen atoms in reactant i. Note that $\Delta_f G''^{\circ}(h2o) = 0$, and so terms for h2o no longer appear in the fundamental equation for G''. Holding the availability of oxygen atoms constant may produce pseudoisomers, as in the case of citrate, isocitrate and cis-aconitate that have compositions that differ only in the numbers of H and O atoms. The standard further transformed Gibbs energy of formation $\Delta_f G_k''^{\circ}$ of a reactant is given in terms of the **standard further transformed Gibbs energies of formation** $\Delta_f G_i''^{\circ}$ of the reactants (pseudoisomers) it contains by

$$\Delta_f G_k''^{\circ} = -RT\ln\sum\exp(-\Delta_f G_i''^{\circ}/RT) \tag{7.3-3}$$

The same result can be obtained by taking the mole-fraction-weighted average of $\Delta_f G_i''^{\circ}$ and adding the further transformed Gibbs energy of mixing. The equilibrium mole fractions r_i of the reactants in the pseudoisomer group are given by

$$r_i = \exp[(\Delta_f G_k''^{\circ} - \Delta_f G_i''^{\circ})/RT] \tag{7.3-4}$$

The **standard further transformed enthalpy of formation** of reactant k at specified $\Delta_f G'^{\circ}(h2o)$ is given by

$$\Delta_f H_k''^{\circ} = \sum r_i\Delta_f H_i''^{\circ} \tag{7.3-5}$$

The **standard further transformed entropy of formation** of reactant k is given by

$$\Delta_f G_k''^{\circ} = \Delta_f H_k''^{\circ} - T\Delta_f S_k''^{\circ} \tag{7.3-6}$$

Thus measurements of $\Delta_f G_k{}''^\circ$ and $\Delta_f H_k{}''^\circ$ at a single temperature yield $\Delta_f S_k{}''^\circ$ at that temperature. If $\Delta_f H_k{}''^\circ$ is known the standard further transformed Gibbs energy of formation can be expressed as a function of temperature, and then all the other thermodynamic properties can be calculated by taking partial derivatives of this function, as we have seen earlier in Chapter 4.

Now we can consider the hydrolysis of glucose 6-phosphate from the viewpoint of G''. The conservation matrix $\mathbf{A}'$ at specified pH (**conmat5**) is converted to $\mathbf{A}''$ by deleting the oxygen row and the h2o column to obtain **conmat6**.

```
In[34]:= conmat6 = {{6, 0, 6}, {1, 1, 0}};

In[35]:= TableForm[conmat6, TableHeadings → {{"C", "P"}, {"glucose6phos", "pi", "glucose"}}]

Out[35]//TableForm=
        glucose6phos    pi    glucose
   C    6               0     6
   P    1               1     0

In[36]:= TableForm[RowReduce[conmat6],
           TableHeadings → {{"glucose6phos", "pi"}, {"glucose6phos", "pi", "glucose"}}]

Out[36]//TableForm=
                    glucose6phos    pi    glucose
   glucose6phos     1               0     1
   pi               0               1     -1
```

A basis for the corresponding stoichiometric number matrix $\boldsymbol{\nu}'$ can be calculated using

$$\mathbf{A}''\boldsymbol{\nu}'' = \mathbf{0} \tag{7.3-7}$$

$\mathbf{A}''$ has the dimensions C''xN'', where C'' is the number of components ($C'' = C - 2$) and N'' is the number of reactants (sum or species). $\boldsymbol{\nu}''$ has the dimensions N''xR''. Note that $N'' = C'' + R''$. The system we are considering is represented by the following reaction:

$$\text{glucose 6-phosphate} = \text{glucose} + \text{P}_1 \tag{7.3-8}$$

When G'' is used, H_2O is omitted from biochemical reactions since it is provided for automatically.

A basis for the stoichiometric number matrix $\boldsymbol{\nu}''$ can be obtained as folllows:

```
In[37]:= TableForm[NullSpace[conmat6], TableHeadings → {{""}, {"glucose6phos", "pi", "glucose"}}]

Out[37]//TableForm=
        glucose6phos    pi    glucose
        -1              1     1
```

The order of multiplication in equation 7.3-7 can be changed by use of the transposes $(\mathbf{A}'')^T$ and $(\boldsymbol{\nu}'')^T$

$$(\boldsymbol{\nu}'')^T(\mathbf{A}'')^T = \mathbf{0} \tag{7.3-9}$$

This makes it possible to obtain a basis for the apparent conservation matrix $\mathbf{A}''$ from the stoichiometric number matrix $\boldsymbol{\nu}''$ for a biochemical reaction system.

It might be thought that the constancy of pH and activity of h2o could be included in the same Legendre transform by use of

$$G' = G - n_c(\text{H})\mu(\text{H}^+) - n_c(\text{O})\mu^\circ(\text{h2o}) \tag{7.3-10}$$

This has been attempted, but it is not possible because $\mu(H^+)$ and $\mu°(h2o)$ are not independent variables since h2o contains two hydrogen atoms. However, after the Legendre transformation defining G ' has been made, the standard transformed chemical potential μ ' °(h2o) can be used in Legendre transform 7.3-1.

7.4 Coupling

An enzyme can couple two or more reactions that could otherwise be catalyzed separately. When this happens, the equilibrium composition reached is different from that when there are separate reactions, as described in Section 7.8. The coupled reaction is the sum of the separate reactions. In adding these separate reactions, one or more reactants may cancel, but that is not necessary. The important point is that all the transformed thermodynamic properties of the coupled reaction are sums of the transformed thermodynamic peoperties of the separate reactions. It is of interest to consider coupling in the context of the six classes of enzyme-catalyzed reactions defined by IUBMB (6). Oxidoreductases (Class 1) all involve coupling because these reactions can be divided into two or more half reactions. This is different from other kinds of coupling, but half reactions do have thermodynamic properties that in principle can be investigated separately. Transferases (Class 2) all involve coupling. Hydrolases (Class 3) do not involve coupling. Lyases (Class 4) generally do not to involve coupling. Isomerases (Class 5) do not involve coupling. Ligases (Class 6) all involve coupling by definition. The IUBMB list indicates that about 60% of enzyme-catalyzed reactions are coupled.

The central concept involved in coupling is the identification of components, which are the things that are conserved in a reaction system. When chemical reactions are studied, atoms of elements are conserved, but some of these conservation equations may not be independent. Redundant conservation equations are not counted as components C. When the pH is specified, the conservation equation for hydrogen atoms is omitted, and so the number of components for a given system is reduced by one: $C' = C - 1$. A test of the conservation matrix A' is that the equation $A'v' = 0$ must yield a suitable basis for the stoichiometric number matrix v'. When it is necessary to recognize that oxygen atoms are available from h2o, A'' must be used, and $C'' = C' - 1$. A test of the conservation matrix A'' is that the equation $A''v'' = 0$ must yield a suitable basis for the stoichiometric number matrix v''.

Three examples of coupled reactions will be considered here; more coupled reactions are discussed in the literature (7). The first is the transferase reaction catalyzed by hexokinase:

$$atp + glucose = glucose6phos + adp \qquad (7.4-1)$$

Row reduction of the conservation matrix A ' for this reaction based on element balances for C, O, and P indicates that there is a single reaction. However, this is a misleading result because a column for h2o is not included in A ', and this excludes the possible hydrolase reactions:

$$atp + h2o = adp + pi \qquad (7.4-2)$$

$$glucose6phos + h2o = glucose + pi \qquad (7.4-3)$$

Including a column for h2o in A ' for a system involving these two reactions, and the use of **RowReduce** indicates that these two reactions can occur. To prevent this, an additional conservation equation has to be put in A ' to tie reactions 7.4-2 and 7.4-3 together. This can be done in several ways, but the conservation equation used here is $n'(atp) + n'(glucose6phos) =$ const. This relation, which is referred to as con1, insures that every time a mole of atp disappears a molecule of glucose6phos appears. The conservation matrix is now

```
In[38]:= conmat1 = {{10, 6, 6, 0, 10}, {13, 6, 9, 1, 10}, {3, 0, 1, 0, 2}, {1, 0, 1, 0, 0}};
```

```
In[39]:= TableForm[conmat1,
            TableHeadings → {{C, O, P, con1}, {"atp", "glucose", "glucose6phos", "h2o", "adp"}}]
```

```
Out[39]//TableForm=
                atp      glucose     glucose6phos     h2o      adp
        C       10       6           6                0        10
        O       13       6           9                1        10
        P       3        0           1                0        2
        con1    1        0           1                0        0
```

The row reduced form of **conmat1** shows that atp, glucose, glucose6phos, and h2o can alternatively be taken as components.

```
In[40]:= TableForm[RowReduce[conmat1],
         TableHeadings → {{"atp", "glucose", "glucose6phos", "h2o"},
            {"atp", "glucose", "glucose6phos", "h2o", "adp"}}]

Out[40]//TableForm=
                    atp      glucose     glucose6phos     h2o      adp
    atp             1        0           0                0        1
    glucose         0        1           0                0        1
    glucose6phos    0        0           1                0        -1
    h2o             0        0           0                1        0
```

This shows that there is now a single biochemical reaction, specifically reaction 7.4-1. The con1 constraint is provided by the mechanism of action of hexokinase.

A basis for the stoichiometric number matrix can be obtained from **conmat 1** by use of **NullSpace** in *Mathematica*.

```
In[41]:= TableForm[RowReduce[NullSpace[conmat1]],
         TableHeadings → {{""}, {"atp", "glucose", "glucose6phos", "h2o", "adp"}}]

Out[41]//TableForm=
            atp      glucose     glucose6phos     h2o      adp
            1        1           -1               0        -1
```

This provides a basis for the stoichiometric number matrix, which correctly indicates that the stoichiometric number for h2o is zero. The actual stoichiometric number matrix for reaction 7.4-1 is

```
In[42]:= TableForm[Transpose[{{-1}, {-1}, {1}, {0}, {1}}],
         TableHeadings → {{""}, {"atp", "glucose", "glucose6phos", "h2o", "adp"}}]

Out[42]//TableForm=
            atp      glucose     glucose6phos     h2o      adp
            -1       -1          1                0        1
```

As a second example, consider the following reaction:

atp + glutamate + ammonia = adp + glutamine + pi   (7.4-4)

In writing out the conservation matrix for this biochemical reaction, the reactants are arbitrarily taken in the order glutamate, atp, pi, ammonia, glutamine, and adp. The elements are taken in the order C, O, N, and P. These elements introduce the following constraints:

```
In[43]:= conmat = {{5, 10, 0, 0, 5, 10}, {4, 13, 4, 0, 3, 10}, {1, 5, 0, 1, 2, 5}, {0, 3, 1, 0, 0, 2}};

In[44]:= TableForm[conmat,
         TableHeadings → {{C, O, N, P}, {"glutamate", "atp", "pi", amm, "glutamine", "adp"}}]
```

Out[44]//TableForm=

	glutamate	atp	pi	amm	glutamine	adp
C	5	10	0	0	5	10
O	4	13	4	0	3	10
N	1	5	0	1	2	5
P	0	3	1	0	0	2

In[45]:= **TableForm[RowReduce[conmat], TableHeadings →**
          {{"glutamate", "atp", "pi", amm}, {"glutamate", "atp", "pi", amm, "glutamine", "adp"}}]

Out[45]//TableForm=

	glutamate	atp	pi	amm	glutamine	adp
glutamate	1	0	0	0	$\frac{5}{7}$	$\frac{2}{7}$
atp	0	1	0	0	$\frac{1}{7}$	$\frac{6}{7}$
pi	0	0	1	0	$-\frac{3}{7}$	$-\frac{4}{7}$
amm	0	0	0	1	$\frac{4}{7}$	$\frac{3}{7}$

The row reduced form shows that this reaction system involves two biochemical reactions. But there is a second way to obtain a conservation matrix, and that is by use of equation 7.2-6. The stoichiometric number matrix for reaction 7.4-4 is

In[46]:= **glutsyn = {{-1}, {-1}, {1}, {-1}, {1}, {1}};**

In[47]:= **TableForm[RowReduce[NullSpace[Transpose[glutsyn]]],**
          TableHeadings → {{"glutamate", "atp", "pi", amm, "glutamine"},
               {"glutamate", "atp", "pi", amm, "glutamine", "adp"}}]

Out[47]//TableForm=

	glutamate	atp	pi	amm	glutamine	adp
glutamate	1	0	0	0	0	1
atp	0	1	0	0	0	1
pi	0	0	1	0	0	-1
amm	0	0	0	1	0	1
glutamine	0	0	0	0	1	-1

It is of interest to note that this row reduced conservation matrix is characteristic of all biochemical reactions of the form A + B + C = D + E + F. The last column indicates that adp = glutamate + atp - pi + ammonia - glutamine, which is equation 7.4-4.

Now the question is "What constraint has to be added to **conmat** to produce this row reduced conservation matrix?" There is no single answer, but the constraint has to tie together the two reactions indicated by **conmat**. We can imagine these are atp + h2o = adp + pi and glutamate + ammonia = glutamine + h2o. One possible additional conservation equation is

$$n(\text{ammonia}) + n(\text{pi}) = \text{constant} \qquad (7.4\text{-}5)$$

This constraint arises in the third step of the following general type of mechanism.

$$E + \text{glutamate} = EG \qquad (7.4\text{-}6)$$

$$EG + \text{atp} = EGP + \text{adp} \qquad (7.4\text{-}7)$$

EGP + ammonia = Eglutamine + pi   (7.4-8)

Eglutamine = E + glutamine   (7.4-9)

The inclusion of conservation equation 7.4-5 yields **conmat2**.

```
In[48]:= conmat2 = {{5, 10, 0, 0, 5, 10}, {4, 13, 4, 0, 3, 10},
              {1, 5, 0, 1, 2, 5}, {0, 3, 1, 0, 0, 2}, {0, 0, 1, 1, 0, 0}};
```

```
In[49]:= TableForm[RowReduce[conmat2],
           TableHeadings → {{"glutamate", "atp", "pi", amm, "glutamine"},
              {"glutamate", "atp", "pi", amm, "glutamine", "adp"}}]
```

Out[49]//TableForm=

	glutamate	atp	pi	amm	glutamine	adp
glutamate	1	0	0	0	0	1
atp	0	1	0	0	0	1
pi	0	0	1	0	0	-1
amm	0	0	0	1	0	1
glutamine	0	0	0	0	1	-1

Thus A'' with the additional constraint 7.4-5 contains the same information as the stoichiometric matrix **glutsyn**. The following shows a more systematic way to show that **conmat2** yields the expected stoichiometric number matrix.

```
In[50]:= TableForm[RowReduce[NullSpace[conmat2]],
           TableHeadings → {{""}, {"glutamate", "atp", "pi", amm, "glutamine", "adp"}}]
```

Out[50]//TableForm=

glutamate	atp	pi	amm	glutamine	adp
1	1	-1	1	-1	-1

As a third example of a coupled reaction consider the following ligase reaction catalyzed by asparagine synthase (glutamine hydrolyzing) that has eight reactants:

atp + aspartate + glutamine + h2o = amp + ppi + asparagine + glutamate   (7.4-10)

(Note that there is an error in reference 6 in that the h2o is missing.) Ligase reactions necessarily involve coupling because they are defined (7) as "joining together of two molecules coupled with the hydrolysis of a pyrophosphate bond in atp of a similar triphosphate." This enzyme couples three hydrolase reactions:

atp + h2o = amp + ppi   (7.4-11)

glutamine + h2o = glutamate + ammonia   (7.4-12)

aspartate + ammonia = asparagineL + h2o   (7.4-13)

The conservation matrix based on the elements C, O, N, and P shows that there would be four independent reactions if these are the only constraints.

To obtain reaction 7.4-4 it is necessary to put in three additional conservation equations that couple the three reactions. There are a number of ways to do this, but the three conservation equations used here are:

con1: $n'(\text{atp}) + n'(\text{asparagine}) = \text{const.}$

con2: $n'(\text{atp}) + n'(\text{glutamate}) = \text{const.}$

con3: $n'(\text{aspartate}) + n'(\text{ppi}) = \text{const.}$

```
In[51]:= a6x3x5x4 = {{10, 4, 5, 0, 10, 0, 4, 5},
            {13, 4, 3, 1, 7, 7, 3, 4}, {5, 1, 2, 0, 5, 0, 2, 1}, {3, 0, 0, 0, 1, 2, 0, 0},
            {1, 0, 0, 0, 0, 0, 1, 0}, {1, 0, 0, 0, 0, 0, 0, 1}, {0, 1, 0, 0, 0, 1, 0, 0}};
```

```
In[52]:= TableForm[a6x3x5x4, TableHeadings → {{C, O, N, P, con1, con2, con3},
            {"atp", "asp", "glutN", "h2o", "amp", "ppi", "aspN", "glut"}}]
```

Out[52]//TableForm=

	atp	asp	glutN	h2o	amp	ppi	aspN	glut
C	10	4	5	0	10	0	4	5
O	13	4	3	1	7	7	3	4
N	5	1	2	0	5	0	2	1
P	3	0	0	0	1	2	0	0
con1	1	0	0	0	0	0	1	0
con2	1	0	0	0	0	0	0	1
con3	0	1	0	0	0	1	0	0

Row reduction shows that there a a single reaction, specifically 7.4-4.

```
In[53]:= TableForm[RowReduce[a6x3x5x4],
            TableHeadings → {{"atp", "asp", "glutN", "h2o", "amp", "ppi", "aspN"},
            {"atp", "asp", "glutN", "h2o", "amp", "ppi", "aspN", "glut"}}]
```

Out[53]//TableForm=

	atp	asp	glutN	h2o	amp	ppi	aspN	glut
atp	1	0	0	0	0	0	0	1
asp	0	1	0	0	0	0	0	1
glutN	0	0	1	0	0	0	0	1
h2o	0	0	0	1	0	0	0	1
amp	0	0	0	0	1	0	0	-1
ppi	0	0	0	0	0	1	0	-1
aspN	0	0	0	0	0	0	1	-1

However, this result is unsatisfactory for some purposes because it assigns a stoichiometric number to h2o, even though [h2o] is not in the expression for the apparent equilibrium constant K'.

Now consider the asparagine synthase (glutamine hydrolyzing) reaction from the viewpoint of the further transformed Gibbs energy G''. The oxygen row and the h2o column are omitted to obtain

```
In[54]:= a6x3x5x4noO = {{10, 4, 5, 10, 0, 4, 5}, {5, 1, 2, 5, 0, 2, 1}, {3, 0, 0, 1, 2, 0, 0},
            {1, 0, 0, 0, 1, 0}, {1, 0, 0, 0, 0, 0, 1}, {0, 1, 0, 0, 1, 0, 0}};
```

This treatment involves $N'' = 7$ reactants and $C'' = 6$ components, 3 of which are element balances (C, N, and P) and 3 of which are conservation equations arising from the enzyme mechanism.

```
In[55]:= TableForm[RowReduce[a6x3x5x4noO],
            TableHeadings → {{"atp", "asp", "glutN", "amp", "ppi", "aspN"},
            {"atp", "asp", "glutN", "amp", "ppi", "aspN", "glut"}}]
```

```
Out[55]//TableForm=
                atp      asp      glutN     amp      ppi      aspN     glut
        atp      1        0        0         0        0        0        1
        asp      0        1        0         0        0        0        1
        glutN    0        0        1         0        0        0        1
        amp      0        0        0         1        0        0       -1
        ppi      0        0        0         0        1        0       -1
        aspN     0        0        0         0        0        1       -1
```

The application of **NullSpace** yields the correct stoichiometric number matrix. Thus the calculation from the standpoint of the further transformed Gibbs energy G'' yields the expected number of constraints introduced by the enzyme mechanism. This reaction is a dramatic example of the difference between chemical reactions and some enzyme-catalyzed reactions. It is the enzymatic mechanism that introduces the three constraints in addition to atom balances.

Other enzyme-catalyzed reactions with two or more constraints in addition to element balances have been discussed (7).

7.5 Components in Reactions Involving Coenzymes

In studying the thermodynamics of systems of biochemical reactions it is desirable to obtain a global view. One way to do that is to assume that coenzymes are in steady states because they are involved in so many different reactions. When coenzyme concentrations are specified, a further transformed Gibbs energy G'' can be defined by (7,8,9)

$$G'' = G' - \sum n_c(\text{coenz}) \, \mu'(\text{coenz}) \tag{7.5-1}$$

This leads to the following equation for the standard further transformed Gibbs energy of formation of a reactant other than a coenzyme:

$$\Delta_f G''^\circ = \Delta_f G'^\circ - \sum N_{\text{coenz}} (\Delta_f G'^\circ (\text{coenz}) + RT \ln[\text{coenz}]) \tag{7.5-2}$$

where N_{coenz} is the number of coenzyme components in a reactant. Row reduction of the conservation matrix for a system of enzyme-catalyzed reactions provides the means for obtaining N_{coenz} for each coenzyme in each reactant. When the concentrations of coenzymes are specified, some reactants may become pseudoisomers. When concentrations of coenzymes in glycolysis are specified, all the thermodynamic properties of this series of ten reactions can be calculated by use of $C_6 = 2C_3$, where C_6 is the sum of reactants with 6 carbon atoms and C_3 is the sum of reactants with 3 carbon atoms.

If all components are included in a Legendre transform, the Gibbs-Duhem equation for a system is obtained. This is useful because it provides a relation between the intensive properties of the system, but to make other calculations at least one component must remain.

7.6 Use of Matrix Multiplication in Calculating Apparent Equilibrium Constants and of Linear-Solve in Calculating Standard Transformed Gibbs Energies of Formation of Reactants

Matrix operations make it possible to carry out calculations on systems of biochemical reactions with equations like 3.1-14. When the transformed Gibbs energy is used the corresponding equation is

$$\Delta_r G_k'^\circ = -RT \ln K_k' = \sum v_{ik}' \Delta_f G_i'^\circ \tag{7.6-1}$$

First, we will use matrix multiplication to calculate K' for the following three reactions at 298.15 K, pH 7, and 0.25 M ionic strength from data on the standard Gibbs energies of formation of the six reactants (9).

atp + h2o = adp + pi (7.6-2)

adp + h2o = amp + pi (7.6-3)

amp + h2o = adenosine + pi (7.6-4)

Second, we will make the reverse calculation and obtain $\Delta_f G'^\circ$ for atp, adp, and amp at 298.15 K, pH 7, and ionic strength 0.25 M from the three K' values and $\Delta_f G'^\circ$ for h2o, pi, and adenosine determined with respect to the elements.

Equation 7.6-1 can be written in matrix notation in two ways:

$$\Delta_r G'^\circ = -RT\ln K' = \Delta_f G'^\circ . v'$$ (7.6-5)

$$\Delta_r G'^\circ = -RT\ln K' = (v')^T . (\Delta_f G'^\circ)^T$$ (7.6-6)

The bold face type indicates matrices. The row matrix of $\Delta_f G'^\circ$ values in kJ mol^{-1} for atp, h2o, adp, pi, amp, and adenosine at 298.15 K and ionic strength 0.25 M is given by

```
In[56]:= Gf,o =
            {{-2292.5, -155.66, -1424.7, -1059.49,
              -554.83, 335.46}};
```

The stoichiometric number matrix for reactions 7.6-2 to 7.6-4 is given by

```
In[57]:= v' = {{-1, 0, 0}, {-1, -1, -1}, {1, -1, 0}, {1, 1, 1}, {0, 1, -1}, {0, 0, 1}};
```

Equation 7.6-5 yields the following row matrix of standard transformed Gibbs energies of reaction for the three reactions:

```
In[58]:= Gf,o.v'
```

```
Out[58]= {{-36.03, -33.96, -13.54}}
```

Equation 7.6-6 yields the same row matrix:

```
In[59]:= Gr,o = Transpose[v'].Transpose[Gf,o]
```

```
Out[59]= {{-36.03}, {-33.96}, {-13.54}}
```

The standard apparent equilibrium constants of reactions 7.6-2 to 7.6-4 are

```
In[60]:= Exp[(-Transpose[v'].Transpose[Gf,o]) / (8.31451 * .29815)]
```

```
Out[60]= {{2.05186×10⁶}, {890227.}, {235.555}}
```

Now we will make the reverse calculation and obtain the standard transformed Gibbs energies of formation of atp, adp, and amp from the three $\Delta_r G'^\circ$ and the $\Delta_f G'^\circ$ values for h2o, pi, and adenosine that have been determined with respect to the elements. This is done by using LinearSolve to solve the three linear equations. Since the $\Delta_f G'^\circ$ values for h2o, pi, and adenosine at 298.15 K, pH 7, and ionic strength 0.25 M are known to be -155.66, -1059.49, and 335.46 kJ mol^{-1}, the $\Delta_r G'^\circ$ values for reactions 7.4-2 to 7.4-4 can be used to calculate $\Delta_r G'^\circ$ for the following partial reactions:

atp = adp (7.6-7)

adp = amp (7.6-8)

amp = 0   (7.6-9)

The adjusted stoichiometric number matrix is

In[61]:= **m = {{-1, 0, 0}, {1, -1, 0}, {0, 1, -1}}**

Out[61]= {{-1, 0, 0}, {1, -1, 0}, {0, 1, -1}}

In[62]:= **Transpose[m]**

Out[62]= {{-1, 1, 0}, {0, -1, 1}, {0, 0, -1}}

The adjusted transformed Gibbs energies for these three reactions are

In[63]:= **ge = {-36.03 + 1059.49 - 155.66, -33.96 + 1059.66 - 155.66, -13.54 - 335.46 + 1059.49 - 155.66}**

Out[63]= {867.8, 870.04, 554.83}

In[64]:= **LinearSolve[Transpose[m], ge]**

Out[64]= {-2292.67, -1424.87, -554.83}

These are the expected the $\Delta_f G'^\circ$ values for atp, adp, and amp. This shows that when the apparent equilibrium constants have been measured for a number of reactions under the same conditions and $\Delta_f G'^\circ$ values are already known for a sufficient number of the reactants, the $\Delta_f G'^\circ$ values for the remaining reactants can be calculated by use of **LinearSolve**, which can handle very large matrices.

7.7 Uses of Matrices in Calculations of Equilibrium Compositions

An important use of matrices and operations of linear algebra is in the *Mathematica* programs for calculating the equilibrium compositions of systems of reactions. The advantage of writing programs in terms of matrices and matrix operations is that such a program can be used for arbitrarily large systems. The equilibrium composition of a chemical reaction system cannot be calculated analytically even when the $\Delta_f G^\circ$ for all the species are known. The equilibrium concentrations for a biochemical reaction system also cannot be calculated analytically. The conservation equations and equilibrium constant expressions have to be satisfied simultaneously. In the Newton-Raphson method, the calculation starts with an estimate of the equilibrium composition, and an iteration is carried out to obtain the solution that satisfies the conservation equations and expressions for apparent equilibrium constants. In 1978 Krambeck (10) wrote a program **equcalc** in APL to calculate the equilibrium composition of a chemical reaction system of ideal gases at a specified total pressure. It was based on the use of the conservation matrix for the system. Then he modified it to **equcalcc** to calculate equilibrium compositions of solutions on the assumption that $\Delta_f G_j = \Delta_f G_j^\circ + RT\ln[j]$. Later Krambeck (11) translated both of these programs into *Mathematica*. This second program is very useful in biochemical thermodynamics, and so it is given here:

```
In[65]:= equcalcc[as_, lnk_, no_] :=
         Module[{l, x, b, ac, m, n, e, k}, (*as=conservation matrix. lnk=-(1/RT)
               (Gibbs energy of formation vector at T). no=initial composition vector.*)
            (*Setup*)
            {m, n} = Dimensions[as];
            b = as.no;
            ac = as;
            (*Initialize*)l = LinearSolve[as.Transpose[as], -as.(lnk + Log[n])];
            (*Solve*)Do[e = b - ac.(x = E^(lnk + l.as));
             If[(10^-10) > Max[Abs[e]], Break[]];
             l = l + LinearSolve[ac.Transpose[as*Table[x, {m}]], e], {k, 100}];
            If[k = 100, Return["Algorithm Failed"]];
            Return[x]]
```

This program has three arguments: (1) as_ is the conservation matrix for the system of reactions. When it is applied to systems of biochemical reactions at specified pH, it has to include constraints in addition to atom balances; this is referred to as the first problem with this program when it is applied to a system of biochemical reactions. (2) lnk_ is equal to $-1/RT$(-Gibbs energy of formation vector at temperature T). At specified pH the standard transformed Gibbs energy of formation of each reactant is included in this vector. (3) no_ is the initial molar composition vector of the species in a chemical reaction and for reactants in a biochemical reaction. The matrix multiplication **as.no** in the program is used to obtain the amounts of components in the system. An estimate of the equilibrium composition is made, and a **Do** loop with **Linear Solve** is used to iterate to the equilibrium composition of the system using the Newton-Raphson method.

The second problem with this program has been discussed in Section 7.3; it is the fact that when h2o is a reactant the stoichiometric number matrix v' is inconsistent with the conservation matrix A' because [h2o] is not included in the expression for the apparent equilibrium constant. A second program **equcalcrx** was developed (4) to solve these problems. This program takes advantage of the fact that a basis for the conservation matrix can be calculated from the stoichiometric number matrix for the system.

```
In[66]:= equcalcrx[nt_, lnkr_, no_] :=
         Module[{as, lnk}, (*nt=transposed stoichiometric number matrix. lnkr=
             ln of equilibrium constants of rxs (vector). no=initial composition vector.*)
            (*Setup*)
            lnk = LinearSolve[nt, lnkr];
            as = NullSpace[nt];
            equcalcc[as, lnk, no]]
```

This program has three arguments: (1) nt_ is the transposed stoichiometric number matrix. (2) lnkr_ is a list of thenatural logarithms of equilibrium constants or apparent equilibrium constants. (3) no_ is the list of initial concentrations of species or reactants. This program uses **LinearSolve** to calculate **lnk** from **nt** and **lnkr**. **NullSpace** is used to calculate a suitable conservation matrix for the system from the stoichiometric number matrix. Now the equilibrium composition can be calculated using the program **equcalcc**. This program can be used to calculate equilibrium compositions of both chemical reaction systems and biochemical reaction systems. This program solves both problems mentioned above. The first problem (failure to take constranints in addition to element balances into account) is solved because the stoichiometric number matrix does include these constraints. The second problem (that h2o is not included in the expression for the apparent equilibrium constant) is solved because the correcct stoichiometric number matrix is used.

7.8 Calculation of Equilibrium Concentrations in a Two-Reaction System

Consider a reaction system at 298.15 K, pH 7, and 0.25 M ionic strength containing atp, adp, pi, glucose and glucose6phos, all initially at 1 mM concentrations. First, we will calculate the equilibrium concentrations when the reactions

atp + h2o = adp + pi   (7.8-1) glucose6phos + h2o = glucose + pi   (7.8-2)

are catalyzed separately. Second, we will calculate the equilibrium concentrations when the reactions are coupled so that the only reaction is the hexokinase reaction:

atp + glucose = adp + glucose6phos   (7.8-3)

The program **equcalcc** cannot be used when h2o is a reactant because when h2o is in the conservation matrix the program trys to calculate [h2o], just like it calculates the equilibrium concentrations of the other reactants. When h2o is a reactant, the program **equcalcrx** is used because it is based on the stoichiometric number matrix, which treats h2o correctly; that is, it leaves it out. This program uses the transposed stoichiometric number matrix **nt** and the vector **lnkr** of the natural logarithms of the apparent equilibrium constants to calculate the **as** and **lnk** needed for **equcalcc**. Then it calls on **equcalcc** to calculate the equilibrium composition using the Newton-Raphson method. To make these calculations we are going to need the functions of pH and ionic strength in BasicBiochemData3.

```
In[67]:=  Off[General::spell1];
          Off[General::spell];
```

```
In[69]:=  << BiochemThermo`BasicBiochemData3`
```

(a) Calculation of the equilibrium composition when the two reactions occur separately

Since the expressions for the apparent equilibrium constants of reactions 7.8-1 and 7.8-2 do not contain [h2o], the program **equcalcrx** is used. The stoichiometric number matrix for this system of two reactions corresponds with

atp = adp + pi   (7.8-4)

glucose6phos = glucose + pi   (7.8-5)

The transposed stoichiometric number matrix **nt** for the reaction system involving reactions 7.8-4 and 7.8-5 is

```
In[70]:=  nt = {{-1, 1, 1, 0, 0}, {0, 0, 1, 1, -1}};
```

```
In[71]:=  TableForm[nt, TableHeadings →
              {{"rx7.8-4", "rx7.8-5"}, {"atp", "adp", "pi", "glucose", "glucose6phos"}}]
```

```
Out[71]//TableForm=
                  atp      adp      pi       glucose      glucose6phos
        rx7.8-4   -1       1        1        0            0
        rx7.8-5   0        0        1        1            -1
```

The apparent equilibrium constants of reactions 7.8-4 and 7.8-5 can be calculated using **calckprime** (13).

```
In[72]:=  calckprime[eq_, pHlist_, islist_] := Module[{energy, dG},(*Calculates the apparent
              equilibrium constant at specified pHs and ionic strengths for a biochemical reaction
              typed in the form atp+h2o+de==adp+pi.  The names of reactants call the appropriate
              functions of pH and ionic strength.  pHlist and is list can be lists.*)
                  energy = Solve[eq, de];
                  dG = energy[[1,1,2]] /. pH -> pHlist /. is -> islist;
                  E^(-(dG/(8.31451*0.29815)))]
```

```
In[73]:=  calckprime[atp + h2o + de == adp + pi, 7, 0.25]
```

```
Out[73]=  2.05626 × 10^6
```

In[74]:= **calckprime[glucose6phos + h2o + de == glucose + pi, 7, 0.25]**

Out[74]= 108.375

Note that Δ_f *G* ' °(h2o) has to be used to calculate these apparent equilibrium constants. The logarithms of the apparent equilibrium constants of reactions 7.8-4 and 7.8-5 are used to specify **lnkr** in the input to **equcalcrx**:

In[75]:= **lnkr = {Log[2.05626 * 10^6], Log[108.375]}**

Out[75]= {14.5364, 4.6856}

Now suppose that initially the concentrations of the five reactants are all 1 mM. The initial composition vector **no** is

In[76]:= **no = {10^-3, 10^-3, 10^-3, 10^-3, 10^-3};**

Note that there is no term for h2o.
      The molar concentrations at equilibrium for the two-reaction system are given by

In[77]:= **eqconc = equcalcrx[nt, lnkr, no]**

Out[77]= {2.91787×10^{-12}, 0.002, 0.00299994, 0.00199994, 5.53608×10^{-8}}

In[78]:= **TableForm[{eqconc},**
      TableHeadings → {{""}, {"atp", "adp", "pi", "glucose", "glucose6phos"}}]

Out[78]//TableForm=

	atp	adp	pi	glucose	glucose6phos
	2.91787×10^{-12}	0.002	0.00299994	0.00199994	5.53608×10^{-8}

We can summarize the equilibrium concentrations by saying that atp and glucose6phos are nearly completely hydrolyzed.

(b) Calculation of the Equilibrium Concentrations When Two Reactions are Coupled

      Since reaction 7.8-3 does not involve h2o, the equilibrium concentrations can be calculated with either **equcalcc** or **equcalcrx**, and so it is done both ways. First, use **equcalcc**.

In[79]:= **as = {{10, 10, 6, 6}, {13, 10, 12, 15}, {3, 2, 0, 1}};**

In[80]:= **lnk = - (1 / (8.31451 * .29815)) * {atp, adp, glucose, glucose6phos} /. pH → 7 /. is → .25**

Out[80]= {924.778, 574.715, 172.131, 532.045}

Assume the initial concentrations are all 1 mM.

In[81]:= **no = {.001, .001, .001, .001};**

The equilibrium composition is given by the following calculation:

In[82]:= **TableForm[{equcalcc[as, lnk, no]},**
      TableHeadings → {{""}, {"atp", "adp", "glucose", "glucose6phos"}}]

Out[82]//TableForm=

	atp	adp	glucose	glucose6phos
	0.0000144149	0.00198559	0.0000144149	0.00198559

We can describe what happens in the reaction atp + glucose = adp + glucose6phos by saying that when the initial concentrations are all 0.001 M, the reaction goes very far to the right so that atp and glucose are essentially used up. These equilibrium concentrations can be compared with the equilibrium concentrations obtained without the constraint provided by the enzyme mechanism. The result of coupling is to produce a lot of glucose6phos. If 0.001 M phosphate is present, its concentration is not affected by reaction 7.8-3.

Second, use **equcalcrx**. The input is as follows:

In[83]:= `nt3 = {{-1, 1, -1, 1}};`

In[84]:= `lnkr3 = {Log[calckprime[atp + glucose + de == adp + glucose6phos, 7, 0.25]]}`

Out[84]= `{9.85081}`

In[85]:= `no3 = {.001, .001, .001, .001};`

In[86]:= `TableForm[{equcalcrx[nt3, lnkr3, no3]},`
`        TableHeadings → {{""}, {"atp", "adp", "glucose", "glucose6phos"}}]`

Out[86]//TableForm=

atp	adp	glucose	glucose6phos
0.0000144149	0.00198559	0.0000144149	0.00198559

This is exactly the result obtained using **equcalcc**.

7.9 Discussion

Chemical equations and biochemical equations are actually matrix equations, and so it is not a surprise that linear algebra is so useful in making thermodynamic calculations on them. More specifics on using matrix notation in writing fundamental equations and equations derived from it are given in Thermodynamics of Biochemical Reactions (2). Matrix operations are very useful in writing computer programs in biochemical thermodynamics, and one advantage of such programs is that the size of the system can be scaled up without rewriting the programs.

Linear algebra clarifies the use of Legendre transforms in biochemical thermodynamics in the sense that when an independent concentration variable is held constant, its row and column are omitted in the conservation matrix and then redundant columns are eliminated because they indicate pseudoisomer groups. By use of **RowReduce** many different choices of components can be found.

Recognition that oxygen atoms are not conserved in biochemical reactions involving h2o raises the question as to whether data bases like BasicBiochemData3 (12) should use G'', rather than G'. The data matrices for species would need to include the number of oxygen atoms in each species, and some new computer programs would have to be written. However, there are two strong arguments against doing this. The first is that biochemists are used to including h2o in biochemical equations, as indicated in Enzyme Nomenclature (6), and the second is that the specification of the availability of oxygen atoms makes citrate, isocitrate, and cis aconitate, for example, pseudoisomers so that only the pseudoisomer group would appear in tables of further transformed thermodynamic properties (13).

Appendix

Printing Enzyme-catalyzed Reactions from Stoichiometric Number Matrices

A system of biochemical reactions is represented by a stoichiometric number matrix. This stoichiometric number matrix can be used to print out the reactions. The programs that can be used to print out the biochemical reactions are **mkeqm** and **nameMatrix**.

```
In[87]:= mkeqm[c_List,s_List]:=(*c_List is the list of stoichiometric numbers for a reaction.
         s_List is a list of the names of species or reactants.  These names have to be put
         in quotation marks.*)Map[Max[#,0]&,-c].s->Map[Max[#,0]&,c].s
```

```
In[88]:= nameMatrix[m_List,s_List]:=(*m_List is the transposed stoichiometric number matrix
         for the system of reactions. s_List is a list of the names of species or reactants.
         These names have to be put in quotation marks.*)Map[mkeqm[#,s]&,m]
```

The first three reactions of glycolysis are

atp + glucose =adp + glucose6phos

glucose6phos = fructose6phos

atp + fructose6phos = adp + fructose16phos

This system of reactions is represented by the following stoichiometric number matrix.

```
In[89]:= nu = {{-1, 0, 0}, {-1, 0, -1}, {1, -1, 0}, {1, 0, 1}, {0, 1, -1}, {0, 0, 1}};
```

```
In[90]:= TableForm[nu]
```

```
Out[90]//TableForm=
      -1      0       0
      -1      0      -1
       1     -1       0
       1      0       1
       0      1      -1
       0      0       1
```

```
In[91]:= names3 = {"glucose", "atp", "glucose6phos", "adp", "fructose6phos", "fructose16phos"};
```

```
In[92]:= TableForm[nu, TableHeadings → {names3, {"rx 13", "rx 14", "rx 15"}}]
```

```
Out[92]//TableForm=
                          rx 13     rx 14     rx 15
        glucose            -1         0         0
        atp                -1         0        -1
        glucose6phos        1        -1         0
        adp                 1         0         1
        fructose6phos       0         1        -1
        fructose16phos      0         0         1
```

The first column can be used to print out the first of the three reactions using **mkeqm**.

```
In[93]:= mkeqm[{-1, -1, 1, 1, 0, 0}, names3]
```

```
Out[93]= atp + glucose → adp + glucose6phos
```

The matrix **nu** can be used to print out the three reactions using **nameMatrix**

```
In[94]:= TableForm[nameMatrix[Transpose[nu], names3]]
```

```
Out[94]//TableForm=
        atp + glucose → adp + glucose6phos
        glucose6phos → fructose6phos
        atp + fructose6phos → adp + fructose16phos
```

References

1. G. Strang, Linear Algebra and its Applications, Harcourt, Brace, Jovanovich, San Diego, 1988.

2. R. A. Alberty, Thermodynamics of Biochemical Reactions, Wiley, Hoboken, NJ, 2003.

3. W. R. Smith and R. W. Missen, Chemical Reaction Equilibrium Analysis: Theory and Algorithms, Wiley-Interscience, New York, 1982.

4. R. A. Alberty, Calculation of equilibrium compositions of biochemical reaction systems involving water as a reactant, J. Phys. Chem. 105B, 1109-1114 (2001).

5, R. A. Alberty, The role of water in the thermodynamics of dilute aqueous solutions, Biophys. Chem. 100, 183-192 (2003).

6. E. C. Webb, Enzyme Nomenclature 1992, Academic Press, New York, 1992.

In[95]:= **http : // www.chem.qmw.ac.uk / iubmb / enzyme /**

7. R. A. Alberty, Components and coupling in enzyme-catalyzed reactions, J. Phys. Chem., 109 B, 2021-2026 (2005).

8. R. A. Alberty, Calculation of equilibrium compositions of large systems of biochemical reactions, J. Phys. Chem. 104 B, 4807-4814 (2000).

9. R. A. Alberty, Equilibrium concentrations for pyruvate dehydrogenase and the citric acid cycle at specified concentrations of certain coenzymes, Biophys. Chem., 109, 73-84 (2004).

10. F. J. Krambeck, Presented at the 71st Annual Meeting of the AIChE, Miami Beach, FL, Nov. 16, 1978.

11. F. J. Krambeck, In Chemical Reactions in Complex Systems; F. J. Krambeck and A. M. Sapre. (Eds.) Van Nostrand Reinhold: New York. 1991.

12. R. A. Alberty, BasicBiochemData3, 2005.

In[95]:= **http : // library.wolfram.com / infocenter / MathSource / 5704**

13. R. A. Alberty, Inverse Legendre Transform in Biochemical Thermodynamics; Applied to the Last Five Reactions of Glycolysis, J. Phys. Chem., 106 B, 6594-6599 (2002).

Chapter 8 Oxidoreductase Reactions (Class 1) at 298.15 K

8.1 Theory

The third largest class of enzymes is the oxidoreductases, which transfer electrons. Oxidoreductase reactions are different from other reactions in that they can be divided into two or more half reactions. Usually there are only two half reactions, but the methane monooxygenase reaction can be divided into three "half reactions." Each chemical half reaction makes an independent contribution to the equilibrium constant K for a chemical redox reaction. For chemical reactions the standard reduction potentials E° can be determined for half reactions by using electrochemical cells, and these measurements have provided most of the information on standard chemical thermodynamic properties of ions. This research has been restricted to rather simple reactions for which electrode reactions are reversible on platinized platinum or other metal electrodes.

When the pH is specified, each biochemical half reaction makes an independent contribution to the apparent equilibrium constant K' for the reaction written in terms of reactants rather than species. The studies of electochemical cells have played an important role in the development of biochemical thermodynamics, as indicated by the outstanding studies by W. Mansfield Clarke (1). The main source of tables of E'° values for biochemical half reactions has been those of Segel (2). Although standard apparent reduction potentials E'° can be measured for some half reactions of biochemical interest, their direct determination is usually not feasible because of the lack of reversibility of the electrode reactions. However, standard apparent reduction potentials can be calculated from K' for oxidoreductase reactions. Goldberg and coworkers (3) have compiled and evaluated the experimental determinations of apparent equilibrium constants and standard transformed enthalpies of oxidoreductase reactions, and their tables have made it possible to calculate E'° values for about 60 half reactions as functions of pH and ionic strength at 298.15 K (4-8).

As an example of an oxidoreductase reaction at a specified pH consider

$$formate + nadpox + h2o = co2tot + nadpred \tag{8.1-1}$$

The half reactions of reaction 8.1-1 written as reduction reactions are

$$nadox + 2\,e = nadred \qquad\qquad (8.1\text{-}2)$$

$$co2tot + 2\,e = formate + h2o \qquad\qquad (8.1\text{-}3)$$

Charges are not shown on electrons because at specified pH, charges are not balanced. It is important to realize that these are not hydrated electrons in water; they are electrons that are transfered and they are refered to as formal electrons. The standard transformed reaction Gibbs energy of an oxidoreductase reaction is given by

$$\Delta_r G'^\circ = -\mid \nu_e \mid FE'^\circ = -\mid \nu_e \mid F\,(E_h{'}^\circ - E_l{'}^\circ) = \Delta_r G_h{'}^\circ - \Delta_r G_l{'}^\circ \qquad\qquad (8.1\text{-}4)$$

where $\mid \nu_e \mid$ is the number of electrons involved in the biochemical reaction, F is the faraday constant (96,485 C mol^{-1}), E'° is the standard apparent electromotive force for the reaction, and $E_h{'}^\circ$ is the standard apparent reduction potential higher in the table given later, and $E_l{'}^\circ$ is the standard apparent reduction potential lower in the table. $\Delta_r G_h{'}^\circ$ and $\Delta_r G_l{'}^\circ$ are the independent contributions of the two half reactions to $\Delta_r G'^\circ$. Standard apparent reduction potentials of half reactions are useful in thinking about the thermodynamics of enzyme-catalyzed reactions in the same sense that $\Delta_r G'^\circ$ values are because they can be used to calculate apparent equilibrium constants of biochemical reactions. The apparent equilibrium constant K' for a biochemical oxidoreductase reaction is given by

$$K' = \exp(-\mid \nu_e \mid FE'^\circ / RT) \qquad\qquad (8.1\text{-}5)$$

The standard transformed reaction enthalpy $\Delta_r H'^\circ$ for a redox reaction is obtained by use of the Gibbs-Helmholtz equation:

$$\Delta_r H'^\circ = T^2 \frac{\partial(\mid \nu_e \mid FE'^\circ / T)}{\partial T} = T^2 \mid \nu_e \mid F\left\{ \frac{\partial(E_h{'}^\circ / T)}{\partial T} - \frac{\partial(E_l{'}^\circ / T)}{\partial T} \right\} = \Delta_r H_h{'}^\circ - \Delta_r H_l{'}^\circ \quad (8.1\text{-}6)$$

Thus $\Delta_r H'^\circ$ is made up of independent contributions from the two half reactions.

The standard transformed reaction entropy $\Delta_r S'^\circ$ of a redox reaction is obtained by use of the following partial derivative:

$$\Delta_r S'^\circ = \frac{\partial(\mid \nu_e \mid FE'^\circ)}{\partial T} = \mid \nu_e \mid F\left\{ \frac{\partial E_h{'}^\circ}{\partial T} - \frac{\partial E_l{'}^\circ}{\partial T} \right\} = \Delta_r S_h{'}^\circ - \Delta_r S_l{'}^\circ \qquad\qquad (8.1\text{-}7)$$

Thus $\Delta_r S'^\circ$ is made up of two independent contributions from the two half reactions.

The change in binding of hydrogen ions $\Delta_r N_H$ in a biochemical redox reaction is given by

$$\Delta_r N_H = -\frac{\mid \nu_e \mid F}{RT\ln(10)} \frac{\partial E'^\circ}{\partial pH} = -\frac{\mid \nu_e \mid F}{RT\ln(10)}\left(\frac{\partial E_h{'}^\circ}{\partial pH} - \frac{\partial E_l{'}^\circ}{\partial pH} \right) = \Delta_r N_{Hh} - \Delta_r N_{Hl} \qquad (8.1\text{-}8)$$

So the change in binding of hydrogen ions in a redox reaction is the difference between the changes in the two half reactions. Higher derivatives yield $\partial \Delta_r H'^\circ / \partial pH$, and $\partial \Delta_r S'^\circ / \partial pH$.

8.2 Tables of Standard Apparent Reduction Potentials of Half Reactions

It is useful to construct a table of E'° values for half reactions arranged according to decreasing values so that apparent equilibrium constants can be calculated for oxidoreductase reactions. **The rule for using such a table is that any oxidized reactant (sum of species in some cases) will react with the reduced reactant (sum of species) in a half reaction with a lower E'° value when the reactant concentrations are all 1 M, except for H_2O.** The half reaction lower in the table is subtracted from the half reaction higher in the table to obtain the oxidoreductase reaction. The electrons in the two half reactions must cancel. The rule in building a table of half reactions is that there should be no fractional coefficients, but the half reactions can be multiplied by an integer or divided by an integer without affecting the standard apparent reduction potential.

Oxidizing agents with standard apparent reduction potentials above 0.807 V at pH 7 tend to oxidize H_2O to $O_2(g)$, as illustrated by:

$F_2(g) + 2e^- = 2F^-$	$E'^{\circ}(\text{pH } 7) = 2.87$ V
$(1/2)O_2(g) + 2e^- = H_2O$	$E'^{\circ}(\text{pH } 7) = 0.81$ V

$$F_2(g) + H_2O = 2F^- + (1/2)O_2(g) \qquad E'^{\circ}(\text{pH } 7) = 2.06 \text{ V}$$

Oxidizing agents with standard apparent reduction potentials below -0.514 V at pH 7 tend to produce molecular hydrogen, as illustrated by

$e^- = (1/2)H_2(g)$	$E'^{\circ}(\text{pH } 7) = -0.574$ V
$Li^+ + 2e^- = Li(s).$	$E'^{\circ}(\text{pH } 7) = -3.045$ V

$$Li(s) = (1/2)H_2(g) + Li^+ \qquad E'^{\circ}(\text{pH } 7) = 2.471 \text{ V}$$

In writing the reduction of hydrogen ions, H^+ is omitted because it is understood that they are supplied or withdrawn at the specified pH as needed. The problem with the usual tables of E'° is that the voltages of biochemical half reactions depend on the temperature, pH, and ionic strength. The effect of temperature cannot be calculated in many cases because of the lack of $\Delta_f H^{\circ}$ data, but the effects of pH and ionic strength can be calculated for more than the 60 half reactions discussed in this chapter. The data file BasicBiochemData3 (9) can be used to calculate the standard apparent reduction potentials of a large number of half reactions. When all the reactants in a half reaction are in BasicBiochemData3, the standard apparent reduction potential at 298.15 K and the desired pH and ionic strength can be calculated by use of the *Mathematica* program **calcappredpot** by typing in the half reactions, specifying the number $|\nu_e|$ of electrons involved and listing the desired pHs and ionic strengths. This program can also be used to make plots. Table 8.1 can be used to calculate apparent equilibrium constants for many more reactions than it took to make BasicBiochemData3 that is given in the Appendix of this book. Apparent equilibrium constants at 298.15 K can be calculated from any pair of half reactions. If the E'° table contains N different half reactions, the apparent equilibrium constants can be calculated for R reactions, where $R = N(N-1)/2$. Thus when the table of standard apparent reduction potentials contains 60 half reactions, it can be used to calculate the apparent equilibrium constants for $60 \times 59/2 = 1770$ oxidoreductase reactions. However, enzymes are not known to exist for all these reactions.

The apparent equilibrium constant K' at 298.15 K for a biochemical reaction obtained by taking the difference between two half reactions is given by

```
In[2]:=  Exp[-nu * 96.485 * ΔE / (8.31451 * .29815)]

Out[2]=  e^{-38.9214 nu ΔE}
```

where **nu** is the absolute value of the number of electrons transferred and ΔE is the difference in standard apparent reduction potentials. Thus if the **nu** are 1, 2, and 3, and the ΔE are -0.1, -0.2, -0.3, -0.4, and -0.5 volts, the apparent equilibrium constants are given by

```
In[3]:=  TableForm[Exp[-nu * 96.485 * ΔE / (8.31451 * .29815)] /. nu → {1, 2, 3} /.
             ΔE → {-0.1, -0.2, -0.3, -0.4, -0.5}, TableHeadings →
                {{"nu=1", "nu=2", "nu=3"}, {"-0.1 V", "-0.2 V", "-0.3 V", "-0.4 V", "-0.5 V"}}]
```

Out[3]//TableForm=

	-0.1 V	-0.2 V	-0.3 V	-0.4 V	-0.5 V
nu=1	49.0156	2402.53	117761.	5.77215×10^6	2.82926×10^8
nu=2	2402.53	5.77215×10^6	1.38678×10^{10}	3.33177×10^{13}	8.00469×10^{16}
nu=3	117761.	1.38678×10^{10}	1.63309×10^{15}	1.92315×10^{20}	2.26473×10^{25}

If the **nu** are 1, 2, and 3, and the ΔE are -0.1, -0.2, -0.3, -0.4, and -0.5 volts, the $\Delta_r G'^\circ$ in kJ mol^{-s} are given by

```
In[4]:=  -nu * 96.485 * ΔE
```

Out[4]= -96.485 nu ΔE

In the following table, the $\Delta_r G'^\circ$ in kJ mol^{-1} are obtained as follows:

```
In[5]:=  TableForm[
            -nu * 96.485 * ΔE /. nu → {1, 2, 3} /. ΔE → {-0.1, -0.2, -0.3, -0.4, -0.5}, TableHeadings →
               {{"nu=1", "nu=2", "nu=3"}, {"-0.1 V", "-0.2 V", "-0.3 V", "-0.4 V", "-0.5 V"}}]
```

Out[5]//TableForm=

	-0.1 V	-0.2 V	-0.3 V	-0.4 V	-0.5 V
nu=1	9.6485	19.297	28.9455	38.594	48.2425
nu=2	19.297	38.594	57.891	77.188	96.485
nu=3	28.9455	57.891	86.8365	115.782	144.728

The standard apparent reduction potential in volts at 298.15 K for a half reaction can be calculated using **calcappredpot**.

```
In[6]:=  Off[General::"spell"];
         Off[General::"spell1"];
```

```
In[8]:=  << BiochemThermo`BasicBiochemData3`
```

```
In[9]:=  calcappredpot[eq_, nu_, pHlist_, islist_] :=
             Module[{energy},(*Calculates the standard apparent reduction potential of a half
             reaction in volts at specified pHs and ionic strengths for a biochemical half
             reaction typed in the form nadox+de==nadred.  The names of the reactants call the
             corresponding functions of pH and ionic strength.  nu is the number of electrons
             involved.  pHlist and islist can be lists.*)
               energy = Solve[eq, de];
                 -(energy[[1,1,2]]/(nu*96.485)) /. pH -> pHlist /.
                   is -> islist]
```

The objective of this section is to prepare a table of E'° for half reactions in volts at specified pH and ionic strength and arrange them in the order of decreasing values. To show how this can be done, consider three half reactions at 298.15 K, pH 6, and 0.25 M ionic strength. The first half reaction is nadox + e = nadred:

```
In[10]:=  calcappredpot[nadox + de == nadred, 2, 6, 0.25]
```

Out[10]= -0.286464

In[11]:= **calcappredpot[co2tot + pyruvate + de == malate + h2o, 2, 6, 0.25]**

Out[11]= -0.25669

In[12]:= **calcappredpot[n2aq + de == 2 * ammonia + h2aq, 8, 6, 0.25]**

Out[12]= -0.245164

This information is summarized in the following table.

In[13]:= **trialdata6 = {{"nadox+2e=nadred", calcappredpot[nadox + de == nadred, 2, 6, 0.25]},**
       {"co2tot+pyruvate+2e=malate+h2o",
       calcappredpot[co2tot + pyruvate + de == malate + h2o, 2, 6, 0.25]},
       {"n2aq+8e=2ammonia+h2aq", calcappredpot[n2aq + de == 2 * ammonia + h2aq, 8, 6, 0.25]}};

In[14]:= **TableForm[trialdata6]**

Out[14]//TableForm=
       nadox+2e=nadred   -0.286464
       co2tot+pyruvate+2e=malate+h2o   -0.25669
       n2aq+8e=2ammonia+h2aq   -0.245164

The entries can be arranged in order of decreasing $E'°$ by use of **Ordering** in *Mathematica*. The ordering of the values from the most positive to the most negative is obtained by using **Reverse[Ordering[Transpose[halfreactiondata][[2]]]]**. The **Reverse** is required because ordering in *Mathematica* gives the $E'°$ in the opposite order. The 2 indicates that the ordering is to be determined by the second column in the table, that is the $E'°$ values. This ordering is then applied to the three half reactions in trialdata6.

In[15]:= **TableForm[trialdata6[[Reverse[Ordering[Transpose[trialdata6][[2]]]]]]]**

Out[15]//TableForm=
       n2aq+8e=2ammonia+h2aq   -0.245164
       co2tot+pyruvate+2e=malate+h2o   -0.25669
       nadox+2e=nadred   -0.286464

A program **redpotsinorder** can be written so that a table can be made for any specified pH and ionic strength at 298.15 K.

In[16]:= **data = {{"nadox+2e=nadred", calcappredpot[nadox + de == nadred, 2, pH, is]},**
       {"co2tot+pyruvate+2e=malate+h2o",
       calcappredpot[co2tot + pyruvate + de == malate + h2o, 2, pH, is]},
       {"n2aq+8e=2ammonia+h2aq", calcappredpot[n2aq + de == 2 * ammonia + h2aq, 8, pH, is]}};

In[17]:= **rxdata = data /. pH → 7 /. is → .25;**

In[18]:= **redpotsinorder[evaldata_] :=**
       Module[{}, (*This program sorts the apparent reduction potentials in decreasing
          order and produces a table. evaldata is in the following form: evaldata=
          {{"nadox+2e=nadred",calcappredpot[nadox+de==nadred,2,pH,is]},
            {"co2tot+pyruvate+2e=malate+h2o",
             calcappredpot[co2tot+pyruvate+de==malate+h2o,2,pH,is]},
             {"n2aq+8e=2ammonia+h2aq",calcappredpot[n2aq+de==2*ammonia+h2aq,
             8,pH,is]}}/.pH->7/.is→.25*)
         TableForm[evaldata[[Reverse[Ordering[Transpose[evaldata][[2]]]]]]]

The apparent reduction potentials are rounded to a maximum of 4 places to the right of the decimal point since 0.02 kJ mol^{-1} corresponds with 0.0002 V when one electron is transferred.

```
In[19]:= PaddedForm[redpotsinorder[rxdata], {5, 4}]

Out[19]//PaddedForm=
      co2tot+pyruvate+2e=malate+h2o        -0.3077
      nadox+2e=nadred                      -0.3160
      n2aq+8e=2ammonia+h2aq                -0.3191
```

Notice that when the pH and ionic strength are changed, the order of the half reactions changes. The program is now applied to a number of half reactions for which there is data in BasicBiochemData3.

```
In[20]:= halfrxdata = {{"nadox+2e=nadred", calcappredpot[nadox + de == nadred, 2, pH, is]},
            {"glyoxylate+2e=glycolate", calcappredpot[glyoxylate + de == glycolate, 2, pH, is]},
            {"pyruvate+2e=lactate", calcappredpot[pyruvate + de == lactate, 2, pH, is]},
            {"oxaloacetate+2e=malate", calcappredpot[oxaloacetate + de == malate, 2, pH, is]},
            {"ketoglotarate+co2tot+2e=citrateiso+h2o",
             calcappredpot[ketoglutarate + co2tot + de == citrateiso + h2o, 2, pH, is]},
            {"fructose+2e=mannitolD", calcappredpot[fructose + de == mannitolD, 2, pH, is]},
            {"glycolate+ammonia+2e=glycine+h2o",
             calcappredpot[glycolate + ammonia + de == glycine + h2o, 2, pH, is]},
            {"acetone+2e=propanol2", calcappredpot[acetone + de == propanol2, 2, pH, is]},
            {"acetate+2e=acetaldehyde+h2o",
             calcappredpot[acetate + de == acetaldehyde + h2o, 2, pH, is]},
            {"acetylcoA+2e=acetaldehyde+coA", calcappredpot[acetylcoA + de == acetaldehyde + coA,
             2, pH, is]}, {"acetylcoA+co2tot+2e=pyruvate+coA+h2o",
             calcappredpot[acetylcoA + co2tot + de == pyruvate + coA + h2o, 2, pH, is]},
            {"fumarate+2e=succinate", calcappredpot[fumarate + de == succinate, 2, pH, is]},
            {"ferredoxinox+e=ferredoxinred",
             calcappredpot[ferredoxinox + de == ferredoxinred, 1, pH, is]},
            {"o2aq+4e=2h2o", calcappredpot[o2aq + de == 2 * h2o, 4, pH, is]},
            {"o2aq+2e=h2o2aq", calcappredpot[o2aq + de == h2o2aq, 2, pH, is]},
            {"acetaldehyde+2e=ethanol", calcappredpot[acetaldehyde + de == ethanol, 2, pH, is]},
            {"o2g+4e=2h2o", calcappredpot[o2g + de == 2 * h2o, 4, pH, is]},
            {"cytochromecox+e=cytochromecred", calcappredpot[cytochromecox + de == cytochromecred,
             1, pH, is]}, {"fmnox+2e=fmnred", calcappredpot[fmnox + de == fmnred, 2, pH, is]},
            {"retinal+2e=retinol", calcappredpot[retinal + de == retinol, 2, pH, is]},
            {"nadpox+2e=nadpred", calcappredpot[nadpox + de == nadpred, 2, pH, is]},
            {"2e=h2g", calcappredpot[de == h2g, 2, pH, is]},
            {"2e=h2aq", calcappredpot[de == h2aq, 2, pH, is]},
            {"methanol+2e=methaneaq+h2o",
             calcappredpot[methanol + de == methaneaq + h2o, 2, pH, is]},
            {"methanol+nadpox+4e=methaneaq+nadpred+h2o",
             calcappredpot[methanol + nadpox + de == methaneaq + nadpred, 2, pH, is]},
            {"co2tot+2e=formate+h2o", calcappredpot[co2tot + de == formate + h2o, 2, pH, is]},
            {"co2g+2e=formate", calcappredpot[co2g + de == formate, 2, pH, is]},
            {"ketoglutarate+ammonia+2e=glutamate+h2o",
             calcappredpot[ketoglutarate + ammonia + de == glutamate + h2o, 2, pH, is]},
            {"pyruvate+ammonia+2e=alanine+h2o", calcappredpot[pyruvate + ammonia + de ==
                alanine + h2o, 2, pH, is]}, {"co2tot+pyruvate+2e=malate+h2o",
             calcappredpot[co2tot + pyruvate + de == malate + h2o, 2, pH, is]},
            {"cystineL+2e=2cysteineL", calcappredpot[cystineL + de == 2 * cysteineL, 2, pH, is]},
            {"citrate+coA+2e=malate+acetylcoA+h2o",
             calcappredpot[citrate + coA + de == malate + acetylcoA + h2o, 2, pH, is]},
            {"glutathioneox+coA+2e=coAglutathione+glutathionered",
             calcappredpot[glutathioneox + coA + de == coAglutathione + glutathionered, 2, pH, is]},
```

```
{"glutathioneox+2e=2*glutathionered",
 calcappredpot[glutathioneox + de == 2 * glutathionered, 2, pH, is]},
{"thioredoxinox+2e=thioredoxinred", calcappredpot[
   thioredoxinox + de == thioredoxinred, 2, pH, is]},
{"n2g+8e=2ammonia+h2g", calcappredpot[n2g + de == 2 * ammonia + h2g, 8, pH, is]},
{"n2aq+8e=2ammonia+h2aq", calcappredpot[n2aq + de == 2 * ammonia + h2aq, 8, pH, is]},
{"oxalylcoA+2e=glyoxylate+coA", calcappredpot[
   oxalylcoA + de == glyoxylate + coA, 2, pH, is]}, {"ribose+2e=deoxyribose+h2o",
 calcappredpot[ribose + de == deoxyribose + h2o, 2, pH, is]},
{"atp+2e=deoxyatp+h2o", calcappredpot[atp + de == deoxyatp + h2o, 2, pH, is]},
{"ethanol+2e=ethaneaq+h2o", calcappredpot[ethanol + de == ethaneaq + h2o, 2, pH, is]},
{"fadox+2e=fadred", calcappredpot[fadox + de == fadred, 2, pH, is]},
{"nitrate+2e=nitrite+h2o", calcappredpot[nitrate + de == nitrite + h2o, 2, pH, is]},
{"sulfate+2e=sulfite+h2o", calcappredpot[sulfate + de == sulfite + h2o, 2, pH, is]},
{"ribose1phos+2e=deoxyribose1phos+h2o",
 calcappredpot[ribose1phos + de == deoxyribose1phos + h2o, 2, pH, is]},
{"ribose5phos+2e=deoxyribose5phos+h2o",
 calcappredpot[ribose5phos + de == deoxyribose5phos + h2o, 2, pH, is]},
{"adenosine+2e=deoxyadenosine+h2o",
 calcappredpot[adenosine + de == deoxyadenosine + h2o, 2, pH, is]},
{"amp+2e=deoxyamp+h2o", calcappredpot[amp + de == deoxyamp + h2o, 2, pH, is]},
{"adp+2e=deoxyadp+h2o", calcappredpot[adp + de == deoxyadp + h2o, 2, pH, is]},
{"atp+2e=deoxyatp+h2o", calcappredpot[atp + de == deoxyatp + h2o, 2, pH, is]},
{"(1/2)*n2oaq+e=(1/2)*n2aq+(1/2)*h2o",
 calcappredpot[(1 / 2) * n2oaq + de == (1 / 2) * n2aq + (1 / 2) * h2o, 1, pH, is]},
{"nitrite+e=noaq+h2o", calcappredpot[nitrite + de == noaq + h2o, 1, pH, is]},
{"glucose+2e=gluconolactone",
 calcappredpot[glucose + de == gluconolactone, 2, pH, is]},
{"nitrite+6e=ammonia+2*h2o", calcappredpot[nitrite + de == ammonia + 2 * h2o,
   6, pH, is]}, {"glucose6phos+2e=gluconolactone6phos",
 calcappredpot[glucose6phos + de == gluconolactone6phos, 2, pH, is]},
{"sorbitol+2e=fructose", calcappredpot[sorbitol + de == fructose, 2, pH, is]},
{"sorbitol6phos+2e=fructose6phos",
 calcappredpot[sorbitol6phos + de == fructose6phos, 2, pH, is]},
{"iditol+2e=sorbose", calcappredpot[iditol + de == sorbose, 2, pH, is]},
{"mannitol1phos+2e=fructose6phos",
 calcappredpot[mannitol1phos + de == fructose6phos, 2, pH, is]},
{"xylitol+2e=xylulose", calcappredpot[xylitol + de == xylulose, 2, pH, is]},
{"ribitol+2e=ribulose", calcappredpot[ribitol + de == ribulose, 2, pH, is]},
{"butanoln+2e=butanal", calcappredpot[butanoln + de == butanal, 2, pH, is]},
{"glycerol+2e=dihydroxyacetone", calcappredpot[
   glycerol + de == dihydroxyacetone, 2, pH, is]}, {"sulfite+4e=sulfurcr+3*h2o",
 calcappredpot[sulfite + de == sulfurcr + 3 * h2o, 4, pH, is]},
{"sulfite+8e=h2saq+3*h2o", calcappredpot[sulfite + de == h2saq + 3 * h2o, 8, pH, is]},
{"sulfite+6e=sulfurcr+3*h2o",
 calcappredpot[sulfite + de == sulfurcr + 3 * h2o, 6, pH, is]}};
```

The program **redpotsinorder** is used to calculate separate tables at 298.15 K, ionic strength 0.25 M, and pHs 6, 7, and 8.

```
In[21]:= halfrxdatapH6is25 = halfrxdata /. pH → 6 /. is → .25;
```

Table 8.1 Standard apparent reduction potentials $E'°$ in volts at 298.15 K, pH 6, and 0.25 M ionic strength

```
In[22]:= PaddedForm[redpotsinorder[halfrxdatapH6is25], {5, 4}]

Out[22]//PaddedForm=
      (1/2)*n2oaq+e=(1/2)*n2aq+(1/2)*h2o              1.3089
      o2aq+4e=2h2o                                    0.9083
```

```
o2g+4e=2h2o                                               0.8658
nitrite+e=noaq+h2o                                        0.4924
nitrate+2e=nitrite+h2o                                    0.4692
nitrite+6e=ammonia+2*h2o                                  0.4163
o2aq+2e=h2o2aq                                            0.4162
glucose+2e=gluconolactone                                0.2991
glucose6phos+2e=gluconolactone6phos                      0.2708
cytochromecox+e=cytochromecred                           0.2121
sorbitol+2e=fructose                                     0.2035
sorbitol6phos+2e=fructose6phos                           0.1943
iditol+2e=sorbose                                        0.1935
mannitol1phos+2e=fructose6phos                           0.1908
xylitol+2e=xylulose                                      0.1667
ribitol+2e=ribulose                                      0.1620
butanoln+2e=butanal                                      0.1459
methanol+2e=methaneaq+h2o                                0.1352
glycerol+2e=dihydroxyacetone                             0.1225
ribose5phos+2e=deoxyribose5phos+h2o                      0.0996
amp+2e=deoxyamp+h2o                                      0.0996
adenosine+2e=deoxyadenosine+h2o                          0.0996
ribose+2e=deoxyribose+h2o                                0.0996
atp+2e=deoxyatp+h2o                                      0.0996
adp+2e=deoxyadp+h2o                                      0.0996
ribose1phos+2e=deoxyribose1phos+h2o                      0.0996
atp+2e=deoxyatp+h2o                                      0.0996
fumarate+2e=succinate                                    0.0969
glycolate+ammonia+2e=glycine+h2o                         0.0270
sulfite+4e=sulfurcr+3*h2o                                0.0187
ethanol+2e=ethaneaq+h2o                                  0.0127
sulfite+6e=sulfurcr+3*h2o                                0.0124
glutathioneox+coA+2e=coAglutathione+glutathionered      -0.0007
glyoxylate+2e=glycolate                                 -0.0402
sulfite+8e=h2saq+3*h2o                                  -0.0449
ketoglutarate+ammonia+2e=glutamate+h2o                 -0.0592
pyruvate+ammonia+2e=alanine+h2o                        -0.0784
oxaloacetate+2e=malate                                 -0.1069
pyruvate+2e=lactate                                    -0.1330
acetaldehyde+2e=ethanol                                -0.1424
fadox+2e=fadred                                        -0.1619
fmnox+2e=fmnred                                        -0.1619
retinal+2e=retinol                                     -0.2187
fructose+2e=mannitolD                                  -0.2229
acetone+2e=propanol2                                   -0.2311
thioredoxinox+2e=thioredoxinred                        -0.2317
glutathioneox+2e=2*glutathionered                      -0.2320
n2aq+8e=2ammonia+h2aq                                  -0.2452
n2g+8e=2ammonia+h2g                                    -0.2466
co2tot+pyruvate+2e=malate+h2o                          -0.2567
nadpox+2e=nadpred                                      -0.2843
ketoglotarate+co2tot+2e=citrateiso+h2o                 -0.2852
nadox+2e=nadred                                        -0.2865
cystineL+2e=2cysteineL                                 -0.3055
citrate+coA+2e=malate+acetylcoA+h2o                    -0.3119
oxalylcoA+2e=glyoxylate+coA                            -0.3297
```

```
         acetylcoA+2e=acetaldehyde+coA                        -0.3297
         2e=h2g                                               -0.3634
         co2tot+2e=formate+h2o                                -0.3662
         co2g+2e=formate                                      -0.4022
         ferredoxinox+e=ferredoxinred                         -0.4030
         acetylcoA+co2tot+2e=pyruvate+coA+h2o                 -0.4244
         sulfate+2e=sulfite+h2o                               -0.4492
         2e=h2aq                                              -0.4546
         acetate+2e=acetaldehyde+h2o                          -0.5139
         methanol+nadpox+4e=methaneaq+nadpred+h2o             -1.0149
```

In[23]:= **halfrxdatapH7is25 = halfrxdata /. pH → 7 /. is → .25;**

Table 8.2 Standard apparent reduction potentials $E'°$ in volts at 298.15 K, pH 7, and 0.25 M ionic strength

In[24]:= **PaddedForm[redpotsinorder[halfrxdatapH7is25], {5, 4}]**

Out[24]//PaddedForm=
```
         (1/2)*n2oaq+e=(1/2)*n2aq+(1/2)*h2o                   1.2497
         o2aq+4e=2h2o                                         0.8491
         o2g+4e=2h2o                                          0.8066
         nitrate+2e=nitrite+h2o                               0.4100
         nitrite+e=noaq+h2o                                   0.3741
         glucose+2e=gluconolactone                            0.3583
         o2aq+2e=h2o2aq                                       0.3570
         nitrite+6e=ammonia+2*h2o                             0.3375
         glucose6phos+2e=gluconolactone6phos                  0.3300
         sorbitol+2e=fructose                                 0.2626
         sorbitol6phos+2e=fructose6phos                       0.2549
         iditol+2e=sorbose                                    0.2526
         mannitol1phos+2e=fructose6phos                       0.2522
         xylitol+2e=xylulose                                  0.2259
         ribitol+2e=ribulose                                  0.2211
         cytochromecox+e=cytochromecred                       0.2121
         butanoln+2e=butanal                                  0.2050
         glycerol+2e=dihydroxyacetone                         0.1816
         methanol+2e=methaneaq+h2o                            0.0761
         ribose5phos+2e=deoxyribose5phos+h2o                  0.0404
         adenosine+2e=deoxyadenosine+h2o                      0.0404
         ribose+2e=deoxyribose+h2o                            0.0404
         amp+2e=deoxyamp+h2o                                  0.0404
         atp+2e=deoxyatp+h2o                                  0.0404
         ribose1phos+2e=deoxyribose1phos+h2o                  0.0404
         atp+2e=deoxyatp+h2o                                  0.0404
         adp+2e=deoxyadp+h2o                                  0.0404
         fumarate+2e=succinate                                0.0366
         glycolate+ammonia+2e=glycine+h2o                     0.0270
         glutathioneox+coA+2e=coAglutathione+glutathionered   0.0002
         sulfite+6e=sulfurcr+3*h2o                           -0.0409
         ethanol+2e=ethaneaq+h2o                             -0.0465
         sulfite+4e=sulfurcr+3*h2o                           -0.0614
         sulfite+8e=h2saq+3*h2o                              -0.0968
         glyoxylate+2e=glycolate                             -0.0994
         ketoglutarate+ammonia+2e=glutamate+h2o              -0.1184
         pyruvate+ammonia+2e=alanine+h2o                     -0.1376
```

oxaloacetate+2e=malate	-0.1666
pyruvate+2e=lactate	-0.1922
acetaldehyde+2e=ethanol	-0.2015
fadox+2e=fadred	-0.2210
fmnox+2e=fmnred	-0.2210
retinal+2e=retinol	-0.2779
fructose+2e=mannitolD	-0.2821
glutathioneox+2e=2*glutathionered	-0.2876
thioredoxinox+2e=thioredoxinred	-0.2888
acetone+2e=propanol2	-0.2902
co2tot+pyruvate+2e=malate+h2o	-0.3077
nadpox+2e=nadpred	-0.3139
nadox+2e=nadred	-0.3160
n2aq+8e=2ammonia+h2aq	-0.3191
n2g+8e=2ammonia+h2g	-0.3205
ketoglotarate+co2tot+2e=citrateiso+h2o	-0.3391
cystineL+2e=2cysteineL	-0.3628
oxalylcoA+2e=glyoxylate+coA	-0.3880
acetylcoA+2e=acetaldehyde+coA	-0.3880
citrate+coA+2e=malate+acetylcoA+h2o	-0.3986
ferredoxinox+e=ferredoxinred	-0.4030
co2tot+2e=formate+h2o	-0.4166
2e=h2g	-0.4225
co2g+2e=formate	-0.4318
acetylcoA+co2tot+2e=pyruvate+coA+h2o	-0.4740
2e=h2aq	-0.5137
sulfate+2e=sulfite+h2o	-0.5256
acetate+2e=acetaldehyde+h2o	-0.6023
methanol+nadpox+4e=methaneaq+nadpred+h2o	-1.0445

In[25]:= **halfrxdatapH8is25 = halfrxdata /. pH → 8 /. is → .25;**

Table 8.3 Standard apparent reduction potentials $E'°$ in volts at 298.15 K, pH 8, and 0.25 M ionic strength

In[26]:= **PaddedForm[redpotsinorder[halfrxdatapH8is25], {5, 4}]**

Out[26]//PaddedForm=

(1/2)*n2oaq+e=(1/2)*n2aq+(1/2)*h2o	1.1906
o2aq+4e=2h2o	0.7900
o2g+4e=2h2o	0.7475
glucose+2e=gluconolactone	0.4174
glucose6phos+2e=gluconolactone6phos	0.3891
nitrate+2e=nitrite+h2o	0.3508
sorbitol+2e=fructose	0.3218
sorbitol6phos+2e=fructose6phos	0.3143
iditol+2e=sorbose	0.3118
mannitol1phos+2e=fructose6phos	0.3118
o2aq+2e=h2o2aq	0.2979
xylitol+2e=xylulose	0.2851
ribitol+2e=ribulose	0.2803
butanoln+2e=butanal	0.2642
nitrite+6e=ammonia+2*h2o	0.2588
nitrite+e=noaq+h2o	0.2558
glycerol+2e=dihydroxyacetone	0.2408
cytochromecox+e=cytochromecred	0.2121

glycolate+ammonia+2e=glycine+h2o	0.0264
methanol+2e=methaneaq+h2o	0.0169
glutathioneox+coA+2e=coAglutathione+glutathionered	0.0043
ribose5phos+2e=deoxyribose5phos+h2o	−0.0187
adp+2e=deoxyadp+h2o	−0.0187
amp+2e=deoxyamp+h2o	−0.0187
adenosine+2e=deoxyadenosine+h2o	−0.0187
ribose+2e=deoxyribose+h2o	−0.0187
ribose1phos+2e=deoxyribose1phos+h2o	−0.0187
atp+2e=deoxyatp+h2o	−0.0187
atp+2e=deoxyatp+h2o	−0.0187
fumarate+2e=succinate	−0.0227
sulfite+6e=sulfurcr+3*h2o	−0.0987
ethanol+2e=ethaneaq+h2o	−0.1057
sulfite+4e=sulfurcr+3*h2o	−0.1480
sulfite+8e=h2saq+3*h2o	−0.1487
glyoxylate+2e=glycolate	−0.1586
ketoglutarate+ammonia+2e=glutamate+h2o	−0.1782
pyruvate+ammonia+2e=alanine+h2o	−0.1974
oxaloacetate+2e=malate	−0.2258
pyruvate+2e=lactate	−0.2513
acetaldehyde+2e=ethanol	−0.2607
fadox+2e=fadred	−0.2802
fmnox+2e=fmnred	−0.2802
glutathioneox+2e=2*glutathionered	−0.3254
thioredoxinox+2e=thioredoxinred	−0.3318
retinal+2e=retinol	−0.3370
fructose+2e=mannitolD	−0.3412
nadpox+2e=nadpred	−0.3435
nadox+2e=nadred	−0.3456
acetone+2e=propanol2	−0.3494
co2tot+pyruvate+2e=malate+h2o	−0.3658
n2aq+8e=2ammonia+h2aq	−0.3927
n2g+8e=2ammonia+h2g	−0.3941
ketoglotarate+co2tot+2e=citrateiso+h2o	−0.3976
ferredoxinox+e=ferredoxinred	−0.4030
cystineL+2e=2cysteineL	−0.4088
oxalylcoA+2e=glyoxylate+coA	−0.4406
acetylcoA+2e=acetaldehyde+coA	−0.4406
co2g+2e=formate	−0.4613
co2tot+2e=formate+h2o	−0.4747
2e=h2g	−0.4817
citrate+coA+2e=malate+acetylcoA+h2o	−0.4936
acetylcoA+co2tot+2e=pyruvate+coA+h2o	−0.5255
2e=h2aq	−0.5729
sulfate+2e=sulfite+h2o	−0.5891
acetate+2e=acetaldehyde+h2o	−0.6911
methanol+nadpox+4e=methaneaq+nadpred+h2o	−1.0740

To obtain an alphabetical list, **Reverse** is removed from **redpotsinorder** and the first column is selected by replacing 2 with 1 in the last line of the program.

```
In[27]:= redpotsinalphaborder[evaldata_] :=
         Module[{}, (*This program sorts the apparent reduction potentials in alphabetical
           order and produces a table.  evaldata is in the following form: evaldata=
           {{"nadox+2e=nadred",calcappredpot[nadox+de==nadred,2,pH,is]},
              {"co2tot+pyruvate+2e=malate+h2o",
               calcappredpot[co2tot+pyruvate+de==malate+h2o,2,pH,is]},
              {"n2aq+8e=2ammonia+h2aq",calcappredpot[n2aq+de==2*ammonia+h2aq,
                8,pH,is]}}/.pH->7/.is→.25*)
          TableForm[evaldata[[Ordering[Transpose[evaldata][[1]]]]]]]
```

Table 8.4 *E'*° for half reactions in alphabetical order at 298.15 K, pH 7, and ionic strength 0.25 M

```
In[28]:= PaddedForm[redpotsinalphaborder[halfrxdatapH7is25], {5, 4}]

Out[28]//PaddedForm=
      (1/2)*n2oaq+e=(1/2)*n2aq+(1/2)*h2o                    1.2497
      2e=h2aq                                              -0.5137
      2e=h2g                                               -0.4225
      acetaldehyde+2e=ethanol                              -0.2015
      acetate+2e=acetaldehyde+h2o                          -0.6023
      acetone+2e=propanol2                                 -0.2902
      acetylcoA+2e=acetaldehyde+coA                        -0.3880
      acetylcoA+co2tot+2e=pyruvate+coA+h2o                 -0.4740
      adenosine+2e=deoxyadenosine+h2o                       0.0404
      adp+2e=deoxyadp+h2o                                   0.0404
      amp+2e=deoxyamp+h2o                                   0.0404
      atp+2e=deoxyatp+h2o                                   0.0404
      atp+2e=deoxyatp+h2o                                   0.0404
      butanoln+2e=butanal                                   0.2050
      citrate+coA+2e=malate+acetylcoA+h2o                  -0.3986
      co2g+2e=formate                                      -0.4318
      co2tot+2e=formate+h2o                                -0.4166
      co2tot+pyruvate+2e=malate+h2o                        -0.3077
      cystineL+2e=2cysteineL                               -0.3628
      cytochromecox+e=cytochromecred                        0.2121
      ethanol+2e=ethaneaq+h2o                              -0.0465
      fadox+2e=fadred                                      -0.2210
      ferredoxinox+e=ferredoxinred                         -0.4030
      fmnox+2e=fmnred                                      -0.2210
      fructose+2e=mannitolD                                -0.2821
      fumarate+2e=succinate                                 0.0366
      glucose+2e=gluconolactone                             0.3583
      glucose6phos+2e=gluconolactone6phos                   0.3300
      glutathioneox+2e=2*glutathionered                    -0.2876
      glutathioneox+coA+2e=coAglutathione+glutathionered    0.0002
      glycerol+2e=dihydroxyacetone                          0.1816
      glycolate+ammonia+2e=glycine+h2o                      0.0270
      glyoxylate+2e=glycolate                              -0.0994
      iditol+2e=sorbose                                     0.2526
      ketoglotarate+co2tot+2e=citrateiso+h2o               -0.3391
      ketoglutarate+ammonia+2e=glutamate+h2o               -0.1184
      mannitol1phos+2e=fructose6phos                        0.2522
      methanol+2e=methaneaq+h2o                             0.0761
      methanol+nadpox+4e=methaneaq+nadpred+h2o             -1.0445
      n2aq+8e=2ammonia+h2aq                                -0.3191
      n2g+8e=2ammonia+h2g                                  -0.3205
      nadox+2e=nadred                                      -0.3160
      nadpox+2e=nadpred                                    -0.3139
      nitrate+2e=nitrite+h2o                                0.4100
      nitrite+6e=ammonia+2*h2o                              0.3375
      nitrite+e=noaq+h2o                                    0.3741
      o2aq+2e=h2o2aq                                        0.3570
      o2aq+4e=2h2o                                          0.8491
      o2g+4e=2h2o                                           0.8066
```

```
oxaloacetate+2e=malate                          -0.1666
oxalylcoA+2e=glyoxylate+coA                     -0.3880
pyruvate+2e=lactate                             -0.1922
pyruvate+ammonia+2e=alanine+h2o                 -0.1376
retinal+2e=retinol                              -0.2779
ribitol+2e=ribulose                              0.2211
ribose1phos+2e=deoxyribose1phos+h2o              0.0404
ribose+2e=deoxyribose+h2o                        0.0404
ribose5phos+2e=deoxyribose5phos+h2o              0.0404
sorbitol+2e=fructose                             0.2626
sorbitol6phos+2e=fructose6phos                   0.2549
sulfate+2e=sulfite+h2o                          -0.5256
sulfite+4e=sulfurcr+3*h2o                       -0.0614
sulfite+6e=sulfurcr+3*h2o                       -0.0409
sulfite+8e=h2saq+3*h2o                          -0.0968
thioredoxinox+2e=thioredoxinred                 -0.2888
xylitol+2e=xylulose                              0.2259
```

Note that in all cases where the deoxy derivative is on the right, the pH dependence of the standard transformed reduction potential of the half reaction is the same because the pKs of reactants and products are the same.

The recognition that a given enzyme-catalyzed reaction is a redox reaction is important from a mechanistic standpoint in that the mechanism of the catalysis may involve separate catalysis of the two half reactions with a redox site in the enzyme facilitating the transfer of electrons between the two half reactions; in other words, atoms do not have to be transfered between the two half reactions.

8.3 Calculation of Effects of pH and Ionic Strength on $E'°$

A more detailed view of the effects of pH and ionic strength on $E'°$ can be obtained by constructing tables for half reactions. The following calculations provide values of $E'°$ in volts at pHs 5, 6, 7, 8, and 9 at ionic strengths of 0, 0.10, and 0.25 M. Values at zero ionic strength are shown in the first row in each table. Tables are given here for only a dozen half reactions, but these calculations can be made for all the half reactions in Tables 8.1 to 8.4.

```
In[29]:= nadoxentry = Join[{{"nadox+2e=nadred"}},
            Transpose[calcappredpot[nadox + de == nadred, 2, {5, 6, 7, 8, 9}, {0, .1, .25}]]];

In[30]:= PaddedForm[TableForm[nadoxentry], {5, 4}]

Out[30]//PaddedForm=
      nadox+2e=nadred
        -0.2653         -0.2949         -0.3244         -0.3540         -0.3836
        -0.2589         -0.2885         -0.3181         -0.3477         -0.3773
        -0.2569         -0.2865         -0.3160         -0.3456         -0.3752

In[31]:= pyruvateentry = Join[{{"pyruvate+2e=lactate"}},
            Transpose[calcappredpot[pyruvate + de == lactate, 2, {5, 6, 7, 8, 9}, {0, .1, .25}]]];

In[32]:= PaddedForm[TableForm[pyruvateentry], {5, 4}]

Out[32]//PaddedForm=
      pyruvate+2e=lactate
        -0.0655             -0.1246         -0.1838         -0.2429         -0.3021
        -0.0718             -0.1310         -0.1901         -0.2493         -0.3084
        -0.0738             -0.1330         -0.1922         -0.2513         -0.3105
```

```
In[33]:= ketoglutarateentry = Join[
             {{"ketoglutarate+co2tot+2e=", "citrateiso", "+h2o"}}, Transpose[calcappredpot[
                ketoglutarate + co2tot + de == citrateiso + h2o, 2, {5, 6, 7, 8, 9}, {0, .1, .25}]]];
```

```
In[34]:= PaddedForm[TableForm[ketoglutarateentry], {5, 4}]
```

Out[34]//PaddedForm=

ketoglutarate+co2tot+2e=	citrateiso	+h2o		
-0.2221	-0.2862	-0.3463	-0.4057	-0.4654
-0.2304	-0.2864	-0.3412	-0.3996	-0.4600
-0.2321	-0.2852	-0.3391	-0.3976	-0.4585

```
In[35]:= PaddedForm[TableForm[
             Join[nadoxentry, pyruvateentry, ketoglutarateentry], TableSpacing → {1, .5}], {5, 4}]
```

Out[35]//PaddedForm=

nadox+2e=nadred				
-0.2653	-0.2949	-0.3244	-0.3540	-0.3836
-0.2589	-0.2885	-0.3181	-0.3477	-0.3773
-0.2569	-0.2865	-0.3160	-0.3456	-0.3752
pyruvate+2e=lactate				
-0.0655	-0.1246	-0.1838	-0.2429	-0.3021
-0.0718	-0.1310	-0.1901	-0.2493	-0.3084
-0.0738	-0.1330	-0.1922	-0.2513	-0.3105
ketoglutarate+co2tot+2e=	citrateiso	+h2o		
-0.2221	-0.2862	-0.3463	-0.4057	-0.4654
-0.2304	-0.2864	-0.3412	-0.3996	-0.4600
-0.2321	-0.2852	-0.3391	-0.3976	-0.4585

In this trial table part of the reaction involving ketoglutarate has been put into the second and third columns to prevent the columns for pHs 6, 7, 8, and 9 from being pushed off the page to the right. Now this trial table is extended to a dozen entries.

```
In[36]:= acetylcoAentry = Join[{{"acetylcoA+2e=", "acetaldehyde+coA"}}, Transpose[
                calcappredpot[acetylcoA + de == acetaldehyde + coA, 2, {5, 6, 7, 8, 9}, {0, .1, .25}]]];
```

```
In[37]:= hydrogenentry = Join[{{"2e=h2aq"}},
             Transpose[calcappredpot[de == h2aq, 2, {5, 6, 7, 8, 9}, {0, .1, .25}]]];
```

```
In[38]:= glutathioneentry = Join[
             {{"glutathioneox+coA+2e=", "coAglutathione", "+glutathionered", "            "}},
             Transpose[calcappredpot[glutathioneox + coA + de == coAglutathione + glutathionered,
                2, {5, 6, 7, 8, 9}, {0, .1, .25}]]];
```

```
In[39]:= atpentry = Join[{{"atp+2e=deoxyatp+h2o"}},
             Transpose[calcappredpot[atp + de == deoxyatp + h2o, 2, {5, 6, 7, 8, 9}, {0, .1, .25}]]];
```

```
In[40]:= nitrateentry = Join[{{"nitrate+2e=nitrite+h2o"}},
             Transpose[calcappredpot[nitrate + de == nitrite + h2o, 2, {5, 6, 7, 8, 9}, {0, .1, .25}]]];
```

```
In[41]:= cystineentry = Join[{{"cystineL+2e=2*cysteineL"}}, Transpose[
                calcappredpot[cystineL + de == 2 * cysteineL, 2, {5, 6, 7, 8, 9}, {0, .1, .25}]]];
```

```
In[42]:= pyrammentry =
             Join[{{"pyruvate+ammonia+2e=", "alanine+h2o"}}, Transpose[calcappredpot[
                pyruvate + ammonia + de == alanine + h2o, 2, {5, 6, 7, 8, 9}, {0, .1, .25}]]];
```

```
In[43]:=  oxygenentry = Join[{{"o2aq+4e=2*h2o"}},
              Transpose[calcappredpot[o2aq + de == 2 * h2o, 4, {5, 6, 7, 8, 9}, {0, .1, .25}]]];

In[44]:=  nitrogenentry = Join[{{"n2aq+8e=2*ammonia+h2aq"}}, Transpose[
              calcappredpot[n2aq + de == 2 * ammonia + h2aq, 8, {5, 6, 7, 8, 9}, {0, .1, .25}]]];
```

Table 8.5 Standard apparent reduction potentials at 298.15 K, pHs 5, 6, 7, 8, and 9 and ionic strengths 0, 0.10, and 0.25 M

```
In[45]:= PaddedForm[
            TableForm[Join[nadoxentry, pyruvateentry, ketoglutarateentry, acetylcoAentry,
                hydrogenentry, glutathioneentry, atpentry, nitrateentry, cystineentry, pyrammentry,
                oxygenentry, nitrogenentry], TableSpacing → {.5, .5, .5, .5, .5}], {5, 4}]

Out[45]//PaddedForm=
    nadox+2e=nadred
        -0.2653             -0.2949         -0.3244         -0.3540         -0.3836
        -0.2589             -0.2885         -0.3181         -0.3477         -0.3773
        -0.2569             -0.2865         -0.3160         -0.3456         -0.3752
    pyruvate+2e=lactate
        -0.0655             -0.1246         -0.1838         -0.2429         -0.3021
        -0.0718             -0.1310         -0.1901         -0.2493         -0.3084
        -0.0738             -0.1330         -0.1922         -0.2513         -0.3105
    ketoglutarate+co2tot+2e= citrateiso      +h2o
        -0.2221             -0.2862         -0.3463         -0.4057         -0.4654
        -0.2304             -0.2864         -0.3412         -0.3996         -0.4600
        -0.2321             -0.2852         -0.3391         -0.3976         -0.4585
    acetylcoA+2e=            acetaldehyde+coA
        -0.2623             -0.3214         -0.3801         -0.4353         -0.4778
        -0.2686             -0.3277         -0.3861         -0.4394         -0.4788
        -0.2707             -0.3297         -0.3880         -0.4406         -0.4790
    2e=h2aq
        -0.3870             -0.4462         -0.5053         -0.5645         -0.6236
        -0.3933             -0.4525         -0.5117         -0.5708         -0.6300
        -0.3954             -0.4546         -0.5137         -0.5729         -0.6320
    glutathioneox+coA+2e=   coAglutathione   +glutathionered
         0.0076              0.0076          0.0076          0.0079          0.0085
         0.0012              0.0013          0.0019          0.0048          0.0080
        -0.0008             -0.0007          0.0002          0.0043          0.0079
    atp+2e=deoxyatp+h2o
         0.1671              0.1080          0.0488         -0.0104         -0.0695
         0.1608              0.1016          0.0425         -0.0167         -0.0759
         0.1587              0.0996          0.0404         -0.0187         -0.0779
    nitrate+2e=nitrite+h2o
         0.5369              0.4776          0.4184          0.3592          0.3001
         0.5305              0.4712          0.4121          0.3529          0.2937
         0.5284              0.4692          0.4100          0.3508          0.2917
    cystineL+2e=2*cysteineL
        -0.2381             -0.2972         -0.3554         -0.4066         -0.4324
        -0.2444             -0.3034         -0.3611         -0.4085         -0.4281
        -0.2465             -0.3055         -0.3628         -0.4088         -0.4265
    pyruvate+ammonia+2e=    alanine+h2o
        -0.0024             -0.0616         -0.1208         -0.1806         -0.2448
        -0.0151             -0.0743         -0.1335         -0.1933         -0.2575
        -0.0192             -0.0784         -0.1376         -0.1974         -0.2616
    o2aq+4e=2*h2o
         0.9758              0.9167          0.8575          0.7984          0.7392
         0.9695              0.9103          0.8512          0.7920          0.7329
         0.9675              0.9083          0.8491          0.7900          0.7308
    n2aq+8e=2*ammonia+h2aq
        -0.1628             -0.2368         -0.3107         -0.3843         -0.4558
        -0.1692             -0.2431         -0.3170         -0.3907         -0.4621
        -0.1712             -0.2452         -0.3191         -0.3927         -0.4642
```

We can look at other aspects of biochemical half reactions by making tables that show the change in the standard transformed Gibbs energy of a half reaction, the change in the binding of hydrogen ions in the half reaction, and the standard apparent reduction potential. The change in binding of hydrogen ions is of interest because it shows the rate of change of the other two properties with pH at specified temperature and ionic strength. The following calculations using **derivefnGNHEM-**

Frx are for the dozen reactions just considered at 0.25 M ionic strength, but these calculations an be made for all the half reactions in Tables 8.1-8.4.

```
In[46]:=  Off[General::"spell"];
          Off[General::"spell1"];
```

```
In[48]:=  derivefnGNHEMFrx[eq_, nu_] :=
            Module[{function, functionG, functionNH, functionEMF}, (*Derives the functions of pH
               and ionic strength that give the standard transformed Gibbs energy of reaction,
               change in number of hydrogen ions bound, and the standard apparent reduction
               potential at 298.15 K for a biochemical half reaction typed in as, for example,
               n2aq+de==2*ammonia+h2aq. nu is the number of formal electrons.  The
                  standard transformed Gibbs energy of reaction is in kJ mol^-1.*)
            function = Solve[eq, de];
            functionG = function[[1, 1, 2]];
            functionNH = (1 / (8.31451 * .29815 * Log[10])) * D[functionG, pH];
            functionEMF = -functionG / (nu * 96.485);
            {functionG, functionNH, functionEMF}]
```

```
In[49]:=  TableForm[derivefnGNHEMFrx[nadox + de == nadred, 2] /. pH → {5, 6, 7, 8, 9} /. is → .25,
            TableHeadings -> {{"ΔᵣG'°", "ΔᵣN_H", "E'°"}, {"pH 5", "pH 6", "pH 7", "pH 8", "pH 9"}}]
```

Out[49]//TableForm=

	pH 5	pH 6	pH 7	pH 8	pH 9
$\Delta_r G'°$	49.5709	55.2789	60.9869	66.695	72.403
$\Delta_r N_H$	1.				
E'°	-0.256884	-0.286464	-0.316044	-0.345624	-0.375204

When the change in binding of hydrogen ions is exactly one independent of pH, *Mathematica* just prints the 1. once.

```
In[50]:=  TableForm[derivefnGNHEMFrx[pyruvate + de == lactate, 2] /. pH → {5, 6, 7, 8, 9} /. is → .25,
            TableHeadings -> {{"ΔᵣG'°", "ΔᵣN_H", "E'°"}, {"pH 5", "pH 6", "pH 7", "pH 8", "pH 9"}}]
```

Out[50]//TableForm=

	pH 5	pH 6	pH 7	pH 8	pH 9
$\Delta_r G'°$	14.2498	25.6658	37.0819	48.498	59.9141
$\Delta_r N_H$	2.				
E'°	-0.0738445	-0.133004	-0.192164	-0.251324	-0.310484

```
In[51]:=  TableForm[derivefnGNHEMFrx[ketoglutarate + co2tot + de == citrateiso + h2o, 2] /.
            pH → {5, 6, 7, 8, 9} /. is → .25,
            TableHeadings -> {{"ΔᵣG'°", "ΔᵣN_H", "E'°"}, {"pH 5", "pH 6", "pH 7", "pH 8", "pH 9"}}]
```

Out[51]//TableForm=

	pH 5	pH 6	pH 7	pH 8	pH 9
$\Delta_r G'°$	44.7831	55.0333	65.4439	76.7224	88.4727
$\Delta_r N_H$	1.97582	1.72024	1.92803	2.00858	2.14667
E'°	-0.232073	-0.285191	-0.33914	-0.397587	-0.458479

```
In[52]:=  TableForm[
            derivefnGNHEMFrx[acetylcoA + de == acetaldehyde + coA, 2] /. pH → {5, 6, 7, 8, 9} /. is → .25,
            TableHeadings -> {{"ΔᵣG'°", "ΔᵣN_H", "E'°"}, {"pH 5", "pH 6", "pH 7", "pH 8", "pH 9"}}]
```

Out[52]//TableForm=

	pH 5	pH 6	pH 7	pH 8	pH 9
$\Delta_r G'°$	52.2278	63.626	74.8706	85.0179	92.4412
$\Delta_r N_H$	1.9992	1.99204	1.92574	1.55487	1.11084
E'°	-0.270652	-0.32972	-0.387991	-0.440576	-0.479044

In[53]:= **TableForm[derivefnGNHEMFrx[de == h2aq, 2] /. pH → {5, 6, 7, 8, 9} /. is → .25,**
        TableHeadings -> {{"Δ_rG'°", "Δ_rN$_H$", "E'°"}, {"pH 5", "pH 6", "pH 7", "pH 8", "pH 9"}}]

Out[53]//TableForm=

	pH 5	pH 6	pH 7	pH 8	pH 9
Δ_rG'°	76.2998	87.7158	99.1319	110.548	121.964
Δ_rN$_H$	2.				
E'°	-0.395397	-0.454557	-0.513717	-0.572877	-0.632037

In[54]:= **TableForm[derivefnGNHEMFrx[glutathioneox + coA + de == coAglutathione + glutathionered, 2] /**
        pH → {5, 6, 7, 8, 9} /. is → .25,
        TableHeadings -> {{"Δ_rG'°", "Δ_rN$_H$", "E'°"}, {"pH 5", "pH 6", "pH 7", "pH 8", "pH 9"}}]

Out[54]//TableForm=

	pH 5	pH 6	pH 7	pH 8	pH 9
Δ_rG'°	0.157159	0.137736	-0.035331	-0.828674	-1.53172
Δ_rN$_H$	-0.000880314	-0.00860976	-0.0699175	-0.182395	-0.0548073
E'°	-0.000814424	-0.000713769	0.000183091	0.00429432	0.00793762

In[55]:= **TableForm[derivefnGNHEMFrx[atp + de == deoxyatp + h2o, 2] /. pH → {5, 6, 7, 8, 9} /. is → .25,**
        TableHeadings -> {{"Δ_rG'°", "Δ_rN$_H$", "E'°"}, {"pH 5", "pH 6", "pH 7", "pH 8", "pH 9"}}]

Out[55]//TableForm=

	pH 5	pH 6	pH 7	pH 8	pH 9
Δ_rG'°	-30.6302	-19.2142	-7.79807	3.61802	15.0341
Δ_rN$_H$	2.	2.	2.	2.	2.
E'°	0.158731	0.0995707	0.0404108	-0.0187491	-0.077909

In[56]:= **TableForm[**
        derivefnGNHEMFrx[nitrate + de == nitrite + h2o, 2] /. pH → {5, 6, 7, 8, 9} /. is → .25,
        TableHeadings -> {{"Δ_rG'°", "Δ_rN$_H$", "E'°"}, {"pH 5", "pH 6", "pH 7", "pH 8", "pH 9"}}]

Out[56]//TableForm=

	pH 5	pH 6	pH 7	pH 8	pH 9
Δ_rG'°	-101.972	-90.5363	-79.1183	-67.702	-56.2859
Δ_rN$_H$	2.00862	2.00087	2.00009	2.00001	2.
E'°	0.528433	0.469173	0.410003	0.350842	0.291682

In[57]:= **TableForm[**
        derivefnGNHEMFrx[cystineL + de == 2 * cysteineL, 2] /. pH → {5, 6, 7, 8, 9} /. is → .25,
        TableHeadings -> {{"Δ_rG'°", "Δ_rN$_H$", "E'°"}, {"pH 5", "pH 6", "pH 7", "pH 8", "pH 9"}}]

Out[57]//TableForm=

	pH 5	pH 6	pH 7	pH 8	pH 9
Δ_rG'°	47.5658	58.9461	70.0179	78.8888	82.3104
Δ_rN$_H$	1.99839	1.98402	1.85092	1.10775	0.220883
E'°	-0.246493	-0.305468	-0.362843	-0.408814	-0.426545

In[58]:= **TableForm[derivefnGNHEMFrx[pyruvate + ammonia + de == alanine + h2o, 2] /.**
        pH → {5, 6, 7, 8, 9} /. is → .25,
        TableHeadings -> {{"Δ_rG'°", "Δ_rN$_H$", "E'°"}, {"pH 5", "pH 6", "pH 7", "pH 8", "pH 9"}}]

Out[58]//TableForm=

	pH 5	pH 6	pH 7	pH 8	pH 9
Δ_rG'°	3.70925	15.1266	26.5551	38.0924	50.4757
Δ_rN$_H$	2.00006	2.00056	2.00557	2.05302	2.35895
E'°	-0.0192219	-0.0783883	-0.137613	-0.197401	-0.261573

```
In[59]:= TableForm[derivefnGNHEMFrx[o2aq + de == 2 * h2o, 4] /. pH → {5, 6, 7, 8, 9} /. is → .25,
            TableHeadings -> {{"ΔrG'°", "ΔrNH", "E'°"}, {"pH 5", "pH 6", "pH 7", "pH 8", "pH 9"}}]
```

Out[59]//TableForm=

	pH 5	pH 6	pH 7	pH 8	pH 9
$\Delta_r G'°$	-373.38	-350.548	-327.716	-304.884	-282.052
$\Delta_r N_H$	4.				
$E'°$	0.967457	0.908297	0.849138	0.789978	0.730818

```
In[60]:= TableForm[
            derivefnGNHEMFrx[n2aq + de == 2 * ammonia + h2aq, 8] /. pH → {5, 6, 7, 8, 9} /. is → .25,
            TableHeadings -> {{"ΔrG'°", "ΔrNH", "E'°"}, {"pH 5", "pH 6", "pH 7", "pH 8", "pH 9"}}]
```

Out[60]//TableForm=

	pH 5	pH 6	pH 7	pH 8	pH 9
$\Delta_r G'°$	132.159	189.237	246.293	303.131	358.277
$\Delta_r N_H$	9.99989	9.99888	9.98886	9.89395	9.2821
$E'°$	-0.171217	-0.245164	-0.319082	-0.392717	-0.464161

Apparent equilibrium constants for biochemical reactions can readily be calculated from standard apparent reduction potentials as illustrated here for

nadox + formate + h2o = nadred + co2tot

at 298.15 K, pH 7, and ionic strength 0.25 M.   Equation 8.5 yields

$$K' = \exp(- \mid \nu_e \mid FE'°/RT) = \exp(-2 \times 96{,}485 \times (-0.316 + 0.417)/(8.31451 \times 0.29815) = 2.51 \times 10^3$$

This result can be obtained more easily by use of the program calckprime (8).

8.4 Plots of Standard Apparent Reduction Potentials Versus pH

Plots can be made of the pH dependencies of all the half reactions in Tables 8.1 to 8.4, but plots are given for only six of these half reactions.

```
In[61]:= plot1=Plot[calcappredpot[acetylcoA+co2tot+de==
            pyruvate+coA+h2o,2,pH,.25],{pH,5,9},AxesOrigin->{5,-.575},AxesLabel->{"pH",""},PlotLa
            bel->"E'°/V   acetylcoA+co2tot+2e=pyruvate+coA+h2o",DisplayFunction->Identity];
```

```
In[62]:= plot2=Plot[calcappredpot[glutathioneox+de==
            2*glutathionered,2,pH,.25],{pH,5,9},AxesOrigin->{5,-0.35},AxesLabel->{"pH",""},PlotLa
            bel->"E'°/V   glutathioneox+2e=2*glutathionered",DisplayFunction->Identity];
```

```
In[63]:= plot3=Plot[calcappredpot[cystineL+de==
            2*cysteineL,2,pH,.25],{pH,5,9},AxesOrigin->{5,-.425},AxesLabel->{"pH",""},PlotLabel->
            "E'°/V   cystineL+2e=2*cysteineL",DisplayFunction->Identity];
```

```
In[64]:= plot4=Plot[calcappredpot[co2tot+de==
            formate+h2o,2,pH,.25],{pH,5,9},AxesOrigin->{5,-.55},AxesLabel->{"pH",""},PlotLabel->"
            E'°/V   co2tot+2e=formate+h2o",DisplayFunction->Identity];
```

```
In[65]:= plot5=Plot[calcappredpot[pyruvate+ammonia+de==
            alanine+h2o,2,pH,.25],{pH,5,9},AxesOrigin->{5,-.26},AxesLabel->{"pH",""},PlotLabel->"
            E'°/V   pyruvate+ammonia+2e=alanine+h2o",DisplayFunction->Identity];
```

```
In[66]:= plot6=Plot[calcappredpot[thioredoxinox+de≈
         2*thioredoxinred,2,pH,.25],{pH,5,9},AxesOrigin->{5,-.7},AxesLabel->{"pH",""},PlotLabe
         l->"E'°/V  thioredoxinox+2e=2*thioredoxinred",DisplayFunction->Identity];

In[67]:= Show[GraphicsArray[{{plot1, plot2}, {plot3, plot4}, {plot5, plot6}}]];
```

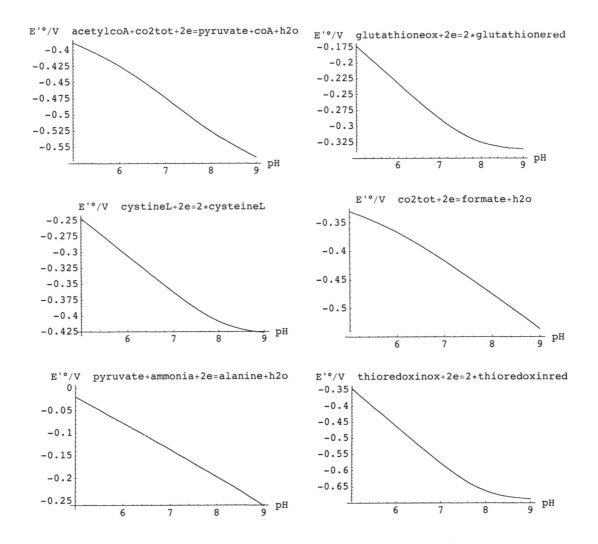

Figure 8.1 Standard apparent reduction potentials in volts of half reactions as functions of pH at 298.15 K and 0.25 M ionic strength.

8.5 Thermodynamic Properties of Enzyme-catalyzed Reactions as Functions of pH

The program **derivefnGNHKprimerx** (10) can be used to calculate tables showing the pH dependencies of the standard transformed Gibbs energy of reaction, the change in the binding of hydrogen ions in the reaction, and the apparent equilibrium constant as functions of pH and ionic strength at 298.15 K. These functions can be used to calculate tables or plots. The following tables are all for 0.25 M ionic strength. Note that $K' = 10^{\log K'}$.

```
In[68]:= derivefnGNHKprimerx[eq_] := Module[{function, functionG, functionNH, functionlogKprime},
         (*Derives the functions of pH and ionic strength that give the standard
           transformed Gibbs energy of reaction, change in number of hydrogen ions bound,
           and the base 10 log of the apparent equilibrium constant at 298.15 K
           for a biochemical reaction typed in as, for example, atp+h2o+de==
           adp+pi. The standard transformed Gibbs energy of reaction is in kJ mol^-1.*)
         function = Solve[eq, de];
         functionG = function[[1, 1, 2]];
         functionNH = (1 / (8.31451 * .29815 * Log[10])) * D[functionG, pH];
         functionlogKprime = -functionG / (8.31451 * .29815 * Log[10]);
         {functionG, functionNH, functionlogKprime}];
```

EC 1.1.1.1   Alcohol dehydrogenase

```
In[69]:= PaddedForm[TableForm[
         derivefnGNHKprimerx[ethanol + nadox + de == acetaldehyde + nadred] /. pH → {5, 6, 7, 8, 9} /.
         is → .25, TableHeadings ->
         {{"ΔrG'°", "ΔrNH", "logK'"}, {"pH 5", "pH 6", "pH 7", "pH 8", "pH 9"}}], {5, 2}]
```

Out[69]//PaddedForm=

	pH 5	pH 6	pH 7	pH 8	pH 9
ΔrG'°	33.51	27.80	22.10	16.39	10.68
ΔrNH	-1.00				
logK'	-5.87	-4.87	-3.87	-2.87	-1.87

EC 1.1.1.26   Glyoxylate reductase

```
In[70]:= PaddedForm[TableForm[
         derivefnGNHKprimerx[glycolate + nadox + de == glyoxylate + nadred] /. pH → {5, 6, 7, 8, 9} /.
         is → .25, TableHeadings ->
         {{"ΔrG'°", "ΔrNH", "logK'"}, {"pH 5", "pH 6", "pH 7", "pH 8", "pH 9"}}], {4, 2}]
```

Out[70]//PaddedForm=

	pH 5	pH 6	pH 7	pH 8	pH 9
ΔrG'°	53.22	47.51	41.81	36.10	30.39
ΔrNH	-1.00				
logK'	-9.32	-8.32	-7.32	-6.32	-5.32

EC 1.1.1.27   L-lactate dehydrogenase

```
In[71]:= PaddedForm[TableForm[
         derivefnGNHKprimerx[lactate + nadox + de == pyruvate + nadred] /. pH → {5, 6, 7, 8, 9} /.
         is → .25, TableHeadings ->
         {{"ΔrG'°", "ΔrNH", "logK'"}, {"pH 5", "pH 6", "pH 7", "pH 8", "pH 9"}}], {4, 2}]
```

```
Out[71]//PaddedForm=
                  pH 5        pH 6        pH 7        pH 8        pH 9
      Δ_rG'°       35.32       29.61       23.91       18.20       12.49
      Δ_rN_H       -1.00
      logK'        -6.19       -5.19       -4.19       -3.19       -2.19
```

EC 1.1.1.37   Malate dehydrogenase

```
In[72]:= PaddedForm[TableForm[
            derivefnGNHKprimerx[malate + acetylcoA + nadox + h2o + de == citrate + coA + nadred] /.
            pH → {5, 6, 7, 8, 9} /. is → .25, TableHeadings ->
            {{"Δ_rG'°", "Δ_rN_H", "logK'"}, {"pH 5", "pH 6", "pH 7", "pH 8", "pH 9"}}], {4, 2}]
```

```
Out[72]//PaddedForm=
                  pH 5        pH 6        pH 7        pH 8        pH 9
      Δ_rG'°        4.15       -4.90      -15.93      -28.55      -43.96
      Δ_rN_H       -1.42       -1.79       -2.05       -2.44       -2.89
      logK'        -0.73        0.86        2.79        5.00        7.70
```

EC 1.1.1.39   Malate dehydrogenase (decarboxylating)

```
In[73]:= PaddedForm[
            TableForm[derivefnGNHKprimerx[ malate + nadox + h2o + de == pyruvate + nadred + co2tot] /.
            pH → {5, 6, 7, 8, 9} /. is → .25, TableHeadings ->
            {{"Δ_rG'°", "Δ_rN_H", "logK'"}, {"pH 5", "pH 6", "pH 7", "pH 8", "pH 9"}}], {4, 2}]
```

```
Out[73]//PaddedForm=
                  pH 5        pH 6        pH 7        pH 8        pH 9
      Δ_rG'°        7.91        5.75        1.61       -3.89       -9.93
      Δ_rN_H       -0.41       -0.50       -0.90       -1.01       -1.15
      logK'        -1.39       -1.01       -0.28        0.68        1.74
```

EC 1.1.1.40   Malate dehydrogenase (decarboxylating) (NADP)

```
In[74]:= PaddedForm[
            TableForm[derivefnGNHKprimerx[malate + nadpox + h2o + de == pyruvate + nadpred + co2tot] /.
            pH → {5, 6, 7, 8, 9} /. is → .25, TableHeadings ->
            {{"Δ_rG'°", "Δ_rN_H", "logK'"}, {"pH 5", "pH 6", "pH 7", "pH 8", "pH 9"}}], {4, 2}]
```

```
Out[74]//PaddedForm=
                  pH 5        pH 6        pH 7        pH 8        pH 9
      Δ_rG'°        7.50        5.33        1.19       -4.31      -10.35
      Δ_rN_H       -0.41       -0.50       -0.90       -1.01       -1.15
      logK'        -1.31       -0.93       -0.21        0.76        1.81
```

EC 1.1.1.42   Isocitrate dehydrogenase (NADP)

```
In[75]:= PaddedForm[TableForm[
            derivefnGNHKprimerx[ citrateiso + nadpox + h2o + de == ketoglutarate + nadpred + co2tot] /.
            pH → {5, 6, 7, 8, 9} /. is → .25, TableHeadings ->
            {{"Δ_rG'°", "Δ_rN_H", "logK'"}, {"pH 5", "pH 6", "pH 7", "pH 8", "pH 9"}}], {4, 2}]
```

```
Out[75]//PaddedForm=
                  pH 5        pH 6        pH 7        pH 8        pH 9
      Δ_rG'°        4.37       -0.17       -4.88      -10.45      -16.49
      Δ_rN_H       -0.98       -0.72       -0.93       -1.01       -1.15
      logK'        -0.77        0.03        0.85        1.83        2.89
```

EC 1.1.1.67   Mannitol 2-dehydrogenase

In[76]:= **PaddedForm[TableForm[**
     derivefnGNHKprimerx[mannitolD + nadox + de == fructose + nadred] /. pH → {5, 6, 7, 8, 9} /.
     is → .25, TableHeadings ->
     {{"$\Delta_r G'°$", "$\Delta_r N_H$", "logK'"}, {"pH 5", "pH 6", "pH 7", "pH 8", "pH 9"}}], {4, 2}]

Out[76]//PaddedForm=

	pH 5	pH 6	pH 7	pH 8	pH 9
$\Delta_r G'°$	17.97	12.26	6.56	0.85	-4.86
$\Delta_r N_H$	-1.00				
logK'	-3.15	-2.15	-1.15	-0.15	0.85

EC 1.1.1.79   Glyoxylate reductase (NADP)

In[77]:= **PaddedForm[TableForm[derivefnGNHKprimerx[glycolate + nadpox + de == glyoxylate + nadpred] /.**
     pH → {5, 6, 7, 8, 9} /. is → .25,
     TableHeadings -> {{"$\Delta_r G'°$", "$\Delta_r N_H$", "logK'"}, {"pH 5", "pH 6", "pH 7", "pH 8", "pH 9"}},
     TableSpacing → {1, 2}], {4, 2}]

Out[77]//PaddedForm=

	pH 5	pH 6	pH 7	pH 8	pH 9
$\Delta_r G'°$	52.80	47.09	41.39	35.68	29.97
$\Delta_r N_H$	-1.00				
logK'	-9.25	-8.25	-7.25	-6.25	-5.25

EC 1.1.1.80   Isopropanol dehydrogenase (NADP)

In[78]:= **PaddedForm[TableForm[**
     derivefnGNHKprimerx[propanol2 + nadox + de == acetone + nadred] /. pH → {5, 6, 7, 8, 9} /.
     is → .25, TableHeadings ->
     {{"$\Delta_r G'°$", "$\Delta_r N_H$", "logK'"}, {"pH 5", "pH 6", "pH 7", "pH 8", "pH 9"}}], {4, 2}]

Out[78]//PaddedForm=

	pH 5	pH 6	pH 7	pH 8	pH 9
$\Delta_r G'°$	16.40	10.69	4.99	-0.72	-6.43
$\Delta_r N_H$	-1.00				
logK'	-2.87	-1.87	-0.87	0.13	1.13

EC 1.1.99.7   Lactate-malate dehydrogenase

In[79]:= **PaddedForm[**
     TableForm[derivefnGNHKprimerx[lactate + oxaloacetate + de == malate + pyruvate] /.
     pH → {5, 6, 7, 8, 9} /. is → .25, TableHeadings ->
     {{"$\Delta_r G'°$", "$\Delta_r N_H$", "logK'"}, {"pH 5", "pH 6", "pH 7", "pH 8", "pH 9"}}], {4, 2}]

Out[79]//PaddedForm=

	pH 5	pH 6	pH 7	pH 8	pH 9
$\Delta_r G'°$	-5.91	-5.04	-4.93	-4.92	-4.92
$\Delta_r N_H$	0.33	0.05	0.00	0.00	0.00
logK'	1.04	0.88	0.86	0.86	0.86

EC 1.1.2.3   L-lactate dehydrogenase (cytochrome)

```
In[80]:= PaddedForm[TableForm[
             derivefnGNHKprimerx[lactate + 2 * cytochromecox + de == pyruvate + 2 * cytochromecred] /.
             pH → {5, 6, 7, 8, 9} /. is → .25,
             TableHeadings -> {{"Δ_rG'°", "Δ_rN_H", "logK'"}, {"pH 5", "pH 6", "pH 7", "pH 8", "pH 9"}},
             TableSpacing → {1, 2}], {4, 2}]
```

Out[80]//PaddedForm=

	pH 5	pH 6	pH 7	pH 8	pH 9
$\Delta_r G'°$	-55.17	-66.59	-78.01	-89.42	-100.80
$\Delta_r N_H$	-2.00				
logK'	9.67	11.67	13.67	15.67	17.67

EC 1.2.1.2   Formate dehydrogenase

```
In[81]:= PaddedForm[TableForm[
             derivefnGNHKprimerx[formate + nadpox + h2o + de == co2tot + nadpred] /. pH → {5, 6, 7, 8, 9} /
             is → .25, TableHeadings ->
             {{"Δ_rG'°", "Δ_rN_H", "logK'"}, {"pH 5", "pH 6", "pH 7", "pH 8", "pH 9"}}], {4, 2}]
```

Out[81]//PaddedForm=

	pH 5	pH 6	pH 7	pH 8	pH 9
$\Delta_r G'°$	-14.50	-15.80	-19.83	-25.32	-31.36
$\Delta_r N_H$	-0.08	-0.45	-0.89	-1.01	-1.15
logK'	2.54	2.77	3.47	4.44	5.49

EC 1.2.1.3   Aldehyde dehydrogenase (NAD)

```
In[82]:= PaddedForm[
             TableForm[derivefnGNHKprimerx[acetaldehyde + nadox + h2o + de == acetate + nadred] /.
             pH → {5, 6, 7, 8, 9} /. is → .25, TableHeadings ->
             {{"Δ_rG'°", "Δ_rN_H", "logK'"}, {"pH 5", "pH 6", "pH 7", "pH 8", "pH 9"}}], {4, 2}]
```

Out[82]//PaddedForm=

	pH 5	pH 6	pH 7	pH 8	pH 9
$\Delta_r G'°$	-33.05	-43.90	-55.25	-66.66	-78.07
$\Delta_r N_H$	-1.77	-1.97	-2.00	-2.00	-2.00
logK'	5.79	7.69	9.68	11.68	13.68

EC 1.2.1.10   Aldehyde dehydrogenase (acetylating)

```
In[83]:= PaddedForm[
             TableForm[derivefnGNHKprimerx[acetaldehyde + coA + nadox + de == acetylcoA + nadred] /.
             pH → {5, 6, 7, 8, 9} /. is → .25, TableHeadings ->
             {{"Δ_rG'°", "Δ_rN_H", "logK'"}, {"pH 5", "pH 6", "pH 7", "pH 8", "pH 9"}}], {4, 2}]
```

Out[83]//PaddedForm=

	pH 5	pH 6	pH 7	pH 8	pH 9
$\Delta_r G'°$	-2.66	-8.35	-13.88	-18.32	-20.04
$\Delta_r N_H$	-1.00	-0.99	-0.93	-0.55	-0.11
logK'	0.47	1.46	2.43	3.21	3.51

EC 1.2.1.12   Glyceraldehyde-3-phosphate dehydrogenase (phosphorylating)

```
In[84]:= PaddedForm[
             TableForm[derivefnGNHKprimerx[glyceraldehydephos + pi + nadox + de == bpg + nadred] /.
             pH → {5, 6, 7, 8, 9} /. is → .25, TableHeadings ->
             {{"Δ_rG'°", "Δ_rN_H", "logK'"}, {"pH 5", "pH 6", "pH 7", "pH 8", "pH 9"}}], {4, 2}]
```

```
Out[84]//PaddedForm=
                  pH 5        pH 6        pH 7        pH 8        pH 9
        Δ_r G'°    14.10       6.72        1.22       -4.22       -9.88
        Δ_r N_H    -1.57      -1.07       -0.92       -0.98       -1.00
        logK'      -2.47      -1.18       -0.21        0.74        1.73
```

EC 1.2.1.17   Glyoxylate dehydrogenase (acylating)

```
In[85]:= PaddedForm[
            TableForm[derivefnGNHKprimerx[glyoxylate + coA + nadpox + de == oxalylcoA + nadpred] /.
                pH → {5, 6, 7, 8, 9} /. is → .25, TableHeadings ->
                {{"Δ_r G'°", "Δ_r N_H", "logK'"}, {"pH 5", "pH 6", "pH 7", "pH 8", "pH 9"}}], {4, 2}]

Out[85]//PaddedForm=
                  pH 5        pH 6        pH 7        pH 8        pH 9
        Δ_r G'°    -3.08      -8.77      -14.30      -18.74      -20.46
        Δ_r N_H    -1.00      -0.99       -0.93       -0.55       -0.11
        logK'       0.54       1.54        2.51        3.28        3.58
```

EC 1.2.1.43   Formate dehydrogenase (NADP)

```
In[86]:= PaddedForm[TableForm[
            derivefnGNHKprimerx[formate + nadpox + h2o + de == co2tot + nadpred] /. pH → {5, 6, 7, 8, 9} /
                is → .25, TableHeadings ->
                {{"Δ_r G'°", "Δ_r N_H", "logK'"}, {"pH 5", "pH 6", "pH 7", "pH 8", "pH 9"}}], {4, 2}]

Out[86]//PaddedForm=
                  pH 5        pH 6        pH 7        pH 8        pH 9
        Δ_r G'°   -14.50     -15.80      -19.83      -25.32      -31.36
        Δ_r N_H    -0.08      -0.45       -0.89       -1.01       -1.15
        logK'       2.54       2.77        3.47        4.44        5.49
```

EC 1.2.1.51   Pyruvate dehydrogenase (NADP)

```
In[87]:= PaddedForm[
            TableForm[derivefnGNHKprimerx[pyruvate + coA + nadpox + de == acetylcoA + co2tot + nadpred] /
                pH → {5, 6, 7, 8, 9} /. is → .25, TableHeadings ->
                {{"Δ_r G'°", "Δ_r N_H", "logK'"}, {"pH 5", "pH 6", "pH 7", "pH 8", "pH 9"}}], {4, 2}]

Out[87]//PaddedForm=
                  pH 5        pH 6        pH 7        pH 8        pH 9
        Δ_r G'°  -204.30    -194.10     -186.60     -179.40     -170.00
        Δ_r N_H     1.92       1.56        1.18        1.44        1.74
        logK'      35.78      34.01       32.68       31.42       29.78
```

EC 1.2.1.52   Oxoglutarate dehydrogenase (NADP)

```
In[88]:= PaddedForm[TableForm[
            derivefnGNHKprimerx[ketoglutarate + coA + nadpox + de == succinylcoA + co2tot + nadpred] /.
                pH → {5, 6, 7, 8, 9} /. is → .25, TableHeadings ->
                {{"Δ_r G'°", "Δ_r N_H", "logK'"}, {"pH 5", "pH 6", "pH 7", "pH 8", "pH 9"}}], {4, 2}]

Out[88]//PaddedForm=
                  pH 5        pH 6        pH 7        pH 8        pH 9
        Δ_r G'°  -211.10    -200.70     -193.20     -186.00     -176.60
        Δ_r N_H     2.00       1.56        1.18        1.44        1.74
        logK'      36.98      35.17       33.84       32.58       30.94
```

EC 1.2.7.1   Pyruvate synthase

```
In[89]:= PaddedForm[
         TableForm[derivefnGNHKprimerx[pyruvate + coA + ferredoxinox + de == acetylcoA + co2tot +
             ferredoxinred] /. pH → {5, 6, 7, 8, 9} /. is → .25, TableHeadings ->
         {{"ΔrG'°", "ΔrNH", "logK'"}, {"pH 5", "pH 6", "pH 7", "pH 8", "pH 9"}}], {4, 2}]
```

Out[89]//PaddedForm=

	pH 5	pH 6	pH 7	pH 8	pH 9
$\Delta_r G'°$	-214.50	-210.10	-208.20	-206.80	-203.10
$\Delta_r N_H$	0.92	0.56	0.18	0.44	0.74
logK'	37.58	36.81	36.48	36.22	35.58

EC 1.3.1.6   Fumarate reductase

```
In[90]:= PaddedForm[TableForm[
         derivefnGNHKprimerx[succinate + nadox + de == fumarate + nadred] /. pH → {5, 6, 7, 8, 9} /.
         is → .25, TableHeadings ->
         {{"ΔrG'°", "ΔrNH", "logK'"}, {"pH 5", "pH 6", "pH 7", "pH 8", "pH 9"}}], {4, 2}]
```

Out[90]//PaddedForm=

	pH 5	pH 6	pH 7	pH 8	pH 9
$\Delta_r G'°$	81.22	73.98	68.05	62.32	56.61
$\Delta_r N_H$	-1.50	-1.10	-1.01	-1.00	-1.00
logK'	-14.23	-12.96	-11.92	-10.92	-9.92

EC 1.4.1.1   Alanine dehydrogenase

```
In[91]:= PaddedForm[
         TableForm[derivefnGNHKprimerx[alanine + nadox + h2o + de == pyruvate + nadred + ammonia] /.
         pH → {5, 6, 7, 8, 9} /. is → .25, TableHeadings ->
         {{"ΔrG'°", "ΔrNH", "logK'"}, {"pH 5", "pH 6", "pH 7", "pH 8", "pH 9"}}], {4, 2}]
```

Out[91]//PaddedForm=

	pH 5	pH 6	pH 7	pH 8	pH 9
$\Delta_r G'°$	45.86	40.15	34.43	28.60	21.93
$\Delta_r N_H$	-1.00	-1.00	-1.01	-1.05	-1.36
logK'	-8.03	-7.03	-6.03	-5.01	-3.84

EC 1.4.1.2   Glutamate dehydrogenase

```
In[92]:= PaddedForm[TableForm[
         derivefnGNHKprimerx[ glutamate + nadox + h2o + de == ketoglutarate + nadred + ammonia] /.
         pH → {5, 6, 7, 8, 9} /. is → .25, TableHeadings ->
         {{"ΔrG'°", "ΔrNH", "logK'"}, {"pH 5", "pH 6", "pH 7", "pH 8", "pH 9"}}], {4, 2}]
```

Out[92]//PaddedForm=

	pH 5	pH 6	pH 7	pH 8	pH 9
$\Delta_r G'°$	49.57	43.86	38.14	32.31	25.64
$\Delta_r N_H$	-1.00	-1.00	-1.01	-1.05	-1.36
logK'	-8.68	-7.68	-6.68	-5.66	-4.49

EC 1.4.1.3   Glutamate dehydrogenase (NADP)

In[93]:= **PaddedForm[TableForm[**
derivefnGNHKprimerx[glutamate + nadpox + h2o + de == ketoglutarate + nadpred + ammonia] /.
pH → {5, 6, 7, 8, 9} /. is → .25, TableHeadings ->
{{"Δ$_r$G'°", "Δ$_r$N$_H$", "logK'"}, {"pH 5", "pH 6", "pH 7", "pH 8", "pH 9"}}], {4, 2}]

Out[93]//PaddedForm=

	pH 5	pH 6	pH 7	pH 8	pH 9
Δ$_r$G'°	49.15	43.44	37.72	31.89	25.22
Δ$_r$N$_H$	-1.00	-1.00	-1.01	-1.05	-1.36
logK'	-8.61	-7.61	-6.61	-5.59	-4.42

EC 1.4.1.10   Glycine dehydrogenase

In[94]:= **PaddedForm[**
TableForm[derivefnGNHKprimerx[glycine + nadox + h2o + de == glyoxylate + nadred + ammonia] /.
pH → {5, 6, 7, 8, 9} /. is → .25, TableHeadings ->
{{"Δ$_r$G'°", "Δ$_r$N$_H$", "logK'"}, {"pH 5", "pH 6", "pH 7", "pH 8", "pH 9"}}], {4, 2}]

Out[94]//PaddedForm=

	pH 5	pH 6	pH 7	pH 8	pH 9
Δ$_r$G'°	58.44	52.73	47.01	41.18	34.51
Δ$_r$N$_H$	-1.00	-1.00	-1.01	-1.05	-1.36
logK'	-10.24	-9.24	-8.24	-7.21	-6.05

EC 1.6.1.1   NAD transhydrogenase

In[95]:= **PaddedForm[TableForm[**
derivefnGNHKprimerx[nadox + nadpred + de == nadred + nadpox] /. pH → {5, 6, 7, 8, 9} /.
is → .25, TableHeadings ->
{{"Δ$_r$G'°", "Δ$_r$N$_H$", "K'"}, {"pH 5", "pH 6", "pH 7", "pH 8", "pH 9"}}], {4, 2}]

Out[95]//PaddedForm=

	pH 5	pH 6	pH 7	pH 8	pH 9
Δ$_r$G'°	0.42	0.42	0.42	0.42	0.42
Δ$_r$N$_H$	0.00				
K'	-0.07	-0.07	-0.07	-0.07	-0.07

EC 1.7.1.1   Nitrate reductase

In[96]:= **PaddedForm[TableForm[**
derivefnGNHKprimerx[nitrite + nadox + h2o + de == nitrate + nadred] /. pH → {5, 6, 7, 8, 9} /.
is → .25, TableHeadings ->
{{"Δ$_r$G'°", "Δ$_r$N$_H$", "logK'"}, {"pH 5", "pH 6", "pH 7", "pH 8", "pH 9"}}], {4, 2}]

Out[96]//PaddedForm=

	pH 5	pH 6	pH 7	pH 8	pH 9
Δ$_r$G'°	151.50	145.80	140.10	134.40	128.70
Δ$_r$N$_H$	-1.01	-1.00	-1.00	-1.00	-1.00
logK'	-26.55	-25.55	-24.55	-23.55	-22.55

EC 1.8.1.7   Glutathione-disulfide reductase

In[97]:= **PaddedForm[**
TableForm[derivefnGNHKprimerx[nadpox + 2 * glutathionered + de == nadpred + glutathioneox] /
pH → {5, 6, 7, 8, 9} /. is → .25, TableHeadings ->
{{"Δ$_r$G'°", "Δ$_r$N$_H$", "logK'"}, {"pH 5", "pH 6", "pH 7", "pH 8", "pH 9"}}], {4, 2}]

Out[97]//PaddedForm=

	pH 5	pH 6	pH 7	pH 8	pH 9
$\Delta_r G'^\circ$	15.72	10.09	5.07	3.49	7.17
$\Delta_r N_H$	-1.00	-0.97	-0.71	0.26	0.89
logK'	-2.75	-1.77	-0.89	-0.61	-1.26

EC 1.8.1.9   Thioredoxin-disulfide reductase

In[98]:= **PaddedForm[**
TableForm[derivefnGNHKprimerx[nadpox + thioredoxinred + de == nadpred + thioredoxinox] /.
pH → {5, 6, 7, 8, 9} /. is → .25, TableHeadings ->
{{"$\Delta_r G'^\circ$", "$\Delta_r N_H$", "logK'"}, {"pH 5", "pH 6", "pH 7", "pH 8", "pH 9"}}], {4, 2}]

Out[98]//PaddedForm=

	pH 5	pH 6	pH 7	pH 8	pH 9
$\Delta_r G'^\circ$	15.83	10.16	4.83	2.25	5.64
$\Delta_r N_H$	-1.00	-0.98	-0.83	0.11	0.88
logK'	-2.77	-1.78	-0.85	-0.39	-0.99

EC 1.8.4.3   Glutathione-coA-glutathione transhydrogenase

In[99]:= **PaddedForm[TableForm[**
derivefnGNHKprimerx[glutathioneox + coA + de == coAglutathione + glutathionered] /.
pH → {5, 6, 7, 8, 9} /. is → .25, TableHeadings ->
{{"$\Delta_r G'^\circ$", "$\Delta_r N_H$", "logK'"}, {"pH 5", "pH 6", "pH 7", "pH 8", "pH 9"}}], {4, 2}]

Out[99]//PaddedForm=

	pH 5	pH 6	pH 7	pH 8	pH 9
$\Delta_r G'^\circ$	0.16	0.14	-0.04	-0.83	-1.53
$\Delta_r N_H$	-0.00	-0.01	-0.07	-0.18	-0.05
logK'	-0.03	-0.02	0.01	0.15	0.27

EC 1.12.1.2   Hydrogen dehydrogenase

In[100]:=
PaddedForm[
TableForm[derivefnGNHKprimerx[nadox + h2aq + de == nadred] /. pH → {5, 6, 7, 8, 9} /. is → .25,
TableHeadings ->
{{"$\Delta_r G'^\circ$", "$\Delta_r N_H$", "logK'"}, {"pH 5", "pH 6", "pH 7", "pH 8", "pH 9"}}], {4, 2}]

Out[100]//PaddedForm=

	pH 5	pH 6	pH 7	pH 8	pH 9
$\Delta_r G'^\circ$	-26.73	-32.44	-38.14	-43.85	-49.56
$\Delta_r N_H$	-1.00				
logK'	4.68	5.68	6.68	7.68	8.68

EC 1.12.7.2   Ferredoxin hydrogenase

In[101]:=
PaddedForm[TableForm[
derivefnGNHKprimerx[2 * ferredoxinred + de == h2aq + 2 * ferredoxinox] /. pH → {5, 6, 7, 8, 9} /.
is → .25, TableHeadings ->
{{"$\Delta_r G'^\circ$", "$\Delta_r N_H$", "logK'"}, {"pH 5", "pH 6", "pH 7", "pH 8", "pH 9"}}], {4, 2}]

```
Out[101]//PaddedForm=
```

	pH 5	pH 6	pH 7	pH 8	pH 9
$\Delta_r G'^\circ$	-1.46	9.96	21.37	32.79	44.20
$\Delta_r N_H$	2.00				
logK'	0.26	-1.74	-3.74	-5.74	-7.74

EC 1.14.13.25 Methane monooxygenase

```
In[102]:=

    PaddedForm[
      TableForm[derivefnGNHKprimerx[methaneaq + o2aq + nadpox + de == methanol + nadpred + h2o] /.
        pH → {5, 6, 7, 8, 9} /. is → .25, TableHeadings ->
        {{"Δ_rG'°", "Δ_rN_H", "logK'"}, {"pH 5", "pH 6", "pH 7", "pH 8", "pH 9"}}], {4, 2}]
```

```
Out[102]//PaddedForm=
```

	pH 5	pH 6	pH 7	pH 8	pH 9
$\Delta_r G'^\circ$	-286.70	-269.60	-252.50	-235.30	-218.20
$\Delta_r N_H$	3.00				
logK'	50.23	47.23	44.23	41.23	38.23

EC 1.17.4.2   Ribonucleoside-triphosphate reductase

```
In[103]:=

    PaddedForm[
      TableForm[derivefnGNHKprimerx[atp + thioredoxinred + de == deoxyatp + h2o + thioredoxinox] /.
        pH → {5, 6, 7, 8, 9} /. is → .25, TableHeadings ->
        {{"Δ_rG'°", "Δ_rN_H", "logK'"}, {"pH 5", "pH 6", "pH 7", "pH 8", "pH 9"}}], {4, 2}]
```

```
Out[103]//PaddedForm=
```

	pH 5	pH 6	pH 7	pH 8	pH 9
$\Delta_r G'^\circ$	-63.96	-63.92	-63.53	-60.41	-51.31
$\Delta_r N_H$	0.00	0.02	0.17	1.11	1.88
logK'	11.20	11.20	11.13	10.58	8.99

EC 1.17.4.2   ATP reductase

```
In[104]:=

    PaddedForm[TableForm[
      derivefnGNHKprimerx[atp + nadred + de == deoxyatp + h2o + nadox] /. pH → {5, 6, 7, 8, 9} /.
      is → .25, TableHeadings ->
        {{"Δ_rG'°", "Δ_rN_H", "logK'"}, {"pH 5", "pH 6", "pH 7", "pH 8", "pH 9"}}], {4, 2}]
```

```
Out[104]//PaddedForm=
```

	pH 5	pH 6	pH 7	pH 8	pH 9
$\Delta_r G'^\circ$	-80.20	-74.49	-68.79	-63.08	-57.37
$\Delta_r N_H$	1.00	1.00	1.00	1.00	1.00
logK'	14.05	13.05	12.05	11.05	10.05

EC 1.18.1.2   Ferredoxin-NADP reductase

```
In[105]:=

    PaddedForm[
      TableForm[derivefnGNHKprimerx[ferredoxinred + nadpox + de == ferredoxinox + nadpred] /.
        pH → {5, 6, 7, 8, 9} /. is → .25, TableHeadings ->
        {{"Δ_rG'°", "Δ_rN_H", "logK'"}, {"pH 5", "pH 6", "pH 7", "pH 8", "pH 9"}}], {4, 2}]
```

Out[105]//PaddedForm=

	pH 5	pH 6	pH 7	pH 8	pH 9
$\Delta_r G'^\circ$	10.27	15.98	21.69	27.40	33.10
$\Delta_r N_H$	1.00				
logK'	-1.80	-2.80	-3.80	-4.80	-5.80

EC 1.18.6.1   Nitrogenase

In[106]:=

```
PaddedForm[TableForm[
   derivefnGNHKprimerx[n2aq + 8 * ferredoxinred + de == 2 * ammonia + h2aq + 8 * ferredoxinox] /.
     pH → {5, 6, 7, 8, 9} /. is → .25, TableHeadings ->
     {{"Δ_rG'°", "Δ_rN_H", "logK'"}, {"pH 5", "pH 6", "pH 7", "pH 8", "pH 9"}}], {4, 2}]
```

Out[106]//PaddedForm=

	pH 5	pH 6	pH 7	pH 8	pH 9
$\Delta_r G'^\circ$	-178.90	-121.80	-64.74	-7.91	47.24
$\Delta_r N_H$	10.00	10.00	9.99	9.89	9.28
logK'	31.34	21.34	11.34	1.39	-8.28

Note the extremely large change in the apparent equilibrium constant with pH. The apparent equilibrium constant decreases by a factor of 10 when the pH is increased 0.10. The hydrolysis of about 16 moles of atp to adp is coupled with this reaction, and so this greatly decreases the dependence of K' on pH (10).

The program **calctrGerx** makes it convenient to calculate $\Delta_r G'^\circ$ for an enzyme-catalyzed reaction at specified pHs and ionic strengths by simply typing the reactions using the names of the functions of pH and ionic strength. The following calculations are all at 298.15 K and 0.25 M ionic strength. Since it is $\Delta_r N_H$ that determines the change in $\Delta_r G'^\circ$ with pH, it is convenient to put these two types of plots together, The program **calctrGerx** is used to construct plots of $\Delta_r G'^\circ$ versus pH.

In[107]:=

```
calctrGerx[eq_, pHlist_, islist_] :=
 Module[{energy}, (*Calculates the standard transformed Gibbs
      energy of reaction in kJ mol^-1 at specified pHs and ionic
      strengths for a biochemical reaction typed in the form atp+h2o+de==
    adp+pi. The names of reactants call the appropriate functions of
      pH and ionic strength. pHlist and is list can be lists.*)
   energy = Solve[eq, de];
   energy[[1, 1, 2]] /. pH → pHlist /. is → islist]
```

The program **calcNHrx** is used to construct plots of $\Delta_r N_H$ versus pH.

In[108]:=

```
calcNHrx[eq_, pHlist_, islist_] := Module[{energy},(*This program calculates the
   change in the binding of hydrogen ions in a biochemical reaction at specified pHs and
   ionic strengths. The reaction is entered in the form atp+h2o+de==adp+pi.*)
      energy = Solve[eq, de];
       D[energy[[1,1,2]], pH]/(8.31451*0.29815*Log[10]) /.
        pH -> pHlist /. is -> islist]
```

In[109]:=

```
plot1 = Plot[calctrGerx[malate + acetylcoA + nadox + h2o + de == citrate + coA + nadred, pH, .25],
    {pH, 5, 9}, AxesLabel → {"pH", ""}, AxesOrigin → {5, -45}, PlotLabel ->
     "malate+acetylcoA+nadox+h2o=citrate+coA+nadred", DisplayFunction → Identity];
```

```
In[110]:=
        plot2 = Plot[calctrGerx[nadpox + thioredoxinred + de == thioredoxinox + nadpred, pH, .25],
           {pH, 5, 9}, AxesLabel → {"pH", ""}, AxesOrigin → {5, 0}, PlotLabel ->
           "nadpox+thioredoxinred=thioredoxinox+nadpred", DisplayFunction → Identity];

In[111]:=
        plot3 = Plot[calctrGerx[methaneaq + o2aq + nadpox + de == methanol + nadpred + h2o, pH, .25],
           {pH, 5, 9}, AxesLabel → {"pH", ""}, AxesOrigin → {5, -290}, PlotLabel ->
           "methaneaq+o2aq+nadpox+de==methanol+nadpred+h2o", DisplayFunction → Identity];

In[112]:=
        plot4 =
          Plot[calctrGerx[n2aq + 8 * ferredoxinred + de == 2 * ammonia + h2aq + 8 * ferredoxinox, pH, .25],
           {pH, 5, 9}, AxesLabel → {"pH", ""}, AxesOrigin → {5, -180},
           PlotLabel -> "    n2aq+8*ferredoxinred=2*ammonia+h2aq+8*ferredoxinox",
           DisplayFunction → Identity];

In[113]:=
        plot11 = Plot[
           Evaluate[calcNHrx[malate + acetylcoA + nadox + h2o + de == citrate + coA + nadred, pH, .25]],
           {pH, 5, 9}, AxesLabel → {"pH", ""}, AxesOrigin → {5, -2.9},
           PlotLabel -> "       ", DisplayFunction → Identity];

In[114]:=
        plot12 =
          Plot[Evaluate[calcNHrx[nadpox + thioredoxinred + de == thioredoxinox + nadpred, pH, .25]],
           {pH, 5, 9}, AxesLabel → {"pH", ""}, AxesOrigin → {5, -1},
           PlotLabel -> "       ", DisplayFunction → Identity];

In[115]:=
        plot13 =
          Plot[Evaluate[calcNHrx[methaneaq + o2aq + nadpox + de == methanol + nadpred + h2o, pH, .25]],
           {pH, 5, 9}, AxesLabel → {"pH", ""}, AxesOrigin → {5, 0},
           PlotRange → {0, 6}, PlotLabel -> "       ", DisplayFunction → Identity];

In[116]:=
        plot14 = Plot[Evaluate[
            calcNHrx[n2aq + 8 * ferredoxinred + de == 2 * ammonia + h2aq + 8 * ferredoxinox, pH, .25]],
           {pH, 5, 9}, AxesLabel → {"pH", ""}, AxesOrigin → {5, 9}, PlotRange → {9, 10},
           PlotLabel -> "       ", DisplayFunction → Identity];

In[117]:=
        Show[GraphicsArray[{{plot1, plot11}, {plot2, plot12}, {plot3, plot13}, {plot4, plot14}}],
           PlotLabel → "        ΔᵣG'°                                    ΔᵣNₕ              "];
```

$\triangle_r G'^\circ$ $\qquad\qquad\qquad\qquad$ $\triangle_r N_H$

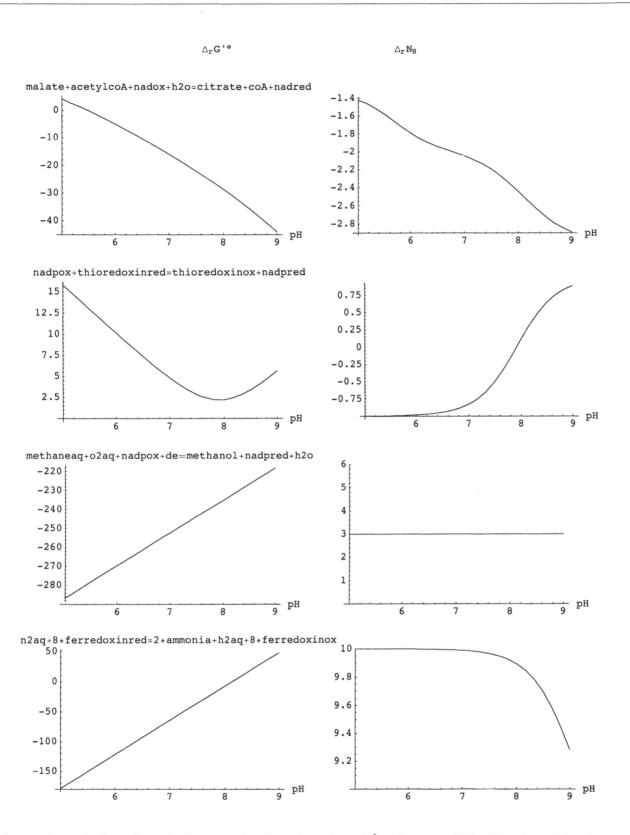

Figure 8.2 Standard transformed Gibbs energies of reaction in kJ mol^{-1} and $\triangle_r$ N_H at 298.15 K and 0.25 M ionic strength for four oxidoreductase reactions.

8.6 Apparent Equilibrium Constants of Oxidoreductase Reactions

Any half reaction in the preceding tables can be combined with any other half reaction to produce a biochemical reaction, but that does not mean that there are enzymes for all of these reactions. For some of the reactions for which there are enzymes, the apparent equilibrium constants are so large that they cannot be determined by direct experiment. Table 8.6 provides apparent equilibrium constants at 298.15 K, pH 7, and 0,25 M ionic strength for a number of oxidoreductase reactions.

The apparent equilibrium constants of the reactions in the preceding section are probably generally of greatest interest at pH 7 and 0.25 M ionic strength. The following table has been prepared to make it easier to compare these reactions.

```
In[118]:=
        reactiondata = {{"1.1.1.1   ethanol+nadox=acetaldehyde+nadred", "1.35×10⁻⁴"},
          {"1.1.1.26   glycolate+nadox=glyoxylate+nadred", "4.74×10⁻⁸"},
          {"1.1.1.27   lactate+nadox=pyruvate+nadred", "6.49×10⁻⁵"},
          {"1.1.1.37   malate+acetylcoA+nadox+h2o=citrate+coA+nadred", "6.02×10²"},
          {"1.1.1.39   malate+nadox+h2o=pyruvate+nadred+co2tot", "0.508"},
          {"1.1.1.40   malate+nadpox+h2o=pyruvate+nadpred+co2tot", "0.488"},
          {"1.1.1.42   citrateiso+nadpox+h2o=ketoglutarate+nadpred+co2tot", "5.80"},
          {"1.1.1.67    mannitolD+nadox=fructose+nadred", "0.071"},
          {"1.1.1.79    glycolate+nadpox=glyoxylate+nadpred", "4.55×10⁻⁸"},
          {"1.1.1.80   propanol2+nadox=acetone+nadred", "0.134"},
          {"1.1.99.7   lactate+oxaloacetate=malate+pyruvate", "7.51"},
          {"1.1.2.3   lactate+2*cytochromecox=pyruvate+2*cytochromecred", "4.63×10¹³"},
          {"1.2.1.2   formate+nadox+h2o=co2tot+nadred", "2.42×10³"},
          {"1.2.1.3   acetaldehyde+nadox+h2o=acetate+nadred", "4.77×10⁹"},
          {"1.2.1.10   acetaldehyde+coA+nadox=acetylcoA+nadred", "2.7×10²"},
          {"1.2.1.12   glyceraldehydephos+pi+nadox=bpg+nadred", "0.612"},
          {"1.2.1.17   glyoxylate+coA+nadpox=oxalylcoA+nadpred", "2.6×10²"},
          {"1.2.1.43   formate+nadpox+h2o=co2tot+nadpred", "2.42×10³"},
          {"1.2.1.51   pyruvate+coA+nadpox=acetylcoA+co2tot+nadpred", "3.9×10³²"},
          {"1.2.1.52   ketoglutarate+coA+nadpox=succinylcoA+co2tot+nadpred", "5.6×10³³"},
          {"1.2.7.1   pyruvate+coA+ferredoxinox=acetylcoA+co2tot+ferrdoxinred", "1.2×10³⁶"},
          {"1.3.1.6   succinate+nadox=fumarate+nadred", "1.20×10⁻¹²"},
          {"1.4.1.1   alanine+nadox+h2o=pyruvate+nadred+ammonia", "9.29×10⁻⁷"},
          {"1.4.1.2   glutamate+nadox+h2o=ketoglutarate+nadred+ammonia", "2.08×10⁻⁷"},
          {"1.4.1.3   glutamate+nadpox+h2o=ketoglutarate+nadpred+ammonia", "2.00×10⁻⁷"},
          {"1.4.1.10    glycine+nadox+h2o=glyoxylate+nadred+ammonia", "5.81×10⁻⁹"},
          {"1.6.1.1    nadox+nadpred=nadred+nadpox", "1.04"},
          {"1.7.1.1    nitrite+nadox+h2o=nitrate+nadred", "2.85×10⁻²⁵"},
          {"1.8.1.7    nadpox+2*glutathionered=nadpred+glutathioneox", "0.105"},
          {"1.8.1.9   nadpox+thioredoxinred=nadpred+thioredoxinox", "0.115"},
          {"1.8.4.3 glutathione+coA=coAglutathione+glutathionered", "1.0"},
          {"1.12.1.2   nadox+h2aq=nadred", "4.82×10⁶"},
          {"1.12.7.2   ferredoxinred+nadpox=ferredoxinox+nadpred", "1.80×10⁻⁴"},
          {"1.14.13.25   methaneaq+o2aq+nadpox=methanol+nadpred+h2o", "1.38×10⁴⁴"},
          {"1.17.4.2    atp+thioredoxinred=deoxyatp + h2o+thioredoxinox", "1.35×10¹¹"},
          {"1.17.4.2   atp+nadred+de=deoxyatp+h2o+nadox", "1.12×10¹²"},
          {"1.18.6.1   n2aq+8*ferredoxinred=2*ammonia+h2aq+8*ferredoxinox", "2.20×10¹¹"}};
```

Table 8.6 Apparent equilibrium constants of redox reactions at 298.15 K, pH 7, and ionic strength 0.25 M

```
In[119]:=
        TableForm[reactiondata]
```

Out[119]//TableForm=

1.1.1.1	ethanol+nadox=acetaldehyde+nadred	1.35×10^{-4}
1.1.1.26	glycolate+nadox=glyoxylate+nadred	4.74×10^{-8}
1.1.1.27	lactate+nadox=pyruvate+nadred	6.49×10^{-5}
1.1.1.37	malate+acetylcoA+nadox+h2o=citrate+coA+nadred	6.02×10^{2}
1.1.1.39	malate+nadox+h2o=pyruvate+nadred+co2tot	0.508
1.1.1.40	malate+nadpox+h2o=pyruvate+nadpred+co2tot	0.488
1.1.1.42	citrateiso+nadpox+h2o=ketoglutarate+nadpred+co2tot	5.80
1.1.1.67	mannitolD+nadox=fructose+nadred	0.071
1.1.1.79	glycolate+nadpox=glyoxylate+nadpred	4.55×10^{-8}
1.1.1.80	propanol2+nadox=acetone+nadred	0.134
1.1.99.7	lactate+oxaloacetate=malate+pyruvate	7.51
1.1.2.3	lactate+2*cytochromecox=pyruvate+2*cytochromecred	4.63×10^{13}
1.2.1.2	formate+nadox+h2o=co2tot+nadred	2.42×10^{3}
1.2.1.3	acetaldehyde+nadox+h2o=acetate+nadred	4.77×10^{9}
1.2.1.10	acetaldehyde+coA+nadox=acetylcoA+nadred	2.7×10^{2}
1.2.1.12	glyceraldehydephos+pi+nadox=bpg+nadred	0.612
1.2.1.17	glyoxylate+coA+nadpox=oxalylcoA+nadpred	2.6×10^{2}
1.2.1.43	formate+nadpox+h2o=co2tot+nadpred	2.42×10^{3}
1.2.1.51	pyruvate+coA+nadpox=acetylcoA+co2tot+nadpred	3.9×10^{32}
1.2.1.52	ketoglutarate+coA+nadpox=succinylcoA+co2tot+nadpred	5.6×10^{33}
1.2.7.1	pyruvate+coA+ferredoxinox=acetylcoA+co2tot+ferrdoxinred	1.2×10^{36}
1.3.1.6	succinate+nadox=fumarate+nadred	1.20×10^{-12}
1.4.1.1	alanine+nadox+h2o=pyruvate+nadred+ammonia	9.29×10^{-7}
1.4.1.2	glutamate+nadox+h2o=ketoglutarate+nadred+ammonia	2.08×10^{-7}
1.4.1.3	glutamate+nadpox+h2o=ketoglutarate+nadpred+ammonia	2.00×10^{-7}
1.4.1.10	glycine+nadox+h2o=glyoxylate+nadred+ammonia	5.81×10^{-9}
1.6.1.1	nadox+nadpred=nadred+nadpox	1.04
1.7.1.1	nitrite+nadox+h2o=nitrate+nadred	2.85×10^{-25}
1.8.1.7	nadpox+2*glutathionered=nadpred+glutathioneox	0.105
1.8.1.9	nadpox+thioredoxinred=nadpred+thioredoxinox	0.115
1.8.4.3	glutathione+coA=coAglutathione+glutathionered	1.0
1.12.1.2	nadox+h2aq=nadred	4.82×10^{6}
1.12.7.2	ferredoxinred+nadpox=ferredoxinox+nadpred	1.80×10^{-4}
1.14.13.25	methaneaq+o2aq+nadpox=methanol+nadpred+h2o	1.38×10^{44}
1.17.4.2	atp+thioredoxinred=deoxyatp + h2o+thioredoxinox	1.35×10^{11}
1.17.4.2	atp+nadred+de=deoxyatp+h2o+nadox	1.12×10^{12}
1.18.6.1	n2aq+8*ferredoxinred=2*ammonia+h2aq+8*ferredoxinox	2.20×10^{11}

8.7 Discussion

Standard apparent reduction potentials are useful because they provide a more global view of reactivity than the standard transformed Gibbs energies of formation of reactants from which they can be calculated. It is also useful to think mechanistically in terms of half reactions because half reactions are connected with other half reaction only through electron transfer.

The tables of standard apparent reduction potentials produced here are different from classical tables (2) of E'° in that they can be reproduced at other pHs in the range 5 to 9 and other ionic strengths in the range 0 to 0.25 M. For some half reactions, standard transformed reduction potenials can be calculated at temperatures other than 298.15 K. These values have all been calculated from the species database BasicBiochemData3 (9) that has been calculated from experimental measurements of apparent equilibrium constants and standard transformed enthalpies of enzyme-catalyzed reactions. The source of each of these values can be traced.

Apparent equilibrium constants at 298.15 K, pH 7, and ionic strength 0.25 M have been calculated for a number of redox reactions, but many more can be calculated from the table of half reactions and the 33 oxidoreductase reactions in Chapter 13.

The effects of temperature have been described for five oxidoreductase reactions in Chapter 4. Two programs have been developed to calculate the effects of temperature on the various standard transformed thermodynamic properties. The first program **derivetrGibbsT** derives the function of T, pH, and ionic strength that yields $\Delta_f G'^{\circ}$ for a biochemical reactant. The functions labelled with speciesnameGT have been given in Section 4.3 for about 60 reactants. The second program given in Chapter 4, **derivefnGHSNHrx** derives $\Delta_r G'^{\circ}$, $\Delta_r H'^{\circ}$, $\Delta_r S'^{\circ}$, and $\Delta_r N_H$ for reactions involving some of these 60 reactants. 2D and 3D plots have been presented for ethanol + nadox = acetaldehyde + nadred and formate + h2o + nadox = co2tot + nadred in Chapter 4. A recent article extends this treatment to more half reactions involving nitrogen and sulfur compounds (7).

References

1. W. M. Clark, Oxidation-Reduction Potentials of Organic Systems, Williams and Wilkins, Baltimore, 1961.

2. I. H. Segel, Biochemical Calculations, Wiley, Hoboken, NJ, 1976.

3. R. N. Goldberg, Y. B. Tewari, D. Bell, and K. Fasio, Thermodynamics of enzyme-catalyzed reactions: Part I. Oxidoreductases, J. Phys. Chem. Ref. Data 22, 515 (1993).

4. R. A. Alberty, Calculation of standard transformed formation properties of biochemical reactants and standard apparent reduction potentials, Arch. Biochem. Biophys. 358, 25-39 (1998).

5. R. A. Alberty, Standard Apparent Reduction Potentials for Biochemical Half Reactions as a Function of pH and Ionic Strength, Arch. Biochem. Biophys., 389, 94-109 (2001).

6. R. A. Alberty, Thermodynamics of Biochemical Reactions, Wiley, Hoboken, NJ, 2003.

7. R. A. Alberty, Standard apparent reduction potentials of biochemical half reactions and thermodynamic data on the species involved, Biophys. Chem., 111, 115-122 (2004).

8. R. A. Alberty, Thermodynamic properties of nucleotide reductase reactions, Biochemistry, 43, 9840-9845 (2004).

9. R. A. Alberty, BasicBiochemData3, 2005.

In[120]:=
    http : // library.wolfram.com / infocenter / MathSource / 5704

10. R. A. Alberty, Thermodynamics of the mechanism of the nitrogenase reaction, Biophysical Chemistry, 114, 115-120 (2005).

Chapter 9 Transferase Reactions (Class 2) at 298.15 K

9.1 Introduction

9.2 Apparent Equilibrium Constants of Transferase Reactions

9.3 Effect of pH on the Standard Transformed Gibbs Energy of Reaction

9.4 Discussion

References

9.1 Introduction

In the IUBMB classification, transferases are enzymes transferring a group, e. g. a methyl group or a glycyl group from a donor to an acceptor. Many, and perhaps all, transferase reactions can be considered to result from the coupling of two oxidoreductase reactions of two hydrolase reactions (1) In thinking about the mechanisms of transferase reactions it is important to understand that enzyme mechanisms must provide this coupling. We have seen an example of this in Section 7.4 where constraints in addition to atom balances are discussed. The fact that two reactions are coupled is very important in thermodynamics. It means that if the transformed thermodynamic properties of two oxidoreductase reactions or two hydrolase reactions are known, the thermodynamic properties of the transferase reaction can be calculated without making any thermodynamic measurements on the transferase reaction. That means that the tables in Chapters 8 and 10 can be used to calculate the standard transformed thermodynamic properties of many hydrolase reactions. In the next section the reactions that can be considered to be coupled are identified. When they are not identified, enzymes may not exist for both reactions or I have not been able to find suitable pairs of reactions.

Goldberg and Tewari (2) have surveyed and evaluated the thermodynamic data on transferase reactions.

9.2 Apparent Equilibrium Constants of Transferase Reactions

The program **derivefnGNHKprime** (3) is used to calculate tables showing the pH dependencies of the standard transformed Gibbs energy of reaction, the change in the number of hydrogen ions bound, and $\log K'$ at 298.15 K and 0.25 M ionic strength.. Note that each transferase reaction has been written in the direction in which it is spontaneous at 298.15 K, pH 7, and 0.25 M ionic strength.

```
In[2]:=  Off[General::"spell"];
         Off[General::"spell1"];

In[4]:=  << BiochemThermo`BasicBiochemData3`
```

```
In[5]:=  derivefnGNHKprimerx[eq_] :=
            Module[{function, functionG, functionNH, functionlogKprime},
              (*Derives the functions of pH and ionic strength that give the standard
                 transformed Gibbs energy of reaction, change in number of hydrogen ions
                 bound, and the base 10 log of the apparent equilibrium constant at 298.15
                 K for a biochemical reaction typed in as, for example, atp+h2o+de==adp+
                 pi. The standard transformed Gibbs energy of reaction is in kJ mol^-1.*)
              function = Solve[eq, de];
              functionG = function[[1, 1, 2]];
              functionNH = (1 / (8.31451 * .29815 * Log[10])) * D[functionG, pH];
              functionlogKprime = -functionG / (8.31451 * .29815 * Log[10]);
              {functionG, functionNH, functionlogKprime}];
```

EC 2.3.1.8 phosphate acetyltransferase can be considered to be 3.1.2.3-3.1.2.1.

```
In[6]:=  PaddedForm[TableForm[
            derivefnGNHKprimerx[acetylphos + coA + de == acetylcoA + pi] /. pH → {5, 6, 7, 8, 9} /. is → .2!
            TableHeadings ->
              {{"Δ_rG'°", "Δ_rN_H", "logK'"}, {"pH 5", "pH 6", "pH 7", "pH 8", "pH 9"}}], {4, 2}]
```

Out[6]//PaddedForm=

	pH 5	pH 6	pH 7	pH 8	pH 9
$\Delta_r G'°$	0.47	-1.05	-3.27	-5.70	-3.39
$\Delta_r N_H$	-0.42	-0.23	-0.55	-0.08	0.78
logK'	-0.08	0.18	0.57	1.00	0.59

EC 2.3.3.1 citrate (*Si*)-synthase

```
In[7]:=  PaddedForm[
            TableForm[derivefnGNHKprimerx[acetylcoA + h2o + oxaloacetate + de == citrate + coA] /.
              pH → {5, 6, 7, 8, 9} /. is → .25, TableHeadings ->
                {{"Δ_rG'°", "Δ_rN_H", "logK'"}, {"pH 5", "pH 6", "pH 7", "pH 8", "pH 9"}}], {4, 2}]
```

Out[7]//PaddedForm=

	pH 5	pH 6	pH 7	pH 8	pH 9
$\Delta_r G'°$	-37.08	-39.56	-44.77	-51.67	-61.36
$\Delta_r N_H$	-0.09	-0.74	-1.04	-1.44	-1.89
logK'	6.50	6.93	7.84	9.05	10.75

2.3.1.54 formate C-acetyltransferase

```
In[8]:=  PaddedForm[TableForm[
            derivefnGNHKprimerx[coA + pyruvate + de == acetylcoA + formate] /. pH → {5, 6, 7, 8, 9} /.
              is → .25, TableHeadings ->
                {{"Δ_rG'°", "Δ_rN_H", "logK'"}, {"pH 5", "pH 6", "pH 7", "pH 8", "pH 9"}}], {4, 2}]
```

Out[8]//PaddedForm=

	pH 5	pH 6	pH 7	pH 8	pH 9
$\Delta_r G'°$	-11.26	-11.24	-11.07	-9.80	-5.81
$\Delta_r N_H$	0.00	0.01	0.07	0.45	0.89
logK'	1.97	1.97	1.94	1.72	1.02

EC 2.4.1.7 sucrose phosphorylase can be considered to be 3.2.1.48-3.1.3.9

```
In[9]:=  PaddedForm[TableForm[
            derivefnGNHKprimerx[sucrose + pi + de == fructose + glucose1phos] /. pH → {5, 6, 7, 8, 9} /.
            is → .25, TableHeadings ->
            {{"ΔrG'°", "ΔrNH", "logK'"}, {"pH 5", "pH 6", "pH 7", "pH 8", "pH 9"}}], {4, 2}]
```

Out[9]//PaddedForm=

	pH 5	pH 6	pH 7	pH 8	pH 9
$\Delta_r G'^{\circ}$	-7.71	-8.90	-10.87	-11.49	-11.57
$\Delta_r N_H$	-0.08	-0.36	-0.23	-0.03	-0.00
logK'	1.35	1.56	1.90	2.01	2.03

2.4.1.8 maltose phosphorylase can be considered to be 3.2.1.20-3.1.3.9

```
In[10]:=  PaddedForm[TableForm[
            derivefnGNHKprimerx[maltose + pi + de == glucose + glucose6phos] /. pH → {5, 6, 7, 8, 9} /.
            is → .25, TableHeadings ->
            {{"ΔrG'°", "ΔrNH", "logK'"}, {"pH 5", "pH 6", "pH 7", "pH 8", "pH 9"}}], {4, 2}]
```

Out[10]//PaddedForm=

	pH 5	pH 6	pH 7	pH 8	pH 9
$\Delta_r G'^{\circ}$	-4.77	-6.17	-8.30	-8.96	-9.04
$\Delta_r N_H$	-0.10	-0.40	-0.24	-0.04	-0.00
logK'	0.84	1.08	1.45	1.57	1.58

2.6.1.1 aspartate transaminase

```
In[11]:=  PaddedForm[
            TableForm[derivefnGNHKprimerx[aspartate + ketoglutarate + de == oxaloacetate + glutamate] /
            pH → {5, 6, 7, 8, 9} /. is → .25, TableHeadings ->
            {{"ΔrG'°", "ΔrNH", "logK'"}, {"pH 5", "pH 6", "pH 7", "pH 8", "pH 9"}}], {4, 2}]
```

Out[11]//PaddedForm=

	pH 5	pH 6	pH 7	pH 8	pH 9
$\Delta_r G'^{\circ}$	-1.47	-1.47	-1.47	-1.47	-1.47
$\Delta_r N_H$	1.24×10^{-15}				
logK'	0.26	0.26	0.26	0.26	0.26

2.6.1.2 alanine transaminase can be considered to be 1.4.1.2-1.4.1.1

```
In[12]:=  PaddedForm[
            TableForm[derivefnGNHKprimerx[alanine + ketoglutarate + de == pyruvate + glutamate] /.
            pH → {5, 6, 7, 8, 9} /. is → .25, TableHeadings ->
            {{"ΔrG'°", "ΔrNH", "logK'"}, {"pH 5", "pH 6", "pH 7", "pH 8", "pH 9"}}], {4, 2}]
```

Out[12]//PaddedForm=

	pH 5	pH 6	pH 7	pH 8	pH 9
$\Delta_r G'^{\circ}$	-3.71	-3.71	-3.71	-3.71	-3.71
$\Delta_r N_H$	-1.24×10^{-15}				
logK'	0.65	0.65	0.65	0.65	0.65

2.6.1.4 glycine trans aminase can be considered to be 1.4.1.10-1.4.1.2

```
In[13]:=  PaddedForm[
            TableForm[derivefnGNHKprimerx[glyoxylate + glutamate + de == glycine + ketoglutarate] /.
            pH → {5, 6, 7, 8, 9} /. is → .25, TableHeadings ->
            {{"ΔrG'°", "ΔrNH", "logK'"}, {"pH 5", "pH 6", "pH 7", "pH 8", "pH 9"}}], {4, 2}]
```

Out[13]//PaddedForm=

	pH 5	pH 6	pH 7	pH 8	pH 9
$\Delta_r G'^\circ$	-8.87	-8.87	-8.87	-8.87	-8.87
$\Delta_r N_H$	0.00				
logK'	1.55	1.55	1.55	1.55	1.55

2.6.1.35 glycine-oxaloacetate transaminase

```
In[14]:=  PaddedForm[
             TableForm[derivefnGNHKprimerx[glyoxylate + aspartate + de == glycine + oxaloacetate] /.
             pH → {5, 6, 7, 8, 9} /. is → .25, TableHeadings ->
             {{"ΔrG'°", "ΔrNH", "logK'"}, {"pH 5", "pH 6", "pH 7", "pH 8", "pH 9"}}], {4, 2}]
```

Out[14]//PaddedForm=

	pH 5	pH 6	pH 7	pH 8	pH 9
$\Delta_r G'^\circ$	-10.34	-10.34	-10.34	-10.34	-10.34
$\Delta_r N_H$	6.22×10^{-16}				
logK'	1.81	1.81	1.81	1.81	1.81

2.7.1.2 hexokinase can be considered to be 3.6.1.3-3.1.3.9

```
In[15]:=  PaddedForm[TableForm[
             derivefnGNHKprimerx[atp + glucose + de == adp + glucose6phos] /. pH → {5, 6, 7, 8, 9} /.
             is → .25, TableHeadings ->
             {{"ΔrG'°", "ΔrNH", "logK'"}, {"pH 5", "pH 6", "pH 7", "pH 8", "pH 9"}}], {4, 2}]
```

Out[15]//PaddedForm=

	pH 5	pH 6	pH 7	pH 8	pH 9
$\Delta_r G'^\circ$	-17.41	-19.47	-24.42	-30.11	-35.82
$\Delta_r N_H$	-0.14	-0.65	-0.98	-1.00	-1.00
logK'	3.05	3.41	4.28	5.28	6.28

2.7.1.6 galactokinase

```
In[16]:=  PaddedForm[TableForm[
             derivefnGNHKprimerx[atp + galactose + de == adp + galactose1phos] /. pH → {5, 6, 7, 8, 9} /.
             is → .25, TableHeadings ->
             {{"ΔrG'°", "ΔrNH", "logK'"}, {"pH 5", "pH 6", "pH 7", "pH 8", "pH 9"}}], {4, 2}]
```

Out[16]//PaddedForm=

	pH 5	pH 6	pH 7	pH 8	pH 9
$\Delta_r G'^\circ$	-15.82	-18.65	-24.06	-29.83	-35.54
$\Delta_r N_H$	-0.23	-0.79	-1.01	-1.00	-1.00
logK'	2.77	3.27	4.22	5.23	6.23

2.7.1.23 NAD kinase

```
In[17]:=  PaddedForm[TableForm[
             derivefnGNHKprimerx[atp + nadox + de == adp + nadpox] /. pH → {5, 6, 7, 8, 9} /. is → .25,
             TableHeadings ->
             {{"ΔrG'°", "ΔrNH", "logK'"}, {"pH 5", "pH 6", "pH 7", "pH 8", "pH 9"}}], {4, 2}]
```

Out[17]//PaddedForm=

	pH 5	pH 6	pH 7	pH 8	pH 9
$\Delta_r G'^\circ$	-2.60	-8.51	-14.63	-20.47	-26.20
$\Delta_r N_H$	-1.02	-1.07	-1.05	-1.01	-1.00
logK'	0.45	1.49	2.56	3.59	4.59

2.7.1.40 pyruvate kinase NADH kinase can be considered to be 3.1.3.60-3.6.1.3

```
In[18]:= PaddedForm[TableForm[
           derivefnGNHKprimerx[adp + pep + de == atp + pyruvate] /. pH → {5, 6, 7, 8, 9} /. is → .25,
           TableHeadings ->
             {{"ΔrG'°", "ΔrNH", "logK'"}, {"pH 5", "pH 6", "pH 7", "pH 8", "pH 9"}}], {4, 2}]
```

Out[18]//PaddedForm=

	pH 5	pH 6	pH 7	pH 8	pH 9
ΔrG'°	-34.47	-33.11	-28.85	-23.29	-17.60
ΔrNH	0.08	0.48	0.93	0.99	1.00
logK'	6.04	5.80	5.05	4.08	3.08

2.7.1.86 NAD$_{red}$ kinase

```
In[19]:= PaddedForm[TableForm[
           derivefnGNHKprimerx[atp + nadred + de == adp + nadpred] /. pH → {5, 6, 7, 8, 9} /. is → .25,
           TableHeadings ->
             {{"ΔrG'°", "ΔrNH", "logK'"}, {"pH 5", "pH 6", "pH 7", "pH 8", "pH 9"}}], {4, 2}]
```

Out[19]//PaddedForm=

	pH 5	pH 6	pH 7	pH 8	pH 9
ΔrG'°	-3.01	-8.93	-15.05	-20.89	-26.62
ΔrNH	-1.02	-1.07	-1.05	-1.01	-1.00
logK'	0.53	1.56	2.64	3.66	4.66

2.7.1.90 diphosphate-fructose-6-phosphate 1-phosphotransferase

```
In[20]:= PaddedForm[TableForm[
           derivefnGNHKprimerx[ppi + fructose6phos + de == pi + fructose16phos] /. pH → {5, 6, 7, 8, 9}
             is → .25, TableHeadings ->
             {{"ΔrG'°", "ΔrNH", "logK'"}, {"pH 5", "pH 6", "pH 7", "pH 8", "pH 9"}}], {4, 2}]
```

Out[20]//PaddedForm=

	pH 5	pH 6	pH 7	pH 8	pH 9
ΔrG'°	-3.17	-6.19	-9.87	-14.11	-16.37
ΔrNH	-0.30	-0.63	-0.73	-0.64	-0.17
logK'	0.55	1.08	1.73	2.47	2.87

2.7.2.1 acetate kinase

```
In[21]:= PaddedForm[TableForm[
           derivefnGNHKprimerx[adp + acetylphos + de == atp + acetate] /. pH → {5, 6, 7, 8, 9} /. is → .2!
           TableHeadings ->
             {{"ΔrG'°", "ΔrNH", "logK'"}, {"pH 5", "pH 6", "pH 7", "pH 8", "pH 9"}}], {4, 2}]
```

Out[21]//PaddedForm=

	pH 5	pH 6	pH 7	pH 8	pH 9
ΔrG'°	2.64	-3.38	-8.60	-12.96	-14.72
ΔrNH	-1.16	-0.96	-0.88	-0.56	-0.12
logK'	-0.46	0.59	1.51	2.27	2.58

2.7.4.3 adenylate kinase can be considered to be 3.6.1.3-3.6.1.5

```
In[22]:= PaddedForm[
            TableForm[derivefnGNHKprimerx[atp + amp + de == 2 * adp] /. pH → {5, 6, 7, 8, 9} /. is → .25,
              TableHeadings ->
                {{"ΔᵣG'°", "ΔᵣNₕ", "logK'"}, {"pH 5", "pH 6", "pH 7", "pH 8", "pH 9"}}], {4, 2}]
```

Out[22]//PaddedForm=

	pH 5	pH 6	pH 7	pH 8	pH 9
$\Delta_r G'°$	-2.23	-2.13	-2.07	-2.09	-2.09
$\Delta_r N_H$	0.01	0.02	-0.00	-0.00	-0.00
logK'	0.39	0.37	0.36	0.37	0.37

2.7.9.1 pyruvate, phosphate dikinase can be considered to be 3.1.3.60-3.6.1.8

```
In[23]:= PaddedForm[TableForm[
            derivefnGNHKprimerx[amp + pep + ppi + de == atp + pyruvate + pi] /. pH → {5, 6, 7, 8, 9} /. is →
            TableHeadings ->
                {{"ΔᵣG'°", "ΔᵣNₕ", "logK'"}, {"pH 5", "pH 6", "pH 7", "pH 8", "pH 9"}}], {4, 2}]
```

Out[23]//PaddedForm=

	pH 5	pH 6	pH 7	pH 8	pH 9
$\Delta_r G'°$	-27.52	-24.47	-17.54	-10.36	-1.20
$\Delta_r N_H$	0.21	0.97	1.26	1.36	1.83
logK'	4.82	4.29	3.07	1.82	0.21

2.7.9.2 pyruvate, water dikinase can be considered to be 3..6.1.5-3.1.3.60

```
In[24]:= PaddedForm[TableForm[
            derivefnGNHKprimerx[atp + pyruvate + h2o + de == amp + pep + pi] /. pH → {5, 6, 7, 8, 9} /. is →
            TableHeadings ->
                {{"ΔᵣG'°", "ΔᵣNₕ", "logK'"}, {"pH 5", "pH 6", "pH 7", "pH 8", "pH 9"}}], {4, 2}]
```

Out[24]//PaddedForm=

	pH 5	pH 6	pH 7	pH 8	pH 9
$\Delta_r G'°$	4.14	2.02	-5.12	-15.70	-27.02
$\Delta_r N_H$	-0.13	-0.75	-1.67	-1.96	-2.00
logK'	-0.73	-0.35	0.90	2.75	4.73

A table is prepared to show the range of apparent equilibrium constants of transferase reactions at 298.15 K, pH 7, and ionic strength 0.25 M.

```
In[25]:= transfKprimedata = {{"acetylphos+coA=acetylcoA+pi", 3.75},
            {"acetylcoA+h2o+oxaloacetate=citrate+coA", 6.97 * 10^7},
            {"coA+pyruvate=acetylcoA+formate", 86.9},
            {"sucrose+pi=fructose+glucose1phos", 80.2},
            {"maltose+pi=glucose+glucose6phos", 28.5},
            {"aspartate+ketoglutarate=oxaloacetate+glutamate", 1.81},
         {"alanine+ketoglutarate=pyruvate+glutamate", 4.47},
         {"glyoxylate+glutamate=glycine+ketoglutarate", 35.8},
         {"glyoxylate+aspartate=glycine+oxaloacetate", 64.8},
         {"atp+glucose=adp+glucose6phos", 1.90 * 10^4},
         {"atp+galactose=adp+galactose1phos", 1.64 * 10^4},
         {"atp+nadox=adp+nadpox", 3.65},
         {"adp+pep=atp+pyruvate", 1.13 * 10^5},
         {"atp+nadred=adp+nadpred", 351},
         {"ppi+fructose6phos=pi+fructose16phos", 53.5},
         {"adp+acetylphos=atp+acetate", 32.1},
         {"atp+amp=2*adp", 2.31},
         {"amp+pep+ppi=atp+pyruvate+pi", 1.18 * 10^3},
         {"atp+pyruvate+h2o=amp+pep+pi", 7.88}};
```

Table 10.1 Apparent equilibrium constants of transferase reactions at 298.15 K, pH 7, and 0.25 M ionic strength

```
In[26]:= TableForm[transfKprimedata[[Ordering[Transpose[transfKprimedata][[2]]]]]]
```

```
Out[26]//TableForm=
```

aspartate+ketoglutarate=oxaloacetate+glutamate	1.81
atp+amp=2*adp	2.31
atp+nadox=adp+nadpox	3.65
acetylphos+coA=acetylcoA+pi	3.75
alanine+ketoglutarate=pyruvate+glutamate	4.47
atp+pyruvate+h2o=amp+pep+pi	7.88
maltose+pi=glucose+glucose6phos	28.5
adp+acetylphos=atp+acetate	32.1
glyoxylate+glutamate=glycine+ketoglutarate	35.8
ppi+fructose6phos=pi+fructose16phos	53.5
glyoxylate+aspartate=glycine+oxaloacetate	64.8
sucrose+pi=fructose+glucose1phos	80.2
coA+pyruvate=acetylcoA+formate	86.9
atp+nadred=adp+nadpred	351
amp+pep+ppi=atp+pyruvate+pi	1180.
atp+galactose=adp+galactose1phos	16400.
atp+glucose=adp+glucose6phos	19000.
adp+pep=atp+pyruvate	113000.
acetylcoA+h2o+oxaloacetate=citrate+coA	6.97×10^7

Note that these apparent equilibrium constants are the products of the apparent equilibrium constants of the reactions that are coupled. The standard transformed Gibbs energies of reaction and change in the binding of hydrogen ions in the reaction are sums of the properties of the reactions being coupled.

9.3 Effect of pH on the Standard Transformed Gibbs Energy of Reaction

The program **calctrGerx** makes it convenient to calculate $\Delta_r G'^\circ$ for an enzyme-catalyzed reaction at specified pHs and ionic strengths by simply typing the reactions using the names of the functions of pH and ionic strength. The following calculations on transferase reactions are all at 298.15 K and 0.25 M ionic strength. Since it is $\Delta_r N_H$ that determines the change in $\Delta_r G'^\circ$ with pH, it is convenient to put these two types of plots together. The program **calctrGerx** is used to construct plots of $\Delta_r G'^\circ$ versus pH.

```
In[27]:= calctrGerx[eq_, pHlist_, islist_] :=
            Module[{energy}, (*Calculates the standard transformed Gibbs
                energy of reaction in kJ mol^-1 at specified pHs and ionic
                strengths for a biochemical reaction typed in the form atp+h2o+de==
            adp+pi.  The names of reactants call the appropriate functions of
                pH and ionic strength.  pHlist and is list can be lists.*)
            energy = Solve[eq, de]; energy[[1, 1, 2]] /. pH → pHlist /. is → islist]
```

The program **calcNHrx** is used to construct plots of $\Delta_r N_H$ versus pH.

```
In[28]:= calcNHrx[eq_, pHlist_, islist_] := Module[{energy},(*This program calculates the
            change in the binding of hydrogen ions in a biochemical reaction at specified pHs
            and ionic strengths.  The reaction is entered in the form atp+h2o+de==adp+pi.*)
            energy = Solve[eq, de];
                D[energy[[1,1,2]], pH]/(8.31451*0.29815*Log[10]) /.
                pH -> pHlist /. is -> islist]
```

```
In[29]:= plot1 = Plot[calctrGerx[acetylcoA + h2o + oxaloacetate + de == citrate + coA, pH, .25],
            {pH, 5, 9}, AxesLabel → {"pH", "Δr G' °"}, PlotLabel ->
              "       acetylcoA+h2o+oxaloacet=citrate+coA", DisplayFunction → Identity];

In[30]:= plot2 = Plot[calctrGerx[coA + pyruvate + de == acetylcoA + formate, pH, .25], {pH, 5, 9},
            AxesLabel → {"pH", "Δr G' °"}, PlotLabel -> "      coA+pyruvate=acetylcoA+formate",
            AxesOrigin → {5, -12}, PlotRange → {-12, -6}, DisplayFunction → Identity];

In[31]:= plot3 = Plot[calctrGerx[atp + glucose + de == adp + glucose6phos, pH, .25],
            {pH, 5, 9}, AxesLabel → {"pH", "Δr G' °"},
            PlotLabel -> "       atp+glucose=adp+glucose6phos", DisplayFunction → Identity];

In[32]:= plot4 = Plot[calctrGerx[atp + nadox + de == adp + nadpox, pH, .25], {pH, 5, 9},
            AxesLabel → {"pH", "Δr G' °"}, PlotLabel -> "atp+nadox=adp+nadpox",
            AxesOrigin → {5, -25}, DisplayFunction → Identity];

In[33]:= plot5 = Plot[calctrGerx[amp + pep + ppi + de == atp + pyruvate + pi, pH, .25], {pH, 5, 9},
            AxesLabel → {"pH", "Δr G' °"}, PlotLabel -> "   amp+pep+ppi=atp+pyruvate+pi",
            AxesOrigin → {5, -28}, DisplayFunction → Identity];

In[34]:= plot6 = Plot[calctrGerx[atp + pyruvate + h2o + de == amp + pep + pi, pH, .25], {pH, 5, 9},
            AxesLabel → {"pH", "Δr G' °"}, PlotLabel -> "   atp+pyruvate+h2o=amp+pep+pi",
            AxesOrigin → {5, -28}, DisplayFunction → Identity];

In[35]:= plot11 = Plot[Evaluate[calcNHrx[acetylcoA + h2o + oxaloacetate + de == citrate + coA, pH, .25]
            {pH, 5, 9}, AxesLabel → {"pH", "Δr NH"}, AxesOrigin → {5, -2}, PlotRange → {-2, 0},
            PlotLabel -> "       acetylcoA+h2o+oxaloacet=citrate+coA",
            DisplayFunction → Identity];

In[36]:= plot12 = Plot[Evaluate[calcNHrx[coA + pyruvate + de == acetylcoA + formate, pH, .25]],
            {pH, 5, 9}, AxesLabel → {"pH", "Δr NH"}, AxesOrigin → {5, 0},
            PlotLabel -> "       coA+pyruvate=acetylcoA+formate", DisplayFunction → Identity];

In[37]:= plot13 = Plot[Evaluate[calcNHrx[atp + glucose + de == adp + glucose6phos, pH, .25]],
            {pH, 5, 9}, AxesLabel → {"pH", "Δr NH"}, AxesOrigin → {5, -1}, PlotRange → {-1, 0},
            PlotLabel -> "       atp+glucose=adp+glucose6phos", DisplayFunction → Identity];

In[38]:= plot14 = Plot[Evaluate[calcNHrx[atp + nadox + de == adp + nadpox, pH, .25]], {pH, 5, 9},
            AxesLabel → {"pH", "Δr NH"}, AxesOrigin → {5, -2}, PlotRange → {-2, 0},
            PlotLabel -> "       atp+nadox=adp+nadpox", DisplayFunction → Identity];

In[39]:= plot15 = Plot[Evaluate[calcNHrx[amp + pep + ppi + de == atp + pyruvate + pi, pH, .25]],
            {pH, 5, 9}, AxesLabel → {"pH", "Δr NH"}, AxesOrigin → {5, 0}, PlotRange → {0, 2},
            PlotLabel -> "       amp+pep+ppi=atp+pyruvate+pi", DisplayFunction → Identity];

In[40]:= plot16 = Plot[Evaluate[calcNHrx[atp + pyruvate + h2o + de == amp + pep + pi, pH, .25]],
            {pH, 5, 9}, AxesLabel → {"pH", "Δr NH"}, AxesOrigin → {5, -2}, PlotRange → {-2, 0},
            PlotLabel -> "       atp+pyruvate+h2o=amp+pep+pi", DisplayFunction → Identity];

In[41]:= Show[GraphicsArray[{{plot1, plot11}, {plot2, plot12}, {plot3, plot13}}]];
```

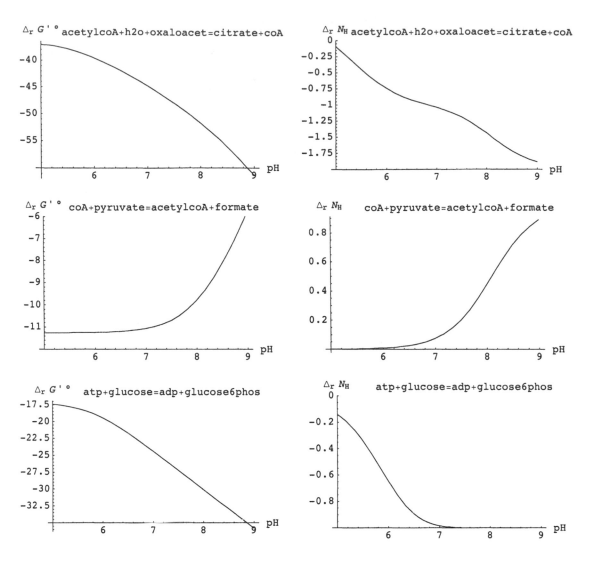

Figure 9.1 Standard transformed Gibbs energies of reaction and changes in the binding of hydrogen ions of three transferase reactions as a function of pH at 298.15 K and 0.25 M ionic strength.

```
In[42]:= Show[GraphicsArray[{{plot4, plot14}, {plot5, plot15}, {plot6, plot16}}]];
```

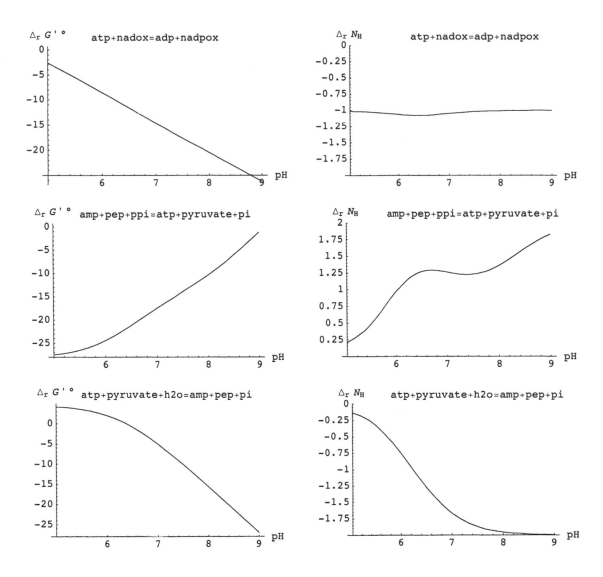

Figure 9.2 Standard transformed Gibbs energies of reaction and changes in the binding of hydrogen ions of three more transferase reactions as a function of pH at 298.15 K and 0.25 M ionic strength.

9.4 Discussion

The standard transformed thermodynamic properties of more transferase reactions are calculated at 289.15 K in Chapter 12. Since $\Delta_f H°$ are known for all species of 20 transferase reactions, these reactions are discussed in Chapter 13.

References

1. R. A. Alberty, Thermodynamic properties of oxidoreductase, transferase, hydrolase, and ligase reactions, Arch. Biochem. Biophys. 435, 363-368 (2005).

2. R. N. Goldberg and Y. B. Tewari, Thermodynamics of Enzyme-catalyzed Reactions: Part 2 Transferases, J. Phys. Chem. 23, 547-617 (1994).

3. R. A. Alberty, Thermodynamics of the mechanism of the nitrogenase reaction, Biophysical Chemistry, 114, 115-120 (2005).

Chapter 10 Hydrolase Reactions (Class 3) at 298.15 K

10.1 Introduction

Hydrolases are enzymes that catalyze the hydrolytic cleavage of C-C, C-N, C-O, and some other bonds, including phosphate ester bonds and phosphoanhydride bonds. Goldberg and Tewari (1,2) have surveyed and evaluated experimental data on the apparent equilibrium constants and heats of hydrolase reactions. The equilibrium constants of hydrolase reactions are greater than unity but range over 32 orders of magnitude. Since apparent equilibrium constants are known for various types of hydrolase reactions, good estimates can be made of apparent equilibrium constants for hydrolase reactions that have not been studied. Hydrolase reactions do not involve coupling.

10.2 Apparent Equilibrium Constants of Hydrolase Reactions

The program **derivefnGNHKprimerx** (3) is used to calculate tables showing the pH dependencies of the standard transformed Gibbs energy of reaction, the change in the number of hydrogen ions bound, and the apparent equilibrium constant at 298.15 K and 0.25 M ionic strength. Since there are problems in rounding exponentials, $\log K'$ is calculated rather than K'.

```
In[2]:=  Off[General::"spell"];
         Off[General::"spell1"];

In[4]:=  << BiochemThermo`BasicBiochemData3`
```

```
In[5]:= derivefnGNHKprimerx[eq_] :=
          Module[{function, functionG, functionNH, functionlogKprime},
            (*Derives the functions of pH and ionic strength that give the standard
              transformed Gibbs energy of reaction, change in number of hydrogen ions
              bound, and the base 10 log of the apparent equilibrium constant at 298.15
              K for a biochemical reaction typed in as, for example, atp+h2o+de==adp+
              pi. The standard transformed Gibbs energy of reaction is in kJ mol^-1.*)
            function = Solve[eq, de];
            functionG = function[[1, 1, 2]];
            functionNH = (1 / (8.31451 * .29815 * Log[10])) * D[functionG, pH];
            functionlogKprime = -functionG / (8.31451 * .29815 * Log[10]);
            {functionG, functionNH, functionlogKprime}];
```

EC 3.1.1.1   Ethylacetase

```
In[6]:= PaddedForm[TableForm[
          derivefnGNHKprimerx[ethylacetate + h2o + de == ethanol + acetate] /. pH → {5, 6, 7, 8, 9} /.
          is → .25, TableHeadings ->
            {{"Δ_rG'°", "Δ_rN_H", "logK'"}, {"pH 5", "pH 6", "pH 7", "pH 8", "pH 9"}}], {4, 2}]
```

Out[6]//PaddedForm=

	pH 5	pH 6	pH 7	pH 8	pH 9
$\Delta_r G'°$	-6.91	-12.05	-17.69	-23.39	-29.10
$\Delta_r N_H$	-0.77	-0.97	-1.00	-1.00	-1.00
logK'	1.21	2.11	3.10	4.10	5.10

EC 3.1.2.1   Acetyl-CoA hydrolase

```
In[7]:= PaddedForm[TableForm[
          derivefnGNHKprimerx[acetylcoA + h2o + de == acetate + coA] /. pH → {5, 6, 7, 8, 9} /. is → .25,
          TableHeadings ->
            {{"Δ_rG'°", "Δ_rN_H", "logK'"}, {"pH 5", "pH 6", "pH 7", "pH 8", "pH 9"}}], {4, 2}]
```

Out[7]//PaddedForm=

	pH 5	pH 6	pH 7	pH 8	pH 9
$\Delta_r G'°$	-30.39	-35.55	-41.36	-48.33	-58.03
$\Delta_r N_H$	-0.77	-0.98	-1.07	-1.44	-1.89
logK'	5.32	6.23	7.25	8.47	10.17

EC 3.1.2.3a   Acetylphosphate hydrolase

```
In[8]:= PaddedForm[TableForm[
          derivefnGNHKprimerx[acetylphos + h2o + de == acetate + pi] /. pH → {5, 6, 7, 8, 9} /. is → .25,
          TableHeadings ->
            {{"Δ_rG'°", "Δ_rN_H", "logK'"}, {"pH 5", "pH 6", "pH 7", "pH 8", "pH 9"}}], {4, 2}]
```

Out[8]//PaddedForm=

	pH 5	pH 6	pH 7	pH 8	pH 9
$\Delta_r G'°$	-29.92	-36.60	-44.64	-54.04	-61.43
$\Delta_r N_H$	-1.20	-1.21	-1.62	-1.53	-1.11
logK'	5.24	6.41	7.82	9.47	10.76

EC 3.1.2.3b   Succinyl-CoA hydrolase

```
In[9]:= PaddedForm[TableForm[
            derivefnGNHKprimerx[succinylcoA + h2o + de == succinate + coA] /. pH → {5, 6, 7, 8, 9} /.
            is → .25, TableHeadings ->
                {{"ΔrG'°", "ΔrNH", "logK'"}, {"pH 5", "pH 6", "pH 7", "pH 8", "pH 9"}}], {4, 2}]
```

Out[9]//PaddedForm=

	pH 5	pH 6	pH 7	pH 8	pH 9
$\Delta_r G'°$	-24.98	-29.12	-34.77	-41.73	-51.42
$\Delta_r N_H$	-0.48	-0.91	-1.06	-1.44	-1.89
logK'	4.38	5.10	6.09	7.31	9.01

EC 3.1.2.3c   Malyl-CoA hydrolase

```
In[10]:= PaddedForm[TableForm[
            derivefnGNHKprimerx[malylcoA + h2o + de == malate + coA] /. pH → {5, 6, 7, 8, 9} /. is → .25,
            TableHeadings ->
                {{"ΔrG'°", "ΔrNH", "logK'"}, {"pH 5", "pH 6", "pH 7", "pH 8", "pH 9"}}], {4, 2}]
```

Out[10]//PaddedForm=

	pH 5	pH 6	pH 7	pH 8	pH 9
$\Delta_r G'°$	-22.43	-27.46	-33.26	-40.22	-49.92
$\Delta_r N_H$	-0.75	-0.97	-1.07	-1.44	-1.89
logK'	3.93	4.81	5.83	7.05	8.75

EC 3.1.2.18   Oxalyl-CoA hydrolase

```
In[11]:= PaddedForm[TableForm[
            derivefnGNHKprimerx[oxalylcoA + h2o + de == oxalate + coA] /. pH → {5, 6, 7, 8, 9} /. is → .25
            TableHeadings ->
                {{"ΔrG'°", "ΔrNH", "logK'"}, {"pH 5", "pH 6", "pH 7", "pH 8", "pH 9"}}], {4, 2}]
```

Out[11]//PaddedForm=

	pH 5	pH 6	pH 7	pH 8	pH 9
$\Delta_r G'°$	-6.49	-12.10	-17.97	-24.94	-34.64
$\Delta_r N_H$	-0.95	-1.00	-1.07	-1.45	-1.89
logK'	1.14	2.12	3.15	4.37	6.07

EC 3.1.3.1a   Ribose 1-phosphate hydrolase

```
In[12]:= PaddedForm[TableForm[
            derivefnGNHKprimerx[ribose1phos + h2o + de == ribose + pi] /. pH → {5, 6, 7, 8, 9} /. is → .25,
            TableHeadings ->
                {{"ΔrG'°", "ΔrNH", "logK'"}, {"pH 5", "pH 6", "pH 7", "pH 8", "pH 9"}}], {4, 2}]
```

Out[12]//PaddedForm=

	pH 5	pH 6	pH 7	pH 8	pH 9
$\Delta_r G'°$	-26.10	-25.33	-23.82	-23.29	-23.22
$\Delta_r N_H$	0.05	0.25	0.19	0.03	0.00
logK'	4.57	4.44	4.17	4.08	4.07

EC 3.1.3.1b   Galactose 1-phosphate hydrolase

```
In[13]:= PaddedForm[TableForm[
            derivefnGNHKprimerx[galactose1phos + h2o + de == galactose + pi] /. pH → {5, 6, 7, 8, 9} /.
            is → .25, TableHeadings ->
                {{"ΔrG'°", "ΔrNH", "logK'"}, {"pH 5", "pH 6", "pH 7", "pH 8", "pH 9"}}], {4, 2}]
```

Out[13]//PaddedForm=

	pH 5	pH 6	pH 7	pH 8	pH 9
$\Delta_r G'^\circ$	-16.74	-14.57	-11.97	-11.25	-11.16
$\Delta_r N_H$	0.19	0.54	0.27	0.04	0.00
logK'	2.93	2.55	2.10	1.97	1.96

EC 3.1.3.60   Phospho*enol*pyruvate phosphohydrolase

In[14]:= **PaddedForm[**
    TableForm[derivefnGNHKprimerx[pep + h2o + de == pyruvate + pi] /. pH → {5, 6, 7, 8, 9} /. is →
    TableHeadings ->
    {{"$\Delta_r G'^\circ$", "$\Delta_r N_H$", "logK'"}, {"pH 5", "pH 6", "pH 7", "pH 8", "pH 9"}}], {4, 2}]

Out[14]//PaddedForm=

	pH 5	pH 6	pH 7	pH 8	pH 9
$\Delta_r G'^\circ$	-67.04	-66.32	-64.88	-64.37	-64.30
$\Delta_r N_H$	0.04	0.23	0.19	0.03	0.00
logK'	11.74	11.62	11.37	11.28	11.26

EC 3.1.3.9   Glucose-6-phosphatase

In[15]:= **PaddedForm[TableForm[**
    derivefnGNHKprimerx[glucose6phos + h2o + de == glucose + pi] /. pH → {5, 6, 7, 8, 9} /. is → .
    TableHeadings ->
    {{"$\Delta_r G'^\circ$", "$\Delta_r N_H$", "logK'"}, {"pH 5", "pH 6", "pH 7", "pH 8", "pH 9"}}], {4, 2}]

Out[15]//PaddedForm=

	pH 5	pH 6	pH 7	pH 8	pH 9
$\Delta_r G'^\circ$	-15.15	-13.75	-11.62	-10.96	-10.88
$\Delta_r N_H$	0.10	0.40	0.24	0.04	0.00
logK'	2.65	2.41	2.03	1.92	1.91

EC 3.1.3.11   Fructose-1,6-bisphosphate 1-phosphohydrolase

In[16]:= **PaddedForm[TableForm[**
    derivefnGNHKprimerx[fructose16phos + h2o + de == fructose6phos + pi] /. pH → {5, 6, 7, 8, 9}
    is → .25, TableHeadings ->
    {{"$\Delta_r G'^\circ$", "$\Delta_r N_H$", "logK'"}, {"pH 5", "pH 6", "pH 7", "pH 8", "pH 9"}}], {4, 2}]

Out[16]//PaddedForm=

	pH 5	pH 6	pH 7	pH 8	pH 9
$\Delta_r G'^\circ$	-20.21	-16.26	-12.79	-11.95	-11.84
$\Delta_r N_H$	0.37	0.84	0.32	0.04	0.00
logK'	3.54	2.85	2.24	2.09	2.07

EC 3.2.1.20   Alpha glucosidase

In[17]:= **PaddedForm[TableForm[**
    derivefnGNHKprimerx[maltose + h2o + de == 2 * glucose] /. pH → {5, 6, 7, 8, 9} /. is → .25,
    TableHeadings ->
    {{"$\Delta_r G'^\circ$", "$\Delta_r N_H$", "logK'"}, {"pH 5", "pH 6", "pH 7", "pH 8", "pH 9"}}], {4, 2}]

Out[17]//PaddedForm=

	pH 5	pH 6	pH 7	pH 8	pH 9
$\Delta_r G'^\circ$	-19.92	-19.92	-19.92	-19.92	-19.92
$\Delta_r N_H$	0.00				
logK'	3.49	3.49	3.49	3.49	3.49

EC 3.2.1.48   Sucrose alpha-glucosidase

```
In[18]:= PaddedForm[TableForm[
            derivefnGNHKprimerx[sucrose + h2o + de == glucose + fructose] /. pH → {5, 6, 7, 8, 9} /.
            is → .25, TableHeadings ->
            {{"ΔᵣG'°", "ΔᵣNₕ", "logK'"}, {"pH 5", "pH 6", "pH 7", "pH 8", "pH 9"}}], {4, 2}]
```

Out[18]//PaddedForm=

	pH 5	pH 6	pH 7	pH 8	pH 9
$\Delta_r G'°$	-29.52	-29.52	-29.52	-29.52	-29.52
$\Delta_r N_H$	0.00				
logK'	5.17	5.17	5.17	5.17	5.17

EC 3.2.2.4   AMP nucleosidase

```
In[19]:= PaddedForm[TableForm[
            derivefnGNHKprimerx[amp + h2o + de == ribose5phos + adenine] /. pH → {5, 6, 7, 8, 9} /. is → .:
            TableHeadings ->
            {{"ΔᵣG'°", "ΔᵣNₕ", "logK'"}, {"pH 5", "pH 6", "pH 7", "pH 8", "pH 9"}}], {4, 2}]
```

Out[19]//PaddedForm=

	pH 5	pH 6	pH 7	pH 8	pH 9
$\Delta_r G'°$	-4.77	-4.64	-4.72	-4.74	-4.74
$\Delta_r N_H$	0.08	-0.01	-0.01	-0.00	-0.00
logK'	0.84	0.81	0.83	0.83	0.83

EC 3.2.2.7   Adenosine nucleosidase

```
In[20]:= PaddedForm[TableForm[
            derivefnGNHKprimerx[adenosine + h2o + de == ribose + adenine] /. pH → {5, 6, 7, 8, 9} /.
            is → .25, TableHeadings ->
            {{"ΔᵣG'°", "ΔᵣNₕ", "logK'"}, {"pH 5", "pH 6", "pH 7", "pH 8", "pH 9"}}], {4, 2}]
```

Out[20]//PaddedForm=

	pH 5	pH 6	pH 7	pH 8	pH 9
$\Delta_r G'°$	-7.20	-6.94	-6.91	-6.91	-6.91
$\Delta_r N_H$	0.11	0.01	0.00	0.00	0.00
logK'	1.26	1.22	1.21	1.21	1.21

EC 3.4.13.18   Glycyl-glycine dipeptidase

```
In[21]:= PaddedForm[TableForm[
            derivefnGNHKprimerx[glycylglycine + h2o + de == 2 * glycine] /. pH → {5, 6, 7, 8, 9} /. is → .:
            TableHeadings ->
            {{"ΔᵣG'°", "ΔᵣNₕ", "logK'"}, {"pH 5", "pH 6", "pH 7", "pH 8", "pH 9"}}], {4, 2}]
```

Out[21]//PaddedForm=

	pH 5	pH 6	pH 7	pH 8	pH 9
$\Delta_r G'°$	-2.43	-2.43	-2.43	-2.43	-2.43
$\Delta_r N_H$	0.00				
logK'	0.43	0.43	0.43	0.43	0.43

EC 3.5.1.1   Asparaginase

In[22]:= **PaddedForm[TableForm[**
        derivefnGNHKprimerx[asparagineL + h2o + de == aspartate + ammonia] /. pH → {5, 6, 7, 8, 9} /.
        is → .25, TableHeadings ->
        {{"Δ_rG'°", "Δ_rN$_H$", "logK'"}, {"pH 5", "pH 6", "pH 7", "pH 8", "pH 9"}}], {4, 2}]

Out[22]//PaddedForm=

	pH 5	pH 6	pH 7	pH 8	pH 9
Δ_rG'°	-13.69	-13.69	-13.70	-13.82	-14.79
Δ_rN$_H$	-0.00	-0.00	-0.01	-0.05	-0.36
logK'	2.40	2.40	2.40	2.42	2.59

EC 3.5.1.2   Glutaminase

In[23]:= **PaddedForm[TableForm[**
        derivefnGNHKprimerx[glutamine + h2o + de == glutamate + ammonia] /. pH → {5, 6, 7, 8, 9} /.
        is → .25, TableHeadings ->
        {{"Δ_rG'°", "Δ_rN$_H$", "logK'"}, {"pH 5", "pH 6", "pH 7", "pH 8", "pH 9"}}], {4, 2}]

Out[23]//PaddedForm=

	pH 5	pH 6	pH 7	pH 8	pH 9
Δ_rG'°	-13.19	-13.19	-13.20	-13.32	-14.29
Δ_rN$_H$	-0.00	-0.00	-0.01	-0.05	-0.36
logK'	2.31	2.31	2.31	2.33	2.50

EC 3.5.1.5   Urease

In[24]:= **PaddedForm[TableForm[**
        derivefnGNHKprimerx[urea + h2o + de == co2tot + 2 * ammonia] /. pH → {5, 6, 7, 8, 9} /. is → .25
        TableHeadings ->
        {{"Δ_rG'°", "Δ_rN$_H$", "logK'"}, {"pH 5", "pH 6", "pH 7", "pH 8", "pH 9"}}], {4, 2}]

Out[24]//PaddedForm=

	pH 5	pH 6	pH 7	pH 8	pH 9
Δ_rG'°	-226.20	-204.60	-185.80	-168.70	-153.90
Δ_rN$_H$	3.92	3.55	3.10	2.89	2.14
logK'	39.62	35.85	32.56	29.56	26.96

EC 3.6.1.1   Inorganic diphosphatase

In[25]:= **PaddedForm[**
        TableForm[derivefnGNHKprimerx[ppi + h2o + de == 2 * pi] /. pH → {5, 6, 7, 8, 9} /. is → .25,
        TableHeadings ->
        {{"Δ_rG'°", "Δ_rN$_H$", "logK'"}, {"pH 5", "pH 6", "pH 7", "pH 8", "pH 9"}}], {4, 2}]

Out[25]//PaddedForm=

	pH 5	pH 6	pH 7	pH 8	pH 9
Δ_rG'°	-23.37	-22.45	-22.66	-26.06	-28.22
Δ_rN$_H$	0.08	0.21	-0.41	-0.60	-0.17
logK'	4.09	3.93	3.97	4.57	4.94

EC 3.6.1.3   Adenosine triphosphatase

In[26]:= **PaddedForm[**
        TableForm[derivefnGNHKprimerx[atp + h2o + de == adp + pi] /. pH → {5, 6, 7, 8, 9} /. is → .25,
        TableHeadings ->
        {{"Δ_rG'°", "Δ_rN$_H$", "logK'"}, {"pH 5", "pH 6", "pH 7", "pH 8", "pH 9"}}], {4, 2}]

Out[26]//PaddedForm=

	pH 5	pH 6	pH 7	pH 8	pH 9
$\Delta_r G'^\circ$	-32.56	-33.22	-36.04	-41.07	-46.70
$\Delta_r N_H$	-0.04	-0.25	-0.74	-0.96	-1.00
logK'	5.70	5.82	6.31	7.20	8.18

EC 3.6.1.5   Adenosine diphosphatase

In[27]:= **PaddedForm[**
         TableForm[derivefnGNHKprimerx[adp + h2o + de == amp + pi] /. pH → {5, 6, 7, 8, 9} /. is → .25,
         TableHeadings ->
            {{"$\Delta_r G'^\circ$", "$\Delta_r N_H$", "logK'"}, {"pH 5", "pH 6", "pH 7", "pH 8", "pH 9"}}], {4, 2}]

Out[27]//PaddedForm=

	pH 5	pH 6	pH 7	pH 8	pH 9
$\Delta_r G'^\circ$	-30.33	-31.09	-33.96	-38.99	-44.61
$\Delta_r N_H$	-0.05	-0.27	-0.74	-0.96	-1.00
logK'	5.31	5.45	5.95	6.83	7.82

EC 3.6.1.8   ATP diphosphatase

In[28]:= **PaddedForm[**
         TableForm[derivefnGNHKprimerx[atp + h2o + de == amp + ppi] /. pH → {5, 6, 7, 8, 9} /. is → .25,
         TableHeadings ->
            {{"$\Delta_r G'^\circ$", "$\Delta_r N_H$", "logK'"}, {"pH 5", "pH 6", "pH 7", "pH 8", "pH 9"}}], {4, 2}]

Out[28]//PaddedForm=

	pH 5	pH 6	pH 7	pH 8	pH 9
$\Delta_r G'^\circ$	-39.52	-41.86	-47.34	-54.00	-63.10
$\Delta_r N_H$	-0.17	-0.73	-1.08	-1.33	-1.83
logK'	6.92	7.33	8.29	9.46	11.05

EC 3.7.1.1   Oxaloacetase

In[29]:= **PaddedForm[TableForm[**
         derivefnGNHKprimerx[oxaloacetate + h2o + de == oxalate + acetate] /. pH → {5, 6, 7, 8, 9} /.
         is → .25, TableHeadings ->
            {{"$\Delta_r G'^\circ$", "$\Delta_r N_H$", "logK'"}, {"pH 5", "pH 6", "pH 7", "pH 8", "pH 9"}}], {4, 2}]

Out[29]//PaddedForm=

	pH 5	pH 6	pH 7	pH 8	pH 9
$\Delta_r G'^\circ$	-43.66	-48.68	-54.31	-60.01	-65.72
$\Delta_r N_H$	-0.72	-0.97	-1.00	-1.00	-1.00
logK'	7.65	8.53	9.52	10.51	11.51

To show the tremendous range of values of apparent equilibrium constants for hydrolase reactions at 298.15 K, pH 7, and 0.25 M ionic strength, a table is prepared in order of increasing values.

```
In[30]:= hydroKprimedata = {{"3.7.1.1  ethylacetate+h2o=ethanol+acetate", 1260},
            {"3.1.2.1  acetylcoA+h2o=acetate+coA", 1.76*10^7},
            {"3.1.2.3a  acetylphos+h2o=acetate+pi", 6.61*10^7},
            {"3.1.2.3b  succinylcoA+h2o=succinate+coA", 1.24*10^6},
            {"3.1.2.3c  malylcoA+h2o=malate+coA", 6.89*10^5},
            {"3.1.2.18  oxalylcoA+h2o=oxalate+coA", 1.41*10^3},
            {"3.1.3.1a  ribose1phos+h2o=ribose+pi", 1.49*10^4},
            {"3.1.3.1b  galactose1phos+h2o=galactose+pi", 1.25*10^2},
            {"3.1.3.60  pep+h2o+de==pyruvate+pi", 2.33*10^11},
            {"3.1.3.9  glucose6phos+h2o=glucose+pi", 1.08*10^2},
            {"3.1.3.11  fructose16phos+h2o=fructose6phos+pi", 1.74*10^2},
            {"3.2.1.20  maltose+h2o=2*glucose", 3.09*10^3},
            {"3.2.1.48  sucrose+h2o=glucose+fructose", 1.48*10^5},
            {"3.2.2.4  amp+h2o=ribose5phos+adenine", 6.70},
            {"3.2.2.7  adenosine+h2o=ribose+adenine", 16.3},
            {"3.4.13.18  glycylglycine+h2o=2*glycine", 2.67},
            {"3.5.1.1  asparagineL+h2o=aspartate+ammonia", 252},
            {"3.5.1.2  glutamine+h2o=glutamate+ammonia", 206},
            {"3.5.1.5  urea+h2o=co2tot+2*ammonia", 3.62*10^32},
            {"3.6.1.1  ppi+h2o=2*pi", 1390},
            {"3.6.1.3  atp+h2o=adp+pi", 2.05*10^6},
            {"3.6.1.5  adp+h2o=amp+pi", 8.90*10^5},
            {"3.6.1.8  atp+h2o=amp+ppi", 1.97*10^9},
            {"3.7.1.1  oxaloacetate+h2o=oxalate+acetate", 3.28*10^9}};
```

Table 10.1 Apparent equilibrium constants of hydrolase reactions at 298.15 K, pH 7, and 0.25 M ionic strength

```
In[31]:= TableForm[hydroKprimedata[[Ordering[Transpose[hydroKprimedata][[2]]]]]]

Out[31]//TableForm=
```

3.4.13.18	glycylglycine+h2o=2*glycine	2.67
3.2.2.4	amp+h2o=ribose5phos+adenine	6.7
3.2.2.7	adenosine+h2o=ribose+adenine	16.3
3.1.3.9	glucose6phos+h2o=glucose+pi	108.
3.1.3.1b	galactose1phos+h2o=galactose+pi	125.
3.1.3.11	fructose16phos+h2o=fructose6phos+pi	174.
3.5.1.2	glutamine+h2o=glutamate+ammonia	206
3.5.1.1	asparagineL+h2o=aspartate+ammonia	252
3.7.1.1	ethylacetate+h2o=ethanol+acetate	1260
3.6.1.1	ppi+h2o=2*pi	1390
3.1.2.18	oxalylcoA+h2o=oxalate+coA	1410.
3.2.1.20	maltose+h2o=2*glucose	3090.
3.1.3.1a	ribose1phos+h2o=ribose+pi	14900.
3.2.1.48	sucrose+h2o=glucose+fructose	148000.
3.1.2.3c	malylcoA+h2o=malate+coA	689000.
3.6.1.5	adp+h2o=amp+pi	890000.
3.1.2.3b	succinylcoA+h2o=succinate+coA	1.24×10^6
3.6.1.3	atp+h2o=adp+pi	2.05×10^6
3.1.2.1	acetylcoA+h2o=acetate+coA	1.76×10^7
3.1.2.3a	acetylphos+h2o=acetate+pi	6.61×10^7
3.6.1.8	atp+h2o=amp+ppi	1.97×10^9
3.7.1.1	oxaloacetate+h2o=oxalate+acetate	3.28×10^9
3.1.3.60	pep+h2o+de≔pyruvate+pi	2.33×10^{11}
3.5.1.5	urea+h2o=co2tot+2*ammonia	3.62×10^{32}

It is evident that very different types of bonds are split. It is important to remember that these apparent equilibrium constants at 298.15 K and ionic strength 0.25 M are a consequence of the equilibrium constants of chemical reference reactions and pKs of reactants. Any attempt to interpret these K' values has to take into account these two different types of effects. The next two sections explore these effects in the hydrolysis of phosphate compounds.

10.3 Effect of pH on the Standard Transformed Gibbs Energy of Hydrolysis of Phosphate Compounds

The program **calctrGerx** makes it convenient to calculate $\Delta_r G'°$ for an enzyme-catalyzed reaction at specified pHs and ionic strengths by simply typing the reactions using the names of the functions of pH and ionic strength. Since it is $\Delta_r N_H$ that determines the change in $\Delta_r G'°$ with pH, it is convenient to put these two types of plots together, The following calculations of plots on the hydrolysis of phosphate compounds are all at 298.15 K and 0.25 M ionic strength. The program **calctrGerx is** used to construct plots of $\Delta_r G'°$ versus pH.

```
In[32]:= calctrGerx[eq_, pHlist_, islist_] :=
            Module[{energy}, (*Calculates the standard transformed Gibbs
                energy of reaction in kJ mol^-1 at specified pHs and ionic
                    strengths for a biochemical reaction typed in the form atp+h2o+de==
                adp+pi.  The names of reactants call the appropriate functions of
                    pH and ionic strength.  pHlist and is list can be lists.*)
                energy = Solve[eq, de]; energy[[1, 1, 2]] /. pH → pHlist /. is → islist]
```

The program **calcNHrx** is used to construct plots of $\Delta_r N_H$ versus pH.

```
In[33]:= calcNHrx[eq_, pHlist_, islist_] := Module[{energy},(*This program calculates the
         change in the binding of hydrogen ions in a biochemical reaction at specified pHs
         and ionic strengths.  The reaction is entered in the form atp+h2o+de==adp+pi.*)
            energy = Solve[eq, de];
             D[energy[[1,1,2]], pH]/(8.31451*0.29815*Log[10]) /.
              pH -> pHlist /. is -> islist]
```

```
In[34]:= plot1 = Plot[calctrGerx[glucose6phos + h2o + de == glucose + pi, pH, .25],
             {pH, 5, 9}, AxesLabel → {"pH", "Δr G' °"},
             PlotLabel -> "glucose6phos+h2o=glucose+pi", DisplayFunction → Identity];
```

```
In[35]:= plot2 = Plot[calctrGerx[amp + h2o + de == adenosine + pi, pH, .25], {pH, 5, 9},
             AxesLabel → {"pH", "Δr G' °"}, PlotLabel -> "amp+h2o=adenosine+pi",
             AxesOrigin -> {5, -16}, DisplayFunction → Identity];
```

```
In[36]:= plot3 =
             Plot[calctrGerx[adp + h2o + de == amp + pi, pH, .25], {pH, 5, 9}, AxesLabel → {"pH", "Δr G' °"}
                PlotLabel -> "adp+h2o=amp+pi", DisplayFunction → Identity];
```

```
In[37]:= plot4 =
             Plot[calctrGerx[atp + h2o + de == adp + pi, pH, .25], {pH, 5, 9}, AxesLabel → {"pH", "Δr G' °"}
                PlotLabel -> "atp+h2o=adp+pi", DisplayFunction → Identity];
```

```
In[38]:= plot5 =
             Plot[calctrGerx[ppi + h2o + de == 2 * pi, pH, .25], {pH, 5, 9}, AxesLabel → {"pH", "Δr G' °"},
                PlotLabel -> "ppi+h2o=2*pi", DisplayFunction → Identity];
```

```
In[39]:= plot6 = Plot[calctrGerx[acetylphos + h2o + de == acetate + pi, pH, .25],
             {pH, 5, 9}, AxesLabel → {"pH", "Δr G' °"},
             PlotLabel -> "acetylphos+h2o=acetate+pi", DisplayFunction → Identity];
```

```
In[40]:= plot11 = Plot[Evaluate[calcNHrx[glucose6phos + h2o + de == glucose + pi, pH, .25]],
            {pH, 5, 9}, AxesLabel → {"pH", "Δr NH"},
            PlotLabel -> "glucose6phos+h2o=glucose+pi", DisplayFunction → Identity];

In[41]:= plot12 = Plot[Evaluate[calcNHrx[amp + h2o + de == adenosine + pi, pH, .25]],
            {pH, 5, 9}, AxesLabel → {"pH", "Δr NH"},
            PlotLabel -> "amp+h2o=adenosine+pi", DisplayFunction → Identity];

In[42]:= plot13 = Plot[Evaluate[calcNHrx[adp + h2o + de == amp + pi, pH, .25]],
            {pH, 5, 9}, AxesLabel → {"pH", "Δr NH"}, PlotLabel -> "adp+h2o=amp+pi",
            AxesOrigin -> {5, -1}, DisplayFunction → Identity];

In[43]:= plot14 = Plot[Evaluate[calcNHrx[atp + h2o + de == adp + pi, pH, .25]],
            {pH, 5, 9}, AxesLabel → {"pH", "Δr NH"}, PlotLabel -> "atp+h2o=adp+pi",
            AxesOrigin -> {5, -1}, DisplayFunction → Identity];

In[44]:= plot15 = Plot[Evaluate[calcNHrx[ppi + h2o + de == 2 * pi, pH, .25]],
            {pH, 5, 9}, AxesLabel → {"pH", "Δr NH"}, PlotLabel -> "ppi+h2o=2*pi",
            AxesOrigin -> {5, -1}, PlotRange → {-1, .3}, DisplayFunction → Identity];

In[45]:= plot16 = Plot[Evaluate[calcNHrx[acetylphos + h2o + de == acetate + pi, pH, .25]],
            {pH, 5, 9}, AxesLabel → {"pH", "Δr NH"},
            PlotLabel -> "acetylphos+h2o=acetate+pi", DisplayFunction → Identity];
```

In[46]:= **Show[GraphicsArray[{{plot1, plot11}, {plot2, plot12}, {plot3, plot13}}]];**

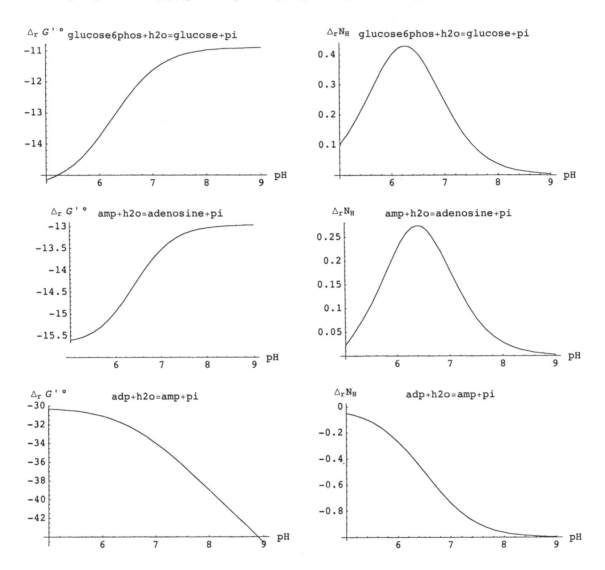

Figure 10.1 Standard transformed Gibbs energies of hydrolysis of phosphate compounds and changes in the binding of hydrogen ions as a function of pH at 298.15 K and 0.25 M ionic strength.

In[47]:= **Show[GraphicsArray[{{plot4, plot14}, {plot5, plot15}, {plot6, plot16}}]];**

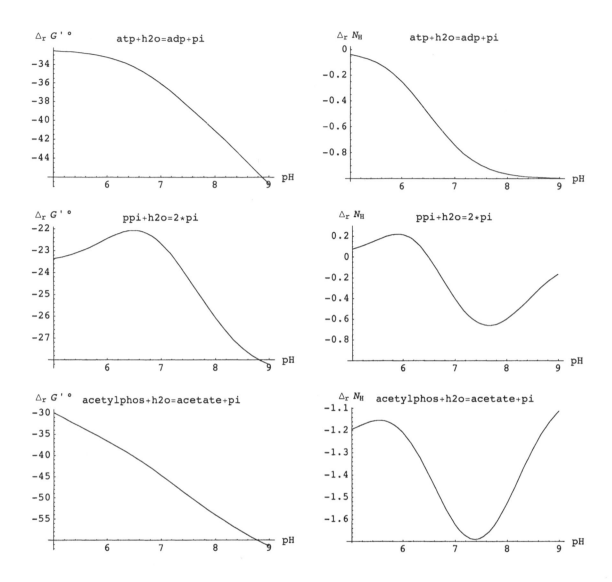

Figure 10.2 Standard transformed Gibbs energies of hydrolysis of three more phosphate compounds and changes in the binding of hydrogen ions as a function of pH at 298.15 K and 0.25 M ionic strength.

   Note that the plots of $\Delta_r N_H$ are directly proportional to the slopes of the plots of $\Delta_r G'^\circ$ versus pH. These plots show that there are two rather different types of phosphate compounds. Glucose 6-phosphate and AMP are phosphate esters. At equilibrium they are hydrolyzed to a greater extent at low pH than high pH and $\Delta_r N_H$ values are relatively low and positive. The other four compounds have phosphoanhydride bonds. At equilibrium they are hydrolyzd to a greater extent at high pH, and $\Delta_r N_H$ values are larger and are negative. The production of H^+ is about the same for ATP and ADP, but the production of H^+ is greater for acetyl phosphate.

10.4 Calculation of p*K*s of Phosphate Compounds at 298.15 K and Zero Ionic Strength

Since standard thermodynamic properties are known for about 17 phosphate ester hydrolyses (4), there is the opportunity to compare their p*K*s and the standard Gibbs energies of reaction for chemical reference reactions at 298.15 K and zero ionic strength. These tables can be used in estimating thermodynamic properties on phosphate hydrolysis reactions that have not yet been determined. The various p*K*s are compared in this section, and then the $\Delta_r G^\circ$ of hydrolysis at high pH is discssed in the next section. The following program is used to calculate the p*K*s for phosphate esters at zero ionic strength.

```
In[48]:= calcpK1is0[speciesmat_] :=
           Module[{lnkzero}, (*This program calculates pK1 at 298.15 K and
              zero ionic stregth for a weak acid.*)
              lnkzero = (speciesmat[[2, 1]] - speciesmat[[1, 1]]) / (8.31451 * .29815);
              N[-lnkzero/Log[10]]]
```

```
In[49]:= listpKs = {adppK, amppK, atppK, dihydroxyacetonephospK,
              fructose16phospK, fructose6phospK, galactose1phospK, gluconolactone6phospK,
              glucose1phospK, glucose6phospK, glyceraldehydephospK, glycerol3phospK,
              mannitol1phospK, peppK, phosphoglycerate2pK, phosphoglycerate3pK,
              ppipK, ribose1phospK, ribose5phospK, sorbitol6phospK};
```

```
In[50]:= listspeciesdata =
           {adpsp, ampsp, atpsp, dihydroxyacetonephossp, fructose16phossp, fructose6phossp,
              galactose1phossp, gluconolactone6phossp, glucose1phossp, glucose6phossp,
              glyceraldehydephossp, glycerol3phossp, mannitol1phossp, pepsp, phosphoglycerate2sp,
              phosphoglycerate3sp, ppisp, ribose1phossp, ribose5phossp, sorbitol6phossp};
```

```
In[51]:= namesofphosphates =
           {"adp", "amp", "atp", "dihydroxyacetonephos", "fructose16phos", "fructose6phos",
              "galactose1phos", "gluconolactone6phos", "glucose1phos", "glucose6phos",
              "glyceraldehydephos", "glycerol3phos", "mannitol1phos", "pep", "phosphoglycerate2",
              "phosphoglycerate3", "ppi", "ribose1phos", "ribose5phos", "sorbitol6phos"};
```

```
In[52]:= Clear[adppK, amppK, atppK, dihydroxyacetonephospK, fructose16phospK, fructose6phospK,
              galactose1phospK, gluconolactone6phospK, glucose1phospK, glucose6phospK,
              glyceraldehydephospK, glycerol3phospK, mannitol1phospK, peppK, phosphoglycerate2pK,
              phosphoglycerate3pK, ppipK, ribose1phospK, ribose5phospK, sorbitol6phospK];
```

```
In[53]:= Evaluate[listpKs] = Map[calcpK1is0, listspeciesdata];
```

Table 10.2 pK_1 for phosphates at 298.15 K and zero ionic strength in alphabetical order

In[54]:= **PaddedForm[TableForm[Transpose[{namesofphosphates, listpKs}]], {3, 2}]**

Out[54]//PaddedForm=

adp	7.18
amp	6.73
atp	7.60
dihydroxyacetonephos	5.70
fructose16phos	6.65
fructose6phos	6.27
galactose1phos	6.15
gluconolactone6phos	6.42
glucose1phos	6.50
glucose6phos	6.42
glyceraldehydephos	5.70
glycerol3phos	6.67
mannitol1phos	6.50
pep	7.00
phosphoglycerate2	7.64
phosphoglycerate3	7.53
ppi	9.46
ribose1phos	6.69
ribose5phos	6.69
sorbitol6phos	6.42

In[55]:= **tablepKs = Transpose[{namesofphosphates, listpKs}];**

Table 10.3 pK_1 for phosphates at 298.15 K and zero ionic strength in order of increasing values

```
In[56]:= PaddedForm[TableForm[tablepKs[[Ordering[Transpose[tablepKs][[2]]]]]], {3, 2}]

Out[56]//PaddedForm=
        dihydroxyacetonephos        5.70
        glyceraldehydephos          5.70
        galactose1phos              6.15
        fructose6phos               6.27
        glucose6phos                6.42
        gluconolactone6phos         6.42
        sorbitol6phos               6.42
        mannitol1phos               6.50
        glucose1phos                6.50
        fructose16phos              6.65
        glycerol3phos               6.67
        ribose1phos                 6.69
        ribose5phos                 6.69
        amp                         6.73
        pep                         7.00
        adp                         7.18
        phosphoglycerate3           7.53
        atp                         7.60
        phosphoglycerate2           7.64
        ppi                         9.46
```

10.5 Calculation of Standard Gibbs Energies of Reaction for Hydrolysis of Phosphate Compounds at 298.15 K and Zero Ionic Strength

The following calculations yield $\Delta_r G^\circ$ for the chemical reaction for the hydrolysis of phosphate esters at zero ionic strength. These chemical reactions are written in terms of species that exist above the pKs in Table 9.2. The properties of these reactants and reactions can be calculated from the values of $\Delta_r G^\circ$ for a reference reaction and the pKs given in the preceding section.

```
In[57]:= calcdGzerohighpH[estersp_, sugarsp_] :=
            Module[{}, (*estersp is the species matrix of the ester. sugar
               is the species matrix of the sugar.  ΔrG° is in kJ mol⁻¹.*)
              (sugarsp[[1]] + pisp[[1]] - (estersp[[1]] + h2osp[[1]]))[[1]]]

In[58]:= dGzero1 = calcdGzerohighpH[glucose6phossp, glucosesp]

Out[58]= -10.87
```

```
In[59]:= hydphosGrx =
           {{"glucose6phos²⁻+h2o=glucose+pi²⁻", calcdGzerohighpH[glucose6phossp, glucosesp]},
            {"glucose1phos²⁻+h2o=glucose+pi²⁻", calcdGzerohighpH[glucose1phossp, glucosesp]},
            {"fructose6phos⁻²+h2o=fructose+pi⁻²", calcdGzerohighpH[fructose6phossp,
              fructosesp]}, {"fructose16phos⁴⁻+h2o=fructose6phos²⁻+pi²⁻",
             calcdGzerohighpH[fructose16phossp, fructose6phossp]},
            {"ribose1phos²⁻+h2o=ribose+pi²⁻", calcdGzerohighpH[ribose1phossp, ribosesp]},
            {"ribose5phos²⁻+h2o=ribose+pi²⁻", calcdGzerohighpH[ribose5phossp, ribosesp]},
            {"glycerol3phos²⁻+h2o=glycerol+pi²⁻", calcdGzerohighpH[
              glycerol3phossp, glycerolsp]}, {"galactose1phos²⁻+h2o=galactose+pi²⁻",
             calcdGzerohighpH[galactose1phossp, galactosesp]},
            {"dihydroxyacetonephos²⁻+h2o=dihydroxyacetone+pi²⁻",
             calcdGzerohighpH[dihydroxyacetonephossp, dihydroxyacetonesp]},
            {"atp⁴⁻+h2o=adp³⁻+pi²⁻+H⁺", calcdGzerohighpH[atpsp, adpsp]},
            {"adp³⁻+h2o=amp²⁻+pi²⁻+H⁺", calcdGzerohighpH[adpsp, ampsp]},
            {"amp²⁻+h2o=adenosine+pi²⁻", calcdGzerohighpH[ampsp, adenosinesp]},
            {"gluconolactone6phos²⁻+h2o=gluconolactone+pi^2-",
             calcdGzerohighpH[gluconolactone6phossp, gluconolactonesp]},
            {"sorbitol6phos²⁻+h2o=sorbitol+pi²⁻",
             calcdGzerohighpH[sorbitol6phossp, sorbitolsp]},
            {"pep³⁻+h2o=pyruvate⁻+pi²⁻", calcdGzerohighpH[pepsp, pyruvatesp]},
            {"mannitol1phos²⁻+h2o=mannitol+pi²⁻",
             calcdGzerohighpH[mannitol1phossp, mannitolDsp]},
            {"ppi⁴⁻+h2o=2 pi²⁻", calcdGzerohighpH[ppisp, pisp]},
            {"acetylphos²⁻+h2o=acetate⁻+pi²⁻+H⁺", calcdGzerohighpH[acetylphossp, acetatesp]}};
```

Table 10.4 Standard Gibbs energies in kJ mol^{-1} of reaction for chemical hydrolysis reactions of phosphates at 298.15 K and zero ionic strength in alphabetical order

In[60]:= **TableForm[hydphosGrx[[Ordering[Transpose[hydphosGrx][[1]]]]]]**

Out[60]//TableForm=

acetylphos^{2-}+h2o=acetate$^-$+pi^{2-}+H$^+$	-8.73
adp^{3-}+h2o=amp^{2-}+pi^{2-}+H$^+$	6.77
amp^{2-}+h2o=adenosine+pi^{2-}	-12.96
atp^{4-}+h2o=adp^{3-}+pi^{2-}+H$^+$	3.06
dihydroxyacetonephos^{2-}+h2o=dihydroxyacetone+pi^{2-}	-13.65
fructose16phos^{4-}+h2o=fructose6phos^{2-}+pi^{2-}	-18.31
fructose6phos^{-2}+h2o=fructose+pi^{-2}	-13.62
galactose1phos^{2-}+h2o=galactose+pi^{2-}	-11.15
gluconolactone6phos^{2-}+h2o=gluconolactone+pi^2-	-16.33
glucose1phos^{2-}+h2o=glucose+pi^{2-}	-17.94
glucose6phos^{2-}+h2o=glucose+pi^{2-}	-10.87
glycerol3phos^{2-}+h2o=glycerol+pi^{2-}	2.57
mannitol1phos^{2-}+h2o=mannitol+pi^{2-}	-7.94
pep^{3-}+h2o=pyruvate$^-$+pi^{2-}	-67.53
ppi^{4-}+h2o=2 pi^{2-}	-35.15
ribose1phos^{2-}+h2o=ribose+pi^{2-}	-23.21
ribose5phos^{2-}+h2o=ribose+pi^{2-}	-15.13
sorbitol6phos^{2-}+h2o=sorbitol+pi^{2-}	-12.17

Table 10.5 Standard Gibbs energies kJ mol^{-1} of reaction for chemical hydrolysis reactions of phosphates at 298.15 K and zero ionic strength in order of increasing values

In[61]:= **TableForm[hydphosGrx[[Ordering[Transpose[hydphosGrx][[2]]]]]]**

```
Out[61]//TableForm=
```

$pep^{3-}+h2o=pyruvate^-+pi^{2-}$	-67.53
$ppi^{4-}+h2o=2\ pi^{2-}$	-35.15
$ribose1phos^{2-}+h2o=ribose+pi^{2-}$	-23.21
$fructose16phos^{4-}+h2o=fructose6phos^{2-}+pi^{2-}$	-18.31
$glucose1phos^{2-}+h2o=glucose+pi^{2-}$	-17.94
$gluconolactone6phos^{2-}+h2o=gluconolactone+pi\char94 2-$	-16.33
$ribose5phos^{2-}+h2o=ribose+pi^{2-}$	-15.13
$dihydroxyacetonephos^{2-}+h2o=dihydroxyacetone+pi^{2-}$	-13.65
$fructose6phos^{-2}+h2o=fructose+pi^{-2}$	-13.62
$amp^{2-}+h2o=adenosine+pi^{2-}$	-12.96
$sorbitol6phos^{2-}+h2o=sorbitol+pi^{2-}$	-12.17
$galactose1phos^{2-}+h2o=galactose+pi^{2-}$	-11.15
$glucose6phos^{2-}+h2o=glucose+pi^{2-}$	-10.87
$acetylphos^{2-}+h2o=acetate^-+pi^{2-}+H^+$	-8.73
$mannitol1phos^{2-}+h2o=mannitol+pi^{2-}$	-7.94
$glycerol3phos^{2-}+h2o=glycerol+pi^{2-}$	2.57
$atp^{4-}+h2o=adp^{3-}+pi^{2-}+H^+$	3.06
$adp^{3-}+h2o=amp^{2-}+pi^{2-}+H^+$	6.77

Note that three of these chemical reactions produce a mole of hydrogen ions; that is the hydrolyses of atp^{4-}, adp^{3-}, and acetyl phosphate^{2-}. Therefore, these three reactions are pulled to the right by specification of pHs in the neutral and alkaline regions.

10.6 Discussion

The effects of temperature on thermodynamic properties of hydrolase reactions have been calculated for six phosphate hydrolysis reactions and three other hydrolase reactions in Chapter 4. The effects of temperature for 20 hydrolase reactions are discussed in Chapter 13.

Appendix

Components in Hydrolase Reactions

Hydrolase reactions do not involve coupling, but they do involve a problem with the treatment of water in biochemical thermodynamics. Therefore, it is of interest to consider their representation by matrices. As a prototypical hydrolase reaction, consider the hydrolysis of glucose 5-phosphate.

glucose6phos + h2o = glucose + pi (A10-1)

The conservation matrix A' for this reaction at specified pH is

```
In[62]:=  conmat1 = {{6, 0, 6, 0}, {9, 1, 6, 4}, {1, 0, 0, 1}};
```

```
In[63]:=  TableForm[conmat1, TableHeadings → {{C, O, P}, {"glucose6phos", "h2o", "glucose", "pi"}}]
```

Out[63]//TableForm=

	glucose6phos	h2o	glucose	pi
C	6	0	6	0
O	9	1	6	4
P	1	0	0	1

The row reduced form shows that the components can be taken as the first three reactants, rather than C, O, and P.

In[64]:= **TableForm[RowReduce[conmat1], TableHeadings →**
      {{"glucose6phos", "h2o", "glucose"}, {"glucose6phos", "h2o", "glucose", "pi"}}]

Out[64]//TableForm=

	glucose6phos	h2o	glucose	pi
glucose6phos	1	0	0	1
h2o	0	1	0	1
glucose	0	0	1	-1

A basis for the conservation matix can be calculated from the stoichiometric number matrix.

In[65]:= **nu1 = {{-1}, {-1}, {1}, {1}};**

In[66]:= **TableForm[nu1, TableHeadings → {{"glucose6phos", "h2o", "glucose", "pi"}, {""}}]**

Out[66]//TableForm=

glucose6phos	-1
h2o	-1
glucose	1
pi	1

In[67]:= **TableForm[RowReduce[NullSpace[Transpose[nu1]]], TableHeadings →**
      {{"glucose6phos", "h2o", "glucose"}, {"glucose6phos", "h2o", "glucose", "pi"}}]

Out[67]//TableForm=

	glucose6phos	h2o	glucose	pi
glucose6phos	1	0	0	1
h2o	0	1	0	1
glucose	0	0	1	-1

However, the concentration of water does not appear in the expression for the equilibrium constant, and so this treatment fails in calculating the equilibrium concentrations using **equcalcc** (see Chapter 7). This calculational problem can be solved by using the further transformed Gibbs energy G'' defined by the Legendre transform (4)

$$G'' = G' - n_c(O) \mu'^\circ(H_2 O) \tag{A10-2}$$

When standard further transformed Gibbs energies of formation $\Delta_r G''$ are used, the row for oxygen and the column for $H_2 O$ are omitted from the conservation matrix.

In[68]:= **conmat2 = {{6, 6, 0}, {1, 0, 1}};**

In[69]:= **TableForm[conmat2, TableHeadings → {{C, P}, {"glucose6phos", "glucose", "pi"}}]**

Out[69]//TableForm=

	glucose6phos	glucose	pi
C	6	6	0
P	1	0	1

The row reduced form of conmat2 shows that the components can be taken as the first two reactants, rather than C and P.

```
In[70]:= TableForm[RowReduce[conmat2],
            TableHeadings → {{"glucose6phos", "glucose"}, {"glucose6phos", "glucose", "pi"}}]
Out[70]//TableForm=
                        glucose6phos      glucose      pi
        glucose6phos    1                 0            1
        glucose         0                 1            -1
```

A basis for the conservation matix can be calculated from the stoichiometric number matrix for the hydrolysis reaction that is now written as

$$glucose6phos = glucose + pi \qquad\qquad (A10-3)$$

```
In[71]:= nu2 = {{-1}, {1}, {1}};

In[72]:= TableForm[nu2, TableHeadings → {{"glucose6phos", "glucose", "pi"}, {""}}]
Out[72]//TableForm=

        glucose6phos    -1
        glucose         1
        pi              1

In[73]:= TableForm[RowReduce[NullSpace[Transpose[nu2]]],
            TableHeadings → {{"glucose6phos", "glucose"}, {"glucose6phos", "glucose", "pi"}}]
Out[73]//TableForm=
                        glucose6phos      glucose      pi
        glucose6phos    1                 0            1
        glucose         0                 1            -1
```

The matrices conmat2 and nu2 now correspond with the way the apparent equilibrium constant is written. Using the further transformed Gibbs energy G'' solves another problem and that is the elimination of the conservation equation for oxygen atoms in conmat1. When $H_2 O$ is a reactant, there is a sense in which oxygen atoms are not conserved. Since the activity of $H_2 O$ remains at unity no matter the extent of reaction, it does not make much sense to write a conservation equation for oxygen atoms with infinity on the right hand side. However, there are good reasons to keep $H_2 O$ in biochemical reactions when it is a reactant and to use G' in spite of this problem. The main point of this Appendix is to show that the mechanisms of hydrolase enzymes do not introduce constraints.

References

1. R. N. Goldberg and Y. B. Tewari, Thermodynamic and transport properties of carbohydrates and their monophosphates: The pentoses and hexoses, J. Phys. Chem. Ref. Data 18, 809-880 (1989).
2. R. N. Goldberg and Y. B. Tewari, Thermodynamics of enzyme-catalyzed reactions: Part 3. Hydrolases, J. Phys. Chem. Ref. Data, 23, 1035-1103 (1994).
3. R. A. Alberty, Thermodynamics of the mechanism of the nitrogenase reaction, Biophysical Chemistry, 114, 115-120 (2005).
4. R. A. Alberty, Role of water in the thermodynamics of dilute aqueous solutions, Biophys. Chem., 100, 183-192 (2003).

Chapter 11 Lyase Reactions (Class 4), Isomerase Reactions (Class 5), and Ligase Reactions (Class 6) at 298.15 K

11.1 Introduction

11.2 Standard Transformed Gibbs Energies of Reaction, Changes in Binding of Hydrogen Ions, and Apparent Equilibrium Constants of Lyase Reactions

11.3 Apparent Equilibrium Constants of Lyase Reactions at 298.15 K

11.4 Standard Transformed Gibbs Energies of Reaction, Changes in Binding of Hydrogen Ions, and Apparent Equilibrium Constants of Isomerase Reactions

11.5 Apparent Equilibrium Constants of Isomerase Reactions at 298.15 K

11.6 Standard Transformed Gibbs Energies of Reaction, Change in the Binding of Hydrogen ions, and Apparent Equilibrium Constants of Ligase Reactions

11.7 Apparent Equilibrium Constants of Ligase Reactions at 298.15 K

11.8 Discussion

Appendix

   Components in Ligase Reactions

References

11.1 Introduction

Lyases are enzymes "cleaving C-C, C-O, C-N, and other bonds by elimination, leaving double bonds or rings, or conversely adding groups to double bonds." Isomerases are enzymes catalyzing isomerizations. Ligases are enzymes catalyzing "the joining together of two molecules coupled with the hydrolysis of a pyrophosphate bond in ATP or a similar triphosphate." (1,2) These reactions are discussed in the same chapter because together they include about only 15% of the enzymes that have neen named. In general, lyase reactions do not involve constraints in addition to element balances; that is they are generally not coupled; but one exception is discussed here. Isomerase reactions have a single component, and so they do not involve other constraints. Lyase reactions are coupled reactions by definition.

Goldberg and Tewari (3,4) have summarized and evaluated experimental thermodynamic data on these three classes of reactions.

11.2 Standard Transformed Gibbs Energies of Reaction, Change in the Binding of Hydrogen ions, and Apparent Equilibrium Constants of Lyase Reactions

The program **derivefnGNHKprimerx** (5) is used to calculate tables showing the pH dependencies of the standard transformed Gibbs energy of reaction, the change in the number of hydrogen ions bound, and the apparent equilibrium constant at 298.15 K and 0.25 M ionic strength.. Note that all the reactions in this chapter have been written in the direction in which they are spontaneous at 298.15 K, pH 7, and 0.25 M ionic strength.

```
In[2]:=  Off[General::"spell"];
         Off[General::"spell1"];

In[4]:=  << BiochemThermo`BasicBiochemData3`

In[5]:=  derivefnGNHKprimerx[eq_] :=
           Module[{function, functionG, functionNH, functionlogKprime},
             (*Derives the functions of pH and ionic strength that give the standard
               transformed Gibbs energy of reaction, change in number of hydrogen ions
               bound, and the base 10 log of the apparent equilibrium constant at 298.15
               K for a biochemical reaction typed in as, for example, atp+h2o+de==adp+
               pi. The standard transformed Gibbs energy of reaction is in kJ mol^-1.*)
             function = Solve[eq, de];
             functionG = function[[1, 1, 2]];
             functionNH = (1 / (8.31451 * .29815 * Log[10])) * D[functionG, pH];
             functionlogKprime = -functionG / (8.31451 * .29815 * Log[10]);
             {functionG, functionNH, functionlogKprime}];
```

EC 4.1.1.12   Aspartate 4-decarboxylase

```
In[6]:=  PaddedForm[TableForm[
           derivefnGNHKprimerx[aspartate + h2o + de == alanine + co2tot] /. pH → {5, 6, 7, 8, 9} /. is → .2
           TableHeadings ->
             {{"ΔrG'°", "ΔrNH", "logK'"}, {"pH 5", "pH 6", "pH 7", "pH 8", "pH 9"}}], {4, 2}]

Out[6]//PaddedForm=
                  pH 5        pH 6        pH 7        pH 8        pH 9
         ΔrG'°   -31.08      -26.67      -24.99      -24.77      -25.10
         ΔrNH      0.92        0.55        0.11       -0.01       -0.15
         logK'     5.44        4.67        4.38        4.34        4.40
```

EC 4.1.1.38   Phospho*enol*pyruvate carboxylate (diphosphate)

```
In[7]:=  PaddedForm[TableForm[
           derivefnGNHKprimerx[pep + co2tot + pi + de == oxaloacetate + ppi + h2o] /. pH → {5, 6, 7, 8, 9} ,
           is → .25, TableHeadings ->
             {{"ΔrG'°", "ΔrNH", "logK'"}, {"pH 5", "pH 6", "pH 7", "pH 8", "pH 9"}}], {4, 2}]

Out[7]//PaddedForm=
                  pH 5        pH 6        pH 7        pH 8        pH 9
         ΔrG'°   -10.34      -14.97      -15.00      -11.29       -8.74
         ΔrNH     -0.95       -0.53        0.48        0.63        0.32
         logK'     1.81        2.62        2.63        1.98        1.53
```

EC 4.1.3.1   Isocitrate lyase

```
In[8]:= PaddedForm[TableForm[
          derivefnGNHKprimerx[succinate + glyoxylate + de == citrateiso] /. pH → {5, 6, 7, 8, 9} /.
          is → .25, TableHeadings ->
             {{"ΔrG'°", "ΔrNH", "logK'"}, {"pH 5", "pH 6", "pH 7", "pH 8", "pH 9"}}], {4, 2}]
```

Out[8]//PaddedForm=

	pH 5	pH 6	pH 7	pH 8	pH 9
$\Delta_r G'°$	-2.21	-0.72	-0.30	-0.24	-0.24
$\Delta_r N_H$	0.30	0.16	0.02	0.00	0.00
logK'	0.39	0.13	0.05	0.04	0.04

EC 4.1.3.6   Citrate (*pro*-3S)-lyase

```
In[9]:= PaddedForm[TableForm[
          derivefnGNHKprimerx[acetate + oxaloacetate + de == citrate] /. pH → {5, 6, 7, 8, 9} /. is → .2
          TableHeadings ->
             {{"ΔrG'°", "ΔrNH", "logK'"}, {"pH 5", "pH 6", "pH 7", "pH 8", "pH 9"}}], {4, 2}]
```

Out[9]//PaddedForm=

	pH 5	pH 6	pH 7	pH 8	pH 9
$\Delta_r G'°$	-6.69	-4.01	-3.41	-3.34	-3.33
$\Delta_r N_H$	0.68	0.24	0.03	0.00	0.00
logK'	1.17	0.70	0.60	0.58	0.58

EC 4.1.99.1   Tryptophanase

```
In[10]:= PaddedForm[
           TableForm[derivefnGNHKprimerx[indole + pyruvate + ammonia + de == tryptophanL + h2o] /.
             pH → {5, 6, 7, 8, 9} /. is → .25, TableHeadings ->
               {{"ΔrG'°", "ΔrNH", "logK'"}, {"pH 5", "pH 6", "pH 7", "pH 8", "pH 9"}}], {4, 2}]
```

Out[10]//PaddedForm=

	pH 5	pH 6	pH 7	pH 8	pH 9
$\Delta_r G'°$	-22.49	-22.49	-22.48	-22.36	-21.39
$\Delta_r N_H$	0.00	0.00	0.01	0.05	0.36
logK'	3.94	3.94	3.94	3.92	3.75

This reaction can be considered to be the result of the coupling tryptophanL + h2o = indole + serine and serine = pyruvate + ammonia, and so it involves one constraint.

EC 4.2.1.2   Fumarate hydratase

```
In[11]:= PaddedForm[
           TableForm[derivefnGNHKprimerx[fumarate + h2o + de == malate] /. pH → {5, 6, 7, 8, 9} /. is → .
             TableHeadings ->
               {{"ΔrG'°", "ΔrNH", "logK'"}, {"pH 5", "pH 6", "pH 7", "pH 8", "pH 9"}}], {4, 2}]
```

Out[11]//PaddedForm=

	pH 5	pH 6	pH 7	pH 8	pH 9
$\Delta_r G'°$	-4.34	-3.69	-3.61	-3.60	-3.60
$\Delta_r N_H$	0.23	0.04	0.00	0.00	0.00
logK'	0.76	0.65	0.63	0.63	0.63

EC 4.2.1.3   Aconitase hydratase

```
In[12]:= PaddedForm[TableForm[
            derivefnGNHKprimerx[aconitatecis + h2o + de == citrate] /. pH → {5, 6, 7, 8, 9} /. is → .25,
            TableHeadings ->
            {{"ΔrG'°", "ΔrNH", "logK'"}, {"pH 5", "pH 6", "pH 7", "pH 8", "pH 9"}}], {4, 2}]
```

Out[12]//PaddedForm=

	pH 5	pH 6	pH 7	pH 8	pH 9
$\Delta_r G'^\circ$	-12.37	-9.12	-8.45	-8.38	-8.37
$\Delta_r N_H$	0.91	0.27	0.03	0.00	0.00
logK'	2.17	1.60	1.48	1.47	1.47

EC 4.3.1.1   Aspartate ammonia-lyase

```
In[13]:= PaddedForm[TableForm[
            derivefnGNHKprimerx[fumarate + ammonia + de == aspartate] /. pH → {5, 6, 7, 8, 9} /. is → .25
            TableHeadings ->
            {{"ΔrG'°", "ΔrNH", "logK'"}, {"pH 5", "pH 6", "pH 7", "pH 8", "pH 9"}}], {4, 2}]
```

Out[13]//PaddedForm=

	pH 5	pH 6	pH 7	pH 8	pH 9
$\Delta_r G'^\circ$	-11.20	-11.43	-11.44	-11.33	-10.36
$\Delta_r N_H$	-0.10	-0.01	0.00	0.05	0.36
logK'	1.96	2.00	2.01	1.98	1.81

12.3 Apparent Equilibrium Constants of Lyase Reactions at 298.15 K

A table is prepared to show the range of apparent equilibrium constants for lyase reactions at 298.15 K, pH 7, and 0.25 M ionic strength.

```
In[14]:= appKdatalyase = {{"aspartate+h2o=alanine+co2tot", 2.39 * 10^4},
            {"pep+co2tot+pi=oxaloacetate+ppi+h2o", 424}, {"succinate+glyoxylate=citrateiso",
            1.13}, {"acetate+oxaloacetate=citrate", 3.95},
            {"indole+pyruvate+ammonia=tryptophaneL+h2o", 8660}, {"fumarate+h2o=malate", 4.41},
            {"aconitatecis+h2o=citrate", 30.3}, {"fumarate+ammonia=aspartate", 101}};
```

Table 11.1 Apparent equilibrium constants of lyase reactions at 298.15 K, pH 7, and ionic strength 0.25 M in order of increasing values

```
In[15]:= TableForm[appKdatalyase[[Ordering[Transpose[appKdatalyase][[2]]]]]]
```

Out[15]//TableForm=

succinate+glyoxylate=citrateiso	1.13
acetate+oxaloacetate=citrate	3.95
fumarate+h2o=malate	4.41
aconitatecis+h2o=citrate	30.3
fumarate+ammonia=aspartate	101
pep+co2tot+pi=oxaloacetate+ppi+h2o	424
indole+pyruvate+ammonia=tryptophaneL+h2o	8660
aspartate+h2o=alanine+co2tot	23900.

The program **calctrGerx** is used to construct plots of $\Delta_r G'^\circ$ versus pH at 298.15 K and 0.25 M ionic strength.

```
In[16]:=  calctrGerx[eq_, pHlist_, islist_] :=
            Module[{energy}, (*Calculates the standard transformed Gibbs
                energy of reaction in kJ mol^-1 at specified pHs and ionic
                strengths for a biochemical reaction typed in the form atp+h2o+de==
            adp+pi.  The names of reactants call the appropriate functions of
                pH and ionic strength.  pHlist and is list can be lists.*)
            energy = Solve[eq, de]; energy[[1, 1, 2]] /. pH → pHlist /. is → islist]
```

The program **calcNHrx** is used to construct plots of $\Delta_r N_H$ versus pH.

```
In[17]:=  calcNHrx[eq_, pHlist_, islist_] := Module[{energy},(*This program calculates the
            change in the binding of hydrogen ions in a biochemical reaction at specified pHs
            and ionic strengths.  The reaction is entered in the form atp+h2o+de==adp+pi.*)
                energy = Solve[eq, de];
                D[energy[[1,1,2]], pH]/(8.31451*0.29815*Log[10]) /.
                pH -> pHlist /. is -> islist]
```

```
In[18]:=  plot1 = Plot[calctrGerx[pep + co2tot + pi + de == oxaloacetate + ppi + h2o, pH, .25],
                {pH, 5, 9}, AxesLabel → {"pH", "Δ_r G' °"},
                PlotLabel -> "           pep+co2tot+pi+oxaloacetate+ppi+h2o",
                AxesOrigin → {5, -16}, DisplayFunction → Identity];
```

```
In[19]:=  plot2 = Plot[calctrGerx[aconitatecis + h2o + de == citrate, pH, .25], {pH, 5, 9},
                AxesLabel → {"pH", "Δ_r G' °"}, PlotLabel -> "           aconitatecis+h2o=citrate",
                AxesOrigin → {5, -12}, PlotRange → {-12, -6}, DisplayFunction → Identity];
```

```
In[20]:=  plot11 = Plot[Evaluate[calcNHrx[pep + co2tot + pi + de == oxaloacetate + ppi + h2o, pH, .25]],
                {pH, 5, 9}, AxesLabel → {"pH", "Δ_r N_H"},
                PlotLabel -> "           pep+co2tot+pi+oxaloacetate+ppi+h2o",
                AxesOrigin → {5, -1}, DisplayFunction → Identity];
```

```
In[21]:=  plot12 = Plot[Evaluate[calcNHrx[aconitatecis + h2o + de == citrate, pH, .25]], {pH, 5, 9},
                AxesLabel → {"pH", "Δ_r N_H"}, PlotLabel -> "           aconitatecis+h2o=citrate",
                AxesOrigin → {5, 0}, DisplayFunction → Identity];
```

```
In[22]:=  Show[GraphicsArray[{{plot1, plot11}, {plot2, plot12}}]];
```

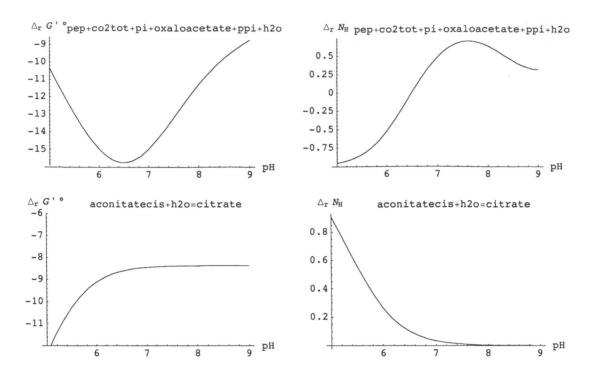

Figure 11.1 $\Delta_r G'°$ and $\Delta_r N_H$ for two lyase reactions at 298.15 K and 0.25 M ionic strength.

The effect of temperature has been calculated earlier for two lyase reactions (4.1.99.1 and 4.3.1.1), and temperature effects are calculated for 7 reactions in Chapter 13.

11.4 Standard Transformed Gibbs Energies of Reaction, Changes in Binding of Hydrogen Ions, and Apparent Equilibrium Constants of Isomerase Reactions

All the catalyzed reactions have been written in the direction in which they are spontaneous at 298.15 K, pH 7, and 0.25 M ionic strength.

EC 5.3.1.5   Xylose isomerase

```
In[23]:= PaddedForm[
            TableForm[derivefnGNHKprimerx[xylulose + de == xylose] /. pH → {5, 6, 7, 8, 9} /. is → .25,
              TableHeadings ->
                {{"ΔrG'°", "ΔrNH", "logK'"}, {"pH 5", "pH 6", "pH 7", "pH 8", "pH 9"}}], {4, 2}]

Out[23]//PaddedForm=
```

	pH 5	pH 6	pH 7	pH 8	pH 9
$\Delta_r G'°$	-4.34	-4.34	-4.34	-4.34	-4.34
$\Delta_r N_H$	0.00				
logK'	0.76	0.76	0.76	0.76	0.76

EC 5.3.1.7   Mannose isomerase

```
In[24]:= PaddedForm[
            TableForm[derivefnGNHKprimerx[mannose + de == fructose] /. pH → {5, 6, 7, 8, 9} /. is → .25,
              TableHeadings ->
                {{"ΔrG'°", "ΔrNH", "logK'"}, {"pH 5", "pH 6", "pH 7", "pH 8", "pH 9"}}], {4, 2}]
```

```
Out[24]//PaddedForm=
                    pH 5        pH 6        pH 7        pH 8        pH 9
        Δ_rG'°      -5.51       -5.51       -5.51       -5.51       -5.51
        Δ_rN_H      0.00
        logK'       0.97        0.97        0.97        0.97        0.97
```

EC 5.3.1.9   Glucose 6-phosphate isomerase

```
In[25]:= PaddedForm[TableForm[
            derivefnGNHKprimerx[fructose6phos + de == glucose6phos] /. pH → {5, 6, 7, 8, 9} /. is → .25,
            TableHeadings ->
              {{"Δ_rG'°", "Δ_rN_H", "logK'"}, {"pH 5", "pH 6", "pH 7", "pH 8", "pH 9"}}], {4, 2}]
```

```
Out[25]//PaddedForm=
                    pH 5        pH 6        pH 7        pH 8        pH 9
        Δ_rG'°      -3.87       -3.46       -3.19       -3.15       -3.14
        Δ_rN_H      0.04        0.08        0.02        0.00        0.00
        logK'       0.68        0.61        0.56        0.55        0.55
```

EC 5.3.1.15   D-lyxose ketol-isomerase

```
In[26]:= PaddedForm[
            TableForm[derivefnGNHKprimerx[xylulose + de == lyxose] /. pH → {5, 6, 7, 8, 9} /. is → .25,
              TableHeadings ->
                {{"Δ_rG'°", "Δ_rN_H", "logK'"}, {"pH 5", "pH 6", "pH 7", "pH 8", "pH 9"}}], {4, 2}]
```

```
Out[26]//PaddedForm=
                    pH 5        pH 6        pH 7        pH 8        pH 9
        Δ_rG'°      -2.99       -2.99       -2.99       -2.99       -2.99
        Δ_rN_H      0.00
        logK'       0.52        0.52        0.52        0.52        0.52
```

EC 5.3.1.20   Ribose isomerase

```
In[27]:= PaddedForm[
            TableForm[derivefnGNHKprimerx[ribulose + de == ribose] /. pH → {5, 6, 7, 8, 9} /. is → .25,
              TableHeadings ->
                {{"Δ_rG'°", "Δ_rN_H", "logK'"}, {"pH 5", "pH 6", "pH 7", "pH 8", "pH 9"}}], {4, 2}]
```

```
Out[27]//PaddedForm=
                    pH 5        pH 6        pH 7        pH 8        pH 9
        Δ_rG'°      -16.06      -16.06      -16.06      -16.06      -16.06
        Δ_rN_H      0.00
        logK'       2.81        2.81        2.81        2.81        2.81
```

EC 5.4.2.7   Phosphopentomatase

```
In[28]:= PaddedForm[TableForm[
            derivefnGNHKprimerx[ribose1phos + de == ribose5phos] /. pH → {5, 6, 7, 8, 9} /. is → .25,
            TableHeadings ->
              {{"Δ_rG'°", "Δ_rN_H", "logK'"}, {"pH 5", "pH 6", "pH 7", "pH 8", "pH 9"}}], {4, 2}]
```

```
Out[28]//PaddedForm=
                    pH 5                pH 6                pH 7        pH 8                pH 9
        Δ_rG'°      -8.08               -8.08               -8.08       -8.08               -8.08
        Δ_rN_H      1.24 × 10^{-15}     -1.24 × 10^{-15}    0.00        1.24 × 10^{-15}     0.00
        logK'       1.42                1.42                1.42        1.42                1.42
```

11.5 Apparent Equilibrium Constants of Isomerase Reactions at 298.15 K

```
In[29]:= appKdataisom = {{"xylulose=xylose", 5.76}, {"mannose=fructose", 9.23},
            {"fructose6phos=glucose6phos", 3.62}, {"xylulose=lyxose", 3.34},
            {"ribulose=ribose", 651},
            {"ribose1phos=ribose5phos", 26.0}};
```

Table 11.2 Apparent equilibrium constants of isomerase reactions at 298.15 K, pH 7, and ionic strength 0.25 M in order of increasing values

```
In[30]:= TableForm[appKdataisom[[Ordering[Transpose[appKdataisom][[2]]]]]]
```

```
Out[30]//TableForm=
        xylulose=lyxose                3.34
        fructose6phos=glucose6phos     3.62
        xylulose=xylose                5.76
        mannose=fructose               9.23
        ribose1phos=ribose5phos        26.
        ribulose=ribose                651
```

The effect of temperature has been calculated earlier for three isomerase reactions (5.1.3.5, 5.3.1.7, and 5.3.1.20), and the effect of temperature is calculated for 6 isomerase reactions in Chapter 13.

11.6 Standard Transformed Gibbs Energies of Reaction, Changes in the Binding of Hydrogen ions, and Apparent Equilibrium Constants of Ligase Reactions

Since all ligase reactions are coupled by definition, the reactions that are coupled are shown.

EC 6.2.1.1   Acetate-coA ligase can be considered to be 3.6.1.8-3.1.2.1.

```
In[31]:= PaddedForm[TableForm[
            derivefnGNHKprimerx[atp + acetate + coA + de == amp + ppi + acetylcoA] /. pH → {5, 6, 7, 8, 9},
            is → .25, TableHeadings ->
            {{"Δ_rG'°", "Δ_rN_H", "logK'"}, {"pH 5", "pH 6", "pH 7", "pH 8", "pH 9"}}], {4, 2}]
```

```
Out[31]//PaddedForm=
            pH 5      pH 6      pH 7      pH 8      pH 9
  Δ_rG'°    -9.12     -6.31     -5.98     -5.67     -5.06
  Δ_rN_H     0.61      0.24     -0.00      0.12      0.06
  logK'      1.60      1.11      1.05      0.99      0.89
```

EC 6.2.1.5   Succinate-coA ligase (ADP forming) can be considered to be 3.6.1.3-3.1.2.3.

```
In[32]:= PaddedForm[TableForm[derivefnGNHKprimerx[atp + succinate + coA + de == adp + pi + succinylcoA]
            pH → {5, 6, 7, 8, 9} /. is → .25, TableHeadings ->
            {{"Δ_rG'°", "Δ_rN_H", "logK'"}, {"pH 5", "pH 6", "pH 7", "pH 8", "pH 9"}}], {4, 2}]
```

```
Out[32]//PaddedForm=
            pH 5      pH 6      pH 7      pH 8      pH 9
  Δ_rG'°    -7.59     -4.09     -1.26      0.65      4.72
  Δ_rN_H     0.44      0.66      0.32      0.48      0.89
  logK'      1.33      0.72      0.22     -0.11     -0.83
```

EC 6.2.1.9   Malate-coA ligase can be considered to be 3.6.1.3-3.1.2.3.

In[33]:= **PaddedForm[TableForm[**
       derivefnGNHKprimerx[atp + malate + coA + de ⩵ adp + pi + malylcoA] /. pH → {5, 6, 7, 8, 9} /.
       is → .25, TableHeadings ->
       {{"Δ$_r$G'°", "Δ$_r$N$_H$", "logK'"}, {"pH 5", "pH 6", "pH 7", "pH 8", "pH 9"}}], {4, 2}]

Out[33]//PaddedForm=

	pH 5	pH 6	pH 7	pH 8	pH 9
Δ$_r$G'°	-10.13	-5.75	-2.78	-0.85	3.22
Δ$_r$N$_H$	0.71	0.72	0.33	0.48	0.89
logK'	1.77	1.01	0.49	0.15	-0.56

EC 6.2.1.13   Acetate-coA ligase (ADP forming) can be considered to be 3.1.2.1-3.6.1.3.

In[34]:= **PaddedForm[TableForm[**
       derivefnGNHKprimerx[adp + pi + acetylcoA + de ⩵ atp + acetate + coA] /. pH → {5, 6, 7, 8, 9} /.
       is → .25, TableHeadings ->
       {{"Δ$_r$G'°", "Δ$_r$N$_H$", "logK'"}, {"pH 5", "pH 6", "pH 7", "pH 8", "pH 9"}}], {4, 2}]

Out[34]//PaddedForm=

	pH 5	pH 6	pH 7	pH 8	pH 9
Δ$_r$G'°	2.17	-2.33	-5.33	-7.26	-11.33
Δ$_r$N$_H$	-0.73	-0.73	-0.33	-0.48	-0.89
logK'	-0.38	0.41	0.93	1.27	1.99

EC 6.3.1.1   Aspartate-ammonia ligase can be considered to be 3.6.1.8-3.5.1.1.

In[35]:= **PaddedForm[**
       TableForm[derivefnGNHKprimerx[atp + aspartate + ammonia + de ⩵ amp + ppi + asparagineL] /.
       pH → {5, 6, 7, 8, 9} /. is → .25, TableHeadings ->
       {{"Δ$_r$G'°", "Δ$_r$N$_H$", "logK'"}, {"pH 5", "pH 6", "pH 7", "pH 8", "pH 9"}}], {4, 2}]

Out[35]//PaddedForm=

	pH 5	pH 6	pH 7	pH 8	pH 9
Δ$_r$G'°	-25.83	-28.17	-33.64	-40.18	-48.30
Δ$_r$N$_H$	-0.17	-0.73	-1.07	-1.28	-1.47
logK'	4.53	4.93	5.89	7.04	8.46

EC 6.3.1.2   Glutamate-ammonia ligase can be considered to be 3.6.1.3-3.5.1.2.

In[36]:= **PaddedForm[**
       TableForm[derivefnGNHKprimerx[atp + glutamate + ammonia + de ⩵ adp + pi + glutamine] /.
       pH → {5, 6, 7, 8, 9} /. is → .25, TableHeadings ->
       {{"Δ$_r$G'°", "Δ$_r$N$_H$", "logK'"}, {"pH 5", "pH 6", "pH 7", "pH 8", "pH 9"}}], {4, 2}]

Out[36]//PaddedForm=

	pH 5	pH 6	pH 7	pH 8	pH 9
Δ$_r$G'°	-19.37	-20.03	-22.83	-27.75	-32.41
Δ$_r$N$_H$	-0.04	-0.25	-0.74	-0.91	-0.64
logK'	3.39	3.51	4.00	4.86	5.68

EC 6.3.5.4   Asparagine synthase (glutamine hydrolyzing) can be considered to be 3.6.1.8-3.5.1.1+3.5.1.2. It is important to note that three reactions are coupled. (Note EC leaves out h2o as a reactant.)

In[37]:= **PaddedForm[**
       TableForm[derivefnGNHKprimerx[atp + aspartate + glutamine + h2o + de ⩵ amp + ppi + asparagine
       glutamate] /. pH → {5, 6, 7, 8, 9} /. is → .25, TableHeadings ->
       {{"Δ$_r$G'°", "Δ$_r$N$_H$", "logK'"}, {"pH 5", "pH 6", "pH 7", "pH 8", "pH 9"}}], {4, 2}]

Out[37]//PaddedForm=

	pH 5	pH 6	pH 7	pH 8	pH 9
$\Delta_r G'^\circ$	-39.02	-41.36	-46.84	-53.50	-62.60
$\Delta_r N_H$	-0.17	-0.73	-1.08	-1.33	-1.83
logK'	6.84	7.25	8.21	9.37	10.97

EC 6.4.1.1   Pyruvate carboxylase

```
In[38]:= PaddedForm[
        TableForm[derivefnGNHKprimerx[atp + pyruvate + co2tot + de == adp + pi + oxaloacetate] /.
          pH → {5, 6, 7, 8, 9} /. is → .25, TableHeadings ->
          {{"ΔrG'°", "ΔrNH", "logK'"}, {"pH 5", "pH 6", "pH 7", "pH 8", "pH 9"}}], {4, 2}]
```

Out[38]//PaddedForm=

	pH 5	pH 6	pH 7	pH 8	pH 9
$\Delta_r G'^\circ$	0.75	-4.31	-8.81	-14.06	-19.36
$\Delta_r N_H$	-0.96	-0.80	-0.85	-0.96	-0.85
logK'	-0.13	0.76	1.54	2.46	3.39

11.7 Apparent Equilibrium Constants of Ligase Reactions at 298.15 K

To show the range of apparent equilibrium constants for ligase reactions at 298.15 K, pH 7, and 0.25 M ionic strength, a table is prepared in order of increasing values.

```
In[39]:= transfKprimedatalig = {{"atp+acetate+coA=amp+ppi+acetylcoA", 11.1},
          {"atp+succinate+coA=adp+pi+succinylcoA", 1.66},
          {"atp+malate+coA=adp+pi+malylcoA", 2.99},
          {"adp+pi+acetylcoA=atp+acetate+coA", 8.58},
          {"atp+aspartate+ammonia=amp+ppi+asparagineL", 7.82 * 10^5},
          {"atp+glutamate+ammonia=adp+pi+glutamine", 1.00 * 10^4},
          {"atp+aspartate+glutamine+h2o=amp+ppi+asparagineL+glutamate", 1.61 * 10^8},
          {"atp+pyruvate+co2tot=adp+pi+oxaloacetate", 34.9}};
```

Table 11.3 Apparent equilibrium constants of ligase reactions at 298.15 K, pH 7, and 0.25 M ionic strength in the order of increasing values

```
In[40]:= TableForm[transfKprimedatalig[[Ordering[Transpose[transfKprimedatalig][[2]]]]]]

Out[40]//TableForm=
        atp+succinate+coA=adp+pi+succinylcoA                        1.66
        atp+malate+coA=adp+pi+malylcoA                              2.99
        adp+pi+acetylcoA=atp+acetate+coA                            8.58
        atp+acetate+coA=amp+ppi+acetylcoA                           11.1
        atp+pyruvate+co2tot=adp+pi+oxaloacetate                     34.9
        atp+glutamate+ammonia=adp+pi+glutamine                      10000.
        atp+aspartate+ammonia=amp+ppi+asparagineL                   782000.
```

atp+aspartate+glutamine+h2o=amp+ppi+asparagineL+glutamate 1.61×10^8

The following plots show the effects of pH on four ligase reactions.

```
In[41]:= plot1 = Plot[calctrGerx[atp + malate + coA + de == adp + pi + malylcoA, pH, .25],
            {pH, 5, 9}, AxesLabel → {"pH", "Δr G'°"}, AxesOrigin → {5, -10},
            PlotLabel -> "        atp+malate+coA=adp+pi+malylcoA", DisplayFunction -> Identity];

In[42]:= plot2 = Plot[calctrGerx[adp + pi + acetylcoA + de == atp + acetate + coA, pH, .25],
            {pH, 5, 9}, AxesLabel → {"pH", "Δr G'°"}, AxesOrigin → {5, -12}, PlotLabel ->
            "        adp+pi+acetylcoA=atp+acetate+coA", DisplayFunction -> Identity];

In[43]:= plot3 = Plot[calctrGerx[atp + aspartate + ammonia + de == amp + ppi + asparagineL, pH, .25],
            {pH, 5, 9}, AxesLabel → {"pH", "Δr G'°"}, AxesOrigin → {5, -48},
            PlotLabel -> "        atp+asp+amm=amp+ppi+aspi", DisplayFunction -> Identity];

In[44]:= plot4 = Plot[
            calctrGerx[atp + aspartate + glutamine + h2o + de == amp + ppi + asparagineL + glutamate, pH, .
            {pH, 5, 9}, AxesLabel → {"pH", "Δr G'°"}, AxesOrigin → {5, -60},
            PlotLabel -> "        atp+asp+gluti+h2o=amp+ppi+aspi+glut",
            DisplayFunction -> Identity];

In[45]:= plot11 = Plot[Evaluate[calcNHrx[atp + malate + coA + de == adp + pi + malylcoA, pH, .25]],
            {pH, 5, 9}, AxesLabel → {"pH", "Δr NH"}, AxesOrigin → {5, 0}, PlotRange -> {0, 1},
            PlotLabel -> "        atp+malate+coA=adp+pi+malylcoA", DisplayFunction -> Identity];

In[46]:= plot12 = Plot[Evaluate[calcNHrx[adp + pi + acetylcoA + de == atp + acetate + coA, pH, .25]],
            {pH, 5, 9}, AxesLabel → {"pH", "ΔrNH"}, AxesOrigin → {5, -1}, PlotRange → {-1, 0},
            PlotLabel -> "        adp+pi+acetylcoA=atp+acetate+coA",
            DisplayFunction -> Identity];

In[47]:= plot13 =
            Plot[Evaluate[calcNHrx[atp + aspartate + ammonia + de == amp + ppi + asparagineL, pH, .25]],
            {pH, 5, 9}, AxesLabel → {"pH", "Δr NH"}, AxesOrigin → {5, -1.5}, PlotRange → {-1.5, 0},
            PlotLabel -> "        atp+asp+amm=amp+ppi+aspi", DisplayFunction -> Identity];

In[48]:= plot14 = Plot[Evaluate[
            calcNHrx[atp + aspartate + glutamine + h2o + de == amp + ppi + asparagineL + glutamate, pH, .2
            {pH, 5, 9}, AxesLabel → {"pH", "Δr NH"}, AxesOrigin → {5, -2}, PlotRange → {-2, 0},
            PlotLabel -> "        atp+asp+gluti+h2o=amp+ppi+aspi+glut",
            DisplayFunction -> Identity];

In[49]:= Show[
            GraphicsArray[{{plot1, plot11}, {plot2, plot12}, {plot3, plot13}, {plot4, plot14}}]];
```

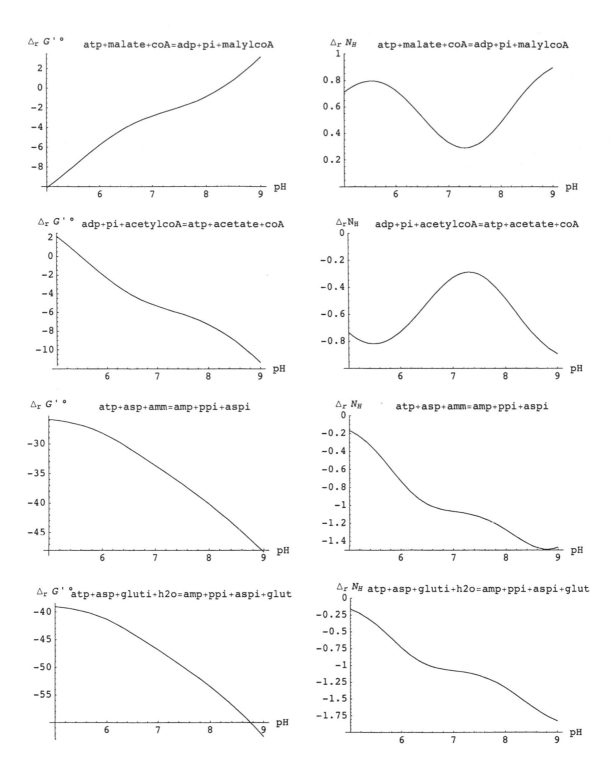

Figure 11.2 $\Delta_r G'^\circ$ and $\Delta_r N_H$ for four ligase reactions at 298.15 K and 0.25 M ionic strength.

The following abbreviations have been used in the lower two plots: asp for aspartate, amm for ammonia, aspi for asparagine, glut for glutamate, and gluti for glutamine.

The effect of temperature on standard transformed thermodynamic properties of ligase reactions have been calculated in Chapter 4 for reactions 6.2.1.1, 6.3.1.2, and 6.4.1.1. The effects of temperature on 5 ligase reactions are shown in Chapter 13.

11.8 Discussion

Ligase reactions provide remarkable examples of enzyme-catalyzed reactions with 6 to 8 reactants. Since $C' = N' - R'$ at specified pH and $C'' = N'' - R''$ at specified pH and specified availability of oxygen atoms, this means there are 5 to 7 components. Three or four conservation equations are provided by elements, and so these enzyme mechanisms introduce 2 to 4 or 1 to 3 conservation equations in addition in coupling. In this respect enzyme-catalyzed reactions can be very different from chemical reactions (see Chapter 7).

Appendix

Components in Ligase Reactions

Ligase reactions necessarily involve coupling. As a first example, consider reaction 6.3.1.2:

glutamate + atp + ammonia = glutamine + adp + pi   (A11-1)

This reaction can be considered to be made up of the following two hydrolase reactions:

EC 3.5.1.2 glutamine + h2o = glutamate + ammonia   (A11-2)

EC 3.6.1.3 atp + h2o = adp + pi   (A11-3)

When this difference is taken h2o cancels. In reaction A11-1 oxygen atoms are conserved. The following conservation matrix balances oxygen. To understand this coupling, a constraint in addition to those for C, O, N, and P has to be added to insure that a molecule of glutamine is produced for each ATP consumed; con1 is n(atp) + n(glutamine) = const. With these five constraints, the conservation matrix is given by

```
In[50]:= a = {{5, 10, 0, 0, 5, 10}, {4, 13, 4, 0, 3, 10},
            {1, 5, 0, 1, 2, 5}, {0, 3, 1, 0, 0, 2}, {0, 1, 0, 0, 1, 0}};
```

```
In[51]:= TableForm[a, TableHeadings →
            {{C, O, N, P, con1}, {"glutamate", "atp", "pi", "amm", "glutamine", "adp"}}]
```

Out[51]//TableForm=

	glutamate	atp	pi	amm	glutamine	adp
C	5	10	0	0	5	10
O	4	13	4	0	3	10
N	1	5	0	1	2	5
P	0	3	1	0	0	2
con1	0	1	0	0	1	0

```
In[52]:= TableForm[RowReduce[a], TableHeadings → {{"glute", "atp", "pi", "amm", "glutne"},
            {"glute", "atp", "pi", "amm", "glutne", "adp"}}]
```

```
Out[52]//TableForm=
                glute     atp      pi      amm     glutne     adp
       glute      1        0       0       0        0          1
       atp        0        1       0       0        0          1
       pi         0        0       1       0        0         -1
       amm        0        0       0       1        0          1
       glutne     0        0       0       0        1         -1
```

When this reaction is considered from the viewpoint of the further transformed Gibbs energy $G\,''$, an additional constraint con2 is needed.

Second, consider reaction 6.3.5.4 that has 8 reactants:

$$atp + aspartate + glutamine + h2o = amp + ppi + asparagineL + glutamate \tag{A11-4}$$

This reaction can be considered to be made up of three hydrolase reactions:

$$atp + h2o = adp + pi \tag{A11-5}$$

$$glutamine + h2o = glutamate + ammonia \tag{A11-6}$$

$$asparagineL + h2o = aspartate + ammonia \tag{A11-7}$$

The conservation matrix for reaction A11-4 is given by

```
In[53]:= a6x3x5x4 = {{10, 4, 5, 0, 10, 0, 4, 5},
             {13, 4, 3, 1, 7, 7, 3, 4}, {5, 1, 2, 0, 5, 0, 2, 1}, {3, 0, 0, 0, 1, 2, 0, 0}};
```

```
In[54]:= TableForm[RowReduce[a6x3x5x4]]
```

```
Out[54]//TableForm=
```

1	0	0	0	$\frac{1}{3}$	$\frac{2}{3}$	0	0
0	1	0	0	$-\frac{10}{9}$	$\frac{10}{9}$	$-\frac{2}{3}$	$\frac{5}{3}$
0	0	1	0	$\frac{20}{9}$	$-\frac{20}{9}$	$\frac{4}{3}$	$-\frac{1}{3}$
0	0	0	1	$\frac{4}{9}$	$\frac{5}{9}$	$\frac{5}{3}$	$-\frac{5}{3}$

This shows that three more constraints are needed. If we view the three reactions as the points of a triangle, we can view the three constraints as the sides of the triangle. There are a number of ways to do this, but the three constraints used here are:
con1: $n(atp) + n(asparagineL)$
con2: $n(atp) + n(glutamate)$
con3: $n(aspartate) + n(ppi)$

```
In[55]:= a6x3x5x4 = {{10, 4, 5, 0, 10, 0, 4, 5},
             {13, 4, 3, 1, 7, 7, 3, 4}, {5, 1, 2, 0, 5, 0, 2, 1}, {3, 0, 0, 0, 1, 2, 0, 0},
             {1, 0, 0, 0, 0, 0, 1, 0}, {1, 0, 0, 0, 0, 0, 0, 1}, {0, 1, 0, 0, 0, 1, 0, 0}};
```

The following shorter names are used to label the columns: asp=aspartate, aspN=aspN=asparagineN, glut=glutamate, glutN=glutamine.

```
In[56]:= TableForm[a6x3x5x4, TableHeadings → {{C, O, N, P, con1, con2, con3},
             {"atp", "asp", "glutN", "h2o", "amp", "ppi", "aspN", "glut"}}]
```

Out[56]//TableForm=

	atp	asp	glutN	h2o	amp	ppi	aspN	glut
C	10	4	5	0	10	0	4	5
O	13	4	3	1	7	7	3	4
N	5	1	2	0	5	0	2	1
P	3	0	0	0	1	2	0	0
con1	1	0	0	0	0	0	1	0
con2	1	0	0	0	0	0	0	1
con3	0	1	0	0	0	1	0	0

In[57]:= **TableForm[RowReduce[a6x3x5x4],**
     TableHeadings → {{"atp", "asp", "glutN", "h2o", "amp", "ppi", "aspN"},
       {"atp", "asp", "glutN", "h2o", "amp", "ppi", "aspN", "glut"}}]

Out[57]//TableForm=

	atp	asp	glutN	h2o	amp	ppi	aspN	glut
atp	1	0	0	0	0	0	0	1
asp	0	1	0	0	0	0	0	1
glutN	0	0	1	0	0	0	0	1
h2o	0	0	0	1	0	0	0	1
amp	0	0	0	0	1	0	0	-1
ppi	0	0	0	0	0	1	0	-1
aspN	0	0	0	0	0	0	1	-1

This conservation matrix yields the correct stoichiometric number matrix, as indicated by the last column.

Now consider reaction 6.3.5.4 from the standpoint of the further transformed Gibbs energy G''. The oxygen row and the h2o column are omitted to give

In[58]:= **a6x3x5x4noO = {{10, 4, 5, 10, 0, 4, 5}, {5, 1, 2, 5, 0, 2, 1}, {3, 0, 0, 1, 2, 0, 0},**
     {1, 0, 0, 0, 0, 1, 0}, {1, 0, 0, 0, 0, 0, 1}, {0, 1, 0, 0, 1, 0, 0}};

In[59]:= **TableForm[RowReduce[a6x3x5x4noO],**
     TableHeadings → {{"atp", "asp", "glutN", "amp", "ppi", "aspN"},
       {"atp", "asp", "glutN", "amp", "ppi", "aspN", "glut"}}]

Out[59]//TableForm=

	atp	asp	glutN	amp	ppi	aspN	glut
atp	1	0	0	0	0	0	1
asp	0	1	0	0	0	0	1
glutN	0	0	1	0	0	0	1
amp	0	0	0	1	0	0	-1
ppi	0	0	0	0	1	0	-1
aspN	0	0	0	0	0	1	-1

The same number of constraints is required. There are now seven reactants, and so there are six constraints: $C'' = 6$ (C,N,P, con1 to con3).

References

1. E. C. Webb, Enzyme Nomenclature 1992, Academic Press (1992).

In[60]:= **http : // www.chem.qmw.ac.uk / iubmb / enzyme /**

2. EC-PDP Enzyme Structures DataBase

In[60]:= **http : // www.ebi.acuk / thornton - srv / databases / enzymes /**

3. R. N. Goldberg and Y. B. Tewari, Thermodynamics of enzyme-catalyzed reactions: Part 4. Lyases, J. Phys. Chem. Ref, Data 24, 1669-1698 (1995).

4. R. N. Goldberg and Y. B. Tewari, Thermodynamics of enzyme-catalyzed reactions: Part 5. Isomerases and ligases, J. Phys. Chem. Ref, Data 24, 1765-1801 (1995).

5. R. A. Alberty, Thermodynamics of the mechanism of the nitrogenase reaction, Biophysical Chemistry, 114, 115-120 (2005).

Chapter 12 Survey of Reactions at 298.15 K

    The objective of this chapter is to demonstrate the breath and depth of the enzyme-catalyzed reactions for which $\Delta_r G'^\circ$ and $\Delta_r N_H$ at 298.15 K can be calculated using the database BasicBiochemData3 (1). This chapter gives these properties for 229 reactions that involve 167 reactants at five pHs and ionic strength 0.25 M. This is not a complete set of reactions for which these properties can be calculated. Chapter 8 showed that these properties can be calculated for 1770 oxidoreductase reactions alone, but enzymes are not known for all of these reactions. Some of the EC enzymes are written in terms of acceptors and reduced acceptors. What is presented here is a list of reactions identified by looking at the reactions in EC (2), EC-PDB (3), and EXPASSY (4) for which the enzymes have been named and $\Delta_f G^\circ$ is known for all the species that are significant in the pH range 5 to 9.

    This list of reactions is useful for estimating $\Delta_r G'^\circ$ and $\Delta_r N_H$ for similar reactions because reactions of a given type are grouped together. However, estimation methods should be based on $\Delta_f G^\circ$ of species and their pKs. An index in the Appendix of this chapter shows the reactions in which each of the reactants is involved. This index shows that if $\Delta_f G^\circ$ values were not known for atp, this list of reactions would be shorter by 45. But if the $\Delta_f G^\circ$ for urea was not known, the list would only be shorter by one. In general, the addition of a reactant to BasicBiochemData3 introduces a number of new reactions, and so the length of such a list of reactions increases exponentially with the number of reactants in the database. There are three ways that $\Delta_f G^\circ$ and $\Delta_f H^\circ$ values of species of reactants can be obtained:

    (1) Some values, like those for co2tot, were obtained from the NBS Tables (5).

    (2) Most of the others were obtained from measurement of apparent equilibrium constants and transformed enthalpies of enzyme-catalyzed reactions. pKs of species are also needed. These calculations, which are described in detail in Chapter 6, have been greatly facilitated by the publications by Goldberg and Tewari of evaluated data from the literature.

(3) $\Delta_f G°$ and $\Delta_f H°$ values of species of reactants can be estimated from tables of these values at zero ionic strength, but this method has not been utilized much as yet. Estimation methods will be useful in the future because molecules of biochemical interest are often large and the various reactive groups may be rather independent of the rest of the molecule.

In the list of reactions discussed here, some reactions can be balanced with gases for which $\Delta_f G°$ values are in the database, but this has not been done here because there is no gas phase in living cells, and so equilibrium concentrations in the aqueous phase are of more interest.

```
In[2]:=  Off[General::"spell"];
         Off[General::"spell1"];


In[4]:=  << BiochemThermo`BasicBiochemData3`
```

A program **trGibbsRxSummary** was written to calculate $\Delta_r G'°$ and $\Delta_r N_H$ at 298.15 K and specified pHs and ionic strengths from the functions (atp, adp,...) of pH and ionic strength in BasicBiochemData3. The use of **round** is discussed in the Appendix.

```
In[5]:=  round[vec_, params_:{6, 2}] :=(*When a list of numbers has more digits to the right
         of the decimal point than you want, say 6, you can request 2 by using
         round[vec,{6,2}],*)
            Flatten[Map[NumberForm[#1, params] & , {vec}, {2}]]

In[6]:=  trGibbsRxSummary[eq_, title_, reaction_, pHlist_, islist_] :=
            Module[{functiom, trGibbse, dvtNH, vectorNH},
               (*When this program is given a reaction equation in the form acetaldehyde+
                  nadred+de==ethanol+nadox,it calculates the standard transformed Gibbs
                  energies of reaction in kJ mol^-1 and the change in binding of hydrogen
                  ions in the reaction at the desired pHs and ionic strengths at 298.15
                  K.  title_ is in the form "EC 1.1.1.1 Alcohol dehydrogenase".  reaction_
                  is in the form "acetaldehyde+nadred=ethanol+nadox".*)
               function = Solve[eq, de];
               trGibbse = round[function[[1, 1, 2]] /. pH -> pHlist /. is -> islist, {4, 2}];
               dvtNH = (1 / (8.31451 * .29815 * Log[10])) * D[function[[1, 1, 2]], pH];
               vectorNH = round[dvtNH /. pH -> pHlist /. is -> islist, {4, 2}];
               Print[title]; Print[reaction]; Print[trGibbse]; Print[vectorNH]]
```

The EC number and name of the enzyme are included in the input. The last two lines of the output give $\Delta_r G'°$ values in kJ mol^{-1} and $\Delta_r N_H$ values at each pH. In each case the reaction equation has been typed in the direction in which the reaction is spontaneous at pH 7. It is more convenient to list $\Delta_r G'°$ values rather than apparent equilibrium constants, but apparent equilibrium constants are readily calculated using $K' = \exp[-\Delta_r G'°/RT]$. The following table gives the K' values that correspond to a range of $\Delta_r G'°$ values.

```
In[7]:=  TableForm[Transpose[
            {{0, 5, 10, 15, 20, 25, 30, 35}, Exp[{0, 5, 10, 15, 20, 25, 30, 35} / (8.31451 * .29815)]}],
            TableHeadings -> {None, {"-Δr G ' °", K'}}]
```

Out[7]//TableForm=

$-\Delta_r G'°$	K'
0	1
5	7.51549
10	56.4825
15	424.494
20	3190.28
25	23976.5
30	180195.
35	1.35425×10^6

The change in binding of hydrogen ions in a biochemical reaction, $\Delta_r N_H$, is given by

$$\Delta_r N_H = (1/(RT\ln(10)))*(\partial\Delta_r G'°/\partial pH) \tag{12.1-1}$$

The rate of change of $\Delta_r G'°$ with pH is given by

$$(\partial\Delta_r G'°/\partial pH) = RT\ln(10)\,\Delta_r N_H \tag{12.1-2}$$

Thus for 298.15 K and ionic strength 0.25 M, the changes in binding of hydrogen ions of 1, 2, 3, 4, 5, 6, 7, and 8 yield the following values of $(\partial\Delta_r G'°/\partial pH)$:

In[8]:= **TableForm[**
      Transpose[{{1, 2, 3, 4, 5, 6, 7, 8}, 8.31451 * .29815 * Log[10] * {1, 2, 3, 4, 5, 6, 7, 8}}],
      TableHeadings → {None, {"$\Delta_r N_H$ ", "$\partial\Delta_r G'° / \partial pH$"}}]

Out[8]//TableForm=

$\Delta_r N_H$	$\partial\Delta_r G'° / \partial pH$
1	5.70804
2	11.4161
3	17.1241
4	22.8322
5	28.5402
6	34.2483
7	39.9563
8	45.6643

A check on the calculations of $\Delta_r N_H$ can be made by writing the reactions in terms of species at high pH to see how many hydrogen ions have to be put on the right or the left to balance charge. It is relatively simple to calculate $\Delta_r N_H$ at high pH by considering the reaction written in terms of the species with the fewest hydrogen atoms. For example, for the reaction

$$\text{acetaldehyde} + \text{nadred} = \text{ethanol} + \text{nadox} \tag{12.1-3}$$

the charges of species at high pH are given by

$$\text{acetaldehyde}^0 + \text{nadred}^{-2} + H^+ = \text{ethanol}^0 + \text{nadox}^{-1} \tag{12.1-4}$$

The H^+ has to be inserted on the left to balance charges. Thus at high pH there is an increase in the binding of hydrogen ions and $\Delta_r N_H = 1.0$. Since none of the four reactants have pKs in the region pH 5 to 9, $\Delta_r N_H$ is independent of pH for this reaction. When $\Delta_r N_H$ is independent of pH *Mathematica* types out this constant value only once.

Another way to calculate $\Delta_r N_H$ at high pH is to count the numbers of hydrogen atoms in the species at high pH.

$$\text{acetaldehydeH}_4 + \text{nadredH}_{27} + H^+ = \text{ethanolH}_6 + \text{nadoxH}_{26} \tag{12.1-5}$$

The H$^+$ has to be added to the left side to balance hydrogen atoms in this chemical reaction. Thus we arrive at the same conclusion that $\Delta_r N_H = 1.0$ independent of pH.

In the following tables for the six classes of enzyme-catalyzed reactions, the reactions are written using the abbreviations for the names of reactants that are used in *Mathematica*. There is a Glossary in the Appendix that gives the full names of reactants. These *Mathematica* names are often transparent like atp, adp, and amp. Names that are not so transparent are pep for phosphoenolpyruvate and prpp for 5-phospho-D-ribose 1-diphosphate. BasicBiochemData3 contains some information on gases, which are indicated by g, as in h2g. When it is important to be clear that the reactant is in aqueous solution, the suffix aq is added in BasicBiochemData3, as in h2aq. Glycerone is referred to as dihydroxyacetone. 2-oxoglutarate is referred to as ketoglutarate. Since names in *Mathematica* cannot start with capital letters or numbers - or contain dots, dashes, or spaces - abbreviations like glycerol3phos are used. In enzyme-catalyzed reactions at specified pH, H$^+$ should **never** be shown. Suffixes ox and red are used, as in nadox and nadred. The sum of the equilibrium concentrations of $CO_2(aq)$, $H_2 CO_3$, $HCO_3{}^-$, and $CO_3{}^{2-}$ is referred to as co2tot. Note that when $CO_2(g)$ is replaced by co2tot in the reaction equation, is is necessary to add h2o to the other side of the reaction.

If $\Delta_r G'^\circ$ in a region of pH becomes more negative with increasing pH (that is K' is increasing), $\Delta_r N_H$ is negative in this region. Conversely, if $\Delta_r G'^\circ$ in a region of pH becomes more positive with increasing pH (that is K' is decreasing), $\Delta_r N_H$ is positive in this region.

The numbers of reactions in each class of the EC list are given in the following table along with the number of reactions for which thermodynamic properties are given in this chapter.

	No. in EC	No. in Ch. 12
EC 1 Oxidoreductases	2487	92
EC 2 Transferases	3273	53
EC 3 Hydrolases	6199	37
EC 4 Lyases	1026	23
EC 5 Isomerases	637	11
EC 6 Ligases	386	13

The following list of tables gives $\Delta_r G'^\circ$ in kJ mol^{-1} and $\Delta_r N_H$ at 298.15 K, 0.25 M ionic strength and pHs 5, 6, 7, 8, and 9.

12.1 Oxidoreductase Reactions

EC 1.1 Acting on the CH-OH group of donors

```
In[9]:= trGibbsRxSummary[acetaldehyde + nadred + de == ethanol + nadox,
          "EC 1.1.1.1a Alcohol dehydrogenase",
          "acetaldehyde+nadred=ethanol+nadox", {5, 6, 7, 8, 9}, .25];

        EC 1.1.1.1a Alcohol dehydrogenase

        acetaldehyde+nadred=ethanol+nadox

        {-33.51, -27.8, -22.1, -16.39, -10.68}

        {1.}

In[10]:= trGibbsRxSummary[butanal + nadred + de == butanoln + nadox,
          "EC 1.1.1.1b Alcohol dehydrognase",
          "butanal+nadred=butanoln+nadox", {5, 6, 7, 8, 9}, .25]
```

EC 1.1.1.1b Alcohol dehydrognase

butanal+nadred=butanoln+nadox

{-32.84, -27.13, -21.43, -15.72, -10.01}

{1.}

In[11]:= **trGibbsRxSummary[acetone + nadred + de == propanol2 + nadox,**
         "EC 1.1.1.1c Alcohol dehydrognase",
         "acetone+nadred=propanol2+nadox", {5, 6, 7, 8, 9}, .25]

EC 1.1.1.1c Alcohol dehydrognase

acetone+nadred=propanol2+nadox

{-16.4, -10.69, -4.99, 0.72, 6.43}

{1.}

In[12]:= **trGibbsRxSummary[dihydroxyacetone + nadred + de == glycerol + nadox,**
         "EC 1.1.1.6 Glycerol dehydrogenase",
         "dihydroxyacetone+nadred=glycerol+nadox", {5, 6, 7, 8, 9}, .25]

EC 1.1.1.6 Glycerol dehydrogenase

dihydroxyacetone+nadred=glycerol+nadox

{-37.35, -31.64, -25.94, -20.23, -14.52}

{1.}

In[13]:= **trGibbsRxSummary[xylulose + nadpred + de == xylitol + nadpox,**
         "EC 1.1.1.10 L-xylulose reductase",
         "xylulose+nadpred=xylitol+nadpox", {5, 6, 7, 8, 9}, .25]

EC 1.1.1.10 L-xylulose reductase

xylulose+nadpred=xylitol+nadpox

{-28.39, -22.68, -16.98, -11.27, -5.56}

{1.}

In[14]:= **trGibbsRxSummary[sorbose + nadred + de == iditol + nadox,**
         "EC 1.1.1.14 L-Iditol 2-dehydrognase",
         "sorbose+nadred=iditol+nadox", {5, 6, 7, 8, 9}, .25]

EC 1.1.1.14 L-Iditol 2-dehydrognase

sorbose+nadred=iditol+nadox

{-23.65, -17.94, -12.24, -6.53, -0.82}

{1.}

In[15]:= **trGibbsRxSummary[fructose6phos + nadred + de == mannitol1phos + nadox,**
         "EC 1.1.1.17 Mannitol-1-phosphate 5-dehydrogenase",
         "fructose6phos+nadred=mannitol1phos+nadox", {5, 6, 7, 8, 9}, .25]

EC 1.1.1.17 Mannitol-1-phosphate 5-dehydrogenase

```
fructose6phos+nadred=mannitol1phos+nadox
```

```
{-24.78, -18.46, -12.32, -6.54, -0.82}
```

```
{1.06, 1.12, 1.03, 1., 1.}
```

In[16]:= **trGibbsRxSummary[glyoxylate + nadred + de == glycolate + nadox,**
        "EC 1.1.1.26 Glycolate reductase",
        "glyoxylate+nadred=glycolate+nadox", {5, 6, 7, 8, 9}, .25]

```
EC 1.1.1.26 Glycolate reductase
```

```
glyoxylate+nadred=glycolate+nadox
```

```
{-53.22, -47.51, -41.81, -36.1, -30.39}
```

```
{1.}
```

In[17]:= **trGibbsRxSummary[pyruvate + nadred + de == lactate + nadox,**
        "EC 1.1.1.27 L-lactate dehydrogenase",
        "pyruvate+nadred=lactate+nadox", {5, 6, 7, 8, 9}, .25]

```
EC 1.1.1.27 L-lactate dehydrogenase
```

```
pyruvate+nadred=lactate+nadox
```

```
{-35.32, -29.61, -23.91, -18.2, -12.49}
```

```
{1.}
```

In[18]:= **trGibbsRxSummary[hydroxypyruvate + nadred + de == glycerate + nadox,**
        "EC 1.1.1.29 Glycerate dehydrogenase",
        "hydroxypyruvate+nadred=glycerate+nadox", {5, 6, 7, 8, 9}, .25]

```
EC 1.1.1.29 Glycerate dehydrogenase
```

```
hydroxypyruvate+nadred=glycerate+nadox
```

```
{-43.33, -37.62, -31.92, -26.21, -20.5}
```

```
{1.}
```

In[19]:= **trGibbsRxSummary[oxaloacetate + nadred + de == malate + nadox,**
        "EC 1.1.1.37 Malate dehydrogenase",
        "oxaloacetate+nadred=malate+nadox", {5, 6, 7, 8, 9}, .25]

```
EC 1.1.1.37 Malate dehydrogenase
```

```
oxaloacetate+nadred=malate+nadox
```

```
{-41.23, -34.65, -28.84, -23.12, -17.41}
```

```
{1.33, 1.05, 1., 1., 1.}
```

In[20]:= **trGibbsRxSummary[pyruvate + nadred + co2tot + de == malate + nadox + h2o,**
        "EC 1.1.1.39 Malate dehydrogenase (decarboxylating)",
        "pyruvate+nadred+co2tot=malate+nadox+h2o", {5, 6, 7, 8, 9}, .25]

```
EC 1.1.1.39 Malate dehydrogenase (decarboxylating)
```

```
pyruvate+nadred+co2tot=malate+nadox+h2o
```

{-7.91, -5.75, -1.61, 3.89, 9.93}

{0.41, 0.5, 0.9, 1.01, 1.15}

For the next three reactions, nadox has to be put on the left.

In[21]:= **trGibbsRxSummary[citrateiso + nadpox + h2o + de == ketoglutarate + nadpred + co2tot,**
     "EC 1.1.1.42 Isocitrate dehydrogenase (NADP)",
     "citrateiso+nadpox+h2o=ketoglutarate+nadpred+co2tot", {5, 6, 7, 8, 9}, .25]

     EC 1.1.1.42 Isocitrate dehydrogenase (NADP)

     citrateiso+nadpox+h2o=ketoglutarate+nadpred+co2tot

     {4.37, -0.17, -4.88, -10.45, -16.49}

     {-0.98, -0.72, -0.93, -1.01, -1.15}

In[22]:= **trGibbsRxSummary[galactose + nadox + de == galactono14lactone + nadred,**
     "EC 1.1.1.48 Galactose 1-dehydrogenase",
     "galactose+nadox=galactono14lactone+nadred", {5, 6, 7, 8, 9}, .25]

     EC 1.1.1.48 Galactose 1-dehydrogenase

     galactose+nadox=galactono14lactone+nadred

     {-5.54, -11.25, -16.95, -22.66, -28.37}

     {-1.}

In[23]:= **trGibbsRxSummary[glucose6phos + nadpox + de == gluconolactone6phos + nadpred,**
     "EC 1.1.1.49 Glucose-6-phosphate 1-dehydrogenase",
     "glucose6phos+nadpox=gluconolactone6phos+nadpred", {5, 6, 7, 8, 9}, .25]

     EC 1.1.1.49 Glucose-6-phosphate 1-dehydrogenase

     glucose6phos+nadpox=gluconolactone6phos+nadpred

     {8.31, 2.6, -3.1, -8.81, -14.52}

     {-1., -1., -1., -1., -1.}

In[24]:= **trGibbsRxSummary[ribulose + nadred + de == ribitol + nadox,**
     "EC 1.1.1.56 Ribitol 2-dehydrogenase",
     "ribulose+nadred=ribitol+nadox", {5, 6, 7, 8, 9}, .25]

     EC 1.1.1.56 Ribitol 2-dehydrogenase

     ribulose+nadred=ribitol+nadox

     {-29.73, -24.02, -18.32, -12.61, -6.9}

     {1.}

In[25]:= **trGibbsRxSummary[fructose + nadred + de == mannitolD + nadox,**
     "EC 1.1.1.67 Mannitol 2-dehydrogenase",
     "fructose+nadred=mannitolD+nadox", {5, 6, 7, 8, 9}, .25]

     EC 1.1.1.67 Mannitol 2-dehydrogenase

     fructose+nadred=mannitolD+nadox

{-17.97, -12.26, -6.56, -0.85, 4.86}

{1.}

In[26]:= **trGibbsRxSummary[glyceraldehyde + nadpred + de == glycerol + nadpox,**
          "EC 1.1.1.72 Glycerol dehydrogenase (NADP)",
          "glyceraldehyde+nadpred=glycerol+nadpox", {5, 6, 7, 8, 9}, .25]

EC 1.1.1.72 Glycerol dehydrogenase (NADP)

glyceraldehyde+nadpred=glycerol+nadpox

{-47.91, -42.2, -36.5, -30.79, -25.08}

{1.}

In[27]:= **trGibbsRxSummary[glyoxylate + nadpred + de == glycolate + nadpox,**
          "EC 1.1.1.79 Glyoxylate reductase (NADP)",
          "glyoxylate+nadpred=glycolate+nadpox", {5, 6, 7, 8, 9}, .25]

EC 1.1.1.79 Glyoxylate reductase (NADP)

glyoxylate+nadpred=glycolate+nadpox

{-52.8, -47.09, -41.39, -35.68, -29.97}

{1.}

In[28]:= **trGibbsRxSummary[acetone + nadpred + de == propanol2 + nadpox,**
          "EC 1.1.1.80 Isopropyl dehydrogenase (NADP)",
          "acetone+nadpred=propanol2+nadpox", {5, 6, 7, 8, 9}, .25]

EC 1.1.1.80 Isopropyl dehydrogenase (NADP)

acetone+nadpred=propanol2+nadpox

{-15.98, -10.27, -4.57, 1.14, 6.85}

{1.}

In[29]:= **trGibbsRxSummary[dihydroxyacetonephos + nadpred + de == glycerol3phos + nadpox,**
          "EC 1.1.1.94 Glycerol-3-phosphate dehydrogenase (NAD(P))",
          "dihydroxyacetonephos+nadpred=glycerol3phos+nadpox", {5, 6, 7, 8, 9}, .25]

EC 1.1.1.94 Glycerol-3-phosphate dehydrogenase (NAD(P))

dihydroxyacetonephos+nadpred=glycerol3phos+nadpox

{-57.51, -49.16, -42., -36.06, -30.32}

{1.35, 1.44, 1.1, 1.01, 1.}

In[30]:= **trGibbsRxSummary[phosphohydroxypyruvate + nadred + de == phosphoglycerate3 + nadox,**
          "EC 1.1.1.95 Phosphoglycerate dehydrogenase",
          "phosphohydroxypyruvate+nadred=phosphoglycerate3+nadox", {5, 6, 7, 8, 9}, .25]

EC 1.1.1.95 Phosphoglycerate dehydrogenase

phosphohydroxypyruvate+nadred=phosphoglycerate3+nadox

{-54.37, -43.38, -34.29, -27.73, -21.92}

{1.98, 1.83, 1.32, 1.05, 1.}

In[31]:= **trGibbsRxSummary[retinal + nadred + de == retinol + nadox,**
"EC 1.1.1.105 Retinol dehydrogenase",
"retinal+nadred=retinol+nadox", {5, 6, 7, 8, 9}, .25]

EC 1.1.1.105 Retinol dehydrogenase

retinal+nadred=retinol+nadox

{-18.78, -13.07, -7.37, -1.66, 4.05}

{1.}

In[32]:= **trGibbsRxSummary[glucose + nadpox + de == gluconolactone + nadpred,**
"EC 1.1.1.119 Glucose 1-dehydrogenase (NADP)",
"glucose+nadpox=gluconolactone+nadpred", {5, 6, 7, 8, 9}, .25]

EC 1.1.1.119 Glucose 1-dehydrogenase (NADP)

glucose+nadpox=gluconolactone+nadpred

{2.85, -2.86, -8.56, -14.27, -19.98}

{-1.}

In[33]:= **trGibbsRxSummary[fructose + nadpred + de == mannitolD + nadpox,**
"EC 1.1.1.138 Mannitol 2-dehydrogenase (NADP)",
"fructose+nadpred=mannitolD+nadpox", {5, 6, 7, 8, 9}, .25]

EC 1.1.1.138 Mannitol 2-dehydrogenase (NADP)

fructose+nadpred=mannitolD+nadpox

{-17.55, -11.84, -6.14, -0.43, 5.28}

{1.}

In[34]:= **trGibbsRxSummary[fructose6phos + nadred + de == sorbitol6phos + nadox,**
"EC 1.1.1.140 Sorbitol-6-phosphate 2-dehydrogenase",
"fructose6phos+nadred=sorbitol6phos+nadox", {5, 6, 7, 8, 9}, .25]

EC 1.1.1.140 Sorbitol-6-phosphate 2-dehydrogenase

fructose6phos+nadred=sorbitol6phos+nadox

{-23.9, -17.78, -11.8, -6.05, -0.34}

{1.04, 1.08, 1.02, 1., 1.}

In[35]:= **trGibbsRxSummary[dihydroxyacetone + nadpred + de == glycerol + nadpox,**
"EC 1.1.1.156 Glycerol 2-dehydrogenase (NADP)",
"dihydroxyacetone+nadpred=glycerol+nadpox", {5, 6, 7, 8, 9}, .25]

EC 1.1.1.156 Glycerol 2-dehydrogenase (NADP)

dihydroxyacetone+nadpred=glycerol+nadpox

{-36.93, -31.22, -25.52, -19.81, -14.1}

{1.}

In[36]:= **trGibbsRxSummary[mannitolD + cytochromecox + de == fructose + cytochromecred,**
"EC 1.1.2.2 Mannitol dehydrogenase (cytochrome)",
"mannitolD+cytochromecox=fructose+cytochromecred", {5, 6, 7, 8, 9}, .25]

EC 1.1.2.2 Mannitol dehydrogenase (cytochrome)

mannitolD+cytochromecox=fructose+cytochromecred

{-52.06, -63.48, -74.89, -86.31, -97.73}

{-2.}

In[37]:= **trGibbsRxSummary[lactate + 2 * cytochromecox + de == pyruvate + 2 * cytochromecred,**
"EC 1.1.2.3 L-Lactate dehydrogenase (cytochrome)",
"lactate+2*cytochromecox=pyruvate+2*cytochromecred", {5, 6, 7, 8, 9}, .25]

EC 1.1.2.3 L-Lactate dehydrogenase (cytochrome)

lactate+2*cytochromecox=pyruvate+2*cytochromecred

{-55.17, -66.59, -78.01, -89.42, -100.8}

{-2.}

In[38]:= **trGibbsRxSummary[malate + o2aq + de == oxaloacetate + h2o2aq,**
"EC 1.1.3.3 Malate oxidase", "malate+o2aq=oxaloacetate+h2o2aq", {5, 6, 7, 8, 9}, .25]

EC 1.1.3.3 Malate oxidase

malate+o2aq=oxaloacetate+h2o2aq

{-100.1, -100.9, -101., -101.1, -101.1}

{-0.33, -0.05, -0.00, -0., -0.}

In[39]:= **trGibbsRxSummary[glucose + o2aq + de == gluconolactone + h2o2aq,**
"EC 1.1.3.4 Glucose oxidase",
"glucose+o2aq=gluconolactone+h2o2aq", {5, 6, 7, 8, 9}, .25]

EC 1.1.3.4 Glucose oxidase

glucose+o2aq=gluconolactone+h2o2aq

{-138., -138., -138., -138., -138.}

{1.24481×10^{-15}}

In[40]:= **trGibbsRxSummary[glycerol3phos + o2aq + de == dihydroxyacetonephos + h2o2aq,**
"EC 1.1.3.21 Glycerol-3-phosphate oxidase",
"glycerol3phos+o2aq=dihydroxyacetonephos+h2o2aq", {5, 6, 7, 8, 9}, .25]

EC 1.1.3.21 Glycerol-3-phosphate oxidase

glycerol3phos+o2aq=dihydroxyacetonephos+h2o2aq

{-83.37, -86.01, -87.47, -87.7, -87.73}

{-0.35, -0.44, -0.1, -0.01, -0.00}

In[41]:= **trGibbsRxSummary[ketoglutarate + nadred + de == hydroxyglutarate + nadox,**
"EC 1.1.99.2 2-Hydroxyglutarate dehydrogenase",
"ketoglutarate+nadred=hydroxyglutarate+nadox", {5, 6, 7, 8, 9}, .25]

EC 1.1.99.2 2-Hydroxyglutarate dehydrogenase

ketoglutarate+nadred=hydroxyglutarate+nadox

{-37.64, -31.93, -26.23, -20.52, -14.81}

{1.}

In[42]:= **trGibbsRxSummary[lactate + oxaloacetate + de == malate + pyruvate,**
"EC 1.1.99.7 Lactate-malate trans-hydrogenase",
"lactate+oxaloacetate=malate+pyruvate", {5, 6, 7, 8, 9}, .25]

EC 1.1.99.7 Lactate-malate trans-hydrogenase

lactate+oxaloacetate=malate+pyruvate

{-5.91, -5.04, -4.93, -4.92, -4.92}

{0.33, 0.05, 0.00, 0., 0.}

EC 1.2 Acting on the aldehyde or oxo group of donors

In[43]:= **trGibbsRxSummary[formate + nadox + h2o + de == co2tot + nadred,**
"EC 1.2.1.2 Formate dehydrogenase",
"formate+nadox+h2o=co2tot+nadred", {5, 6, 7, 8, 9}, .25]

EC 1.2.1.2 Formate dehydrogenase

formate+nadox+h2o=co2tot+nadred

{-14.09, -15.38, -19.41, -24.9, -30.94}

{-0.08, -0.45, -0.89, -1.01, -1.15}

In[44]:= **trGibbsRxSummary[acetaldehyde + nadox + h2o + de == acetate + nadred,**
"EC 1.2.1.3 Aldehyde dehydrogenase",
"acetaldehyde+nadox+h2o=acetate+nadred", {5, 6, 7, 8, 9}, .25]

EC 1.2.1.3 Aldehyde dehydrogenase

acetaldehyde+nadox+h2o=acetate+nadred

{-33.05, -43.9, -55.25, -66.66, -78.07}

{-1.77, -1.97, -2., -2., -2.}

In[45]:= **trGibbsRxSummary[glyceraldehydephos + nadpox + h2o + de == phosphoglycerate3 + nadpred,**
"EC 1.2.1.9 Glyceraldehyde-3-phosphate dehydrogenase (NADP)",
"glyceraldehydephos+nadpox+h2o=phosphoglycerate3+nadpred", {5, 6, 7, 8, 9}, .25]

EC 1.2.1.9 Glyceraldehyde-3-phosphate dehydrogenase (NADP)

glyceraldehydephos+nadpox+h2o=phosphoglycerate3+nadpred

{-27.2, -35.14, -43.46, -54.06, -65.37}

{-1.6, -1.29, -1.69, -1.96, -2.}

In[46]:= **trGibbsRxSummary[acetaldehyde + coA + nadox + de == acetylcoA + nadred,**
"EC 1.2.1.10 Acetaldehyde dehydrogenase (acetylating)",
"acetaldehyde+coA+nadox=acetylcoA+nadred", {5, 6, 7, 8, 9}, .25]

EC 1.2.1.10 Acetaldehyde dehydrogenase (acetylating)

acetaldehyde+coA+nadox=acetylcoA+nadred

{-2.66, -8.35, -13.88, -18.32, -20.04}

{-1., -0.99, -0.93, -0.55, -0.11}

In[47]:= **trGibbsRxSummary[formate + nadpox + h2o + de == co2tot + nadpred,**
"EC 1.2.1.43 Formate dehydrogenase (NADP)",
"formate+nadpox+h2o=co2tot+nadpred", {5, 6, 7, 8, 9}, .25]

EC 1.2.1.43 Formate dehydrogenase (NADP)

formate+nadpox+h2o=co2tot+nadpred

{-14.5, -15.8, -19.83, -25.32, -31.36}

{-0.08, -0.45, -0.89, -1.01, -1.15}

In[48]:= **trGibbsRxSummary[formate + cytochromecox + h2o + de == co2tot + cytochromecred,**
"EC 1.2.2.3 Formate dehydrogenase (cytochrome)",
"formate+cytochromecox+h2o=co2tot+cytochromecred", {5, 6, 7, 8, 9}, .25]

EC 1.2.2.3 Formate dehydrogenase (cytochrome)

formate+cytochromecox+h2o=co2tot+cytochromecred

{-84.12, -91.12, -100.9, -112.1, -123.8}

{-1.08, -1.45, -1.89, -2.01, -2.15}

In[49]:= **trGibbsRxSummary[coaq + 2 * cytochromecox + 2 * h2o + de == co2tot + 2 * cytochromecred,**
"EC 1.2.2.4 Carbon-monoxide dehydrogenase (cytochrome)",
"coaq+2*cytochromecox+2*h2o=co2tot+2*cytochromecred", {5, 6, 7, 8, 9}, .25]

EC 1.2.2.4 Carbon-monoxide dehydrogenase (cytochrome)

coaq+2*cytochromecox+2*h2o=co2tot+2*cytochromecred

{-128.6, -141.4, -156.8, -173.7, -191.2}

{-2.08, -2.45, -2.89, -3.01, -3.15}

In[50]:= **trGibbsRxSummary[pyruvate + pi + o2aq + 2 * h2o + de == acetylphos + co2tot + h2o2aq,**
"EC 1.2.3.3 Pyruvate oxidase",
"pyruvate+pi+o2aq+2*h2o=acetylphos+co2tot+h2o2aq", {5, 6, 7, 8, 9}, .25]

EC 1.2.3.3 Pyruvate oxidase

pyruvate+pi+o2aq+2*h2o=acetylphos+co2tot+h2o2aq

{11.37, 5.91, -1.43, -8.94, -19.}

{-0.65, -1.21, -1.27, -1.48, -2.03}

In[51]:= **trGibbsRxSummary[oxalate + o2aq + 2 * h2o + de == co2tot + h2o2aq, "EC 1.2.3.4 Oxalate oxidase**
"oxalate+o2aq+2*h2o=co2tot+h2o2aq", {5, 6, 7, 8, 9}, .25]

EC 1.2.3.4 Oxalate oxidase

oxalate+o2aq+2*h2o=co2tot+h2o2aq

{377.9, 376.5, 372.5, 367., 360.9}

{-0.13, -0.46, -0.89, -1.01, -1.15}

In[52]:= **trGibbsRxSummary[glyoxylate + o2aq + h2o + de == oxalate + h2o2aq,**
  "EC 1.2.3.5 Glyoxylate oxidase",
  "glyoxylate+o2aq+h2o=oxalate+h2o2aq", {5, 6, 7, 8, 9}, .25]

EC 1.2.3.5 Glyoxylate oxidase

glyoxylate+o2aq+h2o=oxalate+h2o2aq

{-150.4, -156., -161.7, -167.4, -173.2}

{-0.95, -0.99, -1., -1., -1.}

In[53]:= **trGibbsRxSummary[pyruvate + coA + o2aq + h2o + de == acetylcoA + co2tot + h2o2aq,**
  "EC 1.2.3.6 Pyruvate oxidase (CoA-acetylating)",
  "pyruvate+coA+o2aq+h2o=acetylcoA+co2tot+h2o2aq", {5, 6, 7, 8, 9}, .25]

EC 1.2.3.6 Pyruvate oxidase (CoA-acetylating)

pyruvate+coA+o2aq+h2o=acetylcoA+co2tot+h2o2aq

{-166.6, -162.2, -160.4, -158.9, -155.2}

{0.92, 0.56, 0.18, 0.44, 0.74}

In[54]:= **trGibbsRxSummary[pyruvate + coA + ferredoxinox + h2o + de == acetylcoA + co2tot + ferredoxinred**
  "EC 1.2.7.1 Pyruvate synthase",
  "pyruvate+coA+ferredoxinox+h2o=acetylcoA+co2tot+ferredoxinred", {5, 6, 7, 8, 9}, .25]

EC 1.2.7.1 Pyruvate synthase

pyruvate+coA+ferredoxinox+h2o=acetylcoA+co2tot+ferredoxinred

{-36.04, -43.02, -52.59, -62.52, -70.27}

{-1.08, -1.44, -1.82, -1.56, -1.26}

EC 1.4 Acting on the CH-NH(2) group of donors

In[55]:= **trGibbsRxSummary[pyruvate + nadred + ammonia + de == alanine + nadox + h2o,**
  "EC 1.4.1.1 Alanine dehydrogenase",
  "pyruvate+nadred+ammonia=alanine+nadox+h2o", {5, 6, 7, 8, 9}, .25]

EC 1.4.1.1 Alanine dehydrogenase

pyruvate+nadred+ammonia=alanine+nadox+h2o

{-45.86, -40.15, -34.43, -28.6, -21.93}

{1., 1., 1.01, 1.05, 1.36}

In[56]:= **trGibbsRxSummary[ketoglutarate + nadred + ammonia + de == glutamate + nadox + h2o,**
  "EC 1.4.1.2 Glutamate dehydrogenase",
  "ketoglutarate+nadred+ammonia=glutamate+nadox+h2o", {5, 6, 7, 8, 9}, .25]

EC 1.4.1.2 Glutamate dehydrogenase

ketoglutarate+nadred+ammonia=glutamate+nadox+h2o

{-49.57, -43.86, -38.14, -32.31, -25.64}

{1., 1., 1.01, 1.05, 1.36}

In[57]:= **trGibbsRxSummary[ketoglutarate + nadpred + ammonia + de == glutamate + nadpox + h2o,**
     "EC 1.4.1.3 Glutamate dehydrogenase (NAD(P))",
     "ketoglutarate+nadpred+ammonia=glutamate+nadpox+h2o", {5, 6, 7, 8, 9}, .25]

   EC 1.4.1.3 Glutamate dehydrogenase (NAD(P))

   ketoglutarate+nadpred+ammonia=glutamate+nadpox+h2o

   {-49.15, -43.44, -37.72, -31.89, -25.22}

   {1., 1., 1.01, 1.05, 1.36}

In[58]:= **trGibbsRxSummary[hydroxypyruvate + nadred + ammonia + de == serineL + nadox + h2o,**
     "EC 1.4.1.7 Serine 2-dehydrogenase",
     "hydroxypyruvate+nadred+ammonia=serineL+nadox+h2o", {5, 6, 7, 8, 9}, .25]

   EC 1.4.1.7 Serine 2-dehydrogenase

   hydroxypyruvate+nadred+ammonia=serineL+nadox+h2o

   {-49.45, -43.74, -38.02, -32.19, -25.52}

   {1., 1., 1.01, 1.05, 1.36}

In[59]:= **trGibbsRxSummary[methyl2oxopentanoate + nadred + ammonia + de == leucineL + nadox + h2o,**
     "EC 1.4.1.9 Leucine dehydrogenase",
     "methyl2oxopentanoate+nadred+ammonia=leucineL+nadox+h2o", {5, 6, 7, 8, 9}, .25]

   EC 1.4.1.9 Leucine dehydrogenase

   methyl2oxopentanoate+nadred+ammonia=leucineL+nadox+h2o

   {-54.24, -48.53, -42.81, -36.98, -30.31}

   {1., 1., 1.01, 1.05, 1.36}

In[60]:= **trGibbsRxSummary[glyoxylate + nadred + ammonia + de == glycine + nadox + h2o,**
     "EC 1.4.1.10 Glycine dehydrogenase",
     "glyoxylate+nadred+ammonia=glycine+nadox+h2o", {5, 6, 7, 8, 9}, .25]

   EC 1.4.1.10 Glycine dehydrogenase

   glyoxylate+nadred+ammonia=glycine+nadox+h2o

   {-58.44, -52.73, -47.01, -41.18, -34.51}

   {1., 1., 1.01, 1.05, 1.36}

In[61]:= **trGibbsRxSummary[**
     glycine + h2o + 2 * cytochromecox + de == glyoxylate + ammonia + 2 * cytochromecred,
     "EC 1.4.2.1 Glycine dehydrogenase (cytochrome)",
     "glycine+h2o+2*cytochromecox=glyoxylate+ammonia+2*cytochromecred",
     {5, 6, 7, 8, 9}, .25]

   EC 1.4.2.1 Glycine dehydrogenase (cytochrome)

   glycine+h2o+2*cytochromecox=glyoxylate+ammonia+2*cytochromecred

{-32.05, -43.47, -54.9, -66.44, -78.82}

{-2., -2., -2.01, -2.05, -2.36}

In[62]:= **trGibbsRxSummary[glutamate + o2aq + h2o + de == ketoglutarate + ammonia + h2o2aq,**
  "EC 1.4.3.11 L-Glutamate oxidase",
  "glutamate+o2aq+h2o=ketoglutarate+ammonia+h2o2aq", {5, 6, 7, 8, 9}, .25]

  EC 1.4.3.11 L-Glutamate oxidase

  glutamate+o2aq+h2o=ketoglutarate+ammonia+h2o2aq

  {-91.73, -91.73, -91.74, -91.86, -92.83}

  {-0., -0., -0.01, -0.05, -0.36}

In[63]:= **trGibbsRxSummary[aspartate + o2aq + h2o + de == oxaloacetate + ammonia + h2o2aq,**
  "EC 1.4.3.16 L-Aspartate oxidase",
  "aspartate+o2aq+h2o=oxaloacetate+ammonia+h2o2aq", {5, 6, 7, 8, 9}, .25]

  EC 1.4.3.16 L-Aspartate oxidase

  aspartate+o2aq+h2o=oxaloacetate+ammonia+h2o2aq

  {-93.2, -93.2, -93.21, -93.33, -94.3}

  {-0., -0., -0.01, -0.05, -0.36}

EC 1.5 Acting on NADH or NADPH

In[64]:= **trGibbsRxSummary[nadpred + fmnox + de == nadpox + fmnred,**
  "EC 1.5.1.29 FMN reductase", "nadpred+fmnox=nadpox+fmnred", {5, 6, 7, 8, 9}, .25]

  EC 1.5.1.29 FMN reductase

  nadpred+fmnox=nadpox+fmnred

  {-29.33, -23.62, -17.92, -12.21, -6.5}

  {1.}

EC 1.6 Acting on NAD$_{red}$ or NADP$_{red}$

In[65]:= **trGibbsRxSummary[nadpox + nadred + de == nadpred + nadox,**
  "EC 1.6.1.1 NAD(P) transhydrogenase",
  "nadpox+nadred=nadpred+nadox", {5, 6, 7, 8, 9}, .25]

  EC 1.6.1.1 NAD(P) transhydrogenase

  nadpox+nadred=nadpred+nadox

  {-0.42, -0.42, -0.42, -0.42, -0.42}

  {0.}

In[66]:= **trGibbsRxSummary[nadpred + 2 * cytochromecox + de == nadpox + 2 * cytochromecred,**
  "EC 1.6.2.5 NAD(P)-cytochromec reductase",
  "nadpred+2*cytochromecox=nadpox+2*cytochromecred", {5, 6, 7, 8, 9}, .25]

  EC 1.6.2.5 NAD(P)-cytochromec reductase

```
nadpred+2*cytochromecox=nadpox+2*cytochromecred
```

```
{-90.08, -95.78, -101.5, -107.2, -112.9}
```

```
{-1.}
```

In[67]:= **trGibbsRxSummary[nadred + ubiquinoneox + de == nadox + ubiquinonered,**
    "EC 1.6.5.3 NADH dehydrogenase (ubiquinone)",
    "nadred+ubiquinoneox=nadox+ubiquinonered", {5, 6, 7, 8, 9}, .25]

```
EC 1.6.5.3 NADH dehydrogenase (ubiquinone)
```

```
nadred+ubiquinoneox=nadox+ubiquinonered
```

```
{-80.79, -75.08, -69.38, -63.67, -57.96}
```

```
{1.}
```

EC 1.7 Acting on other nitrogenous compounds as donors

In[68]:= **trGibbsRxSummary[nadred + nitrate + de == nadox + nitrite + h2o,**
    "EC 1.7.1.1 Nitrate reductase (NADH)",
    "nadred+nitrate=nadox+nitrite+h2o", {5, 6, 7, 8, 9}, .25]

```
EC 1.7.1.1 Nitrate reductase (NADH)
```

```
nadred+nitrate=nadox+nitrite+h2o
```

```
{-151.5, -145.8, -140.1, -134.4, -128.7}
```

```
{1.01, 1., 1., 1., 1.}
```

In[69]:= **trGibbsRxSummary[nadpred + nitrate + de == nadpox + nitrite + h2o,**
    "EC 1.7.1.3 Nitrate reductase (NADP)",
    "nadpred+nitrate=nadpox+nitrite+h2o", {5, 6, 7, 8, 9}, .25]

```
EC 1.7.1.3 Nitrate reductase (NADP)
```

```
nadpred+nitrate=nadpox+nitrite+h2o
```

```
{-151.1, -145.4, -139.7, -134., -128.3}
```

```
{1.01, 1., 1., 1., 1.}
```

In[70]:= **trGibbsRxSummary[nitrite + 3 * nadpred + de == ammonia + 3 * nadpox + 2 * h2o,**
    "EC 1.7.1.4 Nitrite reductase (NADP)",
    "nitrite+3*nadpred=ammonia+3*nadpox+2*h2o", {5, 6, 7, 8, 9}, .25]

```
EC 1.7.1.4 Nitrite reductase (NADP)
```

```
nitrite+3*nadpred=ammonia+3*nadpox+2*h2o
```

```
{-434.1, -405.6, -377.1, -348.7, -321.1}
```

```
{4.99, 5., 4.99, 4.95, 4.64}
```

In[71]:= **trGibbsRxSummary[nitrite + cytochromecred + de == noaq + cytochromecox + h2o,**
    "EC 1.7.2.1 Nitrite reductase (NO forming)",
    "nitrite+cytochromecred=noaq+cytochromecox+h2o", {5, 6, 7, 8, 9}, .25]

```
EC 1.7.2.1 Nitrite reductase (NO forming)
```

```
            nitrite+cytochromecred=noaq+cytochromecox+h2o

            {-38.45, -27.05, -15.64, -4.22, 7.2}

            {1.99, 2., 2., 2., 2.}

In[72]:=  trGibbsRxSummary[nitrite + 6 * cytochromecred + de == ammonia + 6 * cytochromecox + 2 * h2o,
            "EC 1.7.2.2 Nitrite reductase (cytochrome, ammonia forming)",
            "nitrite+6*cytochromecred=ammonia+6*cytochromecox+2*h2o", {5, 6, 7, 8, 9}, .25]

            EC 1.7.2.2 Nitrite reductase (cytochrome, ammonia forming)

            nitrite+6*cytochromecred=ammonia+6*cytochromecox+2*h2o

            {-163.9, -118.3, -72.61, -27.06, 17.63}

            {7.99, 8., 7.99, 7.95, 7.64}

In[73]:=  trGibbsRxSummary[nitrite + 3 * ferredoxinred + de == ammonia + 3 * ferredoxinox + 2 * h2o,
            "EC 1.7.7.1 Ferredoxin-nitrite reductase",
            "nitrite+3*ferredoxinred=ammonia+3*ferredoxinox+2*h2o", {5, 6, 7, 8, 9}, .25]

            EC 1.7.7.1 Ferredoxin-nitrite reductase

            nitrite+3*ferredoxinred=ammonia+3*ferredoxinox+2*h2o

            {-403.3, -357.7, -312., -266.5, -221.8}

            {7.99, 8., 7.99, 7.95, 7.64}

In[74]:=  trGibbsRxSummary[nitrate + 2 * ferredoxinred + de == nitrite + h2o + 2 * ferredoxinox,
            "EC 1.7.7.2 Ferredoxin-nitrate reductase",
            "nitrate+2*ferredoxinred=nitrite+h2o+2*ferredoxinox", {5, 6, 7, 8, 9}, .25]

            EC 1.7.7.2 Ferredoxin-nitrate reductase

            nitrate+2*ferredoxinred=nitrite+h2o+2*ferredoxinox

            {-179.7, -168.3, -156.9, -145.5, -134.}

            {2.01, 2., 2., 2., 2.}

In[75]:=  trGibbsRxSummary[n2oaq + nadred + de == n2aq + nadox,
            "EC 1.7.99.6 Nitrous oxide reductase", "n2oaq+nadred=n2aq+nadox", {5, 6, 7, 8, 9}, .25]

            EC 1.7.99.6 Nitrous oxide reductase

            n2oaq+nadred=n2aq+nadox

            {-135.1, -140.8, -146.5, -152.2, -157.9}

            {-1.}

In[76]:=  trGibbsRxSummary[2 * noaq + nadred + de == n2oaq + nadox, "EC 1.7.99.7 Nitric oxide reductas(
            "2*noaq+nadred=n2oaq+nadox", {5, 6, 7, 8, 9}, .25]

            EC 1.7.99.7 Nitric oxide reductase

            2*noaq+nadred=n2oaq+nadox

            {-118.5, -124.2, -129.9, -135.6, -141.3}

            {-1.}
```

EC 1.8 Acting on a sulfur group of donors

In[77]:= **trGibbsRxSummary[sulfite + 3 * nadpred + de == h2saq + 3 * nadpox + 3 * h2o,**
   "EC 1.8.1.2 Sulfite reductase",
   "sulfite+3*nadpred=h2saq+3*nadpox+3*h2o", {5, 6, 7, 8, 9}, .25]

   EC 1.8.1.2 Sulfite reductase

   sulfite+3*nadpred=h2saq+3*nadpox+3*h2o

   {-152.8, -129.9, -107., -84.08, -61.24}

   {4., 4.01, 4.02, 4., 4.}

In[78]:= **trGibbsRxSummary[2 * cysteineL + nadox + de == cystineL + nadred,**
   "EC 1.8.1.6 Cystine reductase",
   "2*cysteineL+nadox=cystineL+nadred", {5, 6, 7, 8, 9}, .25]

   EC 1.8.1.6 Cystine reductase

   2*cysteineL+nadox=cystineL+nadred

   {2.01, -3.67, -9.03, -12.19, -9.91}

   {-1., -0.98, -0.85, -0.11, 0.78}

In[79]:= **trGibbsRxSummary[glutathioneox + nadpred + de == 2 * glutathionered + nadpox,**
   "EC 1.8.1.7 Glutathione reductase (NADP)",
   "glutathioneox+nadpred=2*glutathionered+nadpox", {5, 6, 7, 8, 9}, .25]

   EC 1.8.1.7 Glutathione reductase (NADP)

   glutathioneox+nadpred=2*glutathionered+nadpox

   {-15.72, -10.09, -5.07, -3.49, -7.17}

   {1., 0.97, 0.71, -0.26, -0.89}

In[80]:= **trGibbsRxSummary[thioredoxinox + nadpred + de == thioredoxinred + nadpox,**
   "EC 1.8.1.9 Thioredoxin-disulfide reductase (NADP)",
   "thioredoxinox+nadpredthioredoxinred+nadpox", {5, 6, 7, 8, 9}, .25]

   EC 1.8.1.9 Thioredoxin-disulfide reductase (NADP)

   thioredoxinox+nadpredthioredoxinred+nadpox

   {-15.83, -10.16, -4.83, -2.25, -5.64}

   {1., 0.98, 0.83, -0.11, -0.88}

In[81]:= **trGibbsRxSummary[coAglutathione + nadpred + de == coA + glutathionered + nadpox,**
   "EC 1.8.1.10 CoA-glutathione reductase (NADP)",
   "coAglutathione+nadpred=coA+glutathionered+nadpox", {5, 6, 7, 8, 9}, .25]

   EC 1.8.1.10 CoA-glutathione reductase (NADP)

   coAglutathione+nadpred=coA+glutathionered+nadpox

   {-15.88, -10.23, -5.03, -2.66, -5.64}

   {1., 0.98, 0.78, -0.07, -0.83}

In[82]:= **trGibbsRxSummary[sulfite + 2 * cytochromecox + h2o + de == sulfate + 2 * cytochromecred,**
     "EC 1.8.2.1 Sulfite dehydrogenase",
     "sulfite+2*cytochromecox+h2o=sulfate+2*cytochromecred", {5, 6, 7, 8, 9}, .25]

     EC 1.8.2.1 Sulfite dehydrogenase

     sulfite+2*cytochromecox+h2o=sulfate+2*cytochromecred

     {-110.9, -127.6, -142.4, -154.6, -166.1}

     {-2.98, -2.82, -2.32, -2.04, -2.}

In[83]:= **trGibbsRxSummary[sulfite + o2aq + h2o + de == sulfate + h2o2aq,**
     "EC 1.8.3.1 Sulfite oxidase", "sulfite+o2aq+h2o=sulfate+h2o2aq", {5, 6, 7, 8, 9}, .25]

     EC 1.8.3.1 Sulfite oxidase

     sulfite+o2aq+h2o=sulfate+h2o2aq

     {-161.7, -167., -170.3, -171.2, -171.3}

     {-0.98, -0.82, -0.32, -0.04, -0.00}

In[84]:= **trGibbsRxSummary[2 * glutathionered + o2aq + de == glutathioneox + h2o2aq,**
     "EC 1.8.3.3 Glutathione oxidase",
     "2*glutathionered+o2aq=glutathioneox+h2o2aq", {5, 6, 7, 8, 9}, .25]

     EC 1.8.3.3 Glutathione oxidase

     2*glutathionered+o2aq=glutathioneox+h2o2aq

     {-125.2, -125.1, -124.4, -120.3, -110.9}

     {0.00, 0.03, 0.29, 1.26, 1.89}

In[85]:= **trGibbsRxSummary[glutathioneox + 2 * cysteineL + de == 2 * glutathionered + cystineL,**
     "EC 1.8.4.4 Glutathione-cystine transhydrogenase",
     "glutathioneox+2*cysteineL=2*glutathionered+cystineL", {5, 6, 7, 8, 9}, .25]

     EC 1.8.4.4 Glutathione-cystine transhydrogenase

     glutathioneox+2*cysteineL=2*glutathionered+cystineL

     {-14.14, -14.17, -14.52, -16.1, -17.5}

     {-0.00, -0.02, -0.14, -0.36, -0.11}

In[86]:= **trGibbsRxSummary[sulfite + 3 * ferredoxinred + de == h2saq + 3 * ferredoxinox + 3 * h2o,**
     "EC 1.8.7.1 Sulfite reductase (ferredoxin)",
     "sulfite+3*ferredoxinred=h2saq+3*ferredoxinox+3*h2o", {5, 6, 7, 8, 9}, .25]

     EC 1.8.7.1 Sulfite reductase (ferredoxin)

     sulfite+3*ferredoxinred=h2saq+3*ferredoxinox+3*h2o

     {-122., -82., -41.92, -1.89, 38.07}

     {7., 7.01, 7.02, 7., 7.}

EC 1.9 Acting on a heme group of donors

In[87]:= **trGibbsRxSummary[4 * cytochromecred + o2aq + de == 4 * cytochromecox + 2 * h2o,**
       "EC 1.9.3.1 Cytochrome-c oxidase",
       "4*cytochromecred+o2aq=4*cytochromecox+2*h2o", {5, 6, 7, 8, 9}, .25]

   EC 1.9.3.1 Cytochrome-c oxidase

   4*cytochromecred+o2aq=4*cytochromecox+2*h2o

   {-291.5, -268.7, -245.9, -223., -200.2}

   {4.}

In[88]:= **trGibbsRxSummary[nitrate + 2 * cytochromecred + de ==**
       nitrite + 2 * cytochromecox + h2o, "EC 1.9.6.1 Nitrate reductase (cytochrome)",
       "nitrate+2*cytochromecred=nitrite+2*cytochromecox+h2o", {5, 6, 7, 8, 9}, .25]

   EC 1.9.6.1 Nitrate reductase (cytochrome)

   nitrate+2*cytochromecred=nitrite+2*cytochromecox+h2o

   {-61.05, -49.61, -38.2, -26.78, -15.36}

   {2.01, 2., 2., 2., 2.}

In[89]:= **trGibbsRxSummary[ferric + cytochromecred + de == ferrous + cytochromecox,**
       "EC 1.9.99.1 Iron-cytochrome-c reductase",
       "ferric+cytochromecred=ferrous+cytochromecox", {5, 6, 7, 8, 9}, .25]

   EC 1.9.99.1 Iron-cytochrome-c reductase

   ferric+cytochromecred=ferrous+cytochromecox

   {-49.69}

   {0}

EC 1.11 Acting on a peroxide as acceptor

In[90]:= **trGibbsRxSummary[nadred + h2o2aq + de == nadox + 2 * h2o,**
       "EC 1.11.1.1 NADH peroxidase", "nadred+h2o2aq=nadox+2*h2o", {5, 6, 7, 8, 9}, .25]

   EC 1.11.1.1 NADH peroxidase

   nadred+h2o2aq=nadox+2*h2o

   {-331.2, -325.5, -319.8, -314.1, -308.4}

   {1.}

In[91]:= **trGibbsRxSummary[2 * cytochromecred + h2o2aq + de == 2 * cytochromecox + 2 * h2o,**
       "EC 1.11.1.5 Cytochromec peroxidase",
       "2*cytochromecred+h2o2aq=2*cytochromecox+2*h2o", {5, 6, 7, 8, 9}, .25]

   EC 1.11.1.5 Cytochromec peroxidase

   2*cytochromecred+h2o2aq=2*cytochromecox+2*h2o

   {-240.7, -229.3, -217.9, -206.5, -195.1}

   {2.}

In[92]:= **trGibbsRxSummary[2 * h2o2aq + de == o2aq + 2 * h2o,**
   "EC 1.11.1.6 Catalase", "2*h2o2aq=o2aq+2*h2o", {5, 6, 7, 8, 9}, .25]

   EC 1.11.1.6 Catalase

   2*h2o2aq=o2aq+2*h2o

   {-189.9, -189.9, -189.9, -189.9, -189.9}

   {0.}

In[93]:= **trGibbsRxSummary[2 * iodideion + h2o2aq + de == i2cr + 2 * h2o, "EC 1.11.1.8 Iodide peroxidase**
   "2*iodideion+h2o2aq=i2cr+2*h2o", {5, 6, 7, 8, 9}, .25]

   EC 1.11.1.8 Iodide peroxidase

   2*iodideion+h2o2aq=i2cr+2*h2o

   {-176.9, -165.5, -154.1, -142.6, -131.2}

   {2.}

In[94]:= **trGibbsRxSummary[2 * glutathionered + h2o2aq + de == glutathioneox + 2 * h2o,**
   "EC 1.11.1.9 Glutathione peroxidase",
   "2*glutathionered+h2o2aq=glutathioneox+2*h2o", {5, 6, 7, 8, 9}, .25]

   EC 1.11.1.9 Glutathione peroxidase

   2*glutathionered+h2o2aq=glutathioneox+2*h2o

   {-315.1, -315., -314.3, -310.2, -300.8}

   {0.00, 0.03, 0.29, 1.26, 1.89}

EC 1.12 Acting on hydrogen as donor

In[95]:= **trGibbsRxSummary[h2aq + nadox + de == nadred,**
   "EC 1.12.1.2 Hydrogen hydrogenase", "h2aq+nadox=nadred", {5, 6, 7, 8, 9}, .25]

   EC 1.12.1.2 Hydrogen hydrogenase

   h2aq+nadox=nadred

   {-26.73, -32.44, -38.14, -43.85, -49.56}

   {-1.}

In[96]:= **trGibbsRxSummary[.5 * h2aq + cytochromecox + de == cytochromecred,**
   "EC 1.12.2.1 Cytochrome-c hydrogenase",
   ".5*h2aq+cytochromecox=cytochromecred", {5, 6, 7, 8, 9}, .25]

   EC 1.12.2.1 Cytochrome-c hydrogenase

   .5*h2aq+cytochromecox=cytochromecred

   {-58.61, -64.32, -70.03, -75.74, -81.44}

   {-1.}

In[97]:= **trGibbsRxSummary[2 * ferredoxinox + h2aq + de == 2 * ferredoxinred,**
   "EC 1.18.99.1 Hydrogenase",
   "2*ferredoxinox+h2aq=2*ferredoxinred", {5, 6, 7, 8, 9}, .25]

```
EC 1.18.99.1 Hydrogenase

2*ferredoxinox+h2aq=2*ferredoxinred

{1.46, -9.96, -21.37, -32.79, -44.2}

{-2.}
```

EC 1.13 Acting on single donors with incorporation of molecular oxygen

In[98]:= **trGibbsRxSummary[sulfurcr + o2aq + h2o + de == sulfite,**
   "EC 1.13.1.18 Sulfur dioxygenase", "sulfurcr+o2aq+h2o=sulfite", {5, 6, 7, 8, 9}, .25]

```
EC 1.13.1.18 Sulfur dioxygenase

sulfurcr+o2aq+h2o=sulfite

{-337.2, -343.3, -351.4, -362., -373.3}

{-1.02, -1.18, -1.68, -1.96, -2.}
```

In[99]:= **trGibbsRxSummary[lactate + o2aq + de == acetate + co2tot,**
   "EC 1.13.12.4 Lactate 2-monooxygenase",
   "lactate+o2aq=acetate+co2tot", {5, 6, 7, 8, 9}, .25]

```
EC 1.13.12.4 Lactate 2-monooxygenase

lactate+o2aq=acetate+co2tot

{-492.9, -493.7, -497.6, -503.1, -509.1}

{0.15, -0.42, -0.89, -1., -1.15}
```

Note no h2o.

EC 1.14 Acting on paired donors, with incorporation or reduction

In[100]:=
   trGibbsRxSummary[methaneaq + nadpred + o2aq + de == methanol + nadpox + h2o,
    "EC 1.14.13.25 Methane monooxygenase",
    "methaneaq+nadpred+o2aq=methanol+nadpox+h2o", {5, 6, 7, 8, 9}, .25]

```
EC 1.14.13.25 Methane monooxygenase

methaneaq+nadpred+o2aq=methanol+nadpox+h2o

{-385., -379.3, -373.6, -367.9, -362.2}

{1.}
```

EC 1.16 Oxidizing metal ions

In[101]:=
   trGibbsRxSummary[2 * ferric + nadred + de == 2 * ferrous + nadox,
    "EC 1.16.1.7 Ferric chelate reductase",
    "2*ferric+nadred=2*ferrous+nadox", {5, 6, 7, 8, 9}, .25]

```
EC 1.16.1.7 Ferric chelate reductase

2*ferric+nadred=2*ferrous+nadox
```

```
{-189.9, -195.6, -201.3, -207., -212.7}
```

```
{-1.}
```

In[102]:=

```
trGibbsRxSummary[4 * ferrous + o2aq + de == 4 * ferric + 2 * h2o,
  "EC 1.16.3.1 Ferrioxidase", "4*ferrous+o2aq=4*ferric+2*h2o", {5, 6, 7, 8, 9}, .25]
```

```
EC 1.16.3.1 Ferrioxidase
```

```
4*ferrous+o2aq=4*ferric+2*h2o
```

```
{-92.77, -69.94, -47.11, -24.28, -1.45}
```

```
{4.}
```

EC 1.18 Acting on iron-sulfur proteins as donors

In[103]:=

```
g1x18x1x2 = trGibbsRxSummary[ferredoxinox + nadpred + de == ferredoxinred + nadpox,
  "EC 1.18.1.2 Ferredoxin-NADP reductase",
  "ferredoxinox+nadpred=ferredoxinred+nadpox", {5, 6, 7, 8, 9}, .25]
```

```
EC 1.18.1.2 Ferredoxin-NADP reductase
```

```
ferredoxinox+nadpred=ferredoxinred+nadpox
```

```
{-10.27, -15.98, -21.69, -27.4, -33.1}
```

```
{-1.}
```

In[104]:=

```
trGibbsRxSummary[n2aq + 8 * ferredoxinred + 16 * atp + 16 * h2o + de ==
   2 * ammonia + h2aq + 8 * ferredoxinox + 16 * adp + 16 * pi,
  "EC 1.18.6.1 Nitrogenase", "n2aq+8*ferredoxinred+16*atp+16*h2o=2*
    ammonia+h2aq+8*ferredoxinox+16*adp+16*pi", {5, 6, 7, 8, 9}, .25]
```

```
EC 1.18.6.1 Nitrogenase
```

```
n2aq+8*ferredoxinred+16*atp+16*h2o=2*ammonia+h2aq+8*ferredoxinox+16*adp+16*pi
```

```
{-699.9, -653.3, -641.3, -665.1, -700.}
```

```
{9.39, 6.02, -1.88, -5.54, -6.66}
```

12.2 Transferase Reactions

EC 2.1 Transferring one-carbon groups

In[105]:=

```
trGibbsRxSummary[methylmalonylcoA + pyruvate + de == propanoylcoA + oxaloacetate,
  "EC 2.1.3.1 Methylmalonyl-CoA carboxytransferase",
  "methylmalonylcoA+pyruvate=propanoylcoA+oxaloacetate", {5, 6, 7, 8, 9}, .25]
```

```
EC 2.1.3.1 Methylmalonyl-CoA carboxytransferase
```

```
methylmalonylcoA+pyruvate=propanoylcoA+oxaloacetate
```

```
{0.9, 0.72, 0.7, 0.7, 0.7}
```

```
{-0.08, -0.01, -0., -0., -8.44×10^-6}
```

EC 2.3 Acyltransferases

In[106]:=

        trGibbsRxSummary[coA + acetylphos + de == acetylcoA + pi,
         "EC 2.3.1.8 Phosphate acetyltransferase",
         "coA+acetylphos=acetylcoA+pi", {5, 6, 7, 8, 9}, .25]

        EC 2.3.1.8 Phosphate acetyltransferase

        coA+acetylphos=acetylcoA+pi

        {0.47, -1.05, -3.27, -5.7, -3.39}

        {-0.42, -0.23, -0.55, -0.08, 0.78}

In[107]:=

        trGibbsRxSummary[coA + acetoacetylcoA + de == 2 * acetylcoA,
         "EC 2.3.1.9 Acetyl-CoA C-acetyltransferase",
         "coA+acetoacetylcoA+de==2*acetylcoA", {5, 6, 7, 8, 9}, .25]

        EC 2.3.1.9 Acetyl-CoA C-acetyltransferase

        coA+acetoacetylcoA+de==2*acetylcoA

        {-27.57, -27.55, -27.38, -26.11, -22.12}

        {0., 0.01, 0.07, 0.45, 0.89}

In[108]:=

        trGibbsRxSummary[coA + pyruvate + de == acetylcoA + formate,
         "EC 2.3.1.54 Formate C-acetyltransferase",
         "coA+pyruvate=acetylcoA+formate", {5, 6, 7, 8, 9}, .25]

        EC 2.3.1.54 Formate C-acetyltransferase

        coA+pyruvate=acetylcoA+formate

        {-11.26, -11.24, -11.07, -9.8, -5.81}

        {0., 0.01, 0.07, 0.45, 0.89}

In[109]:=

        trGibbsRxSummary[acetylcoA + h2o + oxaloacetate + de == coA + citrate,
         "EC 2.3.3.1 Citrate synthase",
         "acetylcoA+h2o+oxaloacetate=coA+citrate", {5, 6, 7, 8, 9}, .25]

        EC 2.3.3.1 Citrate synthase

        acetylcoA+h2o+oxaloacetate=coA+citrate

        {-37.08, -39.56, -44.77, -51.67, -61.36}

        {-0.09, -0.74, -1.04, -1.44, -1.89}

In[110]:=

        trGibbsRxSummary[adp + pi + acetylcoA + oxaloacetate + de == atp + citrate + coA,
         "EC 2.3.3.8 ATP citrate synthase",
         "adp+pi+acetylcoA+oxaloacetate=atp+citrate+coA", {5, 6, 7, 8, 9}, .25]

        EC 2.3.3.8 ATP citrate synthase

        adp+pi+acetylcoA+oxaloacetate=atp+citrate+coA

{-4.52, -6.34, -8.74, -10.6, -14.66}

{-0.05, -0.49, -0.3, -0.48, -0.89}

In[111]:=

```
trGibbsRxSummary[acetylcoA + h2o + glyoxylate + de == coA + malate,
 "EC 2.3.3.9 Malate synthase",
 "acetylcoA+h2o+glyoxylate=coA+malate", {5, 6, 7, 8, 9}, .25]
```

EC 2.3.3.9 Malate synthase

acetylcoA+h2o+glyoxylate=coA+malate

{-37.11, -41.97, -47.74, -54.7, -64.4}

{-0.67, -0.96, -1.07, -1.44, -1.89}

In[112]:=

```
trGibbsRxSummary[propanoylcoA + h2o + glyoxylate + de == coA + hydroxyglutarate,
 "EC 2.3.3.11 2-Hydroxyglutarate synthase",
 "propanoylcoA+h2o+glyoxylate=coA+hydroxyglutarate", {5, 6, 7, 8, 9}, .25]
```

EC 2.3.3.11 2-Hydroxyglutarate synthase

propanoylcoA+h2o+glyoxylate=coA+hydroxyglutarate

{-34.86, -40.59, -46.47, -53.44, -63.14}

{-1., -1.01, -1.07, -1.45, -1.89}

EC 2.4 Glycosyltransferases

In[113]:=

```
trGibbsRxSummary[sucrose + pi + de == fructose + glucose6phos,
 "EC 2.4.1.7 Sucrose phosphorylase",
 "sucrose+pi=fructose+glucose6phos", {5, 6, 7, 8, 9}, .25]
```

EC 2.4.1.7 Sucrose phosphorylase

sucrose+pi=fructose+glucose6phos

{-14.37, -15.77, -17.9, -18.56, -18.64}

{-0.1, -0.4, -0.24, -0.04, -0.00}

In[114]:=

```
trGibbsRxSummary[maltose + pi + de == glucose + glucose1phos,
 "EC 2.4.1.8 Maltose phosphorylase",
 "maltose+pi=glucose+glucose1phos", {5, 6, 7, 8, 9}, .25]
```

EC 2.4.1.8 Maltose phosphorylase

maltose+pi=glucose+glucose1phos

{1.89, 0.7, -1.27, -1.89, -1.97}

{-0.08, -0.36, -0.23, -0.03, -0.00}

In[115]:=

```
trGibbsRxSummary[glucose + glucose1phos + de == cellobiose + pi,
 "EC 2.4.1.20 Cellosebiose phosphorylase",
 "glucose+glucose1phos=cellobiose+pi", {5, 6, 7, 8, 9}, .25]
```

```
EC 2.4.1.20 Cellosebiose phosphorylase

glucose+glucose1phos=cellobiose+pi

{-13.14, -11.95, -9.98, -9.36, -9.28}

{0.08, 0.36, 0.23, 0.03, 0.00}
```

In[116]:=

trGibbsRxSummary[2 * glucose1phos + de == maltose + 2 * pi,
"EC 2.4.1.39 Maltose synthase", "2*glucose1phos=maltose+2*pi", {5, 6, 7, 8, 9}, .25]

```
EC 2.4.1.39 Maltose synthase

2*glucose1phos=maltose+2*pi

{-202.2, -188.4, -173., -160.4, -148.8}

{2.16, 2.71, 2.46, 2.07, 2.01}
```

In[117]:=

trGibbsRxSummary[adenine + ribose1phos + de == adenosine + pi,
"EC 2.4.2.1 Purine-nucleoside phosphorylase",
"adenine+ribose1phos=adenosine+pi", {5, 6, 7, 8, 9}, .25]

```
EC 2.4.2.1 Purine-nucleoside phosphorylase

adenine+ribose1phos=adenosine+pi

{-18.9, -18.39, -16.91, -16.38, -16.31}

{-0.06, 0.24, 0.19, 0.03, 0.00}
```

In[118]:=

trGibbsRxSummary[adenine + prpp + de == amp + ppi,
"EC 2.4.2.7 Adenine phosphoribosyltransferase",
"adenine+prpp=amp+ppi", {5, 6, 7, 8, 9}, .25]

```
EC 2.4.2.7 Adenine phosphoribosyltransferase

adenine+prpp=amp+ppi

{-37.34, -35.88, -34.22, -34.74, -38.07}

{0.01, 0.45, 0.09, -0.31, -0.82}
```

In[119]:=

trGibbsRxSummary[hypoxanthine + prpp + de == imp + ppi,
"EC 2.4.2.8 Hypoxanthine phosphoribosyltransferase",
"hypoxanthine+prpp=imp+ppi", {5, 6, 7, 8, 9}, .25]

```
EC 2.4.2.8 Hypoxanthine phosphoribosyltransferase

hypoxanthine+prpp=imp+ppi

{-28.67, -26.98, -25.33, -26.18, -31.56}

{0.1, 0.46, 0.08, -0.45, -1.45}
```

EC 2.6 Transferring nitrogenous groups

In[120]:=

    trGibbsRxSummary[aspartate + ketoglutarate + de == oxaloacetate + glutamate,
     "EC 2.6.1.1 Aspartate transaminase",
     "aspartate+ketoglutarate=oxaloacetate+glutamate", {5, 6, 7, 8, 9}, .25]

    EC 2.6.1.1 Aspartate transaminase

    aspartate+ketoglutarate=oxaloacetate+glutamate

    {-1.47, -1.47, -1.47, -1.47, -1.47}

    {1.24481×10^{-15}}

In[121]:=

    trGibbsRxSummary[alanine + ketoglutarate + de == pyruvate + glutamate,
     "EC 2.6.1.2 Alanine transaminase",
     "alanine+ketoglutarate=pyruvate+glutamate", {5, 6, 7, 8, 9}, .25]

    EC 2.6.1.2 Alanine transaminase

    alanine+ketoglutarate=pyruvate+glutamate

    {-3.71, -3.71, -3.71, -3.71, -3.71}

    {-1.24481×10^{-15}}

In[122]:=

    trGibbsRxSummary[glyoxylate + glutamate + de == glycine + ketoglutarate,
     "EC 2.6.1.4 Glycine transaminase",
     "glyoxylate+glutamate=glycine+ketoglutarate", {5, 6, 7, 8, 9}, .25]

    EC 2.6.1.4 Glycine transaminase

    glyoxylate+glutamate=glycine+ketoglutarate

    {-8.87, -8.87, -8.87, -8.87, -8.87}

    {0.}

In[123]:=

    trGibbsRxSummary[methyl2oxopentanoate + glutamate + de == leucineL + ketoglutarate,
     "EC 2.6.1.6 Leucine transaminase",
     "methyl2oxopentanoate+glutamate=leucineL+ketoglutarate", {5, 6, 7, 8, 9}, .25]

    EC 2.6.1.6 Leucine transaminase

    methyl2oxopentanoate+glutamate=leucineL+ketoglutarate

    {-4.67, -4.67, -4.67, -4.67, -4.67}

    {0.}

In[124]:=

    trGibbsRxSummary[glyoxylate + aspartate + de == glycine + oxaloacetate,
     "EC 2.6.1.35 Glycine-oxaloacetate transaminase",
     "glyoxylate+aspartate=glycine+oxaloacetate", {5, 6, 7, 8, 9}, .25]

    EC 2.6.1.35 Glycine-oxaloacetate transaminase

    glyoxylate+aspartate=glycine+oxaloacetate

    {-10.34, -10.34, -10.34, -10.34, -10.34}

$\{6.22405 \times 10^{-16}\}$

In[125]:=

    trGibbsRxSummary[alanine + glyoxylate + de == glycine + pyruvate,
    "EC 2.6.1.44 Alanine-glyoxylate transaminase",
    "alanine+glyoxylate=glycine+pyruvate", {5, 6, 7, 8, 9}, .25]

EC 2.6.1.44 Alanine-glyoxylate transaminase

alanine+glyoxylate=glycine+pyruvate

$\{-12.58, -12.58, -12.58, -12.58, -12.58\}$

$\{0.\}$

In[126]:=

    trGibbsRxSummary[serineL + glyoxylate + de == glycine + hydroxypyruvate,
    "EC 2.6.1.45 Serine-glyoxylate transaminase",
    "serineL+glyoxylate=glycine+hydroxypyruvate", {5, 6, 7, 8, 9}, .25]

EC 2.6.1.45 Serine-glyoxylate transaminase

serineL+glyoxylate=glycine+hydroxypyruvate

$\{-8.99, -8.99, -8.99, -8.99, -8.99\}$

$\{0.\}$

In[127]:=

    trGibbsRxSummary[alanine + hydroxypyruvate + de == serineL + pyruvate,
    "EC 2.6.1.51 Serine-pyruvate transaminase",
    "alanine+hydroxypyruvate=serineL+pyruvate", {5, 6, 7, 8, 9}, .25]

EC 2.6.1.51 Serine-pyruvate transaminase

alanine+hydroxypyruvate=serineL+pyruvate

$\{-3.59, -3.59, -3.59, -3.59, -3.59\}$

$\{0.\}$

EC 2.7 Transferring phosphorous-containing groups

In[128]:=

    trGibbsRxSummary[atp + glucose + de == adp + glucose6phos,
    "EC 2.7.1.2 Glucokinase", "atp+glucose=adp+glucose6phos", {5, 6, 7, 8, 9}, .25]

EC 2.7.1.2 Glucokinase

atp+glucose=adp+glucose6phos

$\{-17.41, -19.47, -24.42, -30.11, -35.82\}$

$\{-0.14, -0.65, -0.98, -1., -1.\}$

In[129]:=

    trGibbsRxSummary[atp + fructose + de == adp + fructose6phos,
    "EC 2.7.1.4 Fructokinase", "atp+fructose=adp+fructose6phos", {5, 6, 7, 8, 9}, .25]

EC 2.7.1.4 Fructokinase

atp+fructose=adp+fructose6phos

{-13.94, -16.4, -21.62, -27.36, -33.07}

{-0.18, -0.73, -1., -1., -1.}

In[130]:=

> **trGibbsRxSummary[atp + galactose + de ⩵ adp + galactose1phos,**
> **"EC 2.7.1.6 Galactokinase", "atp+galactose=adp+galactose1phos", {5, 6, 7, 8, 9}, .25]**

EC 2.7.1.6 Galactokinase

atp+galactose=adp+galactose1phos

{-15.82, -18.65, -24.06, -29.83, -35.54}

{-0.23, -0.79, -1.01, -1., -1.}

In[131]:=

> **trGibbsRxSummary[atp + mannose + de ⩵ adp + mannose6phos,**
> **"EC 2.7.1.7 Mannokinase", "atp+mannose=adp+mannose6phos", {5, 6, 7, 8, 9}, .25]**

EC 2.7.1.7 Mannokinase

atp+mannose=adp+mannose6phos

{-19.34, -21.34, -26.26, -31.94, -37.65}

{-0.13, -0.64, -0.98, -1., -1.}

In[132]:=

> **trGibbsRxSummary[atp + fructose6phos + de ⩵ adp + fructose16phos,**
> **"EC 2.7.1.11 6-phosphofructokinase",**
> **"atp+fructose6phos=adp+fructose16phos", {5, 6, 7, 8, 9}, .25]**

EC 2.7.1.11 6-phosphofructokinase

atp+fructose6phos=adp+fructose16phos

{-12.35, -16.95, -23.25, -29.13, -34.86}

{-0.41, -1.09, -1.07, -1.01, -1.}

In[133]:=

> **trGibbsRxSummary[atp + ribose + de ⩵ adp + ribose5phos,**
> **"EC 2.7.1.15 Ribokinase", "atp+ribose=adp+ribose5phos", {5, 6, 7, 8, 9}, .25]**

EC 2.7.1.15 Ribokinase

atp+ribose=adp+ribose5phos

{-14.54, -15.96, -20.3, -25.87, -31.56}

{-0.09, -0.5, -0.93, -0.99, -1.}

In[134]:=

> **trGibbsRxSummary[atp + adenosine + de ⩵ adp + amp,**
> **"EC 2.7.1.20 Adenosine kinase", "atp+adenosine=adp+amp", {5, 6, 7, 8, 9}, .25]**

EC 2.7.1.20 Adenosine kinase

atp+adenosine=adp+amp

{-16.97, -18.27, -22.49, -28.04, -33.73}

```
                  {-0.06, -0.47, -0.92, -0.99, -1.}
```

In[135]:=

trGibbsRxSummary[atp + nadox + de == adp + nadpox,
   "EC 2.7.1.23 NAD$_{ox}$ kinase", "atp+nadox=adp+nadpox", {5, 6, 7, 8, 9}, .25]

EC 2.7.1.23 NAD$_{ox}$ kinase

atp+nadox=adp+nadpox

{-2.6, -8.51, -14.63, -20.47, -26.2}

{-1.02, -1.07, -1.05, -1.01, -1.}

In[136]:=

trGibbsRxSummary[atp + dihydroxyacetone + de == adp + dihydroxyacetonephos,
   "EC 2.7.1.29 Dihydroxyacetone kinase",
   "atp+dihydroxyacetone=adp+dihydroxyacetonephos", {5, 6, 7, 8, 9}, .25]

EC 2.7.1.29 Dihydroxyacetone kinase

atp+dihydroxyacetone=adp+dihydroxyacetonephos

{-11.57, -15.67, -21.5, -27.32, -33.04}

{-0.44, -0.95, -1.04, -1.01, -1.}

In[137]:=

trGibbsRxSummary[atp + glycerol + de == adp + glycerol3phos,
   "EC 2.7.1.30 Glycerole kinase", "atp+glycerol=adp+glycerol3phos", {5, 6, 7, 8, 9}, .25]

EC 2.7.1.30 Glycerole kinase

atp+glycerol=adp+glycerol3phos

{-32.15, -33.61, -37.99, -43.57, -49.26}

{-0.09, -0.51, -0.94, -1., -1.}

In[138]:=

trGibbsRxSummary[atp + glycerate + de == adp + phosphoglycerate3,
   "EC 2.7.1.31 Glycerate kinase",
   "atp+glycerate=adp+phosphoglycerate3", {5, 6, 7, 8, 9}, .25]

EC 2.7.1.31 Glycerate kinase

atp+glycerate=adp+phosphoglycerate3

{-18.58, -19.21, -21.94, -26.94, -32.56}

{-0.04, -0.24, -0.73, -0.96, -1.}

In[139]:=

trGibbsRxSummary[adp + pep + de == atp + pyruvate,
   "EC 2.7.1.40 Pyruvate kinase", "adp+pep=atp+pyruvate", {5, 6, 7, 8, 9}, .25]

EC 2.7.1.40 Pyruvate kinase

adp+pep=atp+pyruvate

{-34.47, -33.11, -28.85, -23.29, -17.6}

{0.08, 0.48, 0.93, 0.99, 1.}

In[140]:=

> **trGibbsRxSummary[atp + ribulose + de == adp + ribulose5phos,**
> **"EC 2.7.1.47 D-ribulokinase", "atp+ribulose=adp+ribulose5phos", {5, 6, 7, 8, 9}, .25]**

EC 2.7.1.47 D-ribulokinase

atp+ribulose=adp+ribulose5phos

{-30.14, -31.56, -35.9, -41.47, -47.16}

{-0.09, -0.5, -0.93, -0.99, -1.}

In[141]:=

> **trGibbsRxSummary[atp + arabinose + de == adp + arabinose5phos,**
> **"EC 2.7.1.54 D-arabinosekinase",**
> **"atp+arabinose=adp+arabinose5phos", {5, 6, 7, 8, 9}, .25]**

EC 2.7.1.54 D-arabinosekinase

atp+arabinose=adp+arabinose5phos

{-26.88, -28.3, -32.63, -38.2, -43.89}

{-0.09, -0.5, -0.93, -0.99, -1.}

In[142]:=

> **trGibbsRxSummary[atp + inosine + de == adp + imp,**
> **"EC 2.7.1.73 Inosine kinase", "atp+inosine=adp+imp", {5, 6, 7, 8, 9}, .25]**

EC 2.7.1.73 Inosine kinase

atp+inosine=adp+imp

{-17.25, -18.59, -22.8, -28.26, -33.65}

{-0.08, -0.47, -0.92, -0.96, -0.94}

In[143]:=

> **trGibbsRxSummary[atp + deoxyadenosine + de == adp + deoxyamp,**
> **"EC 2.7.1.76 Deoxyadenosine kinase",**
> **"atp+deoxyadenosine=adp+deoxyamp", {5, 6, 7, 8, 9}, .25]**

EC 2.7.1.76 Deoxyadenosine kinase

atp+deoxyadenosine=adp+deoxyamp

{-16.97, -18.27, -22.49, -28.04, -33.73}

{-0.06, -0.47, -0.92, -0.99, -1.}

In[144]:=

> **trGibbsRxSummary[ppi + serineL + de == pi + phosphoserine,**
> **"EC 2.7.1.80 Diphosphate-serine phosphotransferase",**
> **"ppi+serineL=pi+phosphoserine", {5, 6, 7, 8, 9}, .25]**

EC 2.7.1.80 Diphosphate-serine phosphotransferase

ppi+serineL=pi+phosphoserine

{-10.16, -10.59, -12.89, -16.94, -19.18}

{-0.02, -0.18, -0.65, -0.64, -0.17}

In[145]:=

    trGibbsRxSummary[atp + nadred + de == adp + nadpred,
     "EC 2.7.1.86 NAD$_{red}$ kinase", "atp+nadred=adp+nadpred", {5, 6, 7, 8, 9}, .25]

    EC 2.7.1.86 NAD$_{red}$ kinase

    atp+nadred=adp+nadpred

    {-3.01, -8.93, -15.05, -20.89, -26.62}

    {-1.02, -1.07, -1.05, -1.01, -1.}

In[146]:=

    trGibbsRxSummary[ppi + fructose6phos + de == pi + fructose16phos,
     "EC 2.7.1.90 Diphosphate-fructose-6-phosphate 1-phosphotransferase",
     "ppi+fructose6phos=pi+fructose16phos", {5, 6, 7, 8, 9}, .25]

    EC 2.7.1.90 Diphosphate-fructose-6-phosphate 1-phosphotransferase

    ppi+fructose6phos=pi+fructose16phos

    {-3.17, -6.19, -9.87, -14.11, -16.37}

    {-0.3, -0.63, -0.73, -0.64, -0.17}

In[147]:=

    trGibbsRxSummary[glycerol + glucose6phos + de == glycerol3phos + glucose,
     "EC 2.7.1.142 Glycerol-3-phosphate-glucose phosphotransferase",
     "glycerol+glucose6phos=glycerol3phos+glucose", {5, 6, 7, 8, 9}, .25]

    EC 2.7.1.142 Glycerol-3-phosphate-glucose phosphotransferase

    glycerol+glucose6phos=glycerol3phos+glucose

    {-14.73, -14.14, -13.57, -13.45, -13.44}

    {0.05, 0.14, 0.05, 0.01, 0.}

In[148]:=

    trGibbsRxSummary[adp + fructose6phos + de == amp + fructose16phos,
     "EC 2.7.1.146 ADP-specfic phosphofructokinase",
     "adp+fructose6phos=amp+fructose16phos", {5, 6, 7, 8, 9}, .25]

    EC 2.7.1.146 ADP-specfic phosphofructokinase

    adp+fructose6phos=amp+fructose16phos

    {-10.12, -14.83, -21.17, -27.04, -32.77}

    {-0.42, -1.11, -1.06, -1.01, -1.}

In[149]:=

    trGibbsRxSummary[adp + glucose + de == amp + glucose6phos,
     "EC 2.7.1.147 ADP-specific glucokinase",
     "adp+glucose=amp+glucose6phos", {5, 6, 7, 8, 9}, .25]

    EC 2.7.1.147 ADP-specific glucokinase

    adp+glucose=amp+glucose6phos

    {-15.18, -17.34, -22.35, -28.03, -33.73}

```
                {-0.15, -0.67, -0.98, -1., -1.}
```

In[150]:=

```
        trGibbsRxSummary[adp + acetylphos + de == atp + acetate,
          "EC 2.7.2.1 Acetate kinase", "adp+acetylphos=atp+acetate", {5, 6, 7, 8, 9}, .25]
```

```
        EC 2.7.2.1 Acetate kinase

        adp+acetylphos=atp+acetate

        {2.64, -3.38, -8.6, -12.96, -14.72}

        {-1.16, -0.96, -0.88, -0.56, -0.12}
```

In[151]:=

```
        trGibbsRxSummary[pi + acetylphos + de == ppi + acetate,
          "EC 2.7.2.12 Acetate kinase (diphosphate)",
          "pi+acetylphos=ppi+acetate", {5, 6, 7, 8, 9}, .25]
```

```
        EC 2.7.2.12 Acetate kinase (diphosphate)

        pi+acetylphos=ppi+acetate

        {-6.55, -14.15, -21.98, -27.98, -33.21}

        {-1.27, -1.42, -1.22, -0.93, -0.95}
```

In[152]:=

```
        trGibbsRxSummary[atp + amp + de == 2 * adp,
          "EC 2.7.4.3 Phosphotransferases with a phosphate group as acceptor",
          "atp+amp=2*adp", {5, 6, 7, 8, 9}, .25]
```

```
        EC 2.7.4.3 Phosphotransferases with a phosphate group as acceptor

        atp+amp=2*adp

        {-2.23, -2.13, -2.07, -2.09, -2.09}

        {0.01, 0.02, -0.00, -0.00, -0.}
```

In[153]:=

```
        trGibbsRxSummary[atp + ribose5phos + de == amp + prpp,
          "EC 2.7.6.1 Ribose-phosphate diphosphokinase",
          "atp+ribose5phos=amp+prpp", {5, 6, 7, 8, 9}, .25]
```

```
        EC 2.7.6.1 Ribose-phosphate diphosphokinase

        atp+ribose5phos=amp+prpp

        {2.6, -1.34, -8.41, -14.53, -20.28}

        {-0.26, -1.17, -1.16, -1.02, -1.}
```

In[154]:=

```
        trGibbsRxSummary[ppi + nadox + de == atp + nicotinamideribonucleotide,
          "EC 2.7.7.1 nicotinamide-nucleotide adenylyltransferase",
          "ppi+nadox=atp+nicotinamideribonucleotide", {5, 6, 7, 8, 9}, .25]
```

```
        EC 2.7.7.1 nicotinamide-nucleotide adenylyltransferase

        ppi+nadox=atp+nicotinamideribonucleotide
```

{-0.23, -0.72, -3.55, -7.85, -10.12}

{0.03, -0.24, -0.73, -0.65, -0.17}

In[155]:=

```
trGibbsRxSummary[ppi + adenosinephosphosulfate + de == atp + sulfate,
 "EC 2.7.7.4 Sulfate adenylyltransferase",
 "ppi+adenosinephosphosulfate=atp+sulfate", {5, 6, 7, 8, 9}, .25]
```

EC 2.7.7.4 Sulfate adenylyltransferase

ppi+adenosinephosphosulfate=atp+sulfate

{-37.57, -41.97, -46., -50.46, -52.75}

{-0.85, -0.67, -0.8, -0.66, -0.17}

In[156]:=

```
trGibbsRxSummary[amp + pep + ppi + de == atp + pyruvate + pi,
 "EC 2.7.9.1 Pyruvate,phosphate dikinase",
 "amp+pep+ppi=atp+pyruvate+pi", {5, 6, 7, 8, 9}, .25]
```

EC 2.7.9.1 Pyruvate,phosphate dikinase

amp+pep+ppi=atp+pyruvate+pi

{-27.52, -24.47, -17.54, -10.36, -1.2}

{0.21, 0.97, 1.26, 1.36, 1.83}

In[157]:=

```
trGibbsRxSummary[atp + pyruvate + h2o + de == amp + pep + pi,
 "EC 2.7.9.2 Pyruvate,water dikinase",
 "atp+pyruvate+h2o=amp+pep+pi", {5, 6, 7, 8, 9}, .25]
```

EC 2.7.9.2 Pyruvate,water dikinase

atp+pyruvate+h2o=amp+pep+pi

{4.14, 2.02, -5.12, -15.7, -27.02}

{-0.13, -0.75, -1.67, -1.96, -2.}

12.3 Hydrolase Reactions

EC 3.1 Acting on ester bonds

In[158]:=

```
trGibbsRxSummary[ethylacetate + h2o + de == ethanol + acetate,
 "EC 3.1.1.1 Carboxylesterase", "ethylacetate+h2o=ethanol+acetate", {5, 6, 7, 8, 9}, .25]
```

EC 3.1.1.1 Carboxylesterase

ethylacetate+h2o=ethanol+acetate

{-6.91, -12.05, -17.69, -23.39, -29.1}

{-0.77, -0.97, -1., -1., -1.}

In[159]:=

trGibbsRxSummary[acetylcoA + h2o + de == coA + acetate,
"EC 3.1.2.1 Acetyl-CoA hydrolase", "acetylcoA+h2o=coA+acetate", {5, 6, 7, 8, 9}, .25]

EC 3.1.2.1 Acetyl-CoA hydrolase

acetylcoA+h2o=coA+acetate

{-30.39, -35.55, -41.36, -48.33, -58.03}

{-0.77, -0.98, -1.07, -1.44, -1.89}

In[160]:=

trGibbsRxSummary[succinylcoA + h2o + de == coA + succinate,
"EC 3.1.2.3 Succinyl-CoA hydrolase",
"succinylcoA+h2o=coA+succinate", {5, 6, 7, 8, 9}, .25]

EC 3.1.2.3 Succinyl-CoA hydrolase

succinylcoA+h2o=coA+succinate

{-24.98, -29.12, -34.77, -41.73, -51.42}

{-0.48, -0.91, -1.06, -1.44, -1.89}

In[161]:=

trGibbsRxSummary[acetoacetylcoA + h2o + de == coA + acetoacetate,
"EC 3.1.2.11 Acetoacetyl-CoA hydrolase",
"acetoacetylcoA+h2o=coA+acetoacetate", {5, 6, 7, 8, 9}, .25]

EC 3.1.2.11 Acetoacetyl-CoA hydrolase

acetoacetylcoA+h2o=coA+acetoacetate

{-37.16, -42.89, -48.77, -55.74, -65.44}

{-1., -1.01, -1.07, -1.45, -1.89}

In[162]:=

trGibbsRxSummary[oxalylcoA + h2o + de == coA + oxalate,
"EC 3.1.2.18 ADP-dependent short-chain-acyl-CoA hydrolase",
"oxalylcoA+h2o=coA+oxalate", {5, 6, 7, 8, 9}, .25]

EC 3.1.2.18 ADP-dependent short-chain-acyl-CoA hydrolase

oxalylcoA+h2o=coA+oxalate

{-6.49, -12.1, -17.97, -24.94, -34.64}

{-0.95, -1., -1.07, -1.45, -1.89}

In[163]:=

trGibbsRxSummary[phosphoserine + h2o + de == serineL + pi,
"EC 3.1.3.3 Phosphoserine phosphatase",
"phosphoserine+h2o=serineL+pi", {5, 6, 7, 8, 9}, .25]

EC 3.1.3.3 Phosphoserine phosphatase

phosphoserine+h2o=serineL+pi

{-13.21, -11.86, -9.77, -9.12, -9.04}

{0.1, 0.39, 0.24, 0.04, 0.00}

In[164]:=

trGibbsRxSummary[glucose6phos + h2o + de == glucose + pi, "EC 3.1.3.9 Glucose-6-phosphatase"
"glucose6phos+h2o=glucose+pi", {5, 6, 7, 8, 9}, .25]

EC 3.1.3.9 Glucose-6-phosphatase

glucose6phos+h2o=glucose+pi

{-15.15, -13.75, -11.62, -10.96, -10.88}

{0.1, 0.4, 0.24, 0.04, 0.00}

In[165]:=

trGibbsRxSummary[glucose1phos + h2o + de == glucose + pi, "EC 3.1.3.10 Glucose-1-phosphatase
"glucose1phos+h2o=glucose+pi", {5, 6, 7, 8, 9}, .25]

EC 3.1.3.10 Glucose-1-phosphatase

glucose1phos+h2o=glucose+pi

{-21.81, -20.62, -18.65, -18.03, -17.95}

{0.08, 0.36, 0.23, 0.03, 0.00}

In[166]:=

trGibbsRxSummary[fructose16phos + h2o + de == fructose6phos + pi,
"EC 3.1.3.11 Fructose-bisphosphatase",
"fructose16phos+h2o=fructose6phos+pi", {5, 6, 7, 8, 9}, .25]

EC 3.1.3.11 Fructose-bisphosphatase

fructose16phos+h2o=fructose6phos+pi

{-20.21, -16.26, -12.79, -11.95, -11.84}

{0.37, 0.84, 0.32, 0.04, 0.00}

In[167]:=

trGibbsRxSummary[bpg + h2o + de == phosphoglycerate3 + pi,
"EC 3.1.3.13 Bisphosphoglycerate phosphatase",
"bpg+h2o=phosphoglycerate3+pi", {5, 6, 7, 8, 9}, .25]

EC 3.1.3.13 Bisphosphoglycerate phosphatase

bpg+h2o=phosphoglycerate3+pi

{-40.88, -41.44, -44.26, -49.42, -55.07}

{-0.03, -0.23, -0.77, -0.97, -1.}

In[168]:=

trGibbsRxSummary[glycerol + pi + de == glycerol3phos + h2o,
"EC 3.1.3.31 Nucleotidase", "glycerol+pi=glycerol3phos+h2o", {5, 6, 7, 8, 9}, .25]

EC 3.1.3.31 Nucleotidase

glycerol+pi=glycerol3phos+h2o

{0.42, -0.39, -1.95, -2.49, -2.56}

{-0.05, -0.26, -0.2, -0.03, -0.00}

In[169]:=

```
trGibbsRxSummary[phosphoglycerate3 + h2o + de == glycerate + pi,
  "EC 3.1.3.38 3-phosphoglycerate phosphatase",
  "phosphoglycerate3+h2o=glycerate+pi", {5, 6, 7, 8, 9}, .25]
```

EC 3.1.3.38 3-phosphoglycerate phosphatase

phosphoglycerate3+h2o=glycerate+pi

{-13.98, -14.01, -14.09, -14.13, -14.14}

{-0.00, -0.01, -0.01, -0.00, -0.}

In[170]:=

```
trGibbsRxSummary[sorbitol6phos + h2o + de == sorbitol + pi,
  "EC 3.1.3.50 Sorbitol-6-phosphatase",
  "sorbitol6phos+h2o=sorbitol+pi", {5, 6, 7, 8, 9}, .25]
```

EC 3.1.3.50 Sorbitol-6-phosphatase

sorbitol6phos+h2o=sorbitol+pi

{-16.45, -15.05, -12.92, -12.26, -12.18}

{0.1, 0.4, 0.24, 0.04, 0.00}

In[171]:=

```
trGibbsRxSummary[pep + h2o + de == pyruvate + pi,
  "EC 3.1.3.60 Phosphoenolpyruvate phosphatase",
  "pep+h2o=pyruvate+pi", {5, 6, 7, 8, 9}, .25]
```

EC 3.1.3.60 Phosphoenolpyruvate phosphatase

pep+h2o=pyruvate+pi

{-67.04, -66.32, -64.88, -64.37, -64.3}

{0.04, 0.23, 0.19, 0.03, 0.00}

EC 3.2 Glycosylases

In[172]:=

```
trGibbsRxSummary[maltose + h2o + de == 2 * glucose,
  "EC 3.2.1.20 Alpha-glucosidase", "maltose+h2o=2*glucose", {5, 6, 7, 8, 9}, .25]
```

EC 3.2.1.20 Alpha-glucosidase

maltose+h2o=2*glucose

{-19.92, -19.92, -19.92, -19.92, -19.92}

{0.}

In[173]:=

```
trGibbsRxSummary[sucrose + h2o + de == glucose + fructose,
  "EC 3.2.1.48 Sucrose alpha-glucosidase",
  "sucrose+h2o=glucose+fructose", {5, 6, 7, 8, 9}, .25]
```

EC 3.2.1.48 Sucrose alpha-glucosidase

sucrose+h2o=glucose+fructose

{-29.52, -29.52, -29.52, -29.52, -29.52}

{0.}

In[174]:=

    trGibbsRxSummary[lactose + h2o + de == glucose + galactose,
     "EC 3.2.1.108 Lactase", "lactose+h2oglucose+galactose", {5, 6, 7, 8, 9}, .25]

    EC 3.2.1.108 Lactase

    lactose+h2oglucose+galactose

    {-20.31, -20.31, -20.31, -20.31, -20.31}

    {0.}

In[175]:=

    trGibbsRxSummary[inosine + h2o + de == hypoxanthine + ribose,
     "EC 3.2.2.2 Inosine nucleosidase",
     "inosine+h2o=hypoxanthine+ribose", {5, 6, 7, 8, 9}, .25]

    EC 3.2.2.2 Inosine nucleosidase

    inosine+h2o=hypoxanthine+ribose

    {-16.16, -16.15, -16.11, -15.68, -13.34}

    {0., 0.00, 0.02, 0.17, 0.68}

In[176]:=

    trGibbsRxSummary[amp + h2o + de == adenine + ribose5phos,
     "EC 3.2.2.4 AMP nucleosidase", "amp+h2o=adenine+ribose5phos", {5, 6, 7, 8, 9}, .25]

    EC 3.2.2.4 AMP nucleosidase

    amp+h2o=adenine+ribose5phos

    {-4.77, -4.64, -4.72, -4.74, -4.74}

    {0.08, -0.01, -0.01, -0.00, -0.}

In[177]:=

    trGibbsRxSummary[adenosine + h2o + de == adenine + ribose,
     "EC 3.2.2.7 Purinenucleosidase", "adenosine+h2o=adenine+ribose", {5, 6, 7, 8, 9}, .25]

    EC 3.2.2.7 Purinenucleosidase

    adenosine+h2o=adenine+ribose

    {-7.2, -6.94, -6.91, -6.91, -6.91}

    {0.11, 0.01, 0.00, 0., 0.}

In[178]:=

    trGibbsRxSummary[imp + h2o + de == hypoxanthine + ribose5phos,
     "EC 3.2.2.12 Inosinate nucleosidase",
     "imp+h2o=hypoxanthine+ribose5phos", {5, 6, 7, 8, 9}, .25]

    EC 3.2.2.12 Inosinate nucleosidase

    imp+h2o=hypoxanthine+ribose5phos

    {-13.45, -13.53, -13.61, -13.29, -11.26}

{-0.01, -0.02, 0.01, 0.14, 0.62}

EC 3.5 Acting on carbon-nitrogen bonds, other than peptide bnds

In[179]:=

trGibbsRxSummary[asparagineL + h2o + de == aspartate + ammonia,
"EC 3.5.1.1 Asparaginease", "asparagineL+h2o=aspartate+ammonia", {5, 6, 7, 8, 9}, .25]

EC 3.5.1.1 Asparaginease

asparagineL+h2o=aspartate+ammonia

{-13.69, -13.69, -13.7, -13.82, -14.79}

{-0., -0., -0.01, -0.05, -0.36}

In[180]:=

trGibbsRxSummary[glutamine + h2o + de == glutamate + ammonia,
"EC 3.5.1.2 Glutaminase", "glutamine+h2o=glutamate+ammonia", {5, 6, 7, 8, 9}, .25]

EC 3.5.1.2 Glutaminase

glutamine+h2o=glutamate+ammonia

{-13.19, -13.19, -13.2, -13.32, -14.29}

{-0., -0., -0.01, -0.05, -0.36}

In[181]:=

trGibbsRxSummary[urea + 2 * h2o + de == co2tot + 2 * ammonia,
"EC 3.5.1.5 Urease", "urea+2*h2o=co2tot+2*ammonia", {5, 6, 7, 8, 9}, .25]

EC 3.5.1.5 Urease

urea+2*h2o=co2tot+2*ammonia

{-47.67, -37.55, -30.19, -24.51, -21.06}

{1.92, 1.55, 1.1, 0.89, 0.14}

In[182]:=

trGibbsRxSummary[creatine + de == creatinine + h2o,
"EC 3.5.2.10 Creatininase", "creatine=creatinine+h2o", {5, 6, 7, 8, 9}, .25]

EC 3.5.2.10 Creatininase

creatine=creatinine+h2o

{-1.13, -1.13, -1.13, -1.13, -1.13}

{-1.24481 × 10^{-15}}

In[183]:=

trGibbsRxSummary[adenine + h2o + de == hypoxanthine + ammonia, "EC 3.5.4.2 Adenine deaminase"
"adenine+h2o=hypoxanthine+ammonia", {5, 6, 7, 8, 9}, .25]

EC 3.5.4.2 Adenine deaminase

adenine+h2o=hypoxanthine+ammonia

{-37.12, -31.73, -26.07, -20.49, -15.75}

{0.86, 0.98, 0.99, 0.95, 0.64}

In[184]:=

trGibbsRxSummary[adenosine + h2o + de == inosine + ammonia,
  "EC 3.5.4.4 Adenosine deaminase", "adenosine+h2o=inosine+ammonia", {5, 6, 7, 8, 9}, .25]

EC 3.5.4.4 Adenosine deaminase

adenosine+h2o=inosine+ammonia

{-28.16, -22.52, -16.88, -11.72, -9.32}

{0.97, 0.99, 0.97, 0.77, -0.04}

In[185]:=

trGibbsRxSummary[amp + h2o + de == imp + ammonia,
  "EC 3.5.4.6 AMP deaminase", "amp+h2o=imp+ammonia", {5, 6, 7, 8, 9}, .25]

EC 3.5.4.6 AMP deaminase

amp+h2o=imp+ammonia

{-28.44, -22.84, -17.18, -11.93, -9.23}

{0.95, 1., 0.98, 0.81, 0.02}

In[186]:=

trGibbsRxSummary[adp + h2o + de == idp + ammonia,
  "EC 3.5.4.7 ADP deaminase", "adp+h2o=idp+ammonia", {5, 6, 7, 8, 9}, .25]

EC 3.5.4.7 ADP deaminase

adp+h2o=idp+ammonia

{-28.33, -22.75, -17.13, -12.25, -10.59}

{0.94, 0.99, 0.96, 0.67, -0.15}

In[187]:=

trGibbsRxSummary[atp + h2o + de == itp + ammonia,
  "EC 3.5.4.18 ATP deaminase", "atp+h2o=itp+ammonia", {5, 6, 7, 8, 9}, .25]

EC 3.5.4.18 ATP deaminase

atp+h2o=itp+ammonia

{-28.22, -22.66, -17.02, -11.86, -9.48}

{0.94, 0.99, 0.97, 0.77, -0.04}

EC 3.6 Acting on acid anhydrides

In[188]:=

trGibbsRxSummary[ppi + h2o + de == 2 * pi,
  "EC 3.6.1.1 Inorganic diphosphatase", "ppi+h2o=2*pi", {5, 6, 7, 8, 9}, .25]

EC 3.6.1.1 Inorganic diphosphatase

ppi+h2o=2*pi

{-23.37, -22.45, -22.66, -26.06, -28.22}

{0.08, 0.21, -0.41, -0.6, -0.17}

In[189]:=

    trGibbsRxSummary[atp + h2o + de == adp + pi,
     "EC 3.6.1.3 Adenosinetriphosphatase", "atp+h2o=adp+pi", {5, 6, 7, 8, 9}, .25]

    EC 3.6.1.3 Adenosinetriphosphatase

    atp+h2o=adp+pi

    {-32.56, -33.22, -36.04, -41.07, -46.7}

    {-0.04, -0.25, -0.74, -0.96, -1.}

In[190]:=

    trGibbsRxSummary[atp + 2 * h2o + de == amp + 2 * pi,
     "EC 3.6.1.5 Apyrase", "atp+2*h2o=amp+2*pi", {5, 6, 7, 8, 9}, .25]

    EC 3.6.1.5 Apyrase

    atp+2*h2o=amp+2*pi

    {-62.89, -64.31, -70., -80.06, -91.31}

    {-0.09, -0.52, -1.48, -1.93, -1.99}

In[191]:=

    trGibbsRxSummary[atp + h2o + de == amp + ppi,
     "EC 3.6.1.8 ATP diphosphatase", "atp+h2o=amp+ppi", {5, 6, 7, 8, 9}, .25]

    EC 3.6.1.8 ATP diphosphatase

    atp+h2o=amp+ppi

    {-39.52, -41.86, -47.34, -54., -63.1}

    {-0.17, -0.73, -1.08, -1.33, -1.83}

In[192]:=

    trGibbsRxSummary[fadox + h2o + de == amp + fmnox,
     "EC 3.6.1.18 FAD diphophatase", "fadox+h2o=amp+fmnox", {5, 6, 7, 8, 9}, .25]

    EC 3.6.1.18 FAD diphophatase

    fadox+h2o=amp+fmnox

    {-872.1, -878.8, -888.4, -899.5, -910.9}

    {-1.02, -1.4, -1.87, -1.99, -2.}

In[193]:=

    trGibbsRxSummary[nadox + h2o + de == amp + nicotinamideribonucleotide,
     "EC 3.6.1.22 NADox diphosphatase",
     "nadox+h2o=amp+nicotinamideribonucleotide", {5, 6, 7, 8, 9}, .25]

    EC 3.6.1.22 NADox diphosphatase

    nadox+h2o=amp+nicotinamideribonucleotide

    {-39.75, -42.57, -50.89, -61.85, -73.22}

    {-0.13, -0.98, -1.8, -1.98, -2.}

EC 3.7 Acting on carbon-carbon bonds

In[194]:=

> **trGibbsRxSummary[oxaloacetate + h2o + de == oxalate + acetate,**
> **"EC 3.7.1.1 Oxaloacetase", "oxaloacetate=oxalate+acetate", {5, 6, 7, 8, 9}, .25]**

> EC 3.7.1.1 Oxaloacetase

> oxaloacetate=oxalate+acetate

> {-43.66, -48.68, -54.31, -60.01, -65.72}

> {-0.72, -0.97, -1., -1., -1.}

12.4 Lyase Reactions

EC 4.1 Carbon-carbon lyases

In[195]:=

> **trGibbsRxSummary[pyruvate + h2o + de == co2tot + acetaldehyde,**
> **"EC 4.1.1.1 pyruvate decarboxylase",**
> **"pyruvate+h2o=co2tot+acetaldehyde", {5, 6, 7, 8, 9}, .25]**

> EC 4.1.1.1 pyruvate decarboxylase

> pyruvate+h2o=co2tot+acetaldehyde

> {-22.69, -18.28, -16.6, -16.38, -16.71}

> {0.92, 0.55, 0.11, -0.01, -0.15}

In[196]:=

> **trGibbsRxSummary[oxalate + h2o + de == co2tot + formate,**
> **"EC 4.1.1.2 Pyruvate decarboxylase", "oxalate+h2o=co2tot+formate", {5, 6, 7, 8, 9}, .25]**

> EC 4.1.1.2 Pyruvate decarboxylase

> oxalate+h2o=co2tot+formate

> {-31.31, -27.01, -25.35, -25.13, -25.46}

> {0.87, 0.54, 0.11, -0.01, -0.15}

In[197]:=

> **trGibbsRxSummary[acetoacetate + h2o + de == co2tot + acetone,**
> **"EC 4.1.1.4 Acetoacetate decarboxylase",**
> **"acetoacetate+h2o=co2tot+acetone", {5, 6, 7, 8, 9}, .25]**

> EC 4.1.1.4 Acetoacetate decarboxylase

> acetoacetate+h2o=co2tot+acetone

> {-33.98, -29.57, -27.89, -27.67, -28.}

> {0.92, 0.55, 0.11, -0.01, -0.15}

In[198]:=

> **trGibbsRxSummary[aspartate + h2o + de == co2tot + alanine,**
> **"EC 4.1.1.12 Aspartate 4-decarboxylase",**
> **"aspartate+h2o=co2tot+alanine", {5, 6, 7, 8, 9}, .25]**

EC 4.1.1.12 Aspartate 4-decarboxylase

aspartate+h2o=co2tot+alanine

{-31.08, -26.67, -24.99, -24.77, -25.1}

{0.92, 0.55, 0.11, -0.01, -0.15}

In[199]:=

trGibbsRxSummary[pyruvate + oxaloacetate + de == co2tot + pep,
"EC 4.1.1.31 Phosphoenolpyruvate carboxylase",
"pyruvate+oxaloacetate=co2tot+pep", {5, 6, 7, 8, 9}, .25]

EC 4.1.1.31 Phosphoenolpyruvate carboxylase

pyruvate+oxaloacetate=co2tot+pep

{-660.7, -663.2, -671.1, -682., -693.7}

{-0.14, -0.87, -1.77, -1.99, -2.14}

In[200]:=

trGibbsRxSummary[idp + co2tot + pep + de == itp + oxaloacetate + h2o,
"EC 4.1.1.32 Phosphoenolpyruvate carboxykinase",
"idp+co2tot+pep=itp+oxaloacetate+h2o", {5, 6, 7, 8, 9}, .25]

EC 4.1.1.32 Phosphoenolpyruvate carboxykinase

idp+co2tot+pep=itp+oxaloacetate+h2o

{-1.05, -4.11, -1.5, 4.11, 10.86}

{-0.85, -0.07, 0.84, 1.1, 1.25}

In[201]:=

trGibbsRxSummary[glyceraldehydephos + acetaldehyde + de == deoxyribose5phos,
"EC 4.1.2.4 Deoxyribose-phosphate aldolase",
"glyceraldehydephos+acetaldehyde=deoxyribose5phos", {5, 6, 7, 8, 9}, .25]

EC 4.1.2.4 Deoxyribose-phosphate aldolase

glyceraldehydephos+acetaldehyde=deoxyribose5phos

{-24.78, -22.1, -20.6, -20.35, -20.32}

{0.35, 0.45, 0.1, 0.01, 0.00}

In[202]:=

trGibbsRxSummary[dihydroxyacetonephos + glyceraldehydephos + de == fructose16phos,
"EC 4.1.2.13a Fructose-bisphosphate aldolase",
"dihydroxyacetonephos+glyceraldehydephos+de==fructose16phos", {5, 6, 7, 8, 9}, .25]

EC 4.1.2.13a Fructose-bisphosphate aldolase

dihydroxyacetonephos+glyceraldehydephos+de==fructose16phos

{-24.33, -23.18, -23.03, -23.02, -23.02}

{0.29, 0.07, 0.01, 0., 0.}

In[203]:=

trGibbsRxSummary[2 * dihydroxyacetonephos + de == fructose16phos,
 "EC 4.1.2.13b Fructose-bisphosphate aldolase",
 "2*dihydroxyacetonephos=fructose16phos", {5, 6, 7, 8, 9}, .25]

EC 4.1.2.13b Fructose-bisphosphate aldolase

2*dihydroxyacetonephos=fructose16phos

{-16.67, -15.52, -15.37, -15.36, -15.36}

{0.29, 0.07, 0.01, 0., 0.}

In[204]:=

trGibbsRxSummary[formate + acetaldehyde + de == lactate,
 "EC 4.1.2.36 Lactate aldolase", "formate+acetaldehyde=lactate", {5, 6, 7, 8, 9}, .25]

EC 4.1.2.36 Lactate aldolase

formate+acetaldehyde=lactate

{-26.72, -26.72, -26.72, -26.72, -26.72}

{0.}

In[205]:=

trGibbsRxSummary[succinate + glyoxylate + de == citrateiso,
 "EC 4.1.3.1 Isocitratelyase", "succinate+glyoxylate=citrateiso", {5, 6, 7, 8, 9}, .25]

EC 4.1.3.1 Isocitratelyase

succinate+glyoxylate=citrateiso

{-2.21, -0.72, -0.3, -0.24, -0.24}

{0.3, 0.16, 0.02, 0.00, 0.}

In[206]:=

trGibbsRxSummary[acetate + oxaloacetate + de == citrate, "EC 4.1.3.6 Citrate(pro-3S)-lyase"
 "acetate+oxaloacetate=citrate", {5, 6, 7, 8, 9}, .25]

EC 4.1.3.6 Citrate(pro-3S)-lyase

acetate+oxaloacetate=citrate

{-6.69, -4.01, -3.41, -3.34, -3.33}

{0.68, 0.24, 0.03, 0.00, 0.}

In[207]:=

trGibbsRxSummary[pyruvate + glyoxylate + de == hydroxy2oxoglutarate,
 "EC 4.1.3.16 4-hydroxy-2-oxoglutarate aldolase",
 "pyruvate+glyoxylate=hydroxy2oxoglutarate", {5, 6, 7, 8, 9}, .25]

EC 4.1.3.16 4-hydroxy-2-oxoglutarate aldolase

pyruvate+glyoxylate=hydroxy2oxoglutarate

{-12.53, -12.53, -12.53, -12.53, -12.53}

{0.}

In[208]:=

trGibbsRxSummary[tryptophanL + h2o + de == indole + pyruvate,
"EC 4.1.99.1 Tryptophanasease", "tryptophanL+h2o=indole+pyruvate", {5, 6, 7, 8, 9}, .25]

EC 4.1.99.1 Tryptophanasease

tryptophanL+h2o=indole+pyruvate

{-14.79, -37.62, -60.45, -83.29, -106.1}

{-4.}

EC 4.2 Carbon-oxygen lyases

In[209]:=

trGibbsRxSummary[fumarate + h2o + de == malate,
"EC 4.2.1.2 Fumarate hydratase", "fumarate+h2o=malate", {5, 6, 7, 8, 9}, .25]

EC 4.2.1.2 Fumarate hydratase

fumarate+h2o=malate

{-4.34, -3.69, -3.61, -3.6, -3.6}

{0.23, 0.04, 0.00, 0., 0.}

In[210]:=

trGibbsRxSummary[aconitatecis + h2o + de == citrate,
"EC 4.2.1.3 Aconitate hydratase", "aconitatecis+h2o=citrate", {5, 6, 7, 8, 9}, .25]

EC 4.2.1.3 Aconitate hydratase

aconitatecis+h2o=citrate

{-12.37, -9.12, -8.45, -8.38, -8.37}

{0.91, 0.27, 0.03, 0.00, 0.}

In[211]:=

trGibbsRxSummary[phosphoglycerate2 + de == pep + h2o,
"EC 4.2.1.11 Phosphoenolpyruvate hydratase",
"phosphoglycerate2=pep+h2o", {5, 6, 7, 8, 9}, .25]

EC 4.2.1.11 Phosphoenolpyruvate hydratase

phosphoglycerate2=pep+h2o

{-0.94, -1.76, -3.6, -4.35, -4.45}

{-0.05, -0.27, -0.26, -0.04, -0.00}

In[212]:=

trGibbsRxSummary[maleate + h2o + de == malate,
"EC 4.2.1.31 Maleate hydratase", "maleate+h2o=malate", {5, 6, 7, 8, 9}, .25]

EC 4.2.1.31 Maleate hydratase

maleate+h2o=malate

{-14.37, -13.5, -13.39, -13.38, -13.38}

{0.33, 0.05, 0.00, 0., 0.}

In[213]:=

    trGibbsRxSummary[methylmaleate + h2o + de == methylmalate,
    "EC 4.2.1.35 (R)-2methylmalate dehydratase",
    "methylmaleate+h2o=methylmalate", {5, 6, 7, 8, 9}, .25]

    EC 4.2.1.35 (R)-2methylmalate dehydratase

    methylmaleate+h2o=methylmalate

    {-1.55, -5., -5.89, -6., -6.01}

    {-0.83, -0.34, -0.05, -0.01, -0.}

EC 4.3 Carbon-nitrogen lyases

In[214]:=

    trGibbsRxSummary[fumarate + ammonia + de == aspartate, "EC 4.3.1.1 Aspartate ammonia-lyase"
    "aspartate+de==fumarate+ammonia", {5, 6, 7, 8, 9}, .25]

    EC 4.3.1.1 Aspartate ammonia-lyase

    aspartate+de==fumarate+ammonia

    {-11.2, -11.43, -11.44, -11.33, -10.36}

    {-0.1, -0.01, 0.00, 0.05, 0.36}

In[215]:=

    trGibbsRxSummary[serineL + de == pyruvate + ammonia,
    "EC 4.3.1.17 L-serine ammonia-lyase", "serineL=pyruvate+ammonia", {5, 6, 7, 8, 9}, .25]

    EC 4.3.1.17 L-serine ammonia-lyase

    serineL=pyruvate+ammonia

    {-42.33, -42.33, -42.34, -42.46, -43.43}

    {-0., -0., -0.01, -0.05, -0.36}

EC 4.4 Carbon-sulfur lyases

In[216]:=

    trGibbsRxSummary[cysteineL + h2o + de == h2saq + ammonia + pyruvate,
    "EC 4.4.1.15 D-cysteine desulfhydratase",
    "cysteineL+h2o=h2saq+ammonia+pyruvate", {5, 6, 7, 8, 9}, .25]

    EC 4.4.1.15 D-cysteine desulfhydratase

    cysteineL+h2o=h2saq+ammonia+pyruvate

    {-5.07, -5.44, -7.53, -11.19, -13.76}

    {-0.02, -0.16, -0.59, -0.56, -0.46}

EC 4.6 Phosphorus-oxygen lyases

In[217]:=

    trGibbsRxSummary[cyclicamp + ppi + de == atp,
    "EC 4.6.1.1 Adenylate cyclase", "atp=cyclicamp+ppi", {5, 6, 7, 8, 9}, .25]

    EC 4.6.1.1 Adenylate cyclase

```
atp=cyclicamp+ppi
```

{-55.35, -31.21, -6.71, 17.38, 43.63}

{4.15, 4.33, 4.2, 4.34, 4.83}

12.5 Isomerase Reactions

EC 5.2 Cis-trans-isomerases

In[218]:=

trGibbsRxSummary[maleate + de == fumarate,
  "EC 5.2.1.1 Maleate isomerase", "maleate=fumarate", {5, 6, 7, 8, 9}, .25]

EC 5.2.1.1 Maleate isomerase

maleate=fumarate

{-10.04, -9.81, -9.78, -9.78, -9.78}

{0.1, 0.01, 0.00, 0., 0.}

EC 5.3 Intramolecular oxidoreductases

In[219]:=

trGibbsRxSummary[glyceraldehydephos + de == dihydroxyacetonephos,
  "EC 5.3.1.1 Triosephosphate isomerase",
  "glyceraldehydephos=dihydroxyacetonephos", {5, 6, 7, 8, 9}, .25]

EC 5.3.1.1 Triosephosphate isomerase

glyceraldehydephos=dihydroxyacetonephos

{-7.66, -7.66, -7.66, -7.66, -7.66}

$\{-5.79 \times 10^{-14}, -2.3 \times 10^{-14}, -1.87 \times 10^{-15}, 1.24 \times 10^{-15}, 6.22 \times 10^{-16}\}$

In[220]:=

trGibbsRxSummary[xylulose + de == xylose,
  "EC 5.3.1.5 Xylose isomerase", "xylulose=xylose", {5, 6, 7, 8, 9}, .25]

EC 5.3.1.5 Xylose isomerase

xylulose=xylose

{-4.34, -4.34, -4.34, -4.34, -4.34}

{0.}

In[221]:=

trGibbsRxSummary[ribose1phos + de == ribose5phos,
  "EC 5.3.1.6 Ribose-5-phosphate isomerase",
  "ribose1phos=ribose5phos", {5, 6, 7, 8, 9}, .25]

EC 5.3.1.6 Ribose-5-phosphate isomerase

ribose1phos=ribose5phos

{-8.08, -8.08, -8.08, -8.08, -8.08}

$\{1.24 \times 10^{-15}, -1.24 \times 10^{-15}, 0., 1.24 \times 10^{-15}, 0.\}$

In[222]:=

    trGibbsRxSummary[mannose + de == fructose,
     "EC 5.3.1.7 mannose isomerase", "mannose=fructose", {5, 6, 7, 8, 9}, .25]

    EC 5.3.1.7 mannose isomerase

    mannose=fructose

    {-5.51, -5.51, -5.51, -5.51, -5.51}

    {0.}

In[223]:=

    trGibbsRxSummary[mannose6phos + de == fructose6phos,
     "EC 5.3.1.8 Mannose-6-phosphate isomerase",
     "mannose6phos=fructose6phos", {5, 6, 7, 8, 9}, .25]

    EC 5.3.1.8 Mannose-6-phosphate isomerase

    mannose6phos=fructose6phos

    {-0.1, -0.56, -0.87, -0.92, -0.93}

    {-0.05, -0.09, -0.02, -0.00, -0.}

In[224]:=

    trGibbsRxSummary[fructose6phos + de == glucose6phos,
     "EC 5.3.1.9 Glucose-6-phosphate isomerase",
     "fructose6phos=glucose6phos", {5, 6, 7, 8, 9}, .25]

    EC 5.3.1.9 Glucose-6-phosphate isomerase

    fructose6phos=glucose6phos

    {-3.87, -3.46, -3.19, -3.15, -3.14}

    {0.04, 0.08, 0.02, 0.00, 0.}

In[225]:=

    trGibbsRxSummary[xylulose + de == lyxose,
     "EC 5.3.1.15 D-lyxose keto-isomerase", "xylulose=lyxose", {5, 6, 7, 8, 9}, .25]

    EC 5.3.1.15 D-lyxose keto-isomerase

    xylulose=lyxose

    {-2.99, -2.99, -2.99, -2.99, -2.99}

    {0.}

In[226]:=

    trGibbsRxSummary[ribulose + de == ribose,
     "EC 5.3.1.20 Ribose isomerase", "ribulose=ribose", {5, 6, 7, 8, 9}, .25]

    EC 5.3.1.20 Ribose isomerase

    ribulose=ribose

    {-16.06, -16.06, -16.06, -16.06, -16.06}

    {0.}

EC 5.4 Intramolecular transferases (mutases)

In[227]:=

> **trGibbsRxSummary[glucose1phos + de == glucose6phos,**
> **"EC 5.4.2.6 Beta-phosphoglucomutase", "glucose1phos=glucose6phos", {5, 6, 7, 8, 9}, .25]**

EC 5.4.2.6 Beta-phosphoglucomutase

glucose1phos=glucose6phos

{-6.66, -6.87, -7.04, -7.07, -7.07}

{-0.02, -0.05, -0.01, -0.00, -0.}

In[228]:=

> **trGibbsRxSummary[mannose1phos + de == mannose6phos,**
> **"EC 5.4.2.8 Phosphomannomutase", "mannose1phos=mannose6phos", {5, 6, 7, 8, 9}, .25]**

EC 5.4.2.8 Phosphomannomutase

mannose1phos=mannose6phos

{-5.31, -5.31, -5.31, -5.31, -5.31}

$\{2.49 \times 10^{-15}, 2.99 \times 10^{-14}, 1.49 \times 10^{-14}, -1.24 \times 10^{-15}, 1.24 \times 10^{-15}\}$

In[229]:=

> **trGibbsRxSummary[maltose + de == trehalose,**
> **"EC 5.4.99.16 Maltose alpha-D-glucosyltransferase",**
> **"maltose=trehalose", {5, 6, 7, 8, 9}, .25]**

EC 5.4.99.16 Maltose alpha-D-glucosyltransferase

maltose=trehalose

{-8.07, -8.07, -8.07, -8.07, -8.07}

{0.}

12.6 Ligase Reactions

EC 6.2 Forming carbon-sulfur bonds

In[230]:=

> **trGibbsRxSummary[acetate + atp + coA + de == acetylcoA + amp + ppi,**
> **"EC 6.2.1.1 Acetate-CoA ligase",**
> **"acetate+atp+coA=acetylcoA+amp+ppi", {5, 6, 7, 8, 9}, .25]**

EC 6.2.1.1 Acetate-CoA ligase

acetate+atp+coA=acetylcoA+amp+ppi

{-9.12, -6.31, -5.98, -5.67, -5.06}

{0.61, 0.24, -0.00, 0.12, 0.06}

In[231]:=

> **trGibbsRxSummary[succinate + atp + coA + de == succinylcoA + adp + pi,**
> **"EC 6.2.1.5 Succinate-CoA ligase (ADP forming)",**
> **"succinate+atp+coA=succinylcoA+adp+pi", {5, 6, 7, 8, 9}, .25]**

```
EC 6.2.1.5 Succinate-CoA ligase (ADP forming)

succinate+atp+coA=succinylcoA+adp+pi

{-7.59, -4.09, -1.26, 0.65, 4.72}

{0.44, 0.66, 0.32, 0.48, 0.89}
```

In[232]:=

trGibbsRxSummary[oxalate + atp + coA + de == oxalylcoA + adp + pi,
"EC 6.2.1.8 Oxalate-CoA ligase",
"oxalate+atp+coA=oxalylcoA+adp+pi", {5, 6, 7, 8, 9}, .25]

```
EC 6.2.1.8 Oxalate-CoA ligase

oxalate+atp+coA=oxalylcoA+adp+pi

{-26.08, -21.12, -18.07, -16.13, -12.06}

{0.91, 0.75, 0.33, 0.48, 0.89}
```

In[233]:=

trGibbsRxSummary[malate + atp + coA + de == malylcoA + adp + pi, "EC 6.2.1.9 Oxalate-CoA ligase
"malate+atp+coA=malylcoA+adp+pi", {5, 6, 7, 8, 9}, .25]

```
EC 6.2.1.9 Oxalate-CoA ligase

malate+atp+coA=malylcoA+adp+pi

{-10.13, -5.75, -2.78, -0.85, 3.22}

{0.71, 0.72, 0.33, 0.48, 0.89}
```

In[234]:=

trGibbsRxSummary[acetylcoA + adp + pi + de == acetate + atp + coA,
"EC 6.2.1.13 acetate-CoA ligase (ADP forming)",
"acetylcoA+adp+pi=acetate+atp+coA", {5, 6, 7, 8, 9}, .25]

```
EC 6.2.1.13 acetate-CoA ligase (ADP forming)

acetylcoA+adp+pi=acetate+atp+coA

{2.17, -2.33, -5.33, -7.26, -11.33}

{-0.73, -0.73, -0.33, -0.48, -0.89}
```

In[235]:=

trGibbsRxSummary[acetoacetylcoA + amp + ppi + de == acetoacetate + atp + coA,
"EC 6.2.1.16 Acetoacetate-CoA ligase",
"acetoacetylcoA+amp+ppi=acetoacetate+atp+coA", {5, 6, 7, 8, 9}, .25]

```
EC 6.2.1.16 Acetoacetate-CoA ligase

acetoacetylcoA+amp+ppi=acetoacetate+atp+coA

{2.36, -1.03, -1.43, -1.74, -2.35}

{-0.84, -0.27, 0.00, -0.12, -0.06}
```

EC 6.3 Forming carbon-nitrogen bonds

In[236]:=

> **trGibbsRxSummary[aspartate + atp + ammonia + de == asparagineL + amp + ppi,**
> **"EC 6.3.1.1 Aspartate-ammonia ligase",**
> **"aspartate+atp+ammonia=asparagineL+amp+ppi", {5, 6, 7, 8, 9}, .25]**

EC 6.3.1.1 Aspartate-ammonia ligase

aspartate+atp+ammonia=asparagineL+amp+ppi

{-25.83, -28.17, -33.64, -40.18, -48.3}

{-0.17, -0.73, -1.07, -1.28, -1.47}

In[237]:=

> **trGibbsRxSummary[glutamate + atp + ammonia + de == asparagineL + adp + pi,**
> **"EC 6.3.1.2 Glutamate-ammonia ligase",**
> **"glutamate+atp+ammonia=asparagineL+adp+pi", {5, 6, 7, 8, 9}, .25]**

EC 6.3.1.2 Glutamate-ammonia ligase

glutamate+atp+ammonia=asparagineL+adp+pi

{-75.98, -88.05, -102.3, -118.6, -134.7}

{-2.04, -2.25, -2.74, -2.91, -2.64}

In[238]:=

> **trGibbsRxSummary[aspartate + atp + ammonia + de == asparagineL + adp + pi,**
> **"EC 6.3.1.4 Aspartate-ammonia ligase (ADP-forming)",**
> **"aspartate+atp+ammonia=asparagineL+adp+pi", {5, 6, 7, 8, 9}, .25]**

EC 6.3.1.4 Aspartate-ammonia ligase (ADP-forming)

aspartate+atp+ammonia=asparagineL+adp+pi

{-18.87, -19.53, -22.33, -27.25, -31.91}

{-0.04, -0.25, -0.74, -0.91, -0.64}

In[239]:=

> **trGibbsRxSummary[aspartate + atp + glutamine + de == asparagineL + glutamate + amp + ppi,**
> **"EC 6.3.5.4 Asparagine synthase (glutamine-hydrolyzing)",**
> **"aspartate+atp+glutamine=asparagineL+glutamate+amp+ppi", {5, 6, 7, 8, 9}, .25]**

EC 6.3.5.4 Asparagine synthase (glutamine-hydrolyzing)

aspartate+atp+glutamine=asparagineL+glutamate+amp+ppi

{-217.5, -208.4, -202.5, -197.7, -195.4}

{1.83, 1.27, 0.92, 0.67, 0.17}

EC 6.4 Forming carbon-carbon bonds

In[240]:=

> **trGibbsRxSummary[atp + pyruvate + co2tot + de == oxaloacetate + adp + pi,**
> **"EC 6.4.1.1 Pyruvate carboxylase",**
> **"atp+pyruvate+co2tot=oxaloacetate+adp+pi", {5, 6, 7, 8, 9}, .25]**

EC 6.4.1.1 Pyruvate carboxylase

atp+pyruvate+co2tot=oxaloacetate+adp+pi

{0.75, -4.31, -8.81, -14.06, -19.36}

{-0.96, -0.8, -0.85, -0.96, -0.85}

In[241]:=

trGibbsRxSummary[atp + propanoylcoA + co2tot + de == methylmalonylcoA + adp + pi,
 "EC 6.4.1.3 Propanoyl-CoA carboxylase",
 "atp+propanoylcoA+co2tot=methylmalonylcoA+adp+pi", {5, 6, 7, 8, 9}, .25]

EC 6.4.1.3 Propanoyl-CoA carboxylase

atp+propanoylcoA+co2tot=methylmalonylcoA+adp+pi

{-0.15, -5.03, -9.51, -14.76, -20.06}

{-0.88, -0.79, -0.85, -0.96, -0.85}

In[242]:=

trGibbsRxSummary[acetone + co2tot + atp + h2o + de == acetoacetate + amp + 2 * pi,
 "EC 6.4.1.6 Acetone carboxylase",
 "acetone+co2tot+atp+h2o=acetoacetate+amp+2*pi", {5, 6, 7, 8, 9}, .25]

EC 6.4.1.6 Acetone carboxylase

acetone+co2tot+atp+h2o=acetoacetate+amp+2*pi

{-28.92, -34.74, -42.11, -52.39, -63.32}

{-1.01, -1.07, -1.59, -1.92, -1.85}

12.7 Discussion

This chapter demonstrates the usefulness of BasicBiochemData3 (1). In the future this database can be extended and many reactions can be added to this list of 229 enzyme-catalyzed reactions. At the present time $\Delta_f H°$ is known for the species of 94 reactants. Future measurements of $\Delta_r H'°$ and enthalpies of dissociation of weak acids will be used to calculate $\Delta_f H°$ of species of more reactants so that the effects of temperature can be calculated for more reactions. When more heat capacities of species have been determined, it will be possible to make calculations of K' over wider ranges of temperature. BasicBiochemData3 contains functions of pH and ionic strength for $\Delta_r G'°$ of these 229 enzyme-catalyzed reactions so that $\Delta_r G'°$ and $\Delta_r N_H$ can be calculated for these reactions at 298.15 K, pHs in the range 5 to 9, and ionic strengths in the range zero to 0.35 M. The index given in the Appendix lists the EC numbers for reactions that involve these 167 reactants.

Because of coupling (see Chapter 7) there are relationships between the thermodynamic properties of reactions in some of the EC classes. All oxidoreductase reactions can be considered to be coupled reactions because each one can be divided into two, or in a some cases, three half reactions that do not share atoms but are connected by formal electrons. Transferase reactions can each be considered to result from the coupling of two oxidoreductase reactions or two hydrolase reactions. Fifteen examples are discussed in reference (6). Each of the coupled reactions contributes its $\Delta_r G'°$ and $\Delta_r N_H$ to the coupled reaction. Hydrolase reactions and isomerase reactions are never coupled reactions. Some lyase reactions are coupled. Ligase reactions are all coupled by definition because they join together two reactions with the hydrolysis of a pyrophosphate bond in ATP or a similar triphosphate. A spectacular example of coupling is provided by EC 6.3.5.4 because there are seven reactants. This never happens in chemistry.

Appendix

Index of Reactants in Reactions

acetaldehyde   1.1.1.1.a, 1.2.1.3, 1.2.2.3, 1.2.1.10, 4.1.1.1, 4.1.2.4, 4.1.2.36

acetate   1.2.1.3, 1.13.12.4, 2.7.2.1, 2.7.2.12, 3.1.1.1, 3.1.2.1, 3.7.1.1, 2.7.1.1, 4.1.3.6, 6.2.1.1, 6.2.1.13

acetoacetate   3.1.2.11, 4.1.1.4, 6.2.1.16, 6.4.1.6

acetoacetylcoA   2.3.1.9, 3.1.2.11, 6.2.1.16

acetone   1.1.1.1c, 1.1.1.80, 4.1.1.4, 6.4.1.6

acetylcoA   1.2.1.10, 1.2.3.6, 1.2.7.1,2.3.1.8, 2.3.1.9, 2.3.1.54, 2.3.3.8, 2.3.3.9, 2.7.2.1, 3.1.2.1, 6.2.1.1, 6.2.1.13

acetylphos   1.2.3.3, 2.3.1.8, 2.7.2.12

aconitatecis   4.2.1.3

adenine   2.4.2.7, 3.2.2.4, 3.2.2.7, 3.5.4.2

adenosine   2.4.2.1, 2.7.1.20, 3.2.2.7, 3.5.4.4

adeosinephosphosulfate   2.7.7.4

adp   1.18.6.1, 2.3.3.9, 2.7.1.2, 2.7.1.4, 2.7.1.6, 2.7.1.7, 2.7.1.11, 2.7.1.15, 2.7.1.20, 2.7.1.23, 2.7.1.29, 2.7.1.30, 2.7.1.31,
       2.7.1.40, 2.7.1.47, 2.7.1.54, 2.7.1.73, 2.7.1.76, 2.7.1.86, 2.7.1.146, 2.7.1.147, 2.7.2.1, 2.7.2.3,3.5 .4.7, 3.5.4.18,
       3.6.1.3, 6.2.1.5,6.2.1.8, 6.2.1.9, 6.2.1.13, 6.3.1.2, 6.3.1.4, 6.4.1.1, 6.4.1.3

alanine 1.4.1.1, 2.6.1.2, 2.6.1.44, 26.1.51, 4.1.1.12

ammonia   1.4.1.1, 1.4.1.2, 1.4.1.3, 1.4.1.7, 1.4.1.9, 1.4.1.10, 1.4.2.1,1.4.3.11, 1.4.3.16, 1.7.1.4, 1.7.2.2, 1.7.7.1, 1.18.6.1,
       3.5.1.1, 3.5.1.2, 3.5.1.5, 3.5.4.2, 4.1.9.9.1, 6.3.1.4

amp   2.4.2.7, 2.7.1.20, 2.7.1.147, 2.7.2.3, 2.7.6.1, 2.7.9.1, 2.7.9.2, 3.2.2.4, 3.5.4.6, 3.6.1.5, 3.6.1.8, 3.6.1.18, 3.6.1.22,
       6.2.1.1, 6.2.1.16, 6.3.1.1, 6.3.5.4, 6.4.1.6

arabinose   2.7.1.54

arabinose1phos   2.7.1.54

asparagineL   3.5.1.1, 6.3.1.1, 6.3.1.2, 6.3.1.4, 6.3.5.4

aspartate   1.4.3.16, 2.6.1.1, 2.6.1.35, 3.5.1.1, 4.1.1.2, 4.3.1.3, 6.3.1.1, 6.3.1.4, 6.3.5.4

atp   1.18.6.1, 2.3.3.8, 2.7.1.2, 2.7.1.4, 2.7.1.6, 2.7.1.7, 2.7.1.11, 2.7.1.15, 2.7.1.20, 2.7.1.23, 2.7.1.29,2.7.1.30, 2.7.1.31,
       2.7.1.40, 2.7.1.47, 2.7.1.54, 2.7.1.73, 2.7.1.76, 2.7.1.86, 2.7.1.146, 2.7.2.1, 2.7.4.3, 2.7.6.1, 2.7.7.1, 2.7.2.4, 2.7.9.1,
       2.7.9.2, 3.5.4.18, 3.6.1.3, 3.6.1.5, 3.6.1.8, 4.6.1.1, 6.2.1.1, 6.2.1.5, 6.2.1.8, 6.2.1.9, 6.2.1.13, 6.2.1.16, 6.3.1.1, 6.3.1.2,
       6.3.1.4, 6.3.5.4, 6.4.1.1, 6.4.1.3, 6.4.1.6

bpg   3.1.3.13

butanal   1.1.1.1b

butanol   1.1.1.1b

cellobiose   3.4.1.20

citrate   2.3.3.1, 2.3.3.8, 4.1.3.6, 4.2.1.3

citrateiso   1.1.1.42, 4.1.3.1

co2tot 1.1.1.39, 1.1.1.42, 1.2.1.2, 1.2.1.43, 1.2.2.3, 1.2.2.4, 1.2.3.3, 1.2.3.4, 1.2.3.6, 1.2.7.1, 1.13.12.4, 3.5.1.5, 4.1.1.1,
        4.1.1.2, 4.1.1.4, 4.1.112, 4.1.1.31, 4.1.1.32, 6.4.1.1, 6.4.1.3, 6.4.1.6

coA   1.2.1.10, 1.2.3.6, 1.2.7.1, 1.8.1.10, 2.3.1.8, 2.3.1.9, 2.3.1.54, 2.3.3.1, 2.3.3.8, 2.3.3.9, 2.3.3.11, 3.1.2.1, 3.1.2.3,
        3.1.2.11, 3.1.2.18, 6.2.1.1, 6.2.1.5, 6.2.1.8, 6.2.1.9, 6.2.1.13, 6.2.1.16

coAglutathione   1.8.1.10

coaq   1.2.2.4

creatine   3.5.2.10

creatinine   3.5.2.10

cyclicamp   4.6.1.1

cysteineL   1.8.1.6, 1.8.4.4, 4.1.1.15

cystineL   1.8.1.6, 1.8.4.4

cytochromecox   1.1.2.2, 1.1.2.3, 1.2.2.3, 1.2.2.4, 1.4.2.1, 1.6.2.5, 1.7.2.1, 1.7.2.2, 1.8.2.1, 1.9.3.1, 1.9.6.1, 1.9.99.1,
        1.11.1.5, 1.12.2.1

cytochromecored   1.1.2.2, 1.1.2.3, 1.2.2.3, 1.2.2.4, 1.4.2.1, 1.6.2.5, 1.7.2.1, 1.7.2.2, 1.8.2.1, 1.9.3.1, 1.9.6.1, 1.9.99.1,
        1.11.1.5, 1.12.2.1

deoxyadenosine   2.7.1.76

deoxyamp   1.7.1.76

deoxyribose5phos   4.1.2.4

dihydroxyacetone   1.1.1.6, 1.1.1.156, 2.7.1.29

dihydroxyacetonephos 1.1.1.94, 1.1.3.21, 2.7.1.29, 4.1.2.13a, 4.1.2.13b, 5.3.1.1

ethanol   1.1.1.1a, 3.1.1.1

ethylacetate   3.1.1.1

fadox   3.6.1.18

ferredoxinox   1.2.7.1, 1.7.7.1, 1.7.7.12, 1.8.7.1, 1.18.99.1, 1.18.99.1, 1.18.1.2, 1.18.6.1

ferredoxinred 1.2.7.1, 1.7.7.1, 1.7.7.12, 1.8.7.1, 1.18.99.1, 1.18.99.1, 1.18.1.2, 1.18.6.1

ferric 1.9.99.1, 1.16.1.7, 1.16.3.1

ferrous 1.9.99.1, 1.16.1.7, 1.16.3.1

fmnox 1.5.1.29, 3.6.1.8

fmnred 1.5.1.29

formate 1.2.1.3, 1.2.1.43, 1.2.2.3, 2.3.1.54, 4.1.1.2, 4.1.2.36

fructose 1.1.1.67, 1.1.1.138, 1.1.2.2, 2.4.1.7, 3.7.1.4, 3.2.1.48, 5.3.1.7

fructose16phos 2.7.1.11, 2.7.1.90, 2.7.1.148, 3.1.3.11, 4.1.2.13a, 4.1.2.13b,

fructose6phos 1.1.1.16, 1.1.1.140, 2.7.1.4, 2.7.1.11, 2.7.1.90, 2.7.1.146, 3.1.3.11, 5.3.1.8, 5.3.1.9

fumarate 4.2.1.2, 4.3.1.2, 5.2.1.1

galactono14lactone 1.1.1.48

galactose 1.1.1.48, 2.7.1.6, 3.2.1.108

galactose1phos 2.7.1.6

gluconolactone 1.1.1.119, 1.1.3.4

gluconolactone6phos 1.1.1.49

glucose 1.1.1.119, 1.1.3.4, 2.4.1.9, 2.4.1.20, 2.7.1.2, 2.7.1.142, 2.7.1.147,.3.1.3.9, 3.1.3.10, 3.2.1.20, 3.2.1.48, 3.2.1.108

glucose1phos 2.4.1.8, 2.4.1.20, 2.4.1.39, 3.1.3.10, 5.4.2.6

glucose6phos 1.1.1.49, 2.4.1.7, 2.7.1.2, 2.7.1.142, 2.7.1.147, 3.1.3.9, 5.3.1.19, 5.4.2.6

glutamate 1.4.1.2, 1.4.1.3, 1.4.3.11, 2.6.1.1, 2.6.1.2, 2.6.1.4, 2.6.1.6, 3.5.1.2, 6.3.1.2, 6.3.5.4

glutamine 3.5.1.2, 6.3.5.4

glutathioneox 1.8.1.7, 1.8.3.3, 1.8.4.4, 1.11.1.9

glutathionered 1.8.1.7, 1.8.1.10, 1.8.3.3, 1.8.4.4, 1.11.1.9

glyceraldehyde 1.1.1.72

glyceraldehydephos 1.2.1.9, 4.1.2.4, 4.1.2.13a, 5.3.1.1

glycerate 1.1.1.29, 2.7.1.31, 3.1.3.38

glycerol   1.1.1.6, 1.1.1.72, 1.1.1.156, 2.7.1.30, 2.7.1.142, 3.1.3.31

glycerol3phos 1.1.1.94, 1.1.3.21, 2.7.1.30, 2.7.1.142, 3.1.3.31

glycine   1.4.1.10, 1.4.2.1, 2.6.1.4, 2.6.1.35, 2.6.2.44, 2.6.1.45

glycolate   1.1.1.26, 1.1.1.79, 2.3.3.11

glyoxylate   1.1.1.26, 1.1.1.79, 1.2.3.5, 1.4.1.10, 1.4.2.1, 2.3.3.9, 2.6.1.4, 2.6.1.35, 2.6.1.44, 2.6.1.45, 4.1.3.1

h2aq   1.12.1.2, 1.12.2.1, 1.18.99.1, 1.18.6.1

h2o   1.1.1.3, 1.1.1.42, 1.2.1.2, 1.2.1.3, 1.2.1.9, 1.2.1.43, 1.2.2.3, 1.2.2.4, 1.2.3.3, 1.2.3.4, 1.2.3.5, 1.2.3.6, 1.2.7.1, 1.4.1.1,
        1.4.1.2, 1.4.1.3, 1.4.1.7, 1.4.1.9, 1.4.1.10, 1.4.2.1, 1.4.3.11, 1.4.3.16, 1.7.1.1, 1.7.1.3, 1.7.1.4, 1.7.2.1, 1.7.2.3,
        1.7.7.1, 1.7.7.2, 1.7.99.6, 1.7.99.7, 1.8.1.2, 1.8.2.1, 1.8.3.1, 1.8.7.1, 1.9.3.1, 1.9.6.1, 1.11.1.1, 1.11.1.5, 1.11.1.6,
        1.11.1.8, 1.11.1.9, 1.13.1.18, 1.16.3.1, 1.18.6.1, 2.3.3.1, 2.3.3.9, 2.3.3.11, 2.7.9.2, 3.1.1.1, 3.1.2.1, 3.1.2.3, 3.1.2.11,
        3.1.2.18, 3.1.3.3, 3.1.3.9, 3.1.3.10, 3.1.3.11, 3.1.3.13, 3.1.3.31, 3.1.3.38, 3.1.3.50, 3.1.6.0, 3.2.1.20, 3.2.1.48,
        3.2.1.108, 3.2.2.2, 3.2.2.4, 3.2.2.7, 3.2.2.12, 3.5.1.1, 3.5.1.2, 3.5.1.5, 3.5.2.10, 3.5.4.2, 2.5.4.4, 3.5.4.6, 3.5.4.7,
        3.5.4.18, 3.6.1.1, 3.6.1.3, 3.6.1.5, 3.6.1.8, 3.6.1.18, 3.6.1.22, 4.1.1.1, 4.1.1.2, 4.1.1.4, 4.1.1.12, 4.1.1.32, 4.1.99.1,
        4.2.1.2, 4.2.1.3, 4.2.1.11, 4.2.1.31, 4.2.1.35, 4.4.1.15, 6.4.1.6

h2o2aq 1.1.3.3, 1.1.3.4, 1.1.3.21, 1.2.3.3, 1.2.3.4, 1.2.3.5, 1.2.3.6, 1.4.3.11, 1.4.3.16, 1.8.3.1, 1.8.3.3, 1.11.1.1, 1.11.1.5,
        1.11.1.6, 1.11.1.8, 1.11.1.9

h2saq 1.8.1.2, 1.8.7.1, 4.4.1.45

hydroxy2oxoglutarate 4.1.3.16

hydroxyglutarate   1.1.99.1, 2.3.3.11

hydroxypyruvate   1.1.1.29, 1.4.1.7, 2.6.1.45, 2.6..51

hypoxanthine   2.4.2.8, 3.2.2.2, 3.2.2.12, 3.5.4.2

i2cr   1.11.1.8

iditol   1.1.1.14

idp   3.5.4.7, 4.1.1.32

iodideion   1.11.1.8

imp   2.4.2.8, 2.7.1.73, 3.2.2.12, 3.5.4.6

indole 4.1.99.1

inosine 2.7.1.73, 3.2.2.2, 3.5.4.4

itp   3.5.4.18, 4.1.1.32

ketoglutaramate   1.1.1.42

ketoglutarate   1.1.99.2, 1.4.1.2, 1.4.1.3, 1.4.3.11, 2.6.1.1, 2.6.1.2, 2.6.1.4, 2.6.1.6

lactate 1.1.1.27, 1.1.2.3, 1.1.99.7, 1.13.12.4, 4.1.2.36

lactose 3.2.1.108

leucineL   1.4.1.9, 2.6.1.6

lyxose   5.3.1.15

malate   1.1.1.37, 1.1.1.39, 1.1.3.3, 1.1.99.7, 2.3.3.9, 4.2.1.2, 4.2.1.31, 6.2.1.9

maleate   4.2.1.31, 5.2.1.1

maltose   2.4.1.8, 2.4.1.39, 3.2.1.20, 5.4.99.16

malylcoA   6.2.1.9

mannitol1phos 1.1.1.13

mannitolD   1.1.1.67, 1.1.1.138, 1.1.2.2

mannose   2.7.1.7, 5.3.1.7

mannose1phos 5.4.2.8

mannose6phos 2.7.1.7, 5.3.1.8, 5.4.2.8

methaneaq   1.14.13.25

methanol   1.14.13.25

methyl2oxopentanoate 1.4.1.9, 2.6.1.6

methylmalate   4.2.1.35

methylmaleate 4.2.1.35

methylmalonylcoA   2.1.3.1, 6.4.1.3

n2aq   1.7.99.6, 1.18.6.1

n2oaq   1.7.99.6, 1.7.99.7

nadox   1.1.1.1a, 1.1.1.1b, 1.1.1.1c, 1.1.1.6, 1.1.1.14, 1.1.1.17, 1.1.1.26, 1.1.1.27, 1.1.1.29, 1.1.1.37, 1.1.1.39, 1.1.1.48, 1.1.1.56, 1.1.1.95, 1.1.1.105, 1.1.1.140, 1.2.1.2, 1.2.1.3, 1.2.1.10, 1.6.1.1, 1.6.5.3, 1.7.1.1, 1.7.99.6, 1.7.99.7, 1.8.1.6, 1.11.1.1, 1.12.1.2, 1.16.1.7, 2.7.1.23, 2.7.7.1, 3.6.1.22

nadpox   1.1.1.10, 1.1.1.42, 1.1.1.49, 1.1.1.72, 1.1.1.79,1.1.1.80, 1.1.1.94, 1.1.1.195, 1.1.1.138, 1.1.1.156, 1.1.99.2, 1.2.1.9, 1.2.1.43, 1.4.1.3, 1.5.1.29, 1.6.1.1, 1.6.2.5, 1.7.1.3, 1.7.1.4, 1.8.1.2, 1.8.1.7, 1.8.1.9, 1.8.1.10, 1.14.13.25, 1.18.1.2, 2.7.1.23

nadpred   1.1.1.10, 1.1.1.42, 1.1.1.49, 1.1.1.72, 1.1.1.79,1.1.1.80, 1.1.1.94, 1.1.1.195, 1.1.1.138, 1.1.1.156, 1.1.99.2,
           1.2.1.9, 1.2.1.43, 1.4.1.3, 1.5.1.29, 1.6.1.1, 1.6.2.5, 1.7.1.3, 1.7.1.4, 1.8.1.2, 1.8.1.7, 1.8.1.9, 1.8.1.10,
           1.14.13.25, 1.18.1.2, 2.7.1.86

nadred   1.1.1.1a, 1.1.1.1b, 1.1.1.1c, 1.1.1.6, 1.1.1.14, 1.1.1.17, 1.1.1.26, 1.1.1.27, 1.1.1.29, 1.1.1.37, 1.1.1.39,
           1.1.1.48, 1.1.1.56, 1.1.1.95, 1.1.1.105, 1.1.1.140, 1.2.1.2, 1.2.1.3, 1.2.1.10, 1.6.1.1, 1.6.5.3, 1.7.1.1, 1.7.99.6,
           1.7.99.7, 1.8.1.6, 1.11.1.1, 1.12.1.2, 1.16.1.7, 2.7.1.86

nicotinamideribonucleotide   2.7.7.1, 3.6.1.22

nitrate 1.7.1.1, 1.7.1.3, 1.7.7.2, 1.9.6.1

nitrite 1.7.1.1, 1.7.1.3, 1.7.1.4, 1.7.2.1, 1.7.2.2, 1.7.7.1, 1.7.7.2, 1.9.6.1

noaq   1.7.2.1, 1.7.99.7

o2aq   1.1.3.3, 1.1.3.4, 1.1.3.21, 1.2.2.3, 1.2.3.4, 1.2.3.6, 1.4.3.1, 1.4.3.11, 1.4.3.16, 1.8.3.1, 1.8.3.3, 1.9.3.1, 1.11.1.6,
        1.13.12.4, 1.14,13,25, 1.16.3.1

oxalate 1.2.3.4, 1.2.3.5, 3.1.2.11, 3.7.1.1, 4.1.1.2, 6.2.1.8

oxaloacetate   1.1.1.37, 1.1.3.3, 1.99.7, 1.4.3.16, 2.3.3.1, 2.3.3.8, 2.6.1.1, 2.6.1.35, 3.7.1.1, 4.1.1.31, 4.1.1.32, 4.1.3.6, 6.4.1.1

oxaloauccinate 3.1.2.18

oxalylcoA   6.2.1.8

pep   2.7.1.40, 2.7.9.1, 2.7.9.2, 3.1.3.60, 4.1.1.32, 4.2.1.11

phosphoglycerate2   3.1.3.38, 4.2.1.11

phosphoglycerate3   1.1.1.95, 1.2.1.9, 2.7.1.31, 3.1.3.13

phosphohydroxypyruvate   1.1.1.95

phosphoserine 2.7.1.80, 3.1.3.3

pi   1.2.3.3, 1.18.6.1, 2.3.1.8, 2.3.3.8, 2.4.1.7, 2.4.1.8, 2.4.1.20, 2.4.1.39, 2.4.1.39, 2.4.2.1, 2.7.1.80, 2.7.1.90, 2.7.2.12,
      2.7.9.1, 2.7.9.2, 3.1.3.3, 3.1.3.9, 3.1.3.14, 3.1.3.15, 3.1.3.11, 3.1.3.31, 3.1.3.38, 3.1.3.50, 3.1.3.60, 3.6.1.1, 3.6.1.3,
      3.6.1.5, 6.2.1.5,6.2.1.8, 6.2.1.13, 6.3.1.2, 6.3.1.4, 6.4.1.1, 6.4.1.3, 6.4.1.6

ppi   2.4.2.7, 2.4.2.8, 2.7.1.80, 2.7.1.90, 2.7.2.12, 2.7.7.1, 2.7.7.4, 2.7.9.1, 3.6.1.1, 3.6.1.8, 4.6.1.1, 6.2.1.1, 6.2.1.16, 6.3.1.1,
       6.3.5.4

propanol2   1.1.1.1c, 1.1.1.80

propanoylcoA 2.1.3.1, 2.3.3.11, 6.4.1.3

prpp   2.4.2.7, 2.4.2.8, 2.7.6.1

pyruvate   1.1.1.27, 1.1.1.39, 1.1.2.3, 1.1.99.7, 1.2.3.3, 1.2.3.6, 1.2.7.1, 1.4.1.1, 2.1.3.1, 2.3.1.54, 2.6.1.2, 2.6.1.44, 2.6.1.51, 2.7.1.40, 2.7.9.1, 2.7.9.2, 3.1.3.60, 4.1.1.1, 4.1.1.31, 4.1.3.16, 4.1.99.1, 4.3.1.17, 4.4.1.15, 6.4.1.1

retinal 1.1.1.105

retinol 1.1.1.105

ribitol 1.1.1.56

ribose 2.7.1.15, 3.2.2.2, 3.2.2.7, 5.3.1.20

ribose1phos   2.4.2.1, 5.3.1.6

ribose5phos   2.7.1.15, 2.7.6.1, 3.2.2.4, 3.2.2.12, 5.3.1.6

ribulose   1.1.1.56, 2.7.1.87, 5.3.1.20

ribulose5phos 2.7.1.47

serineL   1,4,1,7, 2,6,1,45, 2.6.1.51, 2.7.1.80, 3.1.3.3, 4.3.1.17

sorbitol   3.1.3.50

sorbitol6phos 1.1.1.140, 3.1.3.50

sorbose   1.1.1.14

succinate   3.1.2.3, 4.1.3.1, 6.2.1.5

succinylcoA   3.1.2.3, 6.2.1.5

sucrose   2.4.1.7, 3.2.1.48

sulfate   1.8.2.1, 1.8.3.1, 2.7.7.4

sulfite   1.8.1.2, 1.8.2.1, 1.8.3.1, 1.8.7.1, 1.13.1.18

sulfurcr   1.13.1.18

thioredoxinox 1.8.1.9

thioredoxinred 1.8.1.9

trehalose   5.4.99.16

tryptophanL   4.1.99.1

ubiquinoneox 1.6.5.3

ubiquinonered 1.6.5.3

urea	3.5.1.5
xylitol	1.1.1.10
xylose	5.3.1.5
xylulose	1.1.110, 5.3.1.5, 5.3.1.15

Rounding

Mathematica carries out numerical calculations to more digits than are justified by experimental data, but it is not convenient to continually round calculated properties. It is convenient to use PaddedForm in making tables. There are other situations where the program **round** can be used.

```
In[243]:=
    round[vec_, params_:{4, 2}] :=(*When a list of numbers has more digits to the right of
    the decimal point than you want, say 5, you can request 2 by using round[vec,{5,2}],*)
      Flatten[Map[NumberForm[#1, params] & , {vec}, {2}]]
```

```
In[247]:=
 calcpK[speciesmat_, no_, is_] :=
  Module[{lnkzero, sigmanuzsq, lnK}, (*Calculates pKs for a weak acid at 298.15 K at specified
    ionic strengths (is) when the number no of the pK is specified. pKs are numbered 1,
    2,3,... from the highest pK to the lowest pK,
    but the highest pK for a weak acid may be omitted if it is outside
    of the range 5 t0 9. For H3PO4,pK1=calcpK[pisp,1,{0}]=7.22.*)
   lnkzero = (speciesmat[[no + 1, 1]] - speciesmat[[no, 1]]) / (8.31451 * 0.29815);
   sigmanuzsq = speciesmat[[no, 3]]^2 - speciesmat[[no + 1, 3]]^2 + 1;
   lnK = lnkzero + (1.17582 * is^0.5 * sigmanuzsq) / (1 + 1.6 * is^0.5);
   N[-(lnK / Log[10])]]]
```

```
In[248]:=
    calcpK[acetatesp, 1, {0, .1, .25}]
```

```
Out[248]=
    {4.7547, 4.54024, 4.471}
```

```
In[249]:=
    roundedpKs = round[calcpK[acetatesp, 1, {0, .1, .25}], {5, 2}]
```

```
Out[249]=
    {4.75, 4.54, 4.47}
```

This looks like a vector, but it has a different character, which means the rounded values cannot be used in calculations.

```
In[250]:=
    2 * roundedpKs
```

```
Out[250]=
    {2 4.75, 2 4.54, 2 4.47}
```

References

1. R. A. Alberty, BasicBiochemData3, 2005.

    http : // library.wolfram.com / infocenter / MathSource / 5704

2. E. C. Webb, Enzyme Nomenclature 1992, Academic Press, New York (1992).

\http : // www.chem.qmw.ac.uk / iubmb / enzyme /

3. EC-PDP Enzyme Structure Database

\http : // www.ebi.acuk / thornton – srv / databases / enzymes /

4. Swissprot Enzyme (Enzyme nomenclature database)

\http : // us.expasy.org / enzyme /

5. D. D. Wagman, W. H. Evans, V. B. Parker, R. H. Schumm, I. Halow, S. M. Bailey, K. L. Churney, and R. L. Nuttall, The NBS tables of chemical thermodynamic properties, J. Phys. Chem. Ref. Data, 11, Supplement 2 (1982).

6. R. A. Alberty, Thermodynamic properties of oxidoreductase, transferase, hydrolase, and ligase reactions, Arch. Biochem. Biophys. 435, 363-368 (2005).

Chapter 13 Survey of Reactions at Various Temperatures

13.1 Introduction

13.2 Tables of Transformed Thermodynamic Properties at 298.15 K for Reactions in Which $\Delta_f H°$ are Known for All Species

13.3 Calculation of Transformed Reaction Properties at Other Temperatures

13.4 Three Dimensional Plots

13.5 Discussion

References

13.1 Introduction

The objective of this chapter is to calculate transformed thermodynamic properties of 90 reactions for which standard enthalpies of formation of species are known for all species. The calculations of $\Delta_r H'°$ and $\Delta_r S'°$ for these reactions makes it possible to see the relative contributions these properties make to $\Delta_r G'°$. For these reactions it is possible to calculate thermodynamic properties over a range of temperatures where $\Delta_f H°$ is essentially independent of temperature. The first set of calculations is for 298.15 K to check that the programs used here and in the previous chapter give the same results. In making these tables, $\log K'$ is given, rather than K', because of the difficulty in rounding numbers involving exponents; of course, $K' = 10^{\log K'}$. Tables given here show that $\Delta_r H'°$ does not change much with pH, and so it is the dependence of $\Delta_r S'°$ on pH that is primarily responsible for the pH dependence of K'.

We have seen earlier in Chapter 3 that the various transformed thermodynamic properties of a reaction are interrelated by Maxwell relations. The Maxwell relations used in this chapter are

$$\Delta_r N_H = \frac{1}{RT\ln(10)} \frac{\partial \Delta_r G'°}{\partial \text{pH}} \tag{13.1-1}$$

$$\Delta_r S'° = -\frac{\partial \Delta_r G'°}{\partial T} \tag{13.1-2}$$

$$\Delta_r H'° = -T^2 \frac{\partial(\Delta_r G'°/T)}{\partial T} \tag{13.1-3}$$

$$\frac{\partial \Delta_r S'°}{\partial \text{pH}} = -R\ln(10)\frac{\partial(T\Delta_r N_H)}{\partial T} \tag{13.1-4}$$

$$\frac{\partial \Delta_r H'°}{\partial \text{pH}} = -RT^2\ln(10)\frac{\partial \Delta_r N_H}{\partial T} \tag{13.1-5}$$

In the previous chapter equation 13.1-1 was used a lot, but now that temperature is available as an independent variable, there are four more Maxwell relations. Note that the right hand side of equation 13.1-3 minus T times the right hand side of equation 13.1-2 yields $\Delta_r G'^\circ$ as it must. Also notice that the dependencies of $\Delta_r S'^\circ$ and $\Delta_r H'^\circ$ on pH both involve the change in $\Delta_r N_H$ with temperature. The right side of equation 13.1-5 minus the right side of equation 13.1-4 yields $\partial \Delta_r G'^\circ / \partial$ pH as it must.

The standard formation properties of the species involved in atp + h2o = adp + pi are known at 298.15 K (1), and so it is possible to calculate $\Delta_r G'^\circ$, $\Delta_r H'^\circ$, $\Delta_r S'^\circ$, $\Delta_r N_H$, K', $\partial \Delta_r H'^\circ / \partial$pH, and $\partial \Delta_r S'^\circ / \partial$pH over a range of temperatures (2). This reaction is remarkable because the thermodynamic properties of the complex ions with magnesium are also known. This made it possible to calculate ten properties of this hydrolysis of atp as functions of temperature, pH, pMg, and ionic strength (3). Note that the number of Maxwell relations increases rapidly with the number of independent variables.

Since $\Delta_r H'^\circ$, $\Delta_r S'^\circ$, $\Delta_r N_H$, and $\log K'$ can be calculated from $\Delta_r G'^\circ$ expressed as a function of temperature, pH, and ionic strength, a program **rxSummaryT** has been written to calculate these properties at specified temperatures, pHs, and ionic strengths. Notice that this program uses the functions like atpGT that are available in BasicBiochemData3 (4).

```
In[2]:=  Off[General::"spell"];
         Off[General::"spell1"];

In[4]:=  << BiochemThermo`BasicBiochemData3`

In[5]:=  rxSummaryT[title_, reaction_, mathfunct_, temp_, pHlist_, islist_] :=
           Module[{trGibbse, trenthalpy, trentropy, logK, vectorNH, table},
             (*title_ is in the form "EC 1.1.1.1 Alcohol dehydrogenase".
               reaction_ is in the form "acetaldehyde+nadred=ethanol+nadox".
               mathfunct is of the form ethanolGT+nadoxGT-(acetaldehydeGT+nadredGT).
               temp is of the form 298.15 or other in the range 273.15 to 313.15.
               pHlist is of the form {5,6,7,8,9} or other.
               islist is of the form 0.25 or other.
               This program uses the mathfunction of T, pH, and ionic strength to calculate
                the standard transformed Gibbs energies of reaction in kJ mol^-1,
                the standard transformed enthalpies of formation in kJ mol^-1,
                the standard transformed entropy of reaction in J K^-1 mol^-1,
                the change in the binding of hydrogen ions,
                and logK' at the desired temperature, pHs, and ionic strength.*)
             trGibbse = mathfunct /. t -> temp /. pH -> pHlist /. is -> islist;
             trenthalpy = -t^2*D[mathfunct/t, t] /. t -> temp /. pH -> pHlist /. is -> islist;
             trentropy = -D[mathfunct, t] *1000 /. t -> temp /. pH -> pHlist /. is -> islist;
             vectorNH = (1 / (8.31451* (t / 1000) *Log[10])) *D[mathfunct, pH] /. t -> temp /. pH -> pHlist,
               is -> islist;
             logK = -mathfunct / (8.31452*temp*Log[10] / 1000) /. t -> temp /. pH -> pHlist /. is -> islist
             table = PaddedForm[TableForm[{trGibbse, trenthalpy, trentropy, vectorNH, logK},
               TableHeadings -> {{"Δr G'°", "Δr H'°", "Δr S'°", "ΔrNH", "logK '"},
                 {" pH 5", " pH 6", " pH 7", " pH 8", " pH 9"}}], {4, 2}];
             Print[title]; Print[reaction]; Print[table]]
```

This program is applied to the 90 reactions for which $\Delta_f H^\circ$ are known for all species. The reactions have all been written in the direction in which they are spontaneous at 298.15 K, pH 7, and 0.25 M ionic strength. The standard transformed Gibbs energies of formation and standard transformed enthalpies of formation are in kJ mol^{-1} and the standard transformed entropies of formation are in J K^{-1} mol^{-1}.

13.2 Tables of Transformed Thermodynamic Properties at 298.15 K for Reactions in Which $\Delta_f H°$ are Known for All Species

Class 1: Oxidoreductases

In[6]:= **rxSummaryT["EC 1.1.1.1a Alcohol dehydrogenase", "acetaldehyde+nadred=ethanol+nadox", ethanolGT + nadoxGT - (acetaldehydeGT + nadredGT), 298.15, {5, 6, 7, 8, 9}, .25]**

EC 1.1.1.1a Alcohol dehydrogenase

acetaldehyde+nadred=ethanol+nadox

	pH 5	pH 6	pH 7	pH 8	pH 9
$\Delta_r G'°$	-33.51	-27.80	-22.09	-16.39	-10.68
$\Delta_r H'°$	-45.78	-45.78	-45.78	-45.78	-45.78
$\Delta_r S'°$	-41.14	-60.29	-79.43	-98.58	-117.70
$\Delta_r N_H$	1.00				
log$K$ '	5.87	4.87	3.87	2.87	1.87

In[7]:= **rxSummaryT["EC 1.1.1.1c Alcohol dehydrogenase", "acetone+nadred=propanol2+nadox", propanol2GT + nadoxGT - (acetoneGT + nadredGT), 298.15, {5, 6, 7, 8, 9}, .25]**

EC 1.1.1.1c Alcohol dehydrogenase

acetone+nadred=propanol2+nadox

	pH 5	pH 6	pH 7	pH 8	pH 9
$\Delta_r G'°$	-16.40	-10.69	-4.98	0.72	6.43
$\Delta_r H'°$	-78.83	-78.83	-78.83	-78.83	-78.83
$\Delta_r S'°$	-209.40	-228.50	-247.70	-266.80	-286.00
$\Delta_r N_H$	1.00				
log$K$ '	2.87	1.87	0.87	-0.13	-1.13

In[8]:= **rxSummaryT["EC 1.1.1.27 Lactate dehydrogenase", "pyruvate+nadred=lactate+nadox", lactateGT + nadoxGT - (pyruvateGT + nadredGT), 298.15, {5, 6, 7, 8, 9}, .25]**

EC 1.1.1.27 Lactate dehydrogenase

pyruvate+nadred=lactate+nadox

	pH 5	pH 6	pH 7	pH 8	pH 9
$\Delta_r G'°$	-35.32	-29.61	-23.90	-18.20	-12.49
$\Delta_r H'°$	-60.13	-60.13	-60.13	-60.13	-60.13
$\Delta_r S'°$	-83.20	-102.30	-121.50	-140.60	-159.80
$\Delta_r N_H$	1.00				
log$K$ '	6.19	5.19	4.19	3.19	2.19

In[9]:= **rxSummaryT["EC 1.1.1.37 Malate dehydrogenase", "oxaloacetate+nadred=malate+nadox", malateGT + nadoxGT - (oxaloacetateGT + nadredGT), 298.15, {5, 6, 7, 8, 9}, .25]**

EC 1.1.1.37 Malate dehydrogenase

oxaloacetate+nadred=malate+nadox

	pH 5	pH 6	pH 7	pH 8	pH 9
$\Delta_r G'°$	-41.23	-34.65	-28.84	-23.12	-17.41
$\Delta_r H'°$	-90.19	-89.68	-89.61	-89.60	-89.60
$\Delta_r S'°$	-164.20	-184.60	-203.80	-223.00	-242.10
$\Delta_r N_H$	1.33	1.05	1.00	1.00	1.00
log$K$ '	7.22	6.07	5.05	4.05	3.05

In[10]:= **rxSummaryT["EC 1.1.1.39 Malate dehydrogenase (decarboxylating)",**
        "pyruvate+nadred+co2tot=malate+nadox+h2o",
        malateGT + nadoxGT + h2oGT - (pyruvateGT + nadredGT + co2totGT), 298.15, {5, 6, 7, 8, 9}, .25]

EC 1.1.1.39 Malate dehydrogenase (decarboxylating)

pyruvate+nadred+co2tot=malate+nadox+h2o

	pH 5	pH 6	pH 7	pH 8	pH 9
$\Delta_r G'^{\circ}$	-7.91	-5.75	-1.61	3.89	9.93
$\Delta_r H'^{\circ}$	-39.07	-41.75	-45.42	-46.47	-48.72
$\Delta_r S'^{\circ}$	-104.50	-120.70	-146.90	-168.90	-196.70
$\Delta_r N_H$	0.41	0.50	0.90	1.01	1.15
$\log K'$	1.39	1.01	0.28	-0.68	-1.74

In[11]:= **rxSummaryT["EC 1.1.1.80 Isopropyl dehydrogenase (NADP)",**
        "acetone+nadpred=propanol2+nadpox",
        propanol2GT + nadpoxGT - (acetoneGT + nadpredGT), 298.15, {5, 6, 7, 8, 9}, .25]

EC 1.1.1.80 Isopropyl dehydrogenase (NADP)

acetone+nadpred=propanol2+nadpox

	pH 5	pH 6	pH 7	pH 8	pH 9
$\Delta_r G'^{\circ}$	-15.98	-10.27	-4.57	1.14	6.85
$\Delta_r H'^{\circ}$	-83.24	-83.24	-83.24	-83.24	-83.24
$\Delta_r S'^{\circ}$	-225.60	-244.70	-263.90	-283.00	-302.10
$\Delta_r N_H$	1.00				
$\log K'$	2.80	1.80	0.80	-0.20	-1.20

In[12]:= **rxSummaryT["EC 1.1.3.3 Malate oxidase", "malate+o2ox=oxaloacetate+h2o2aq",**
        oxaloacetateGT + h2o2aqGT - (malateGT + o2aqGT), 298.15, {5, 6, 7, 8, 9}, .25]

EC 1.1.3.3 Malate oxidase

malate+o2ox=oxaloacetate+h2o2aq

	pH 5	pH 6	pH 7	pH 8	pH 9
$\Delta_r G'^{\circ}$	-100.10	-100.90	-101.00	-101.10	-101.10
$\Delta_r H'^{\circ}$	-58.98	-59.50	-59.57	-59.58	-59.58
$\Delta_r S'^{\circ}$	137.80	139.00	139.10	139.10	139.10
$\Delta_r N_H$	-0.33	-0.05	-0.00	-0.00	-0.00
$\log K'$	17.53	17.68	17.70	17.70	17.70

EC 1.1.3.3 Malate oxidase

In[13]:= **rxSummaryT["EC 1.1.99.7 Lactate-malate trans-hydrogenase",**
        "lactate+oxaloacetate=malate+pyruvate",
        malateGT + pyruvateGT - (lactateGT + oxaloacetateGT), 298.15, {5, 6, 7, 8, 9}, .25]

EC 1.1.99.7 Lactate-malate trans-hydrogenase

lactate+oxaloacetate=malate+pyruvate

	pH 5	pH 6	pH 7	pH 8	pH 9
$\Delta_r G'^{\circ}$	-5.91	-5.04	-4.93	-4.92	-4.92
$\Delta_r H'^{\circ}$	-30.07	-29.55	-29.48	-29.47	-29.47
$\Delta_r S'^{\circ}$	-81.01	-82.23	-82.33	-82.34	-82.34
$\Delta_r N_H$	0.33	0.05	0.00	0.00	0.00
$\log K'$	1.04	0.88	0.86	0.86	0.86

In[14]:= **rxSummaryT["EC 1.2.1.2 Formate dehydrogenase", "formate+nadox+h2o=co2tot+nadred",**
   co2totGT + nadredGT - (formateGT + nadoxGT + h2oGT), 298.15, {5, 6, 7, 8, 9}, .25]

EC 1.2.1.2 Formate dehydrogenase

formate+nadox+h2o=co2tot+nadred

	pH 5	pH 6	pH 7	pH 8	pH 9
$\Delta_r G'^\circ$	-14.09	-15.38	-19.41	-24.90	-30.94
$\Delta_r H'^\circ$	-18.72	-15.53	-11.79	-10.72	-8.48
$\Delta_r S'^\circ$	-15.54	-0.51	25.57	47.56	75.33
$\Delta_r N_H$	-0.08	-0.45	-0.89	-1.01	-1.15
logK '	2.47	2.69	3.40	4.36	5.42

In[15]:= **rxSummaryT["EC 1.2.1.3 Aldehyde dehydrogenase",**
   "acetaldehyde+nadox+h2o=acetate+nadred",
   acetateGT + nadredGT - (acetaldehydeGT + nadoxGT + h2oGT), 298.15, {5, 6, 7, 8, 9}, .25]

EC 1.2.1.3 Aldehyde dehydrogenase

acetaldehyde+nadox+h2o=acetate+nadred

	pH 5	pH 6	pH 7	pH 8	pH 9
$\Delta_r G'^\circ$	-33.05	-43.90	-55.25	-66.66	-78.07
$\Delta_r H'^\circ$	-17.55	-17.43	-17.42	-17.42	-17.42
$\Delta_r S'^\circ$	51.99	88.75	126.90	165.10	203.40
$\Delta_r N_H$	-1.77	-1.97	-2.00	-2.00	-2.00
logK '	5.79	7.69	9.68	11.68	13.68

In[16]:= **rxSummaryT["EC 1.2.1.43 Formate dehydrogenase", "formate+nadox+h2o=co2tot+nadred",**
   co2totGT + nadredGT - (formateGT + nadoxGT + h2oGT), 298.15, {5, 6, 7, 8, 9}, .25]

EC 1.2.1.43 Formate dehydrogenase

formate+nadox+h2o=co2tot+nadred

	pH 5	pH 6	pH 7	pH 8	pH 9
$\Delta_r G'^\circ$	-14.09	-15.38	-19.41	-24.90	-30.94
$\Delta_r H'^\circ$	-18.72	-15.53	-11.79	-10.72	-8.48
$\Delta_r S'^\circ$	-15.54	-0.51	25.57	47.56	75.33
$\Delta_r N_H$	-0.08	-0.45	-0.89	-1.01	-1.15
logK '	2.47	2.69	3.40	4.36	5.42

In[17]:= **rxSummaryT["EC 1.4.1.1 Alanine dehydrogenase",**
   "pyruvate+nadred+ammonia=alanine+nadox+h2o",
   alanineGT + nadoxGT + h2oGT - (pyruvateGT + nadredGT + ammoniaGT), 298.15, {5, 6, 7, 8, 9}, .2

EC 1.4.1.1 Alanine dehydrogenase

pyruvate+nadred+ammonia=alanine+nadox+h2o

	pH 5	pH 6	pH 7	pH 8	pH 9
$\Delta_r G'^\circ$	-45.86	-40.15	-34.43	-28.60	-21.93
$\Delta_r H'^\circ$	-82.43	-82.46	-82.72	-85.20	-101.20
$\Delta_r S'^\circ$	-122.70	-141.90	-162.00	-189.80	-265.80
$\Delta_r N_H$	1.00	1.00	1.01	1.05	1.36
logK '	8.03	7.03	6.03	5.01	3.84

In[18]:= **rxSummaryT["EC 1.4.1.2 Glutamate dehydrogenase",**
   "ketoglutarate+nadred+ammonia=glutamate+nadox+h2o", glutamateGT + nadoxGT +
   h2oGT - (ketoglutarateGT + nadredGT + ammoniaGT), 298.15, {5, 6, 7, 8, 9}, .25]

EC 1.4.1.2 Glutamate dehydrogenase

ketoglutarate+nadred+ammonia=glutamate+nadox+h2o

	pH 5	pH 6	pH 7	pH 8	pH 9
$\Delta_r G'^o$	-49.57	-43.86	-38.14	-32.31	-25.64
$\Delta_r H'^o$	-60.51	-60.53	-60.80	-63.27	-79.25
$\Delta_r S'^o$	-36.68	-55.92	-75.98	-103.80	-179.80
$\Delta_r N_H$	1.00	1.00	1.01	1.05	1.36
$\log K'$	8.68	7.68	6.68	5.66	4.49

In[19]:= **rxSummaryT["EC 1.4.1.3 Glutamate dehydrogenase",**
     "ketoglutarate+nadpred+ammonia=glutamate+nadpox+h2o", glutamateGT + nadpoxGT +
     h2oGT - (ketoglutarateGT + nadpredGT + ammoniaGT), 298.15, {5, 6, 7, 8, 9}, .25]

EC 1.4.1.3 Glutamate dehydrogenase

ketoglutarate+nadpred+ammonia=glutamate+nadpox+h2o

	pH 5	pH 6	pH 7	pH 8	pH 9
$\Delta_r G'^o$	-49.15	-43.44	-37.72	-31.89	-25.22
$\Delta_r H'^o$	-64.92	-64.94	-65.20	-67.68	-83.66
$\Delta_r S'^o$	-52.87	-72.11	-92.17	-120.00	-196.00
$\Delta_r N_H$	1.00	1.00	1.01	1.05	1.36
$\log K'$	8.61	7.61	6.61	5.59	4.42

In[20]:= **rxSummaryT["EC 1.4.3.11 L-Glutamate oxidase",**
     "glutamate+o2aq+h2o=ketoglutarate+ammonia+h2o2aq", ketoglutarateGT +
     ammoniaGT + h2o2aqGT - (glutamateGT + h2oGT + o2aqGT), 298.15, {5, 6, 7, 8, 9}, .25]

EC 1.4.3.11 L-Glutamate oxidase

glutamate+o2aq+h2o=ketoglutarate+ammonia+h2o2aq

	pH 5	pH 6	pH 7	pH 8	pH 9
$\Delta_r G'^o$	-91.73	-91.73	-91.74	-91.86	-92.83
$\Delta_r H'^o$	-88.67	-88.64	-88.38	-85.90	-69.93
$\Delta_r S'^o$	10.26	10.35	11.27	19.99	76.82
$\Delta_r N_H$	-0.00	-0.00	-0.01	-0.05	-0.36
$\log K'$	16.07	16.07	16.07	16.09	16.26

In[21]:= **rxSummaryT["EC 1.4.3.16 L-aspartate oxidase",**
     "aspartate+o2aq+h2o=oxaloacetate+ammonia+h2o2aq", oxaloacetateGT +
     ammoniaGT + h2o2aqGT - (aspartateGT + h2oGT + o2aqGT), 298.15, {5, 6, 7, 8, 9}, .25]

EC 1.4.3.16 L-aspartate oxidase

aspartate+o2aq+h2o=oxaloacetate+ammonia+h2o2aq

	pH 5	pH 6	pH 7	pH 8	pH 9
$\Delta_r G'^o$	-93.20	-93.20	-93.21	-93.33	-94.30
$\Delta_r H'^o$	-40.99	-40.96	-40.70	-38.22	-22.25
$\Delta_r S'^o$	175.10	175.20	176.10	184.80	241.70
$\Delta_r N_H$	-0.00	-0.00	-0.01	-0.05	-0.36
$\log K'$	16.33	16.33	16.33	16.35	16.52

In[22]:= **rxSummaryT["EC 1.6.1.1 NAD(P) transhydrogenase", "nadpox+nadred=nadpred+nadox",**
     nadpredGT + nadoxGT - (nadpoxGT + nadredGT), 298.15, {5, 6, 7, 8, 9}, .25]

EC 1.6.1.1 NAD(P) transhydrogenase

nadpox+nadred=nadpred+nadox

	pH 5	pH 6	pH 7	pH 8	pH 9
$\Delta_r G'^\circ$	-0.42	-0.42	-0.42	-0.42	-0.42
$\Delta_r H'^\circ$	4.41	4.41	4.41	4.41	4.41
$\Delta_r S'^\circ$	16.19	16.19	16.19	16.19	16.19
$\Delta_r N_H$	0.00				
$\log K'$	0.07	0.07	0.07	0.07	0.07

In[23]:= **rxSummaryT["EC 1.7.1.1 Nitrate reductase (NADH)", "nadred+nitrate=nitrite+nadox+h2o",**
   nitriteGT + nadoxGT + h2oGT - (nadredGT + nitrateGT), 298.15, {5, 6, 7, 8, 9}, .25]

EC 1.7.1.1 Nitrate reductase (NADH)

nadred+nitrate=nitrite+nadox+h2o

	pH 5	pH 6	pH 7	pH 8	pH 9
$\Delta_r G'^\circ$	-151.50	-145.80	-140.10	-134.40	-128.70
$\Delta_r H'^\circ$	-155.30	-155.20	-155.10	-155.10	-155.10
$\Delta_r S'^\circ$	-12.50	-31.31	-50.42	-69.57	-88.71
$\Delta_r N_H$	1.01	1.00	1.00	1.00	1.00
$\log K'$	26.55	25.55	24.55	23.55	22.55

In[24]:= **rxSummaryT["EC 1.7.1.3 Nitrate reductase (NADPH)",**
   "nadpred+nitrate=nitrite+nadpox+h2o",
   nitriteGT + nadpoxGT + h2oGT - (nadpredGT + nitrateGT), 298.15, {5, 6, 7, 8, 9}, .25]

EC 1.7.1.3 Nitrate reductase (NADPH)

nadpred+nitrate=nitrite+nadpox+h2o

	pH 5	pH 6	pH 7	pH 8	pH 9
$\Delta_r G'^\circ$	-151.10	-145.40	-139.70	-134.00	-128.30
$\Delta_r H'^\circ$	-159.70	-159.60	-159.50	-159.50	-159.50
$\Delta_r S'^\circ$	-28.69	-47.50	-66.61	-85.75	-104.90
$\Delta_r N_H$	1.01	1.00	1.00	1.00	1.00
$\log K'$	26.48	25.47	24.47	23.47	22.47

In[25]:= **rxSummaryT["EC 1.7.1.4 Nitrite reductase (NADP)",**
   "nitrite+3*nadpred=ammonia+3*nadpox+2*h2o",
   ammoniaGT + 3 * nadpoxGT + 2 * h2oGT - (nitriteGT + 3 * nadpredGT), 298.15, {5, 6, 7, 8, 9}, .25]

EC 1.7.1.4 Nitrite reductase (NADP)

nitrite+3*nadpred=ammonia+3*nadpox+2*h2o

	pH 5	pH 6	pH 7	pH 8	pH 9
$\Delta_r G'^\circ$	-434.10	-405.60	-377.10	-348.70	-321.10
$\Delta_r H'^\circ$	-522.60	-522.70	-522.40	-520.00	-504.00
$\Delta_r S'^\circ$	-296.80	-392.70	-487.60	-574.60	-613.50
$\Delta_r N_H$	4.99	5.00	4.99	4.95	4.64
$\log K'$	76.05	71.06	66.06	61.08	56.25

In[26]:= **rxSummaryT["EC 1.7.99.6 Nitrous oxide reductase", "n2oaq+nadred=n2aq+nadox+h2o",**
   n2aqGT + nadoxGT + h2oGT - (n2oaqGT + nadredGT), 298.15, {5, 6, 7, 8, 9}, .25]

EC 1.7.99.6 Nitrous oxide reductase

n2oaq+nadred=n2aq+nadox+h2o

	pH 5	pH 6	pH 7	pH 8	pH 9
$\Delta_r G'^\circ$	-313.60	-307.90	-302.10	-296.40	-290.70
$\Delta_r H'^\circ$	-348.10	-348.10	-348.10	-348.10	-348.10
$\Delta_r S'^\circ$	-115.90	-135.10	-154.20	-173.40	-192.50
$\Delta_r N_H$	1.00				
$\log K'$	54.93	53.93	52.93	51.93	50.93

In[27]:= **rxSummaryT["EC 1.8.1.2 Sulfite reductase", "sulfite+3*nadpred=h2saq+3*nadpox+3*h2o",**
h2saqGT + 3 * nadpoxGT + 3 * h2oGT - (sulfiteGT + 3 * nadpredGT), 298.15, {5, 6, 7, 8, 9}, .25]

EC 1.8.1.2 Sulfite reductase

sulfite+3*nadpred=h2saq+3*nadpox+3*h2o

	pH 5	pH 6	pH 7	pH 8	pH 9
$\Delta_r G'^\circ$	-152.80	-129.90	-107.00	-84.08	-61.24
$\Delta_r H'^\circ$	-193.50	-189.00	-173.70	-165.00	-163.70
$\Delta_r S'^\circ$	-136.60	-198.20	-223.90	-271.50	-343.70
$\Delta_r N_H$	4.00	4.01	4.02	4.00	4.00
$\log K'$	26.77	22.76	18.74	14.73	10.73

In[28]:= **rxSummaryT["EC 1.8.3.1 Sulfite oxidase", "sulfite+o2aq+h2o=sulfate+h2o2aq",**
sulfateGT + h2o2aqGT - (sulfiteGT + o2aqGT + h2oGT), 298.15, {5, 6, 7, 8, 9}, .25]

EC 1.8.3.1 Sulfite oxidase

sulfite+o2aq+h2o=sulfate+h2o2aq

	pH 5	pH 6	pH 7	pH 8	pH 9
$\Delta_r G'^\circ$	-161.70	-167.00	-170.30	-171.20	-171.30
$\Delta_r H'^\circ$	-174.90	-173.70	-169.80	-167.70	-167.40
$\Delta_r S'^\circ$	-44.13	-22.45	1.70	11.44	12.79
$\Delta_r N_H$	-0.98	-0.82	-0.32	-0.04	-0.00
$\log K'$	28.33	29.26	29.84	29.99	30.00

In[29]:= **rxSummaryT["EC 1.11.1.1 NADH peroxidase", "nadred+h2o2aq=nadox+2*h2o",**
nadoxGT + 2 * h2oGT - (nadredGT + h2o2aqGT), 298.15, {5, 6, 7, 8, 9}, .25]

EC 1.11.1.1 NADH peroxidase

nadred+h2o2aq=nadox+2*h2o

	pH 5	pH 6	pH 7	pH 8	pH 9
$\Delta_r G'^\circ$	-331.20	-325.50	-319.80	-314.10	-308.40
$\Delta_r H'^\circ$	-350.20	-350.20	-350.20	-350.20	-350.20
$\Delta_r S'^\circ$	-63.65	-82.79	-101.90	-121.10	-140.20
$\Delta_r N_H$	1.00				
$\log K'$	58.03	57.03	56.03	55.03	54.03

In[30]:= **rxSummaryT["EC 1.11.1.6 Catalase", "2*h2o2aq=o2aq+2*h2o",**
o2aqGT + 2 * h2oGT - (2 * h2o2aqGT), 298.15, {5, 6, 7, 8, 9}, .25]

EC 1.11.1.6 Catalase

2*h2o2aq=o2aq+2*h2o

	pH 5	pH 6	pH 7	pH 8	pH 9
$\Delta_r G'^\circ$	-189.90	-189.90	-189.90	-189.90	-189.90
$\Delta_r H'^\circ$	-201.00	-201.00	-201.00	-201.00	-201.00
$\Delta_r S'^\circ$	-37.23	-37.23	-37.23	-37.23	-37.23
$\Delta_r N_H$	0.00				
$\log K'$	33.27	33.27	33.27	33.27	33.27

In[31]:= **rxSummaryT["EC 1.11.1.8 Iodide peroxidase", "2*iodideion+h2o2aq=i2cr+2*h2o",**
        i2crGT + 2 * h2oGT - (2 * iodideionGT + h2o2aqGT), 298.15, {5, 6, 7, 8, 9}, .25]

EC 1.11.1.8 Iodide peroxidase

2*iodideion+h2o2aq=i2cr+2*h2o

	pH 5	pH 6	pH 7	pH 8	pH 9
$\Delta_r G'^\circ$	-176.90	-165.50	-154.10	-142.60	-131.20
$\Delta_r H'^\circ$	-271.80	-271.80	-271.80	-271.80	-271.80
$\Delta_r S'^\circ$	-318.20	-356.50	-394.80	-433.10	-471.30
$\Delta_r N_H$	2.00				
$\log K'$	30.99	28.99	26.99	24.99	22.99

In[32]:= **rxSummaryT["EC 1.12.1.2 Hydrogen hydrogenase",**
        "h2aq+nadox=nadred", nadredGT - (h2aqGT + nadoxGT), 298.15, {5, 6, 7, 8, 9}, .25]

EC 1.12.1.2 Hydrogen hydrogenase

h2aq+nadox=nadred

	pH 5	pH 6	pH 7	pH 8	pH 9
$\Delta_r G'^\circ$	-26.73	-32.44	-38.15	-43.85	-49.56
$\Delta_r H'^\circ$	-26.09	-26.09	-26.09	-26.09	-26.09
$\Delta_r S'^\circ$	2.14	21.28	40.43	59.57	78.71
$\Delta_r N_H$	-1.00				
$\log K'$	4.68	5.68	6.68	7.68	8.68

In[33]:= **rxSummaryT["EC 1.13.1.18 Sulfur dioxygenase", "sulfurcr+o2aq+h2o=sulfite",**
        sulfiteGT - (sulfurcrGT + o2aqGT + h2oGT), 298.15, {5, 6, 7, 8, 9}, .25]

EC 1.13.1.18 Sulfur dioxygenase

sulfurcr+o2aq+h2o=sulfite

	pH 5	pH 6	pH 7	pH 8	pH 9
$\Delta_r G'^\circ$	-337.20	-343.30	-351.40	-362.00	-373.30
$\Delta_r H'^\circ$	-328.00	-329.20	-333.10	-335.20	-335.50
$\Delta_r S'^\circ$	30.77	47.36	61.50	90.05	127.00
$\Delta_r N_H$	-1.02	-1.18	-1.68	-1.96	-2.00
$\log K'$	59.07	60.15	61.57	63.42	65.40

In[34]:= **rxSummaryT["EC 1.13.12.4 Lactate 2-monooxygenase", "lactate+o2aq=acetate+co2tot",**
        acetateGT + co2totGT - (lactateGT + o2aqGT), 298.15, {5, 6, 7, 8, 9}, .25]

EC 1.13.12.4 Lactate 2-monooxygenase

lactate+o2aq=acetate+co2tot

	pH 5	pH 6	pH 7	pH 8	pH 9
$\Delta_r G'^\circ$	-492.90	-493.70	-497.60	-503.10	-509.20
$\Delta_r H'^\circ$	-486.80	-483.50	-479.70	-478.70	-476.40
$\Delta_r S'^\circ$	20.64	34.15	60.06	82.03	109.80
$\Delta_r N_H$	0.15	-0.42	-0.89	-1.00	-1.15
$\log K'$	86.36	86.49	87.18	88.14	89.20

In[35]:= **rxSummaryT["EC 1.14.13.25 Methane monooxygenase",**
        "methaneaq+nadpred+o2aq=methanol+nadpox+h2o",
        methanolGT + nadpoxGT + h2oGT - (methaneaqGT + nadpredGT + o2aqGT), 298.15, {5, 6, 7, 8, 9}, .

EC 1.14.13.25 Methane monooxygenase

```
methaneaq+nadpred+o2aq=methanol+nadpox+h2o
```

	pH 5	pH 6	pH 7	pH 8	pH 9
$\Delta_r G'°$	-385.00	-379.30	-373.60	-367.90	-362.20
$\Delta_r H'°$	-405.10	-405.10	-405.10	-405.10	-405.10
$\Delta_r S'°$	-67.46	-86.60	-105.70	-124.90	-144.00
$\Delta_r N_H$	1.00				
$\log K'$	67.45	66.45	65.45	64.45	63.45

In[36]:= **rxSummaryT["EC 1.16.1.7 Ferric chelate reductase", "2*ferric+nadred=2*ferrous+nadox",**
2 * ferrousGT + nadoxGT - (2 * ferricGT + nadredGT), 298.15, {5, 6, 7, 8, 9}, .25]

```
EC 1.16.1.7 Ferric chelate reductase

2*ferric+nadred=2*ferrous+nadox
```

	pH 5	pH 6	pH 7	pH 8	pH 9
$\Delta_r G'°$	-189.90	-195.60	-201.30	-207.00	-212.70
$\Delta_r H'°$	-54.20	-54.20	-54.20	-54.20	-54.20
$\Delta_r S'°$	455.00	474.20	493.30	512.50	531.60
$\Delta_r N_H$	-1.00				
$\log K'$	33.26	34.26	35.26	36.26	37.26

In[37]:= **rxSummaryT["EC 1.16.3.1 Ferrioxidase", "4*ferrous+o2aq=4*ferric+2*h2o",**
4 * ferricGT + 2 * h2oGT - (4 * ferrousGT + o2aqGT), 298.15, {5, 6, 7, 8, 9}, .25]

```
EC 1.16.3.1 Ferrioxidase

4*ferrous+o2aq=4*ferric+2*h2o
```

	pH 5	pH 6	pH 7	pH 8	pH 9
$\Delta_r G'°$	-92.77	-69.94	-47.11	-24.28	-1.45
$\Delta_r H'°$	-391.00	-391.00	-391.00	-391.00	-391.00
$\Delta_r S'°$	-1000.00	-1077.00	-1153.00	-1230.00	-1306.00
$\Delta_r N_H$	4.00				
$\log K'$	16.25	12.25	8.25	4.25	0.25

Class 2: Transferases

In[38]:= **rxSummaryT["EC 2.4.1.7 Sucrose phosphorylase", "sucrose+pi=fructose+glucose6phos",**
fructoseGT + glucose6phosGT - (sucroseGT + piGT), 298.15, {5, 6, 7, 8, 9}, .25]

```
EC 2.4.1.7 Sucrose phosphorylase

sucrose+pi=fructose+glucose6phos
```

	pH 5	pH 6	pH 7	pH 8	pH 9
$\Delta_r G'°$	-14.37	-15.77	-17.90	-18.56	-18.64
$\Delta_r H'°$	-31.68	-32.60	-35.32	-36.72	-36.93
$\Delta_r S'°$	-58.07	-56.44	-58.41	-60.93	-61.33
$\Delta_r N_H$	-0.10	-0.40	-0.24	-0.04	-0.00
$\log K'$	2.52	2.76	3.14	3.25	3.27

In[39]:= **rxSummaryT["EC 2.4.2.1 Purine-nucleoside phosphorylase",**
"adenine+ribose1phos=adenosine+pi",
adenosineGT + piGT - (adenineGT + ribose1phosGT), 298.15, {5, 6, 7, 8, 9}, .25]

```
EC 2.4.2.1 Purine-nucleoside phosphorylase

adenine+ribose1phos=adenosine+pi
```

	pH 5	pH 6	pH 7	pH 8	pH 9
$\Delta_r G'^o$	-18.90	-18.39	-16.91	-16.38	-16.31
$\Delta_r H'^o$	-20.15	-17.84	-11.04	-8.67	-8.36
$\Delta_r S'^o$	-4.17	1.85	19.67	25.85	26.67
$\Delta_r N_H$	-0.06	0.24	0.19	0.03	0.00
log$K$ '	3.31	3.22	2.96	2.87	2.86

In[40]:= **rxSummaryT["EC 2.6.1.1 Aspartate transaminase",**
   "aspartate+ketoglutarate=oxaloacetate+glutamate",
   oxaloacetateGT + glutamateGT - (aspartateGT + ketoglutarateGT), 298.15, {5, 6, 7, 8, 9}, .2!

EC 2.6.1.1 Aspartate transaminase

aspartate+ketoglutarate=oxaloacetate+glutamate

	pH 5	pH 6	pH 7	pH 8	pH 9
$\Delta_r G'^o$	-1.47	-1.47	-1.47	-1.47	-1.47
$\Delta_r H'^o$	47.68	47.68	47.68	47.68	47.68
$\Delta_r S'^o$	164.80	164.80	164.80	164.80	164.80
$\Delta_r N_H$	-1.45×10^{-15}				
log$K$ '	0.26	0.26	0.26	0.26	0.26

In[41]:= **rxSummaryT["EC 2.6.1.2 Alanine transaminase",**
   "alanine+ketoglutarate=pyruvate+glutamate",
   pyruvateGT + glutamateGT - (alanineGT + ketoglutarateGT), 298.15, {5, 6, 7, 8, 9}, .25]

EC 2.6.1.2 Alanine transaminase

alanine+ketoglutarate=pyruvate+glutamate

	pH 5	pH 6	pH 7	pH 8	pH 9
$\Delta_r G'^o$	-3.71	-3.71	-3.71	-3.71	-3.71
$\Delta_r H'^o$	21.93	21.93	21.93	21.93	21.93
$\Delta_r S'^o$	85.99	85.99	85.99	85.99	85.99
$\Delta_r N_H$	-1.45×10^{-15}				
log$K$ '	0.65	0.65	0.65	0.65	0.65

In[42]:= **rxSummaryT["EC 2.7.1.2 Glucokinase", "atp+glucose=adp+glucose6phos",**
   adpGT + glucose6phosGT - (atpGT + glucoseGT), 298.15, {5, 6, 7, 8, 9}, .25]

EC 2.7.1.2 Glucokinase

atp+glucose=adp+glucose6phos

	pH 5	pH 6	pH 7	pH 8	pH 9
$\Delta_r G'^o$	-17.41	-19.47	-24.42	-30.11	-35.82
$\Delta_r H'^o$	-22.20	-22.45	-22.52	-22.42	-22.41
$\Delta_r S'^o$	-16.04	-9.99	6.36	25.79	45.00
$\Delta_r N_H$	-0.14	-0.65	-0.98	-1.00	-1.00
log$K$ '	3.05	3.41	4.28	5.28	6.28

In[43]:= **rxSummaryT["EC 2.7.1.4 Fructokinase", "atp+fructose=adp+fructose6phos",**
   adpGT + fructose6phosGT - (atpGT + fructoseGT), 298.15, {5, 6, 7, 8, 9}, .25]

EC 2.7.1.4 Fructokinase

atp+fructose=adp+fructose6phos

	pH 5	pH 6	pH 7	pH 8	pH 9
$\Delta_r G'^\circ$	-13.94	-16.40	-21.62	-27.36	-33.07
$\Delta_r H'^\circ$	-13.74	-14.00	-14.07	-13.96	-13.95
$\Delta_r S'^\circ$	0.65	8.05	25.34	44.93	64.15
$\Delta_r N_H$	-0.18	-0.73	-1.00	-1.00	-1.00
$\log K'$	2.44	2.87	3.79	4.79	5.79

In[44]:= **rxSummaryT["EC 2.7.1.7 Mannokinase", "atp+mannose=adp+mannose6phos",**
        adpGT + mannose6phosGT - (atpGT + mannoseGT), 298.15, {5, 6, 7, 8, 9}, .25]

EC 2.7.1.7 Mannokinase

atp+mannose=adp+mannose6phos

	pH 5	pH 6	pH 7	pH 8	pH 9
$\Delta_r G'^\circ$	-19.34	-21.34	-26.26	-31.94	-37.65
$\Delta_r H'^\circ$	-22.99	-23.24	-23.32	-23.22	-23.21
$\Delta_r S'^\circ$	-12.23	-6.36	9.84	29.25	48.46
$\Delta_r N_H$	-0.13	-0.64	-0.98	-1.00	-1.00
$\log K'$	3.39	3.74	4.60	5.60	6.60

In[45]:= **rxSummaryT["EC 2.7.1.11 6-phosphofructokinase",**
        "atp+fructose6phos=adp+fructose16phos",
        adpGT + fructose16phosGT - (atpGT + fructose6phosGT), 298.15, {5, 6, 7, 8, 9}, .25]

EC 2.7.1.11 6-phosphofructokinase

atp+fructose6phos=adp+fructose16phos

	pH 5	pH 6	pH 7	pH 8	pH 9
$\Delta_r G'^\circ$	-12.35	-16.95	-23.25	-29.13	-34.86
$\Delta_r H'^\circ$	-84.70	-83.53	-83.13	-82.96	-82.94
$\Delta_r S'^\circ$	-242.60	-223.30	-200.80	-180.60	-161.30
$\Delta_r N_H$	-0.41	-1.09	-1.07	-1.01	-1.00
$\log K'$	2.16	2.97	4.07	5.10	6.11

In[46]:= **rxSummaryT["EC 2.7.1.15 Ribokinase", "atp+ribose=adp+ribose5phos",**
        adpGT + ribose5phosGT - (atpGT + riboseGT), 298.15, {5, 6, 7, 8, 9}, .25]

EC 2.7.1.15 Ribokinase

atp+ribose=adp+ribose5phos

	pH 5	pH 6	pH 7	pH 8	pH 9
$\Delta_r G'^\circ$	-14.54	-15.96	-20.30	-25.87	-31.56
$\Delta_r H'^\circ$	-6.59	-10.24	-14.64	-15.53	-15.62
$\Delta_r S'^\circ$	26.68	19.18	18.98	34.68	53.47
$\Delta_r N_H$	-0.09	-0.50	-0.93	-0.99	-1.00
$\log K'$	2.55	2.80	3.56	4.53	5.53

In[47]:= **rxSummaryT["EC 2.7.1.20 Adenosine kinase", "atp+adenosine=adp+amp",**
        adpGT + ampGT - (atpGT + adenosineGT), 298.15, {5, 6, 7, 8, 9}, .25]

EC 2.7.1.20 Adenosine kinase

atp+adenosine=adp+amp

	pH 5	pH 6	pH 7	pH 8	pH 9
$\Delta_r G'^{\circ}$	-16.97	-18.27	-22.49	-28.04	-33.73
$\Delta_r H'^{\circ}$	-19.02	-20.11	-21.88	-22.19	-22.22
$\Delta_r S'^{\circ}$	-6.89	-6.19	2.08	19.63	38.62
$\Delta_r N_H$	-0.06	-0.47	-0.92	-0.99	-1.00
$\log K'$	2.97	3.20	3.94	4.91	5.91

In[48]:= **rxSummaryT["EC 2.7.1.23 NAD$_{ox}$ kinase", "atp+nadox=adp+nadpox",**
     adpGT + nadpoxGT - (atpGT + nadoxGT), 298.15, {5, 6, 7, 8, 9}, .25]

EC 2.7.1.23 NAD$_{ox}$ kinase

atp+nadox=adp+nadpox

	pH 5	pH 6	pH 7	pH 8	pH 9
$\Delta_r G'^{\circ}$	-2.60	-8.51	-14.63	-20.47	-26.20
$\Delta_r H'^{\circ}$	-13.91	-14.09	-14.12	-14.01	-13.99
$\Delta_r S'^{\circ}$	-37.95	-18.71	1.71	21.69	40.96
$\Delta_r N_H$	-1.02	-1.07	-1.05	-1.01	-1.00
$\log K'$	0.45	1.49	2.56	3.59	4.59

In[49]:= **rxSummaryT["EC 2.7.1.30 Glycerol kinase", "atp+glycerol=adp+glycerol3phos",**
     adpGT + glycerol3phosGT - (atpGT + glycerolGT), 298.15, {5, 6, 7, 8, 9}, .25]

EC 2.7.1.30 Glycerol kinase

atp+glycerol=adp+glycerol3phos

	pH 5	pH 6	pH 7	pH 8	pH 9
$\Delta_r G'^{\circ}$	-32.15	-33.61	-37.99	-43.57	-49.26
$\Delta_r H'^{\circ}$	-55.98	-56.21	-56.31	-56.21	-56.20
$\Delta_r S'^{\circ}$	-79.93	-75.83	-61.45	-42.41	-23.25
$\Delta_r N_H$	-0.09	-0.51	-0.94	-1.00	-1.00
$\log K'$	5.63	5.89	6.65	7.63	8.63

In[50]:= **rxSummaryT["EC 2.7.1.40 Pyruvate kinase", "adp+pep=atp+pyruvate",**
     atpGT + pyruvateGT - (adpGT + pepGT), 298.15, {5, 6, 7, 8, 9}, .25]

EC 2.7.1.40 Pyruvate kinase

adp+pep=atp+pyruvate

	pH 5	pH 6	pH 7	pH 8	pH 9
$\Delta_r G'^{\circ}$	-34.47	-33.11	-28.85	-23.29	-17.60
$\Delta_r H'^{\circ}$	32.22	32.16	31.88	31.70	31.67
$\Delta_r S'^{\circ}$	223.70	218.90	203.70	184.40	165.20
$\Delta_r N_H$	0.08	0.48	0.93	0.99	1.00
$\log K'$	6.04	5.80	5.05	4.08	3.08

In[51]:= **rxSummaryT["EC 2.7.1.73 Inosine kinase", "atp+inosine=adp+imp",**
     adpGT + impGT - (atpGT + inosineGT), 298.15, {5, 6, 7, 8, 9}, .25]

EC 2.7.1.73 Inosine kinase

atp+inosine=adp+imp

	pH 5	pH 6	pH 7	pH 8	pH 9
$\Delta_r G'^{\circ}$	-17.25	-18.59	-22.80	-28.26	-33.65
$\Delta_r H'^{\circ}$	-19.44	-20.94	-22.73	-22.63	-18.62
$\Delta_r S'^{\circ}$	-7.35	-7.89	0.23	18.89	50.41
$\Delta_r N_H$	-0.08	-0.47	-0.92	-0.96	-0.94
$\log K'$	3.02	3.26	3.99	4.95	5.89

In[52]:= **rxSummaryT["EC 2.7.1.86 NAD$_{red}$ kinase", "atp+nadred=adp+nadpred",**
         adpGT + nadpredGT - (atpGT + nadredGT), 298.15, {5, 6, 7, 8, 9}, .25]

EC 2.7.1.86 NAD$_{red}$ kinase

atp+nadred=adp+nadpred

	pH 5	pH 6	pH 7	pH 8	pH 9
$\Delta_r G'^{\circ}$	-3.01	-8.93	-15.05	-20.89	-26.62
$\Delta_r H'^{\circ}$	-9.50	-9.68	-9.71	-9.60	-9.58
$\Delta_r S'^{\circ}$	-21.76	-2.52	17.90	37.88	57.15
$\Delta_r N_H$	-1.02	-1.07	-1.05	-1.01	-1.00
$\log K'$	0.53	1.56	2.64	3.66	4.66

In[53]:= **rxSummaryT["EC 2.7.1.90 Diphosphate-fructose-6-phosphate 1-phosphotransferase",**
         "ppi+fructose6phos=pi+fructose16phos",
         piGT + fructose16phosGT - (ppiGT + fructose6phosGT), 298.15, {5, 6, 7, 8, 9}, .25]

EC 2.7.1.90 Diphosphate-fructose-6-phosphate 1-phosphotransferase

ppi+fructose6phos=pi+fructose16phos

	pH 5	pH 6	pH 7	pH 8	pH 9
$\Delta_r G'^{\circ}$	-3.17	-6.19	-9.87	-14.11	-16.37
$\Delta_r H'^{\circ}$	-83.26	-82.45	-80.59	-80.63	-82.81
$\Delta_r S'^{\circ}$	-268.70	-255.80	-237.20	-223.10	-222.80
$\Delta_r N_H$	-0.30	-0.63	-0.73	-0.64	-0.17
$\log K'$	0.55	1.08	1.73	2.47	2.87

In[54]:= **rxSummaryT["EC 2.7.1.142 Glycerol-3-phosphate-glucosephosphotransferase",**
         "glycerol+glucose6phos=glycerol3phos+glucose",
         glycerol3phosGT + glucoseGT - (glycerolGT + glucose6phosGT), 298.15, {5, 6, 7, 8, 9}, .25]

EC 2.7.1.142 Glycerol-3-phosphate-glucosephosphotransferase

glycerol+glucose6phos=glycerol3phos+glucose

	pH 5	pH 6	pH 7	pH 8	pH 9
$\Delta_r G'^{\circ}$	-14.73	-14.14	-13.57	-13.45	-13.44
$\Delta_r H'^{\circ}$	-33.78	-33.77	-33.78	-33.79	-33.79
$\Delta_r S'^{\circ}$	-63.89	-65.84	-67.81	-68.21	-68.25
$\Delta_r N_H$	0.05	0.14	0.05	0.01	0.00
$\log K'$	2.58	2.48	2.38	2.36	2.35

In[55]:= **rxSummaryT["EC 2.7.1.146 ADP-specific phosphofructokinase",**
         "adp+fructose6phos=amp+fructose16phos",
         ampGT + fructose16phosGT - (adpGT + fructose6phosGT), 298.15, {5, 6, 7, 8, 9}, .25]

EC 2.7.1.146 ADP-specific phosphofructokinase

adp+fructose6phos=amp+fructose16phos

	pH 5	pH 6	pH 7	pH 8	pH 9
$\Delta_r G'^{\circ}$	-10.12	-14.83	-21.17	-27.04	-32.77
$\Delta_r H'^{\circ}$	-84.70	-83.98	-83.74	-83.63	-83.62
$\Delta_r S'^{\circ}$	-250.20	-232.00	-209.90	-189.80	-170.50
$\Delta_r N_H$	-0.42	-1.11	-1.06	-1.01	-1.00
$\log K'$	1.77	2.60	3.71	4.74	5.74

In[56]:= **rxSummaryT["EC 2.7.1.147 ADP-specific glucokinasekinase",**
   "adp+glucose=amp+glucose6phos",
   ampGT + glucose6phosGT - (adpGT + glucoseGT), 298.15, {5, 6, 7, 8, 9}, .25]

EC 2.7.1.147 ADP-specific glucokinasekinase

adp+glucose=amp+glucose6phos

	pH 5	pH 6	pH 7	pH 8	pH 9
$\Delta_r G'^\circ$	-15.18	-17.34	-22.35	-28.03	-33.73
$\Delta_r H'^\circ$	-22.20	-22.90	-23.14	-23.09	-23.08
$\Delta_r S'^\circ$	-23.56	-18.63	-2.67	16.55	35.73
$\Delta_r N_H$	-0.15	-0.67	-0.98	-1.00	-1.00
$\log K'$	2.66	3.04	3.91	4.91	5.91

In[57]:= **rxSummaryT["EC 2.7.4.3 Phosphotransferase with a phosphate group as acceptor",**
   "atp+amp=2*adp", 2 * adpGT - (atpGT + ampGT), 298.15, {5, 6, 7, 8, 9}, .25]

EC 2.7.4.3 Phosphotransferase with a phosphate group as acceptor

atp+amp=2*adp

	pH 5	pH 6	pH 7	pH 8	pH 9
$\Delta_r G'^\circ$	-2.23	-2.13	-2.07	-2.09	-2.09
$\Delta_r H'^\circ$	0.01	0.45	0.62	0.67	0.68
$\Delta_r S'^\circ$	7.52	8.65	9.02	9.24	9.28
$\Delta_r N_H$	0.01	0.02	-0.00	-0.00	-0.00
$\log K'$	0.39	0.37	0.36	0.37	0.37

Class 3: Hydrolases

In[58]:= **rxSummaryT["EC 3.1.1.1 Carboxylesterase", "ethylacetate+h2o=ethanol+acetate",**
   ethanolGT + acetateGT - (ethylacetateGT + h2oGT), 298.15, {5, 6, 7, 8, 9}, .25]

EC 3.1.1.1 Carboxylesterase

ethylacetate+h2o=ethanol+acetate

	pH 5	pH 6	pH 7	pH 8	pH 9
$\Delta_r G'^\circ$	-6.91	-12.05	-17.69	-23.39	-29.10
$\Delta_r H'^\circ$	-5.79	-5.67	-5.66	-5.66	-5.66
$\Delta_r S'^\circ$	3.77	21.39	40.37	59.49	78.64
$\Delta_r N_H$	-0.77	-0.97	-1.00	-1.00	-1.00
$\log K'$	1.21	2.11	3.10	4.10	5.10

In[59]:= **rxSummaryT["EC 3.1.3.9 Glucose-6-phosphatase", "glucose6phos+h2o=glucose+pi",**
   glucoseGT + piGT - (glucose6phosGT + h2oGT), 298.15, {5, 6, 7, 8, 9}, .25]

EC 3.1.3.9 Glucose-6-phosphatase

glucose6phos+h2o=glucose+pi

	pH 5	pH 6	pH 7	pH 8	pH 9
$\Delta_r G'^\circ$	-15.15	-13.75	-11.62	-10.96	-10.88
$\Delta_r H'^\circ$	-4.19	-3.27	-0.55	0.85	1.06
$\Delta_r S'^\circ$	36.77	35.14	37.11	39.63	40.03
$\Delta_r N_H$	0.10	0.40	0.24	0.04	0.00
$\log K'$	2.65	2.41	2.03	1.92	1.91

In[60]:= **rxSummaryT["EC 3.1.3.11 Fructose-bisphosphatase",**
  "fructose16phos+h2o=fructose6phos+pi",
  fructose6phosGT + piGT - (fructose16phosGT + h2oGT), 298.15, {5, 6, 7, 8, 9}, .25]

EC 3.1.3.11 Fructose-bisphosphatase

fructose16phos+h2o=fructose6phos+pi

	pH 5	pH 6	pH 7	pH 8	pH 9
$\Delta_r G'^o$	-20.21	-16.26	-12.79	-11.95	-11.84
$\Delta_r H'^o$	58.31	57.81	60.05	61.40	61.59
$\Delta_r S'^o$	263.40	248.50	244.30	246.00	246.30
$\Delta_r N_H$	0.37	0.84	0.32	0.04	0.00
log$K$ '	3.54	2.85	2.24	2.09	2.07

In[61]:= **rxSummaryT["EC 3.1.3.31 Nucleotidease", "glycerol+pi=glycerol3phos+h2o",**
  glycerol3phosGT + h2oGT - (glycerolGT + piGT), 298.15, {5, 6, 7, 8, 9}, .25]

EC 3.1.3.31 Nucleotidease

glycerol+pi=glycerol3phos+h2o

	pH 5	pH 6	pH 7	pH 8	pH 9
$\Delta_r G'^o$	0.42	-0.39	-1.95	-2.49	-2.56
$\Delta_r H'^o$	-29.60	-30.50	-33.23	-34.64	-34.85
$\Delta_r S'^o$	-100.70	-101.00	-104.90	-107.80	-108.30
$\Delta_r N_H$	-0.05	-0.26	-0.20	-0.03	-0.00
log$K$ '	-0.07	0.07	0.34	0.44	0.45

In[62]:= **rxSummaryT["EC 3.2.1.20 Alpha-glucosidase", "maltose+h2o=2*glucose",**
  2 * glucoseGT - (maltoseGT + h2oGT), 298.15, {5, 6, 7, 8, 9}, .25]

EC 3.2.1.20 Alpha-glucosidase

maltose+h2o=2*glucose

	pH 5	pH 6	pH 7	pH 8	pH 9
$\Delta_r G'^o$	-19.92	-19.92	-19.92	-19.92	-19.92
$\Delta_r H'^o$	-0.49	-0.49	-0.49	-0.49	-0.49
$\Delta_r S'^o$	65.17	65.17	65.17	65.17	65.17
$\Delta_r N_H$	2.90×10^{-15}				
log$K$ '	3.49	3.49	3.49	3.49	3.49

In[63]:= **rxSummaryT["EC 3.2.1.48 Sucrose alpha-glucosidase", "sucrose+h2o=glucose+fructose",**
  glucoseGT + fructoseGT - (sucroseGT + h2oGT), 298.15, {5, 6, 7, 8, 9}, .25]

EC 3.2.1.48 Sucrose alpha-glucosidase

sucrose+h2o=glucose+fructose

	pH 5	pH 6	pH 7	pH 8	pH 9
$\Delta_r G'^o$	-29.52	-29.52	-29.52	-29.52	-29.52
$\Delta_r H'^o$	-35.87	-35.87	-35.87	-35.87	-35.87
$\Delta_r S'^o$	-21.30	-21.30	-21.30	-21.30	-21.30
$\Delta_r N_H$	2.90×10^{-15}				
log$K$ '	5.17	5.17	5.17	5.17	5.17

In[64]:= **rxSummaryT["EC 3.2.1.108 Lactase", "lactose+h2o=glucose+galactose",**
  glucoseGT + galactoseGT - (lactoseGT + h2oGT), 298.15, {5, 6, 7, 8, 9}, .25]

EC 3.2.1.108 Lactase

```
lactose+h2o=glucose+galactose
```

	pH 5	pH 6	pH 7	pH 8	pH 9
$\Delta_r G'^{\circ}$	-20.31	-20.31	-20.31	-20.31	-20.31
$\Delta_r H'^{\circ}$	1.52	1.52	1.52	1.52	1.52
$\Delta_r S'^{\circ}$	73.22	73.22	73.22	73.22	73.22
$\Delta_r N_H$	2.90×10^{-15}				
$\log K'$	3.56	3.56	3.56	3.56	3.56

In[65]:= **rxSummaryT["EC 3.2.2.4 AMP nucleosidase", "amp+h2o=adenine+ribose5phos",**
  adenineGT + ribose5phosGT - (ampGT + h2oGT), 298.15, {5, 6, 7, 8, 9}, .25]

```
EC 3.2.2.4 AMP nucleosidase

amp+h2o=adenine+ribose5phos
```

	pH 5	pH 6	pH 7	pH 8	pH 9
$\Delta_r G'^{\circ}$	-4.77	-4.64	-4.72	-4.74	-4.74
$\Delta_r H'^{\circ}$	12.78	12.23	9.84	9.29	9.23
$\Delta_r S'^{\circ}$	58.89	56.59	48.82	47.04	46.84
$\Delta_r N_H$	0.08	-0.01	-0.01	-0.00	-0.00
$\log K'$	0.84	0.81	0.83	0.83	0.83

In[66]:= **rxSummaryT["EC 3.2.2.7 Purinenucleosidase", "adenosine+h2o=adenine+ribose",**
  adenineGT + riboseGT - (adenosineGT + h2oGT), 298.15, {5, 6, 7, 8, 9}, .25]

```
EC 3.2.2.7 Purinenucleosidase

adenosine+h2o=adenine+ribose
```

	pH 5	pH 6	pH 7	pH 8	pH 9
$\Delta_r G'^{\circ}$	-7.20	-6.94	-6.91	-6.91	-6.91
$\Delta_r H'^{\circ}$	0.35	2.36	2.60	2.63	2.63
$\Delta_r S'^{\circ}$	25.33	31.21	31.92	31.99	32.00
$\Delta_r N_H$	0.11	0.01	0.00	0.00	0.00
$\log K'$	1.26	1.22	1.21	1.21	1.21

In[67]:= **rxSummaryT["EC 3.5.1.1 Asparaginease", "asparagineL+h2o=aspartate+ammonia",**
  aspartateGT + ammoniaGT - (asparagineLGT + h2oGT), 298.15, {5, 6, 7, 8, 9}, .25]

```
EC 3.5.1.1 Asparaginease

asparagineL+h2o=aspartate+ammonia
```

	pH 5	pH 6	pH 7	pH 8	pH 9
$\Delta_r G'^{\circ}$	-13.69	-13.69	-13.70	-13.82	-14.79
$\Delta_r H'^{\circ}$	-23.17	-23.15	-22.89	-20.41	-4.43
$\Delta_r S'^{\circ}$	-31.81	-31.72	-30.80	-22.08	34.75
$\Delta_r N_H$	-0.00	-0.00	-0.01	-0.05	-0.36
$\log K'$	2.40	2.40	2.40	2.42	2.59

In[68]:= **rxSummaryT["EC 3.5.1.2 Glutaminease", "glutamine+h2o=glutamate+ammonia",**
  glutamateGT + ammoniaGT - (glutamineGT + h2oGT), 298.15, {5, 6, 7, 8, 9}, .25]

```
EC 3.5.1.2 Glutaminease

glutamine+h2o=glutamate+ammonia
```

	pH 5	pH 6	pH 7	pH 8	pH 9
$\Delta_r G'^o$	-13.19	-13.19	-13.20	-13.32	-14.29
$\Delta_r H'^o$	-20.74	-20.72	-20.46	-17.98	-2.00
$\Delta_r S'^o$	-25.34	-25.24	-24.32	-15.61	41.22
$\Delta_r N_H$	-0.00	-0.00	-0.01	-0.05	-0.36
$\log K'$	2.31	2.31	2.31	2.33	2.50

In[69]:= **rxSummaryT["EC 3.5.1.5 Urease", "urea+2*h2o=co2tot+2*ammonia",**
     co2totGT + 2 * ammoniaGT - (ureaGT + 2 * h2oGT), 298.15, {5, 6, 7, 8, 9}, .25]

EC 3.5.1.5 Urease

urea+2*h2o=co2tot+2*ammonia

	pH 5	pH 6	pH 7	pH 8	pH 9
$\Delta_r G'^o$	-47.67	-37.55	-30.19	-24.51	-21.06
$\Delta_r H'^o$	-74.69	-71.45	-67.18	-61.16	-26.97
$\Delta_r S'^o$	-90.63	-113.70	-124.10	-122.90	-19.81
$\Delta_r N_H$	1.92	1.55	1.10	0.89	0.14
$\log K'$	8.35	6.58	5.29	4.29	3.69

In[70]:= **rxSummaryT["EC 3.5.4.4 Adenosinedeaminase", "adenosine+h2o=inosine+ammonia",**
     inosineGT + ammoniaGT - (adenosineGT + h2oGT), 298.15, {5, 6, 7, 8, 9}, .25]

EC 3.5.4.4 Adenosinedeaminase

adenosine+h2o=inosine+ammonia

	pH 5	pH 6	pH 7	pH 8	pH 9
$\Delta_r G'^o$	-28.16	-22.52	-16.88	-11.72	-9.32
$\Delta_r H'^o$	-44.75	-45.08	-44.34	-37.57	-7.50
$\Delta_r S'^o$	-55.63	-75.68	-92.12	-86.70	6.09
$\Delta_r N_H$	0.97	0.99	0.97	0.77	-0.04
$\log K'$	4.93	3.95	2.96	2.05	1.63

In[71]:= **rxSummaryT["EC 3.5.4.6 AMP deaminase", "amp+h2o=imp+ammonia",**
     impGT + ammoniaGT - (ampGT + h2oGT), 298.15, {5, 6, 7, 8, 9}, .25]

EC 3.5.4.6 AMP deaminase

amp+h2o=imp+ammonia

	pH 5	pH 6	pH 7	pH 8	pH 9
$\Delta_r G'^o$	-28.44	-22.84	-17.18	-11.93	-9.23
$\Delta_r H'^o$	-45.17	-45.91	-45.20	-38.00	-3.90
$\Delta_r S'^o$	-56.10	-77.37	-93.97	-87.44	17.88
$\Delta_r N_H$	0.95	1.00	0.98	0.81	0.02
$\log K'$	4.98	4.00	3.01	2.09	1.62

In[72]:= **rxSummaryT["EC 3.5.4.7 ADP deaminase", "adp+h2o=idp+ammonia",**
     idpGT + ammoniaGT - (adpGT + h2oGT), 298.15, {5, 6, 7, 8, 9}, .25]

EC 3.5.4.7 ADP deaminase

adp+h2o=idp+ammonia

	pH 5	pH 6	pH 7	pH 8	pH 9
$\Delta_r G'^o$	-28.33	-22.75	-17.13	-12.25	-10.59
$\Delta_r H'^o$	-44.95	-45.88	-44.78	-34.67	-2.03
$\Delta_r S'^o$	-55.75	-77.56	-92.73	-75.20	28.73
$\Delta_r N_H$	0.94	0.99	0.96	0.67	-0.15
$\log K'$	4.96	3.99	3.00	2.15	1.86

In[73]:= **rxSummaryT["EC 3.5.4.18 ATP deaminase", "atp+h2o=itp+ammonia",**
    itpGT + ammoniaGT - (atpGT + h2oGT), 298.15, {5, 6, 7, 8, 9}, .25]

EC 3.5.4.18 ATP deaminase

atp+h2o=itp+ammonia

	pH 5	pH 6	pH 7	pH 8	pH 9
$\Delta_r G'°$	-28.22	-22.66	-17.02	-11.86	-9.48
$\Delta_r H'°$	-44.96	-45.91	-45.32	-38.49	-8.24
$\Delta_r S'°$	-56.13	-77.96	-94.93	-89.34	4.16
$\Delta_r N_H$	0.94	0.99	0.97	0.77	-0.04
$logK'$	4.94	3.97	2.98	2.08	1.66

In[74]:= **rxSummaryT["EC 3.6.1.1 Inorganic diphosphatase",**
    "ppi+h2o=2*pi", 2 * piGT - (ppiGT + h2oGT), 298.15, {5, 6, 7, 8, 9}, .25]

EC 3.6.1.1 Inorganic diphosphatase

ppi+h2o=2*pi

	pH 5	pH 6	pH 7	pH 8	pH 9
$\Delta_r G'°$	-23.37	-22.45	-22.66	-26.06	-28.22
$\Delta_r H'°$	-24.95	-24.63	-20.55	-19.23	-21.22
$\Delta_r S'°$	-5.29	-7.32	7.08	22.89	23.47
$\Delta_r N_H$	0.08	0.21	-0.41	-0.60	-0.17
$logK'$	4.09	3.93	3.97	4.57	4.94

In[75]:= **rxSummaryT["EC 3.6.1.3 Adenosinetriphosphatase",**
    "atp+h2o=adp+pi", adpGT + piGT - (atpGT + h2oGT), 298.15, {5, 6, 7, 8, 9}, .25]

EC 3.6.1.3 Adenosinetriphosphatase

atp+h2o=adp+pi

	pH 5	pH 6	pH 7	pH 8	pH 9
$\Delta_r G'°$	-32.56	-33.22	-36.04	-41.07	-46.70
$\Delta_r H'°$	-26.38	-25.72	-23.08	-21.57	-21.35
$\Delta_r S'°$	20.73	25.15	43.47	65.42	85.03
$\Delta_r N_H$	-0.04	-0.25	-0.74	-0.96	-1.00
$logK'$	5.70	5.82	6.31	7.20	8.18

In[76]:= **rxSummaryT["EC 3.6.1.5 Apyrase", "atp+2*h2o=amp+2*pi",**
    ampGT + 2 * piGT - (atpGT + 2 * h2oGT), 298.15, {5, 6, 7, 8, 9}, .25]

EC 3.6.1.5 Apyrase

atp+2*h2o=amp+2*pi

	pH 5	pH 6	pH 7	pH 8	pH 9
$\Delta_r G'°$	-62.89	-64.31	-70.00	-80.06	-91.31
$\Delta_r H'°$	-52.77	-51.89	-46.77	-43.80	-43.37
$\Delta_r S'°$	33.94	41.66	77.91	121.60	160.80
$\Delta_r N_H$	-0.09	-0.52	-1.48	-1.93	-1.99
$logK'$	11.02	11.27	12.26	14.03	16.00

In[77]:= **rxSummaryT["EC 3.6.1.8 ATP diphosphatase", "atp+h2o=amp+ppi",**
    ampGT + ppiGT - (atpGT + h2oGT), 298.15, {5, 6, 7, 8, 9}, .25]

EC 3.6.1.8 ATP diphosphatase

atp+h2o=amp+ppi

	pH 5	pH 6	pH 7	pH 8	pH 9
$\Delta_r G'^o$	-39.52	-41.86	-47.34	-54.00	-63.10
$\Delta_r H'^o$	-27.82	-27.25	-26.22	-24.57	-22.15
$\Delta_r S'^o$	39.23	48.98	70.83	98.71	137.30
$\Delta_r N_H$	-0.17	-0.73	-1.08	-1.33	-1.83
$\log K'$	6.92	7.33	8.29	9.46	11.05

Class 4: Lyases

In[78]:= **rxSummaryT["EC 4.1.1.1 Pyruvate decarboxylase", "pyruvate+h2o=co2tot+acetaldehyde",**
   co2totGT + acetaldehydeGT - (pyruvateGT + h2oGT), 298.15, {5, 6, 7, 8, 9}, .25]

   EC 4.1.1.1 Pyruvate decarboxylase

   pyruvate+h2o=co2tot+acetaldehyde

	pH 5	pH 6	pH 7	pH 8	pH 9
$\Delta_r G'^o$	-22.69	-18.28	-16.60	-16.38	-16.71
$\Delta_r H'^o$	-29.99	-26.80	-23.06	-21.99	-19.75
$\Delta_r S'^o$	-24.49	-28.60	-21.67	-18.82	-10.19
$\Delta_r N_H$	0.92	0.55	0.11	-0.01	-0.15
$\log K'$	3.97	3.20	2.91	2.87	2.93

In[79]:= **rxSummaryT["EC 4.1.1.12 Aspartate 4-decarboxylase", "aspartate+h2o=co2tot+alanine",**
   co2totGT + alanineGT - (aspartateGT + h2oGT), 298.15, {5, 6, 7, 8, 9}, .25]

   EC 4.1.1.12 Aspartate 4-decarboxylase

   aspartate+h2o=co2tot+alanine

	pH 5	pH 6	pH 7	pH 8	pH 9
$\Delta_r G'^o$	-31.08	-26.67	-24.99	-24.77	-25.10
$\Delta_r H'^o$	-25.37	-22.18	-18.44	-17.37	-15.13
$\Delta_r S'^o$	19.15	15.04	21.97	24.82	33.44
$\Delta_r N_H$	0.92	0.55	0.11	-0.01	-0.15
$\log K'$	5.44	4.67	4.38	4.34	4.40

In[80]:= **rxSummaryT["EC 4.1.2.36 Lactate aldolase", "formate+acetaldehyde=lactate",**
   lactateGT - (formateGT + acetaldehydeGT), 298.15, {5, 6, 7, 8, 9}, .25]

   EC 4.1.2.36 Lactate aldolase

   formate+acetaldehyde=lactate

	pH 5	pH 6	pH 7	pH 8	pH 9
$\Delta_r G'^o$	-26.72	-26.72	-26.72	-26.72	-26.72
$\Delta_r H'^o$	-48.86	-48.86	-48.86	-48.86	-48.86
$\Delta_r S'^o$	-74.26	-74.26	-74.26	-74.26	-74.26
$\Delta_r N_H$	0.00				
$\log K'$	4.68	4.68	4.68	4.68	4.68

In[81]:= **rxSummaryT["EC 4.1.3.6 Citrate(pro-3S)-lyase", "acetate+oxaloacetate=citrate",**
   citrateGT - (acetateGT + oxaloacetateGT), 298.15, {5, 6, 7, 8, 9}, .25]

   EC 4.1.3.6 Citrate(pro-3S)-lyase

   acetate+oxaloacetate=citrate

	pH 5	pH 6	pH 7	pH 8	pH 9
$\Delta_r G'^{\circ}$	-6.69	-4.01	-3.41	-3.34	-3.33
$\Delta_r H'^{\circ}$	-72.52	-69.07	-67.75	-67.57	-67.55
$\Delta_r S'^{\circ}$	-220.80	-218.20	-215.80	-215.40	-215.40
$\Delta_r N_H$	0.68	0.24	0.03	0.00	0.00
$logK'$	1.17	0.70	0.60	0.58	0.58

In[82]:= **rxSummaryT["EC 4.1.99.1 Tryptophanase", "tryptophanL+h2o=indole+pyruvate",**
indoleGT + pyruvateGT - (tryptophanLGT + h2oGT), 298.15, {5, 6, 7, 8, 9}, .25]

EC 4.1.99.1 Tryptophanase

tryptophanL+h2o=indole+pyruvate

	pH 5	pH 6	pH 7	pH 8	pH 9
$\Delta_r G'^{\circ}$	-14.79	-37.62	-60.45	-83.29	-106.10
$\Delta_r H'^{\circ}$	194.40	194.40	194.40	194.40	194.40
$\Delta_r S'^{\circ}$	701.50	778.10	854.70	931.30	1008.00
$\Delta_r N_H$	-4.00				
$logK'$	2.59	6.59	10.59	14.59	18.59

In[83]:= **rxSummaryT["EC 4.2.1.2 Fumarate hydratase", "fumarate+h2o=malate",**
malateGT - (fumarateGT + h2oGT), 298.15, {5, 6, 7, 8, 9}, .25]

EC 4.2.1.2 Fumarate hydratase

fumarate+h2o=malate

	pH 5	pH 6	pH 7	pH 8	pH 9
$\Delta_r G'^{\circ}$	-4.34	-3.69	-3.61	-3.60	-3.60
$\Delta_r H'^{\circ}$	-17.29	-16.67	-16.58	-16.57	-16.57
$\Delta_r S'^{\circ}$	-43.46	-43.52	-43.50	-43.50	-43.50
$\Delta_r N_H$	0.23	0.04	0.00	0.00	0.00
$logK'$	0.76	0.65	0.63	0.63	0.63

In[84]:= **rxSummaryT["EC 4.3.1.1 Aspartate ammonia-lyase", "fumarate+ammonia=aspartate",**
aspartateGT - (fumarateGT + ammoniaGT), 298.15, {5, 6, 7, 8, 9}, .25]

EC 4.3.1.1 Aspartate ammonia-lyase

fumarate+ammonia=aspartate

	pH 5	pH 6	pH 7	pH 8	pH 9
$\Delta_r G'^{\circ}$	-11.20	-11.43	-11.44	-11.33	-10.36
$\Delta_r H'^{\circ}$	-35.29	-35.20	-35.45	-37.93	-53.90
$\Delta_r S'^{\circ}$	-80.77	-79.72	-80.51	-89.22	-146.00
$\Delta_r N_H$	-0.10	-0.01	0.00	0.05	0.36
$logK'$	1.96	2.00	2.01	1.98	1.81

Class 5: Isomerases

In[85]:= **rxSummaryT["EC 5.3.1.5 Xylose isomerase",**
"xylulose=xylose", xyloseGT - xyluloseGT, 298.15, {5, 6, 7, 8, 9}, .25]

EC 5.3.1.5 Xylose isomerase

xylulose=xylose

	pH 5	pH 6	pH 7	pH 8	pH 9
$\Delta_r G'^\circ$	-4.34	-4.34	-4.34	-4.34	-4.34
$\Delta_r H'^\circ$	-16.29	-16.29	-16.29	-16.29	-16.29
$\Delta_r S'^\circ$	-40.08	-40.08	-40.08	-40.08	-40.08
$\Delta_r N_H$	0.00				
$\log K'$	0.76	0.76	0.76	0.76	0.76

In[86]:= **rxSummaryT["EC 5.3.1.6 Ribose-5-phosphate isomerase",**
   "ribose1phos=ribose5phos", ribose5phosGT - ribose1phosGT, 298.15, {5, 6, 7, 8, 9}, .25]

EC 5.3.1.6 Ribose-5-phosphate isomerase

ribose1phos=ribose5phos

	pH 5	pH 6	pH 7	pH 8	pH 9
$\Delta_r G'^\circ$	-8.08	-8.08	-8.08	-8.08	-8.08
$\Delta_r H'^\circ$	5.06×10^{-14}	-4.41×10^{-13}	3.53×10^{-13}	-8.38×10^{-13}	-8.19×10^{-14}
$\Delta_r S'^\circ$	27.10	27.10	27.10	27.10	27.10
$\Delta_r N_H$	1.24×10^{-15}	-2.61×10^{-14}	6.22×10^{-15}	1.24×10^{-15}	-1.24×10^{-15}
$\log K'$	1.42	1.42	1.42	1.42	1.42

In[87]:= **rxSummaryT["EC 5.3.1.7 Mannose isomerase",**
   "mannose=fructose", fructoseGT - mannoseGT, 298.15, {5, 6, 7, 8, 9}, .25]

EC 5.3.1.7 Mannose isomerase

mannose=fructose

	pH 5	pH 6	pH 7	pH 8	pH 9
$\Delta_r G'^\circ$	-5.51	-5.51	-5.51	-5.51	-5.51
$\Delta_r H'^\circ$	-0.72	-0.72	-0.72	-0.72	-0.72
$\Delta_r S'^\circ$	16.07	16.07	16.07	16.07	16.07
$\Delta_r N_H$	0.00				
$\log K'$	0.97	0.97	0.97	0.97	0.97

In[88]:= **rxSummaryT["EC 5.3.1.8 Mannose-6-phosphate isomerase", "mannose6phos=fructose6phos",**
   fructose6phosGT - mannose6phosGT, 298.15, {5, 6, 7, 8, 9}, .25]

EC 5.3.1.8 Mannose-6-phosphate isomerase

mannose6phos=fructose6phos

	pH 5	pH 6	pH 7	pH 8	pH 9
$\Delta_r G'^\circ$	-0.10	-0.56	-0.87	-0.92	-0.93
$\Delta_r H'^\circ$	8.52	8.52	8.54	8.54	8.54
$\Delta_r S'^\circ$	28.94	30.48	31.56	31.74	31.76
$\Delta_r N_H$	-0.05	-0.09	-0.02	-0.00	-0.00
$\log K'$	0.02	0.10	0.15	0.16	0.16

In[89]:= **rxSummaryT["EC 5.3.1.9 Glucose-6-phosphate isomerase", "fructose6phos=glucose6phos",**
   glucose6phosGT - fructose6phosGT, 298.15, {5, 6, 7, 8, 9}, .25]

EC 5.3.1.9 Glucose-6-phosphate isomerase

fructose6phos=glucose6phos

	pH 5	pH 6	pH 7	pH 8	pH 9
$\Delta_r G'^{\,o}$	-3.87	-3.46	-3.19	-3.15	-3.14
$\Delta_r H'^{\,o}$	-11.26	-11.26	-11.27	-11.27	-11.27
$\Delta_r S'^{\,o}$	-24.80	-26.16	-27.10	-27.25	-27.27
$\Delta_r N_H$	0.04	0.08	0.02	0.00	0.00
$\log K'$	0.68	0.61	0.56	0.55	0.55

In[90]:= **rxSummaryT["EC 5.3.1.20 Ribose isomerase",**
"ribulose=ribose", riboseGT - ribuloseGT, 298.15, {5, 6, 7, 8, 9}, .25]

EC 5.3.1.20 Ribose isomerase

ribulose=ribose

	pH 5	pH 6	pH 7	pH 8	pH 9
$\Delta_r G'^{\,o}$	-16.06	-16.06	-16.06	-16.06	-16.06
$\Delta_r H'^{\,o}$	-10.98	-10.98	-10.98	-10.98	-10.98
$\Delta_r S'^{\,o}$	17.04	17.04	17.04	17.04	17.04
$\Delta_r N_H$	0.00				
$\log K'$	2.81	2.81	2.81	2.81	2.81

Class 6: Ligases

In[91]:= **rxSummaryT["EC 6.3.1.1 Aspartate-ammonia ligase",**
"aspartate+atp+ammonia=asparagineL+amp+ppi",
asparagineLGT + ampGT + ppiGT - (aspartateGT + atpGT + ammoniaGT), 298.15, {5, 6, 7, 8, 9}, .2

EC 6.3.1.1 Aspartate-ammonia ligase

aspartate+atp+ammonia=asparagineL+amp+ppi

	pH 5	pH 6	pH 7	pH 8	pH 9
$\Delta_r G'^{\,o}$	-25.83	-28.17	-33.64	-40.18	-48.30
$\Delta_r H'^{\,o}$	-4.65	-4.11	-3.34	-4.16	-17.72
$\Delta_r S'^{\,o}$	71.04	80.70	101.60	120.80	102.60
$\Delta_r N_H$	-0.17	-0.73	-1.07	-1.28	-1.47
$\log K'$	4.53	4.93	5.89	7.04	8.46

In[92]:= **rxSummaryT["EC 6.3.1.2 Glutamate-ammonia ligase",**
"glutamate+atp+ammonia=glutamine+adp+pi",
glutamineGT + adpGT + piGT - (glutamateGT + atpGT + ammoniaGT), 298.15, {5, 6, 7, 8, 9}, .25]

EC 6.3.1.2 Glutamate-ammonia ligase

glutamate+atp+ammonia=glutamine+adp+pi

	pH 5	pH 6	pH 7	pH 8	pH 9
$\Delta_r G'^{\,o}$	-19.37	-20.03	-22.83	-27.75	-32.41
$\Delta_r H'^{\,o}$	-5.64	-5.00	-2.62	-3.59	-19.35
$\Delta_r S'^{\,o}$	46.07	50.39	67.79	81.03	43.81
$\Delta_r N_H$	-0.04	-0.25	-0.74	-0.91	-0.64
$\log K'$	3.39	3.51	4.00	4.86	5.68

In[93]:= **rxSummaryT["EC 6.3.1.4 Aspartate-ammonia ligase (ADP-forming)",**
"aspartate+atp+ammonia=asparagineL+adp+pi",
asparagineLGT + adpGT + piGT - (aspartateGT + atpGT + ammoniaGT), 298.15, {5, 6, 7, 8, 9}, .25

EC 6.3.1.4 Aspartate-ammonia ligase (ADP-forming)

aspartate+atp+ammonia=asparagineL+adp+pi

	pH 5	pH 6	pH 7	pH 8	pH 9
$\Delta_r G'^\circ$	-18.87	-19.53	-22.33	-27.25	-31.91
$\Delta_r H'^\circ$	-3.21	-2.57	-0.19	-1.16	-16.92
$\Delta_r S'^\circ$	52.54	56.87	74.26	87.50	50.29
$\Delta_r N_H$	-0.04	-0.25	-0.74	-0.91	-0.64
$\log K'$	3.31	3.42	3.91	4.77	5.59

In[94]:= **rxSummaryT["EC 6.3.5.4 Asparagine synthase (glutamine-hydrolyzing)",**
     "aspartate+atp+glutamine=asparagineL+glutamate+amp+ppi",
     asparagineLGT + glutamateGT + ampGT + ppiGT - (aspartateGT + atpGT + glutamineGT),
     298.15, {5, 6, 7, 8, 9}, .25]

EC 6.3.5.4 Asparagine synthase (glutamine-hydrolyzing)

aspartate+atp+glutamine=asparagineL+glutamate+amp+ppi

	pH 5	pH 6	pH 7	pH 8	pH 9
$\Delta_r G'^\circ$	-217.50	-208.40	-202.50	-197.70	-195.40
$\Delta_r H'^\circ$	-312.00	-311.50	-310.40	-308.80	-306.40
$\Delta_r S'^\circ$	-317.10	-345.60	-362.10	-372.50	-372.10
$\Delta_r N_H$	1.83	1.27	0.92	0.67	0.17
$\log K'$	38.11	36.52	35.48	34.64	34.24

In[95]:= **rxSummaryT["EC 6.4.1.1 Pyruvate carboxylate",**
     "atp+pyruvate+co2tot=oxaloacetate+adp+pi",
     oxaloacetateGT + adpGT + piGT - (atpGT + pyruvateGT + co2totGT), 298.15, {5, 6, 7, 8, 9}, .25]

EC 6.4.1.1 Pyruvate carboxylate

atp+pyruvate+co2tot=oxaloacetate+adp+pi

	pH 5	pH 6	pH 7	pH 8	pH 9
$\Delta_r G'^\circ$	0.75	-4.31	-8.81	-14.06	-19.36
$\Delta_r H'^\circ$	24.74	22.22	21.11	21.56	19.53
$\Delta_r S'^\circ$	80.45	88.98	100.40	119.50	130.50
$\Delta_r N_H$	-0.96	-0.80	-0.85	-0.96	-0.85
$\log K'$	-0.13	0.76	1.54	2.46	3.39

Since all these enzyme-catalyzed reactions have been written in the direction with $K' > 1$ at 298.15 K, pH 7, and 0.25 M ionic strength, we can see the separate contributions of $\Delta_r H'^\circ$ and $\Delta_r S'^\circ$. These contributions come from different sources. $\Delta_r H'^\circ$ measures energy and a negative value indicates that heat is produced. $\Delta_r S'^\circ$ is configurational. If $\Delta_r H'^\circ < 0$ and $\Delta_r S'^\circ > 0$ at specified temperature, pH, and ionic strength these properties both contribute to making $K' > 1$. When this is not the case, either $\Delta_r H'^\circ$ or $\Delta_r S'^\circ$ will predominate in determining K'. If $\Delta_r H'^\circ < 0$ and $\Delta_r H'^\circ < T\Delta_r S'^\circ$, then $\Delta_r H'^\circ$ predominates. Otherwise, $\Delta_r S'^\circ$ predominates. For this set of 90 enzyme-catalyzed reactions, $\Delta_r H'^\circ$ and $\Delta_r S'^\circ$ both contribute to making $K' > 1$ in 31 cases. $\Delta_r H'^\circ$ predominates in 45 cases, and $\Delta_r S'^\circ$ predominates in 14 cases.

The values of the reaction properties depend on the transformed properties of the reactants, and the transformed properties of the reactants depend on the species properties in the database. Hopefully, someday these species properties will be understood in terms of quantum mechanics and statistical mechanics, but the hydration of these species in aqueous solution has to be taken into account.

13.3 Calculations of Transformed Reaction Properties at Other Temperatures

The calculations in the preceding section can readily be repeated at other temperatures in the range 273.15 K to about 313.15 K using **rxSummaryT**, but there are two other ways that the effects of temperature, pH, and ionic strength on various standard transformed properties of enzyme-catalyzed reactions can be calculated. The first is to use the functions like atpHT, atpST, and atpNHT that are available in BasicBiochemData3 (4). The second is to use partial derivatives to obtain this information from the functions of temperature, pH, and ionic strength that yield $\Delta_r G'^\circ$ for reactions. These two methods are demonstrated for the hydrolysis of atp to adp. The function for the transformed reaction Gibbs energy is named as follows for the hydrolysis of atp:

In[96]:= **ec3x6x1x3GT = adpGT + piGT - (atpGT + h2oGT);**

In[97]:= **ec3x6x1x3GT /. t → 298.15 /. pH → 7 /. is → .25**

Out[97]= -36.0353

The function for the standard reaction enthalpy is named as follows:

In[98]:= **ec3x6x1x3HT = adpHT + piHT - (atpHT + h2oHT);**

In[99]:= **ec3x6x1x3HT /. t → 298.15 /. pH → 7 /. is → .25**

Out[99]= -23.0761

The function for the standard reaction entropy in kJ K^{-1} mol^{-1} is named as follows:

In[100]:=
      ec3x6x1x3ST = adpST + piST - (atpST + h2oST);

In[101]:=
      ec3x6x1x3ST /. t → 298.15 /. pH → 7 /. is → .25

Out[101]=
      0.0434652

The change in the binding of hydrogen ions is obtained as follows:

In[102]:=
      ec3x6x1x3NHT = adpNHT + piNHT - (atpNHT + h2oNHT);

In[103]:=
      ec3x6x1x3NHT /. t → 298.15 /. pH → 7 /. is → .25

Out[103]=
      -0.741953

The second method is to use partial derivatives. The standard transformed reaction enthalpy is obtained by use of the Gibbs-Helmholtz equation:

In[104]:=
      -t^2 * D[((ec3x6x1x3GT / t)), t] /. t → 298.15 /. pH → 7 /. is → .25

Out[104]=
      -23.0761

The standard transformed reaction entropy is obtained as follows:

```
In[105]:=
        -D[ec3x6x1x3GT, t] /. t → 298.15 /. pH → 7 /. is → .25

Out[105]=
        0.0434652
```

This value is in kJ K^{-1} mol^{-1}. The change in binding of hydrogen ions is obtained as follows:

```
In[106]:=
        (1 / (8.31451 * (t / 1000) *Log[10])) *D[ec3x6x1x3GT, pH] /. t → 298.15 /. pH → 7 /. is → .25

Out[106]=
        -0.741953
```

It is also convenient to calculate logK'.

```
In[107]:=
        -ec3x6x1x3GT / (8.31451 * (t / 1000) *Log[10]) /. t → 298.15 /. pH → 7 /. is → .25

Out[107]=
        6.31307
```

The functions of temperature, pH, and ionic strength for the reactions discussed in the preceding section are now calculated. They can be used to calculate these thermodynamic properties for the other 89 reactions, or to make tables or plots.

Class 1: Oxidoreductases

```
In[108]:=
        ec1x1x1x1aGT = ethanolGT + nadoxGT - (acetaldehydeGT + nadredGT);

In[109]:=
        ec1x1x1x1cGT = propanol2GT + nadoxGT - (acetoneGT + nadredGT);

In[110]:=
        ec1x1x1x27GT = lactateGT + nadoxGT - (pyruvateGT + nadredGT);

In[111]:=
        ec1x1x1x37GT = malateGT + nadoxGT - (oxaloacetateGT + nadredGT);

In[112]:=
        ec1x1x1x39GT = malateGT + nadoxGT + h2oGT - (pyruvateGT + co2totGT + nadredGT);

In[113]:=
        ec1x1x1x80GT = propanol2GT + nadpoxGT - (acetoneGT + nadpredGT);

In[114]:=
        ec1x1x3x3GT = oxaloacetateGT + h2o2aqGT - (malateGT + o2aqGT);

In[115]:=
        ec1x1x99x7GT = malateGT + pyruvateGT - (lactateGT + oxaloacetateGT);

In[116]:=
        ec1x2x1x2GT = co2totGT + nadredGT - (formateGT + nadoxGT + h2oGT);

In[117]:=
        ec1x2x1x3GT = acetateGT + nadredGT - (acetaldehydeGT + nadoxGT + h2oGT);
```

```
In[118]:=
        ec1x2x1x43GT = co2totGT + nadredGT - (formateGT + nadoxGT + h2oGT);

In[119]:=
        ec1x4x1x1GT = alanineGT + nadoxGT + h2oGT - (pyruvateGT + nadredGT + ammoniaGT);

In[120]:=
        ec1x4x1x2GT = glutamateGT + nadoxGT + h2oGT - (ketoglutarateGT + nadredGT + ammoniaGT);

In[121]:=
        ec1x4x1x3GT = glutamateGT + nadpoxGT + h2oGT - (ketoglutarateGT + nadpredGT + ammoniaGT);

In[122]:=
        ec1x4x3x11GT = ketoglutarateGT + h2o2aqGT + ammoniaGT - (glutamateGT + o2aqGT + h2oGT);

In[123]:=
        ec1x4x3x16GT = oxaloacetateGT + h2o2aqGT + ammoniaGT - (aspartateGT + o2aqGT + h2oGT);

In[124]:=
        ec1x6x1x1GT = nadpredGT + nadoxGT - (nadpoxGT + nadredGT);

In[125]:=
        ec1x7x1x1GT = nitriteGT + nadoxGT + h2oGT - (nitrateGT + nadredGT);

In[126]:=
        ec1x7x1x3GT = nitriteGT + nadpoxGT + h2oGT - (nitrateGT + nadpredGT);

In[127]:=
        ec1x7x1x4GT = ammoniaGT + 3 * nadpoxGT + 2 * h2oGT - (nitriteGT + 3 * nadpredGT);

In[128]:=
        ec1x7x99x6GT = n2aqGT + nadoxGT + h2oGT - (n2oaqGT + nadredGT);

In[129]:=
        ec1x8x1x2GT = h2saqGT + 3 * nadpoxGT + 3 * h2oGT - (sulfiteGT + 3 * nadpredGT);

In[130]:=
        ec1x8x3x1GT = sulfateGT + h2o2aqGT - (sulfiteGT + o2aqGT + h2oGT);

In[131]:=
        ec1x11x1x1GT = nadoxGT + 2 * h2oGT - (nadredGT + h2o2aqGT);

In[132]:=
        ec1x11x1x6GT = o2aqGT + 2 * h2oGT - (2 * h2o2aqGT);

In[133]:=
        ec1x11x1x8GT = i2crGT + 2 h2oGT - (2 * iodideionGT + h2o2aqGT);

In[134]:=
        ec1x12x1x2GT = nadredGT - (h2aqGT + nadoxGT);

In[135]:=
        ec1x13x1x18GT = sulfiteGT - (sulfurcrGT + o2aqGT + h2oGT);

In[136]:=
        ec1x13x12x4GT = acetateGT + co2totGT - (lactateGT + o2aqGT);

In[137]:=
        ec1x14x13x25GT = methanolGT + nadpoxGT + h2oGT - (methaneaqGT + nadpredGT + o2aqGT);
```

In[138]:=
```
ec1x16x1x7GT = 2 * ferrousGT + nadoxGT - (2 * ferricGT + nadredGT) ;
```

In[139]:=
```
ec1x16x3x1GT = 4 * ferricGT + 2 * h2oGT - (4 * ferrousGT + o2aqGT) ;
```

Class 2: Transferases

In[140]:=
```
ec2x4x1x7GT = fructoseGT + glucose6phosGT - (sucroseGT + piGT) ;
```

In[141]:=
```
ec2x4x2x1GT = adenosineGT + piGT - (adenineGT + ribose1phosGT) ;
```

In[142]:=
```
ec2x6x1x1GT = oxaloacetateGT + glutamateGT - (aspartateGT + ketoglutarateGT) ;
```

In[143]:=
```
ec2x6x1x2GT = pyruvateGT + glutamateGT - (alanineGT + ketoglutarateGT) ;
```

In[144]:=
```
ec2x7x1x2GT = adpGT + glucose6phosGT - (atpGT + glucoseGT) ;
```

In[145]:=
```
ec2x7x1x4GT = adpGT + fructose6phosGT - (atpGT + fructoseGT) ;
```

In[146]:=
```
ec2x7x1x7GT = adpGT + mannose6phosGT - (atpGT + mannoseGT) ;
```

In[147]:=
```
ec2x7x1x11GT = adpGT + fructose16phosGT - (atpGT + fructose6phosGT) ;
```

In[148]:=
```
ec2x7x1x13GT = adpGT + ribose5phosGT - (atpGT + riboseGT) ;
```

In[149]:=
```
ec2x7x1x20GT = adpGT + ampGT - (atpGT + adenosineGT) ;
```

In[150]:=
```
ec2x7x1x23GT = adpGT + nadpoxGT - (atpGT + nadoxGT) ;
```

In[151]:=
```
ec2x7x1x30GT = adpGT + glycerol3phosGT - (atpGT + glycerolGT) ;
```

In[152]:=
```
ec2x7x1x40GT = atpGT + pyruvateGT - (adpGT + pepGT) ;
```

In[153]:=
```
ec2x7x1x73GT = adpGT + impGT - (atpGT + inosineGT) ;
```

In[154]:=
```
ec2x7x1x86GT = adpGT + nadpredGT - (atpGT + nadredGT) ;
```

In[155]:=
```
ec2x7x1x90GT = piGT + fructose16phosGT - (ppiGT + fructose6phosGT) ;
```

In[156]:=
```
ec2x7x1x142GT = glycerol3phosGT + glucoseGT - (glycerolGT + glucose6phosGT) ;
```

```
In[157]:=
        ec2x7x1x146GT = ampGT + fructose16phosGT - (adpGT + fructose6phosGT);

In[158]:=
        ec2x7x1x147GT = ampGT + glucose6phosGT - (adpGT + glucoseGT);

In[159]:=
        ec2x7x4x3GT = 2 * adpGT - (atpGT + ampGT);
```

Class 3: Hydrolases

```
In[160]:=
        ec3x1x1x1GT = ethanolGT + acetateGT - (ethylacetateGT + h2oGT);

In[161]:=
        ec3x1x3x9GT = glucoseGT + piGT - (glucose6phosGT + h2oGT);

In[162]:=
        ec3x1x3x11GT = fructose6phosGT + piGT - (fructose16phosGT + h2oGT);

In[163]:=
        ec3x1x3x31GT = glycerol3phosGT + h2oGT - (glycerolGT + piGT);

In[164]:=
        ec3x2x1x20GT = 2 * glucoseGT - (maltoseGT + h2oGT);

In[165]:=
        ec3x2x1x48GT = glucoseGT + fructoseGT - (sucroseGT + h2oGT);

In[166]:=
        ec3x2x1x108GT = glucoseGT + galactoseGT - (lactoseGT + h2oGT);

In[167]:=
        ec3x2x2x4GT = adenineGT + ribose5phosGT - (ampGT + h2oGT);

In[168]:=
        ec3x2x2x7GT = adenineGT + riboseGT - (adenosineGT + h2oGT);

In[169]:=
        ec3x5x1x1GT = aspartateGT + ammoniaGT - (asparagineLGT + h2oGT);

In[170]:=
        ec3x5x1x2GT = glutamateGT + ammoniaGT - (glutamineGT + h2oGT);

In[171]:=
        ec3x5x1x5GT = co2totGT + 2 * ammoniaGT - (ureaGT + 2 * h2oGT);

In[172]:=
        ec3x5x4x4GT = inosineGT + ammoniaGT - (adenosineGT + h2oGT);

In[173]:=
        ec3x5x4x6GT = impGT + ammoniaGT - (ampGT + h2oGT);

In[174]:=
        ec3x5x4x7GT = idpGT + ammoniaGT - (adpGT + h2oGT);

In[175]:=
        ec3x5x4x18GT = itpGT + ammoniaGT - (atpGT + h2oGT);
```

In[176]:=
```
ec3x6x1x1GT = 2 * piGT - (ppiGT + h2oGT);
```

In[177]:=
```
ec3x6x1x3GT = adpGT + piGT - (atpGT + h2oGT);
```

In[178]:=
```
ec3x6x1x5GT = ampGT + 2 * piGT - (atpGT + 2 * h2oGT);
```

In[179]:=
```
ec3x6x1x8GT = ampGT + ppiGT - (atpGT + h2oGT);
```

Class 4 Lyases

In[180]:=
```
ec4x1x1x1GT = co2totGT + acetaldehydeGT - (pyruvateGT + h2oGT);
```

In[181]:=
```
ec4x1x1x2GT = co2totGT + alanineGT - (aspartateGT + h2oGT);
```

In[182]:=
```
ec4x1x2x36GT = lactateGT - (formateGT + acetaldehydeGT);
```

In[183]:=
```
ec4x1x3x36GT = citrateGT - (acetateGT + oxaloacetateGT);
```

In[184]:=
```
ec4x1x99x1GT = indoleGT + pyruvateGT - (tryptophanLGT + h2oGT);
```

In[185]:=
```
ec4x2x1x2GT = malateGT - (fumarateGT + h2oGT);
```

In[186]:=
```
ec4x3x1x1GT = aspartateGT - (fumarateGT + ammoniaGT);
```

5. Isomerases

In[187]:=
```
ec5x3x1x5GT = xyloseGT - xyluloseGT;
```

In[188]:=
```
ec5x3x1x6GT = ribose5phosGT - ribose1phosGT;
```

In[189]:=
```
ec5x3x1x7GT = fructoseGT - mannoseGT;
```

In[190]:=
```
ec5x3x1x8GT = fructose6phosGT - mannose6phosGT;
```

In[191]:=
```
ec5x3x1x9GT = glucose6phosGT - fructose6phosGT;
```

In[192]:=
```
ec5x3x1x20GT = riboseGT - ribuloseGT;
```

6. Ligases

In[193]:=
```
ec6x3x1x1GT = asparagineLGT + ampGT + ppiGT - (aspartateGT + atpGT + ammoniaGT);
```

```
In[194]:=
        ec6x3x1x2GT = glutamineGT + adpGT + piGT - (glutamateGT + atpGT + ammoniaGT);

In[195]:=
        ec6x3x1x4GT = asparagineLGT + adpGT + piGT - (aspartateGT + atpGT + ammoniaGT);

In[196]:=
        ec6x3x5x4GT = asparagineLGT + glutamateGT + ampGT + ppiGT - (aspartateGT + atpGT + glutamineGT);

In[197]:=
        ec6x4x1x1GT = oxaloacetateGT + adpGT + piGT - (pyruvateGT + atpGT + co2totGT);
```

13.4 Three Dimensional Plots

The best way to get an overview of the effects of temperature and pH is to construct 3D plots at specified ionic strength. That is done here for six properties of atp+h2o=adp+pi (EC 3.6.1.3).

```
In[198]:=
        plot1 = Plot3D[(ec3x6x1x3GT) /. is → .25,
          {pH, 5, 9}, {t, 273.15, 313.15}, AxesLabel -> {"pH", "T/K", " "},
          PlotLabel → "Δ_rG'°/kJ mol⁻¹", DisplayFunction → Identity];

In[199]:=
        plot2 = Plot3D[Evaluate[-t^2 * D[(ec3x6x1x3GT) /t, t] /. is → .25],
          {pH, 5, 9}, {t, 273.15, 313.15}, AxesLabel -> {"pH", "T/K", " "},
          PlotLabel → "Δ_rH'°/kJ mol⁻¹", DisplayFunction → Identity];

In[200]:=
        plot3 = Plot3D[Evaluate[-1000 * D[(ec3x6x1x3GT), t] /. is → .25],
          {pH, 5, 9}, {t, 273.15, 313.15}, AxesLabel -> {"pH", "T/K", " "},
          PlotLabel → "Δ_rS'°/J K⁻¹ mol⁻¹", DisplayFunction → Identity];

In[201]:=
        plot4 = Plot3D[Evaluate[(1 / (8.31451 * (t / 1000) *Log[10])) *D[(ec3x6x1x3GT), pH] /. is → .25]
          {pH, 5, 9}, {t, 273.15, 313.15}, AxesLabel -> {"pH", "T/K", " "},
          PlotLabel → "Δ_rN_H", DisplayFunction → Identity];

In[202]:=
        plot5 = Plot3D[Evaluate[D[-t^2 * D[(ec3x6x1x3GT) /t, t], pH] /. is → .25],
          {pH, 5, 9}, {t, 273.15, 313.15}, AxesLabel -> {"pH", "T/K", " "},
          PlotLabel → "(∂Δ_rH'°/∂pH)/kJ mol⁻¹", DisplayFunction → Identity];

In[203]:=
        plot6 = Plot3D[Evaluate[D[-1000 * D[(ec3x6x1x3GT), t], pH] /. is → .25],
          {pH, 5, 9}, {t, 273.15, 313.15}, AxesLabel -> {"pH", "T/K", " "},
          PlotLabel → "(∂Δ_rS'°/∂pH)/J K⁻¹ mol⁻¹", DisplayFunction → Identity];

In[204]:=
        Show[
          GraphicsArray[{{plot1, plot2}, {plot3, plot4}, {plot5, plot6}}, GraphicsSpacing -> .1]];
```

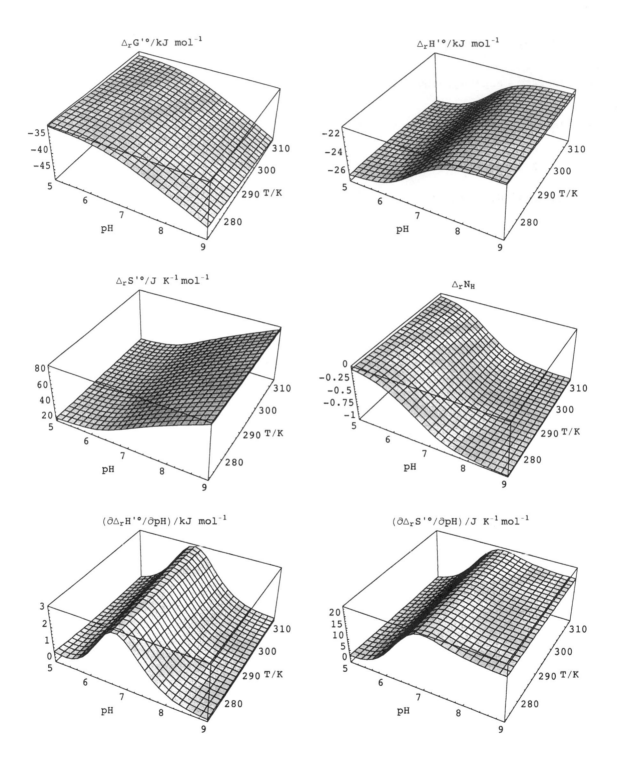

Figure 13.1 $\Delta_r G'^\circ$, $\Delta_r H'^\circ$, $\Delta_r S'^\circ$, $\Delta_r N_H$, $\partial\Delta_r H'^\circ/\partial pH$, and $\partial\Delta_r S'^\circ/\partial pH$ for the reaction atp+ h2o = adp + pi at 0.25 M ionic strength.

13.5 Discussion

This chapter has emphasized again the advantage of having $\Delta_r G'^o$ for an enzyme-catalyzed reaction as a function of temperature, pH, and ionic strength. If magnesium ions or other ions are bound by reactants, the free concentrations of more ions can be included as independent variables. This chapter has also emphasized the value of calorimetric data. More standard transformed enthalpies of reaction need to be measured so that temperature effects can be calculated for more reactions. The database can also be extended by use of reliable estimation methods based on species properties. This may be especially useful with larger biochemical reactants where reactive sites are nearly independent.

Calculations on a number of reactions have already revealed some errors in the database, which have been corrected. There are higher derivatives that are also interesting thermodynamic properties, but taking second derivatives with respect to the same variable is to be avoided because the functions of temperature, pH, and ionic strength are not known sufficiently accurately.

References

1. R. A. Alberty and R. N. Goldberg, Calculation of thermodynamic formation properties for the ATP series at specified pH and pMg, Biochem. 31, 10610-10615 (1992).
2. R. A. Alberty, Effect of temperature on the standard transformed thermodynamic properties of biochemical reactions with emphasis on the Maxwell equations, J. Phys. Chem. 107 B, 3631-3635 (2003). (Supplementary Information is available.)
3. R. A. Alberty, Thermodynamics of the hydrolysis of adenosine triphosphate as a function of temperature, pH pMg, and ionic strength, J. Phys. Chem. 107 B, 12324-12330 (2003).
4. R. A. Alberty, BasicBiochemData3, 2005.

In[205]:=

    http : // library.wolfram.com / infocenter / MathSource / 5704

In[205]:=

Chapter 14   Thermodynamics of the Binding of Ligands by Proteins

14.1 Introduction

    Chapter 7 of Thermodynamics of Biochemical Reactions (1) has treated a number of aspects of the binding of oxygen by hemoglobin from the point of view of the transformed Gibbs energy to make the pH an independent variable and the further transformed Gibbs energy to make pH and $[O_2]$ independent variables. The standard transformed Gibbs energies of formation of the various oxygenated forms of the tetramer of hemoglobin can be calculated, but in some ways it is more useful to consider the tetramer of hemoglobin as an entity at a specified concentration of molecular oxygen, just as atp is considered as an entity at a specified pH. In order to introduce the concentration of molecular oxygen as a natural variable, the following Legendre transform is used to define a further transformed Gibbs energy G'' (2):

$$G'' = G' - n_c(O)\,\mu(O_2) \tag{14.1-1}$$

where $n_c(O)$ is the amount of the oxygen component and $\mu(O_2)$ is the chemical potential of dissolved molecular oxygen. At specified concentration of molecular oxygen, the five forms of hemoglobin are pseudoisomers, and they have the same further transformed Gibbs energy of formation at equilibrium. The further transformed thermodynamic properties of the tetramer can be calculated from experimental measurements of the fractional saturation, but in order to interpret experimental data, it is necessary to provide for the partial dissociation of tetramer $\alpha_2\,\beta_2$ into dimers $\alpha\beta$ (3). Seven apparent equilibrium constants are required to describe experimental data, and it is shown that all seven can be determined using limiting forms (4).

    The previous book (1) has discussed a number of aspects of protein-ligand binding that are not treated in this chapter, which is oriented toward a simple system in which one molecule of ligand is bound.

14.2 Fundamental Equation the Transformed Gibbs Energy

The following simple reaction system has some of the characteristics of the hemoglobin-oxygen system that is much more complex because four molecules of oxygen are bound by the tetramer and the tetramer is partly dissociated. The system discussed here is like myoglobin except that myoglobin does not have a Bohr effect.

$$
\begin{array}{l}
pK_{1M} \quad pK_{2M} \\
M = HM = H_2 M \\
\| \\
MO_2 = HMO_2 = H_2 M O_2 \\
\quad pK_{1MO2} \quad pK_{2MO2}
\end{array}
\qquad (14.2\text{-}1)
$$

The dissociation constant involving the most basic forms is given by $K = [M][O_2]/[MO_2]$. The thermodynamics of this simple system can be treated in three ways:
(1) It can be treated in terms of species using the Gibbs energy G.
(2) At a specified pH it can be treated in terms of the reaction $H_{av} MO_2 = H_{av} M + O_2$ using the transformed Gibbs energy G'. $H_{av} MO_{2\,av}$ is the pseudoisomer group MO_2, HMO_2, and $H_2 M O_2$. $H_{av} M$ is the pseudoisomer group M, HM, and $H_2 M$.
(3) At a specified pH and specified concentration of molecular oxygen, the system in equation 14.2-1 can be treated in terms of the reactant $H_{av} MO_{2\,av}$ using the further transformed Gibbs energy G''. The subscripts av indicate average numbers of hydrogen atoms and oxygen molecules bound at equilibrium.

The possibility of using this third type of treatment led Wyman (5,6,7) to introduce the binding potential that he represented by the Russian L (see Appendix to this chapter). The second way to treat the reaction system 14.2-1 will be used in this section, and the third way will be discussed in the next section. Three proteins are included in BasicBiochemData3 (cytochrome c, ferredoxin, and thioredoxin), but they are not known to have acid groups linked to the redox site.

Reaction system 14.2-1 involving eight species and the five reactions can be discussed in terms of the fundamental equation for G for chemical reactions that was discussed in Section 3.1. In the absence of experimental means to determine the equilibrium concentrations of these eight species, the standard Gibbs energies of formation of the six species in addition to H^+ and $O_2(aq)$ cannot be determined directly (8). However, they can in principle be determined by the methods based on the transformed Gibbs energy G' and the further transformed Gibbs energy G''.

When the pH is specified, the criterion for spontaneous change and equilibrium is provided by the transformed Gibbs energy and the reactants can be taken to be $H_{av} M$, $H_{av} MO_2$, and O_2 that are involved in the following biochemical reaction.

$$
H_{av} MO_2 = H_{av} M + O_2 \qquad (14.2\text{-}2)
$$

The fundamental equation for the system involving this reaction is given by (9)

$$
dG' = -S'dT + VdP + \Delta_f G'(H_{av} M)\,dn'(H_{av} M) + \Delta_f G'(H_{av} MO_2)\,dn'(H_{av} MO_2) + \Delta_f G'(O_2)\,dn'(M) + RT\ln(10)\,n_c(H)\,pH
$$
$$
(14.2\text{-}3)
$$

Introducing the extent ξ of reaction yields

$$
dG' = -S'dT + VdP + \Delta_r G'\,d\xi + RT\ln(10)\,n_c(H)\,dpH \qquad (14.2\text{-}4)
$$

This fundamental equation shows that $S' = -\partial G'/\partial T$, $\Delta_r G' = \partial G'/\partial\xi$, and $RT\ln(10)n_c(H) = \partial G'/\partial pH$. These equations are not directly useful because there is no experimental way to determine G'; **but Maxwell relations for this and other fundamental equations do provide equations for experimental determinable properties.** For the system being discussed the standard transformed Gibbs energy of reaction is given by

$$\Delta_r G'^\circ = -RT\ln K'$$ (14.2-5)

where the apparent equilibrium constant K' is given by

$$K' = \frac{[H_{av} M][O_2]}{[H_{av} MO_2]} = K\frac{1+10^{pK1M-pH}+10^{pK1M+pK2M-2pH}}{1+10^{pK1MO2-pH}+10^{pK1+pK2MO2-2pH}}$$ (14.2-6)

The apparent equilibrium constant at very high pH is equal to $K = [M][O_2]/[MO_2]$, $pK_1(MO_2)$ is the first pK of the oxygenated macromolecule, and $pK_2(MO_2)$ is the second.

Maxwell equations provide the connection with experimentally determined properties and relations between the properties. Ignoring the VdP term, fundamental equation 14.2-4 leads to five Maxwell equations. As we have seen earlier (Chapter 3), these Maxwell equations can be written as

$$\Delta_r N_H = \frac{1}{RT\ln(10)}\frac{\partial\Delta_r G'^\circ}{\partial pH}$$ (14.2-7)

$$\Delta_r S'^\circ = -\frac{\partial\Delta_r G'^\circ}{\partial T}$$ (14.2-8)

$$\Delta_r H'^\circ = -T^2\frac{\partial(\Delta_r G'^\circ/T)}{\partial T}$$ (14.2-9)

$$\frac{\partial\Delta_r S'^\circ}{\partial pH} = -R\ln(10)\frac{\partial(T\Delta_r N_H)}{\partial T}$$ (14.2-10)

$$\frac{\partial\Delta_r H'^\circ}{\partial pH} = -RT^2\ln(10)\frac{\partial\Delta_r N_H}{\partial T}$$ (14.2-11)

Equation 14.2-7 is of special interest because of the connection it provides between experimental measurements and standard thermodynamic properties. The substitution of equations 14.2-5 and 14.2-6 into equation 14.2-7 yields

$$\Delta_r N_H = -\frac{1}{\ln(10)}\frac{\partial\ln P(H_{av} M)}{\partial pH} + \frac{1}{\ln(10)}\frac{\partial\ln P(H_{av} MO_2)}{\partial pH} = \overline{N}_H(H_{av} M) - \overline{N}_H(H_{av} MO_2)$$ (14.2-12)

were $P(H_{av} M)$ and $P(H_{av} MO_2)$ are binding polynomials for hydrogen ions. The average bindings of hydrogen ions in $H_{av} M$ and $H_{av} MO_2$ can be determined by titrating the macromolecule in the absence of oxygen and in the presence of a high enough concentration of molecular oxygen to essentially saturate it. It is of interest to note that German and Wyman (10) showed how it is possible to obtain relative values of the oxygen affinity of hemoglobin as a function of pH from titration curves for oxygenated and deoxygenated hemoglobin. This type of integration was also used by Antonini, et al (11). The two equations inherent in equation 14.2-12 can be integrated.

$$\int \mathrm{d}\ln P(\mathrm{H_{av}\,M}) - \ln(10)\int \overline{N}_\mathrm{H}(\mathrm{H_{av}\,M})\mathrm{dpH} \tag{14.2-13}$$

$$\int \mathrm{d}\ln P(\mathrm{H_{av}\,MO_2}) - \ln(10)\int \overline{N}_\mathrm{H}(\mathrm{H_{av}\,MO_2})\mathrm{dpH} \tag{14.2-14}$$

These integrations yield $P(\mathrm{H_{av}\,M})$ and $P(\mathrm{H_{av}\,MO_2})$ as functions of pH, but with integration constants. This is the first example of a Maxwell equation providing an equation for an experimental property in protein-ligand binding. The next section deals with a second example.

14.3 Fundamental Equation the Further Transformed Gibbs Energy

When the pH and $[\mathrm{O_2}]$ are specified, the criterion for spontaneous change and equilibrium is provided by G'', and the reaction system 4.2-1 is represented by a single reactant $\mathrm{H_{av}\,MO_{2\,av}}$, that is the pseudoisomer group $\mathrm{H_{av}\,M}$ and $\mathrm{H_{av}\,MO_2}$. The fundamental equation for the further transformed Gibbs energy is given by

$$\mathrm{d}G'' = -S''\mathrm{d}T + V\mathrm{d}P + \Delta_\mathrm{f} G''(\mathrm{H_{av}\,MO_{2\,av}})\,\mathrm{d}n''(\mathrm{H_{av}\,MO_{2\,av}}) + RT\ln(10)\,n_c(\mathrm{H})\,\mathrm{dpH} - \{RTn_c(\mathrm{O_2})/[\mathrm{O_2}]\}\mathrm{d}[\mathrm{O_2}] \tag{14.3-1}$$

At constant T, P, pH, and $[\mathrm{O_2}]$ this equation integrates to $G'' = \Delta_\mathrm{f} G''(\mathrm{H_{av}\,MO_{2\,av}})\,n''(\mathrm{H_{av}\,MO_{2\,av}})$, and so there is a single reactant. Ignoring the $V\mathrm{d}P$ term this fundamental equation leads to the following eight Maxwell equations:

$$\overline{N}_{\mathrm{O_2}} = -\frac{[\mathrm{O_2}]}{RT}\frac{\partial \Delta_\mathrm{f} G''^\circ}{\partial[\mathrm{O_2}]} \tag{14.3-2}$$

$$\overline{N}_\mathrm{H} = \frac{1}{RT\ln(10)}\frac{\partial \Delta_\mathrm{f} G''^\circ}{\partial \mathrm{pH}} \tag{14.3-3}$$

$$\Delta_\mathrm{f} S''^\circ = -\frac{\partial \Delta_\mathrm{f} G''^\circ}{\partial T} \tag{14.3-4}$$

$$[\mathrm{O_2}]\frac{\partial \Delta_\mathrm{f} S''^\circ}{\partial[\mathrm{O_2}]} = -R\ln(10)\frac{\partial(T\,\overline{N}_\mathrm{H})}{\partial T} \tag{14.3-5}$$

$$\frac{\partial \Delta_\mathrm{f} S''^\circ}{\partial \mathrm{pH}} = -R\ln(10)\frac{\partial(T\,\overline{N}_\mathrm{H})}{\partial T} \tag{14.3-6}$$

$$\Delta_\mathrm{f} H''^\circ = -T^2\frac{\partial(\Delta_\mathrm{f} G''^\circ/T)}{\partial T} \tag{14.3-7}$$

$$\frac{\partial \Delta_\mathrm{f} H''^\circ}{\partial \mathrm{pH}} = -RT^2\ln(10)\frac{\partial \overline{N}_\mathrm{H}}{\partial T} \tag{14.3-8}$$

$$\ln(10)[\mathrm{O_2}]\frac{\partial \overline{N}_\mathrm{H}}{\partial[\mathrm{O_2}]} = -\frac{\partial \overline{N}_{\mathrm{O_2}}}{\partial \mathrm{pH}} \tag{14.3-9}$$

This last equation is referred to as a **reciprocal relation**.

The first of these Maxwell equations is especially important because it provides the connection between $\Delta_f G''^\circ$ and the experimental quantities $\overline{N}_{O_2}$ and $[O_2]$ at equilibrium as a function of temperature and pH.

The standard further transformed Gibbs energy of formation of $H_{av} MO_{2\,av}$ is given by $\Delta_f G''^\circ(H_{av} MO_{2\,av})$.

$$\Delta_f G''^\circ(H_{av} MO_{2\,av}) = \Delta_f G''^\circ(H_{av} M) - RT\ln(1 + [O_2]/K') \qquad (14.3\text{-}10)$$

$\Delta_f G''^\circ(H_{av} M)$ is equal to $\Delta_f G'^\circ(H_{av} M)$ because this reactant does not contain oxygen. The standard transformed Gibbs energy of formation of $\Delta_f G'^\circ(H_{av} M)$ is given by

$$\Delta_f G'^\circ(H_{av} M) = \Delta_f G_1{}'^\circ - RT\ln(1 + 10^{pK1M-pH} + 10^{pK1M+pK2M-2\,pH}) \qquad (14.3\text{-}11)$$

where $\Delta_f G_1{}'^\circ$ is for M. Since $\Delta_f G^\circ(M) = 0$ by convention, and M does not contain any dissociable hydrogen atoms, $\Delta_f G_1{}'^\circ = 0$. Substituting this equation into equation 14.3-10 yields

$$\Delta_f G''^\circ(H_{av} MO_{2\,av}) = -RT\ln(1 + 10^{pK1M-pH} + 10^{pK1M+pK2M-2\,pH}) - RT\ln(1 + [O_2]/K') \qquad (14.3\text{-}12)$$

Taking the derivative of $\Delta_f G''^\circ(H_{av} MO_{2\,av})$ with respect to the concentration of molecular oxygen yields

$$\frac{\partial \Delta_f G''^\circ}{\partial [O_2]} = -\frac{RT/K'}{1 + [O_2]/K'} \qquad (14.3\text{-}13)$$

Substituting this equation into Maxwell equation 14.3-2 yields

$$\overline{N}_{O_2} = \frac{[O_2]/K'}{1 + [O_2]/K'}$$

$$(14.3\text{-}14)$$

Thus experimental determinations of $\overline{N}_{O_2}$ as a function of $[O_2]$ yields K' as a function of $[O_2]$. The expression for K' given by equation 14.2-6 shows that its dependence on temperature and pH contains all the thermodynamic information on the system. If the apparent equilibrium constant can be determined as a function of pH spectrophometrically, for example, the values of K and the four pKs can be calculated (12).

14.4 Three-dimensional Plots

Since the thermodynamic properties for the binding of a ligand by a protein are functions of temperature, pH, and the concentration of the ligand, a useful way to visualize these functions is to use three-dimensional plots at desired temperatures. The standard further transformed Gibbs energy of $H_{av} MO_{2\,av}$ is given by equation 14.3-12. The following program (13) derives the function of temperature, pH, and concentration of molecular oxygen that yields $\Delta_f G''^\circ$.

```
In[2]:= calcstdfrtrGe := Module[{pK1M, pK2M, pK1MO2, pK2MO2, k, pM, pMO2, kprimeO2},
          (*This program derives the function of temperature, pH,
           and concentration of molecular oxygen that yields the standard further transformed
           Gibbs energy of formation of HavMO2av.  Energies are in joules per mole.*)
          pK1M = 7.85 - (37.7 * 10^3 / (8.3145 * Log[10])) * (1 / 293.15 - 1 / t);
          pK2M = 5.46 - (-6.3 * 10^3 / (8.3145 * Log[10])) * (1 / 293.15 - 1 / t);
          pK1MO2 = 6.67 - (37.7 * 10^3 / (8.3145 * Log[10])) * (1 / 293.15 - 1 / t);
          pK2MO2 = 6.04 - (-6.3 * 10^3 / (8.3145 * Log[10])) * (1 / 293.15 - 1 / t);
          k = 10^-5 * Exp[(60.7 * 10^3 / 8.3145) * (1 / 293.15 - 1 / t)];
          pM = (1 + 10^(-pH + pK1M) + 10^(-2 * pH + pK1M + pK2M));
          pMO2 = (1 + 10^(-pH + pK1MO2) + 10^(-2 * pH + pK1MO2 + pK2MO2));
          kprimeO2 = k * pM / pMO2;
          -8.3145 * t * Log[pM] - 8.3145 * t * Log[1 + co2 / kprimeO2]]
```

The standard enthalpies of the dissociation reactions are those determined by Antonini, et al (11) for horse hemoglobin. Their pKs are based on the assumption that the oxygen-linked groups are independent. However, pK_1 and pK_2 in equation 14.2-1 are the usual thermodynamic pKs independent of any assumptions. Actually some of the acid dissociations are cooperative, which indicates that they are not independent. The pKs used in the current calculations were obtained using

$$pK_1 = \log(10^{pK_\alpha} + 10^{pK_\beta}) \tag{14.4-1}$$

$$pK_2 = pK_\alpha + pK_\beta - pK_1 \tag{14.4-2}$$

where pK_α and pK_β are the pKs based on the assumption that the oxygen-linked groups are independent (14). When the pKs for a dibasic acid are far apart, $pK_1 = pK_\alpha$ and $pK_2 = pK_\beta$, but pK_1 may be quite different from pK_α and pK_2 may be quite different from pK_β when pK_1 and pK_2 ae close. If pK_1 and pK_2 are closer than 0.60, the acid dissociation is cooperative (9).

The program **calcstdfrtrGe** can be used to make tables or 2D plots or 3D plots. Other standard further transformed properties can be calculated by use of Maxwell equations 14.3-2 to 14.3-9.

```
In[3]:= plot1 = Plot3D[Evaluate[(calcstdfrtrGe / 1000) /. t -> 293.15],
          {co2, 10^-10, .0001}, {pH, 5, 9}, AxesLabel → {"[O2]", "pH", ""},
          PlotLabel -> "ΔfG''°/kJ mol^-1", DisplayFunction → Identity];

In[4]:= plot2 = Plot3D[Evaluate[-(co2 / (8.31452 * t)) * D[calcstdfrtrGe, co2] /. t -> 293.15],
          {co2, 10^-10, .0001}, {pH, 5, 9}, AxesLabel → {"[O2]", "pH", ""},
          PlotLabel -> "N̄o2", DisplayFunction → Identity];

In[5]:= plot3 = Plot3D[Evaluate[(1 / (8.31452 * t * Log[10])) * D[calcstdfrtrGe, pH] /. t -> 293.15],
          {co2, 10^-10, .0001}, {pH, 5, 9}, AxesLabel → {"[O2]", "pH", ""},
          PlotLabel -> "N̄H", DisplayFunction → Identity];

In[6]:= plot4 = Plot3D[Evaluate[-t^2 * D[(calcstdfrtrGe / (1000 * t)), t] /. t -> 293.15],
          {co2, 10^-10, .0001}, {pH, 5, 9}, AxesLabel → {"[O2]", "pH", ""},
          PlotLabel -> "ΔfH''°/kJ mol^-1", DisplayFunction → Identity];

In[7]:= plot5 = Plot3D[Evaluate[-D[calcstdfrtrGe / 1000, t] /. t -> 293.15],
          {co2, 10^-10, .0001}, {pH, 5, 9}, AxesLabel → {"[O2]", "pH", ""},
          PlotLabel -> "ΔfS''°/kJ K^-1 mol^-1", DisplayFunction → Identity];

In[8]:= plot6 = Plot3D[Evaluate[D[(co2 / (8.31452 * t)) * D[calcstdfrtrGe, co2], pH] /. t -> 293.15],
          {co2, 10^-10, .0001}, {pH, 5, 9}, AxesLabel → {"[O2]", "pH", ""},
          PlotLabel -> "-∂N̄o2/∂pH", DisplayFunction → Identity];
```

The same plot is obtained by using the other side of the reciprocal relation 14.3-9.

```
In[9]:=  Plot3D[Evaluate[
           Log[10] * co2 * D[(1 / (8.31452 * t * Log[10])) * D[calcstdfrtrGe, pH], co2] /. t -> 293.15],
           {co2, 10^-10, .0001}, {pH, 5, 9}, AxesLabel -> {"[O₂]", "pH", ""},
           PlotLabel -> "ln(10)[O2]∂Ñ_H/∂[O2]", DisplayFunction -> Identity];

In[10]:=  Show[GraphicsArray[{{plot1, plot2}, {plot3, plot4}, {plot5, plot6}}]];
```

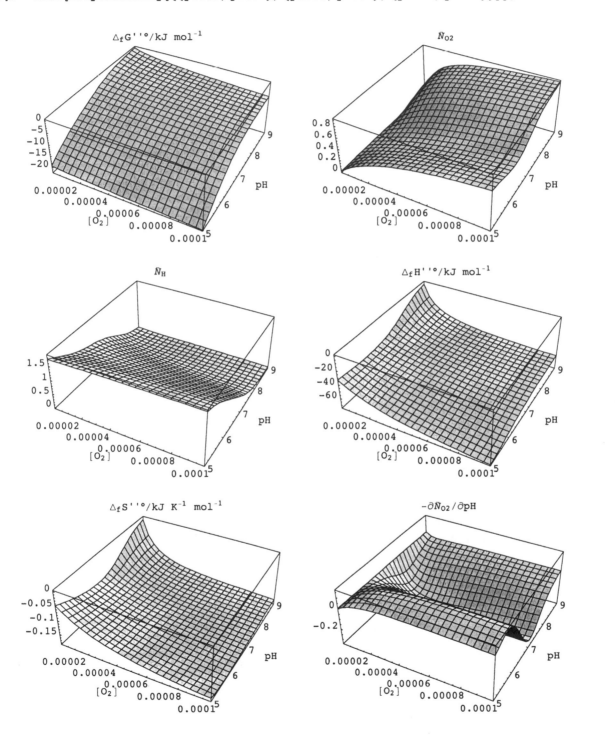

Figure 14.1 Plots of thermodynamic properties versus $[O_2]$ and pH at 298.15 K.

       With this program plots can be readily made at other temperatures in the range 273.15 K to about 313.15 K. Tables of these properties are available on the web (13).

Appendix

Interpretation of Acid Titration Curves for Proteins in the Presence and Absence of Ligand

In general the binding of ligands by proteins depends on the pH, and so the thermodynamics of the binding has to be treated with transformed thermodynamic properties. The binding of oxygen by hemoglobin is the prime example because the Bohr effect (7) plays a role in facilitating the "discharge" of gaseous carbon dioxide. This is also a good example because the binding of oxygen at the iron atom of heme is affected by two acidic groups in the binding site. It is an approximation to ignore the effects from the rest of the hemoglobin molecule, but it is good approximation in this case. However, the study of the Bohr effect is difficult because the binding of four molecules by the hemoglobin molecule is cooperative, which makes it difficult to determine the four binding binding constants and their pH dependencies. This chapter has concentrated on the binding of a ligand by a single binding site.

There are two ways to represent the titration curve for a protein or other polyprotic weak acid. The usual thermodynamic way used in Chapter 2 is to represent the average number of hydrogen ions bound with

$$\overline{N}_H = \frac{10^{pK1-pH} + 2 \times 10^{pK1+pK2-2\,pH} + ...}{1 + 10^{pK1-pH} + 10^{pK1+pK2-2\,pH} + ...} \tag{14A-1}$$

where the pKs are negative base-10 logarithms of acid dissociation constants. The other way is to assume that the acid dissociation constants are independent with $pK_1, pK_2,...$ Kappa's are used to indicate that the values are different. When the acid dissociations are independent, the expression for the average number of hydrogen ions bound is given by

$$\overline{N}_H = \left(\frac{10^{pK1-pH}}{1 + 10^{pK1-pH}} \right) \left(\frac{10^{pK1-pH}}{1 + 10^{pK1-pH}} \right) ... \tag{14A-2}$$

An acid titration curve can be represented by either one of these equations (14) This second form is useful in thinking about the titration curve of a protein that binds a ligand. The pKs of these independent groups can be divided into two classes, those that are not affected by the binding of the ligand and those that are.

The Binding Potential of Wyman

Wyman (5,6,7) introduced the binding potential, which he represented by the Russian L for linkage. This is a molar thermodynamic property that is defined by a Legendre transform that introduces the chemical potential of the ligand as an independent intensive property. The binding potential is given by

$$L = RT\ln\sum \nu_i \, e^{L_i/RT} \tag{14A-3}$$

where L_i is the binding potential of the ith form and ν_i is the equilibrium mole fraction of the ith form. The thermodynamic property used in this book that is comparable to L is $\Delta_f G''(H_{av} MO_{2\,av})$.

$$\Delta_f G''(H_{av} MO_{2\,av}) = - RT\ln\sum e^{-\Delta_f G_i''^\circ/RT} + RT\ln[H_{av} MO_{2\,av}] \tag{14A-4}$$

It may appear puzzling that the summation in this equation does not involve the equilibrium mole fractions that appear in equation 14A-3. This is because the exponential terms in equation 14A-4 involve the standard further transformed Gibbs energies of formation rather than $\Delta_f G_i''$. The standard further transformed Gibbs energies of formation $\Delta_f G_i''^\circ$ of the various forms are given by

$$\Delta_f G_i''^\circ = \Delta_f G_i'' - RT\ln[M_i] = \Delta_f G_i'' - RT\ln(r_i[H_{av} MO_{2\,av}]) \tag{14A-5}$$

where r_i is the equilibrium mole fraction of $H_{av} MO_{2\,av}$. Substituting equation 14A-5 in equation 14A-4 and setting $[H_{av} MO_{2\,av}] = 1$ M yields

$$\Delta_f G''(H_{av} MO_{2\,av}) = -RT \ln \sum r_i e^{-\Delta_f G_i''/RT} \qquad (14A\text{-}6)$$

that is the same as equation 14A-3 except for the sign difference. In the Wyman approach, a normalization is used to eliminate the term for the macromolecule in the fundamental equation, and so the concept of standard state for the macromolecule does not arise. This does not cause a problem in deriving equation 14A-3, but the concentration of the macromolecule is needed in treating an enzyme-catalyzed reaction involving a macromolecule.

References

1. R. A. Alberty, Thermodynamics of Biochemical Reactions, Wiley, Hoboken, NJ, 2003.
2. R. A. Alberty, Thermodynamics of the binding of ligands by macromolecules, Biophys. Chem. 62, 141-159 (1996).
3. F. C. Mills, M. L. Johnson, and J. K Ackers, Oxygen-linked subunit interactions in human hemoglobin: Experimental studies on the concentratioin dependence of oxygenation curves, Biochemistry 15, 5350 (1976).
4. R. A. Alberty, Determination of the seven apparent equilibrium constants for the binding of oxygen by hemoglobin from measured fractional saturations, Biophys. Chem. 63, 119-132 (1997).
5. J. Wyman, Heme Proteins, Adv. Protein Chem. 4, 407-531 (1948).
6. J. Wyman, Linked functions and reciprocal effects in hemoglobin, Adv. Protein Chem. 19, 223-286 (1964).
7. J. Wyman and S. J. Gill, Binding and Linkage, Wiley, Hoboken, NJ (1990).
8. R. A. Alberty, Effect of pH on protein-ligand equilibria, J. Phys. Chem. 104 B, 9929-9934 (2000).
9. R. A. Alberty, Fundamental equation of thermodynamics for protein-ligand binding, Biophys. Chem. 104, 543-559 (2003).
10. B. German, and J. Wyman, The titration curves for oxygenated and reduced hemoglobin, J. Biol. Chem. 117, 533-550 (1937).
11. E. Antonini, J. Wyman, M. Brunori, C. Fronticelli, E. Bicci, and A. Rossi-Fanelli, Studies of the relations between molecular and functional properties of hemoglobin, J. Biol. Chem. 240, 1090-1103 (1965).
12. R. A. Alberty, Calculation of thermodynamic properties of species from binding of a ligand by a macromolecule, Biophys. Chem. 105, 45-58 (2003).
13. R. A. Alberty, ProteinLigandProg, 2003.

In[11]:= **http : // library.wolfram.com / infocenter / MathSource / 4808**

14. I. M. Klotz, Ligand-Receptor Energetics, Wiley, Hoboken, NJ (1997).

Chapter 15 Calorimetry of Biochemical Reactions

15.1 Importance of the Standard Transformed Enthapy of Reaction

The enthalpy H of a chemical reaction system is of special interest because when a reaction occurs at constant temperature and pressure, the change in enthalpy $\Delta_r H$ is equal to the heat q of reaction. The change in enthalpy in a chemical reaction is also of interest because it determines the change in the equilibrium constant K with temperature. Similarly, the transformed enthalpy $\Delta_r H$ ' of an enzyme-catalyzed reaction is of special interest because when the reaction occurs at constant temperature, pressure, and pH, the change in transformed enthalpy $\Delta_r H$ ' is equal to the heat q of the enzyme-catalyzed reaction.

The change in entropy S in a chemical reaction is important because the change in Gibbs energy is made up of contributions from $\Delta_r H°$ and $\Delta_r S°$, which result from different causes. Similarly, the change in transformed entropy in an enzyme-catalyzed reaction is important because the change in transformed Gibbs energy $\Delta_r G$ '° is made up of contributions from $\Delta_r H$ '° and $\Delta_r S$ '°, which result from different causes. The $\Delta_r H$ '° for an enzyme-catalyzed reaction does not change much with pH, and so $\Delta_r S$ '° is largely responsible for the change in K ' with pH.

The determination of the transformed enthalpy of an enzyme-catalyzed reaction is complicated by the fact that the enzyme-catalyzed reaction may produce or consume hydrogen ions that react with the buffer to produce an additional heat effect that depends on the enthalpy of dissociation of the buffer. When metal ions are bound by species of reactants, a similar effect depends on the enthalpy of dissociation of complex ions involving species of the reactants. When adjustments are applied for these buffer effects, the $\Delta_r H$ '° determined from calorimetric measurements and determined from the temperature dependence of K ' should agree. Enthalpies of reaction may be rather constant over narrow temperature ranges, but over wider temperature ranges it is necessary to take account of the change of heat capacity in the reaction. The molar heat capacity of a species C_{Pm} ° is defined by equation 4.1-1, and the effect of ionic strength is described by equation A4-1. The standard transformed heat capacity of a reactant C_{Pm} '° is not simply a mole-fraction-weighted average, but contains an additional term due to the Le Chatelier effect, as shown by equation A4-2. There is not much information on heat capacities of species in aqueous solution and chemical reactions in aqueous solutions in the literature, but Goldberg and Tewari (1) summarized values known in 1989.

15.2 Calorimetric Determination of the Standard Transformed Enthalpy of Reaction

When the apparent equilibrium constant K' for an enzyme-catalyzed reaction depends on pH and pMg, the calorimetric enthalpy of reaction $\Delta_r H$ (cal) is given by (2)

$$\Delta_r H \text{ (cal)} = \Delta_r H'^{\circ} + \Delta_r N_H \Delta_r H^{\circ} \text{ (Buff)} + \Delta_r N_{Mg} \Delta_r H^{\circ} \text{ (MgBuff)} \tag{15.2-1}$$

where $\Delta_r H^{\circ}$ (Buff) is the standard enthalpy for the acid dissociation of the buffer and $\Delta_r H^{\circ}$ (MgBuff) is the standard enthalpy for the dissociation of the magnesium complex ion of the buffer. Thus if hydrogen ions or metal ions are produced or consumed in the enzyme-catalyzed reaction, it is necessary to know $\Delta_r H^{\circ}$ (Buff) and $\Delta_r H^{\circ}$ (MgBuff) at the temperature and ionic strength of the calorimetric experiment. The values of $\Delta_r N_H$ and $\Delta_r N_{Mg}$ can be calculated if the pKs and pK_{Mg}s of all the reactants are known, as described earlier (Sections 2.7 and 2.8), or they can be calculated from measurements of K' at pHs and pMgs in the range of conditions used in the calorimetric experiment. Alberty and Goldberg (2) calculated these effects for the hydrolysis of ATP to ADP at 298.15 K, pH 7.0, pMg 5.0, and 0.25 M ionic strength. The changes in the binding of hydrogen ions and magnesium ions under these conditions are -0.62 and -0.49. Under these conditions, $\Delta_r H$ (cal) = 177.74 kJ mol^{-1}, and so the standard transformed enthalpy of the hydrolysis of ATP under these conditions is given by

$$\Delta_r H'^{\circ} = 177.74 - (-0.62)(0.41) - (-0.49)(-465.36) = -30.76 \tag{15.2-2}$$

Note that the third term is larger than the first term and has the opposite sign.

15.3 Third Law of Thermodynamics

As mentioned in Sections 1.1 and 2.9, the third law of thermodynamics makes it possible to obtain the standard Gibbs energy of formation of species in aqueous solution from measurements of the heat capacity of the crystalline reactant down to about 10 K, its solubility in water and heat of solution, the heat of combustion, and the enthalpy of solution. According to the third law, the standard molar entropy of a pure crystalline substance at zero Kelvin is equal to zero. Therefore, the standard molar entropy of the crystalline substance at temperature T is given by

$$S_m^{\circ}(T) = \int_0^T (C_{Pm}^{\circ}/T)\, dT \tag{15.3-1}$$

Molar heat capacities C_{Pm}° (crystalline reactant) can be determined down to about 10 K, and the Debye equation that applies at very low temperatures can be used to estimate heat capacities below 10 K. The Debye equation is $C_{Pm}^{\circ} = kT^3$.

Heat of combustion measurements can be used to obtain $\Delta_f H^{\circ}$ (298.15 K) of the crystalline substance at 298.15 K, and the heat of solution makes it possible to calculate $\Delta_f H^{\circ}$ (aq soln,298.15 K). When third law measurements have been made, the standard Gibbs energy of formation of the substance in dilute aqueous solution can be calculated using

$$\Delta_f G^{\circ}(\text{aq soln,298.15 K}) = \Delta_f H^{\circ}(\text{aq soln,298.15 K}) - 298.15\, \Delta_f S^{\circ}(\text{aq soln,298.15 K}) \tag{15.3-2}$$

Thus $\Delta_f G^{\circ}$ for a species in aqueous solution can be determined calorimetrically. The standard entropy of formation of a species at 298.15 K is related to its standard molar entropy at 298.15 K by

$$\Delta_f S^{\circ}(\text{aq soln,298.15 K}) = S_m^{\circ}(\text{aq soln298.15 K}) - \sum N_{elei}(S_m^{\circ}(298.15 \text{ K}))_{elei} \tag{15.3-3}$$

N_{elei} is the number of atoms of element i in the crystalline substance and $(S_m\,°(298.15\,K))_{\mathrm{elei}}$ is the standard molar entropy of element i in its thermodynamic reference state. This equation makes it possible to calculate $\Delta_f S\,°$ for a species when $S_m\,°$ has been determined by the third law method. Then $\Delta_f G°$ for the species in dilute aqueous solution can be calculated using equation 15.3-2. Measurements of pKs, pK_{Mg}s, and enthalpies of dissociation make it possible to calculate $\Delta_f G°$ and $\Delta_f H°$ for the other species of a reactant that are significant in the pH range of interest (usually pH 5 to 9). When this can be done, the species properties of solutes in aqueous solution are obtained with respect to the elements in their reference states, just like other species in the NBS Tables (3).

This is important for biochemistry because when it can be done, calculations can be made on the thermodynamics of the formation of a reactant all the way back to the elements it contains. Thanks to the research of Boeiro-Goates and coworkers, $\Delta_f G°$ and $\Delta_f H°$ are now known for adenosine (aq) (4), adenine (aq) (5), and inosine (aq) (6). This has made it possible to calculate the standard thermodynamic properties of species of AMP, ADP, ATP, IMP, IDP, and ITP as well.

15.4 Calculation of Standard Entropies of Formation of Species

When $\Delta_f G_j\,°$ and $\Delta_f H_j\,°$ are known for a species, the value of $\Delta_f S_j\,°$ can be calculated using

$$\Delta_f S_j\,° = (\Delta_f H_j\,° - \Delta_f G_j\,°)/T \tag{15.4-1}$$

Thus it is possible to calculate $\Delta_f S_j\,°$ for all the species of about 94 reactants from the small data matrices in BasicBiochemData3 (7). To make a table of $\Delta_f S_j\,°$ of the species of a reactant, different programs are written for reactants with different numbers of species. The table of $\Delta_f S_j\,°$ prepared here includes z_j and $N_H(j)$ so that the species can be readily identified.

```
In[2]:=
  Off[General::"spell"];
  Off[General::"spell1"];

In[4]:=
  << BiochemThermo`BasicBiochemData3`

In[5]:=
  calcSform1sp[reactantname_, speciesmat_] :=
  Module[{dGzero, dHzero, zi, nH}, (*This program is used to calculate the
    standard entropy of formation of the single species of a reactant in
    kJ K^-1 mol^-1 at 298.15 K and zero ionic strength.  The reactant
    name should be in quotation marks.  The output is a 1 x4 matrix.*)
  {dGzero, dHzero, zi, nH} = Transpose[speciesmat];
  Transpose[{{reactantname}, (dHzero - dGzero) / 298.15, zi, nH}]]

In[6]:=
  calcSform2sp[reactantname_, speciesmat_] :=
  Module[{dGzero, dHzero, zi, nH}, (*This program is used to calculate the
    standard entropies of formation of the two species of a reactant in
    kJ K^-1 mol^-1 at 298.15 K and zero ionic strength.  The reactant
    name should be put in quotation marks.  The output is a 2 x4 matrix.*)
  {dGzero, dHzero, zi, nH} = Transpose[speciesmat];
  Transpose[{{reactantname, ""}, (dHzero - dGzero) / 298.15, zi, nH}]]
```

In[7]:=

```
calcSform3sp[reactantname_, speciesmat_] :=
  Module[{dGzero, dHzero, zi, nH}, (*This program is used to calculate the
      standard entropies of formation of the three species of a reactant
      in kJ K^-1 mol^-1 at 298.15 K and zero ionic strength.  The reactant
      name should be put in quotation marks.  The output is a 3 x4 matrix.*)
    {dGzero, dHzero, zi, nH} = Transpose[speciesmat];
    Transpose[{{reactantname, "", ""}, (dHzero - dGzero) / 298.15, zi, nH}]]
```

The calculation of $\Delta_f S°$ for species is shown for twelve reactants of biochemical interest.

In[8]:=

```
glucoseSsp = calcSform1sp["glucose", glucosesp];
```

In[9]:=

```
piSsp = calcSform2sp["phosphate", pisp];
```

In[10]:=

```
adpSsp = calcSform3sp["adenosine diphosphate", adpsp];
```

In[11]:=

```
ampSsp = calcSform3sp["adenosine monophosphate", ampsp];
```

In[12]:=

```
atpSsp = calcSform3sp["adenosine triphosphate", atpsp];
```

In[13]:=

```
formateSsp = calcSform1sp["formate", formatesp];
```

In[14]:=

```
glucose6phosSsp = calcSform2sp["glucose 6-phosphate", glucose6phossp];
```

In[15]:=

```
h2oSsp = calcSform1sp["h2o", h2osp];
```

In[16]:=

```
acetaldehydeSsp = calcSform1sp["acetaldehyde", acetaldehydesp];
```

In[17]:=

```
adenosineSsp = calcSform2sp["adenosine", adenosinesp];
```

In[18]:=

```
co2totSsp = calcSform3sp["co2tot", co2totsp];
```

In[19]:=

```
ethanolSsp = calcSform1sp["ethanol", ethanolsp];
```

The standard entropies of formation for species of these reactants are shown in Table 15.1.

Table 15.1 $\Delta_f S_j^{\,\circ}$ in kJ K^{-1} mol^{-1} for species in dilute aqueous solution at 298.15 K and zero ionic strength

In[20]:=

```
TableForm[Join[acetaldehydeSsp, adenosineSsp, adpSsp, ampSsp, atpSsp, co2totSsp,
   ethanolSsp, formateSsp, glucoseSsp, h2oSsp, glucose6phosSsp, piSsp],
  TableHeadings -> {None, {"Reactant", "ΔfSj °", "zj", "NH(j)"}}]
```

Out[20]//TableForm=

Reactant	$\Delta_f S_j^{\,\circ}$	z_j	$N_H(j)$
acetaldehyde	-0.245615	0	4
adenosine	-1.43149	0	13
	-1.42016	1	14
adenosine diphosphate	-2.41627	-3	12
	-2.26007	-2	13
	-2.23565	-1	14
adenosine monophosphate	-1.99537	-2	12
	-1.84843	-1	13
	-1.83277	0	14
adenosine triphosphate	-2.85464	-4	12
	-2.68801	-3	13
	-2.64877	-2	14
co2tot	-0.500855	-2	0
	-0.35291	-1	1
	-0.256649	0	2
ethanol	-0.357739	0	6
formate	-0.250042	-1	1
glucose	-1.16146	0	12
h2o	-0.163139	0	2
glucose 6-phosphate	-1.71893	-2	11
	-1.58997	-1	12
phosphate	-0.68053	-2	1
	-0.554419	-1	2

The values of z_j and $N_H(j)$ identify the species. A longer list of standard entropies of formation is given in reference (8).

15.5 Calculation of Standard Molar Entropies of Species

Whenever the standard entropy of formation of a species is known, its standard molar entropy can be calculated using equation 15.3-3. This calculation is carried out using **calcentropy298** (8).

In[21]:=

```
calcentropy298[entropyform_, nH_, nC_, nN_, nO_, nP_, nS_] :=
 Module[{}, (*This program calculates the standard
    molar entropy of a species in dilute aqueous solution at T=
   298.15 K and zero ionic strength from the standard entropy of formation of the
    species (entropyform_) at T=298.15 K and zero ionic strength.The entropies are in J
     K^-1 mol^-1. The numbers are numbers of atoms in the molecule or ion.*)entropyform +
  nH * 130.684 / 2 + nC * 5.740 + nN * 191.61 / 2 + nO * 205.138 / 2 + 41.09 * nP + 31.80 * nS]
```

The coefficients in this calculation are from the NBS Tables (3). The calculation of $S_m{}^\circ$ for a neutral species is straight foward. The formation of acetic acid in dilute aqueous solution from its elements is represented by

$$2C(\text{graphite}) + 2H_2(g) + O_2(g) = CH_3CO_2H(aq) \tag{15.5-1}$$

The standard molar entropy of acetic acid at 298.15 K can be calculated using the program **calcentropy298**. To calculate $S_m{}^\circ$ for the acetate ion in dilute aqueous solution at 298.15 K and zero ionic strength, the formation reaction is balanced by adding $H^+(aq)$ on the right side.

$$2C(\text{graphite}) + 2H_2(g) + O_2(g) = CH_3CO_2{}^-(aq) + H^+(aq) \tag{15.5-2}$$

Note that $S_m{}^\circ$ for $H^+(aq) = 0$ by definition. For a polyprotic weak acid, the number of hydrogen atoms used in the computer program is the same for all the species of the polyprotic acid.

The formation reaction for the ammonium ion is

$$(1/2)N_2(g) + (3/2)H_2(g) + H^+(aq) = NH_4{}^+(aq) \tag{15.5-3}$$

In this case the hydrogen ion is added to the left side to balance the formation reaction. For more complicated organic weak acids it is useful to remember that the atomic composition to be used in **calcentropy298** for each species is that shown for the uncharged form in the Merck Index (9). The standard molar entropies of species in $kJ\,K^{-1}\,mol^{-1}$ are calculated using the $\Delta_f S_j{}^\circ$ in Table 15.1.

```
In[22]:=
 acetaldehydesm = calcentropy298[-245.6, 4, 2, 0, 1, 0, 0] / 1000;

In[23]:=
 adenosinesm1 = calcentropy298[-1431.5, 13, 10, 5, 4, 0, 0] / 1000;

In[24]:=
 adenosinesm2 = calcentropy298[-1420.2, 13, 10, 5, 4, 0, 0] / 1000;

In[25]:=
 adpsm1 = calcentropy298[-2416.3, 15, 10, 5, 10, 2, 0] / 1000;

In[26]:=
 adpsm2 = calcentropy298[-2260.1, 15, 10, 5, 10, 2, 0] / 1000;

In[27]:=
 adpsm3 = calcentropy298[-2235.7, 15, 10, 5, 10, 2, 0] / 1000;

In[28]:=
 ampsm1 = calcentropy298[-1995.4, 14, 10, 5, 7, 1, 0] / 1000;

In[29]:=
 ampsm2 = calcentropy298[-1848.4, 14, 10, 5, 7, 1, 0] / 1000;

In[30]:=
 ampsm3 = calcentropy298[-1832.8, 14, 10, 5, 7, 1, 0] / 1000;

In[31]:=
 atpsm1 = calcentropy298[-2854.6, 16, 10, 5, 13, 3, 0] / 1000;

In[32]:=
 atpsm2 = calcentropy298[-2688.0, 16, 10, 5, 13, 3, 0] / 1000;
```

```
In[33]:=
 atpsm3 = calcentropy298[-2648.8, 16, 10, 5, 13, 3, 0] / 1000;

In[34]:=
 co2totsm1 = calcentropy298[-500.9, 2, 1, 0, 3, 0, 0] / 1000;

In[35]:=
 co2totsm2 = calcentropy298[-352.9, 2, 1, 0, 3, 0, 0] / 1000;

In[36]:=
 co2totsm3 = calcentropy298[-256.6, 2, 1, 0, 3, 0, 0] / 1000;

In[37]:=
 ethanolsm = calcentropy298[-357.7, 6, 2, 0, 1, 0, 0] / 1000;

In[38]:=
 formatesm = calcentropy298[-250.0, 2, 1, 0, 2, 0, 0] / 1000;

In[39]:=
 glucosesm = calcentropy298[-1161.5, 12, 6, 0, 6, 0, 0] / 1000;

In[40]:=
 glucose6phossm1 = calcentropy298[-1718.9, 13, 6, 0, 9, 1, 0] / 1000;

In[41]:=
 glucose6phossm2 = calcentropy298[-1590.0, 13, 6, 0, 9, 1, 0] / 1000;

In[42]:=
 h2osm = calcentropy298[-163.1, 2, 0, 0, 1, 0, 0] / 1000;

In[43]:=
 pism1 = calcentropy298[-680.5, 3, 0, 0, 4, 1, 0] / 1000;

In[44]:=
 pism2 = calcentropy298[-554.4, 3, 0, 0, 4, 1, 0] / 1000;

In[45]:=
 entropiesform = {-245.6, -1431.5, -1420.2, -2416.3, -2260.1,
     -2235.7, -1995.4, -1848.4, -1832.8, -2854.6, -2688.0, -2648.8, -500.9, -352.9,
     -256.6, -357.7, -250.0, -1161.5, -1718.9, -1590.0, -163.1, -680.5, -554.4} / 1000;

In[46]:=
 entropies = {acetaldehydesm, adenosinesm1, adenosinesm2, adpsm1, adpsm2, adpsm3, ampsm1,
     ampsm2, ampsm3, atpsm1, atpsm2, atpsm3, co2totsm1, co2totsm2, co2totsm3, ethanolsm,
     formatesm, glucosesm, glucose6phossm1, glucose6phossm2, h2osm, pism1, pism2};

In[47]:=
 namesentropies = {"acetaldehydesm", "adenosinesm1", "adenosinesm2",
     "adpsm1", "adpsm2", "adpsm3", "ampsm1", "ampsm2", "ampsm3", "atpsm1", "atpsm2",
     "atpsm3", "co2totsm1", "co2totsm2", "co2totsm3", "ethanolsm", "formatesm",
     "glucosesm", "glucose6phossm1", "glucose6phossm2", "h2osm", "pism1", "pism2"};
```

Table 15.2 $\Delta_f S_j{}^\circ$ and $S_m{}^\circ$ in kJ K^{-1} mol^{-1} for species at 298.15 K and zero ionic strength

In[48]:=

```
PaddedForm[TableForm[Transpose[{entropiesform, entropies}],
   TableHeadings → {namesentropies, {"  ΔfSj°", "  Sm°"}}], {5, 4}]
```

Out[48]//PaddedForm=

	$\Delta_f S_j{}^\circ$	$S_m{}^\circ$
acetaldehydesm	-0.2456	0.1298
adenosinesm1	-1.4315	0.3646
adenosinesm2	-1.4202	0.3759
adpsm1	-2.4163	0.2081
adpsm2	-2.2601	0.3643
adpsm3	-2.2357	0.3887
ampsm1	-1.9954	0.2149
ampsm2	-1.8484	0.3619
ampsm3	-1.8328	0.3775
atpsm1	-2.8546	0.1840
atpsm2	-2.6880	0.3506
atpsm3	-2.6488	0.3898
co2totsm1	-0.5009	-0.0568
co2totsm2	-0.3529	0.0912
co2totsm3	-0.2566	0.1875
ethanolsm	-0.3577	0.1484
formatesm	-0.2500	0.0916
glucosesm	-1.1615	0.2725
glucose6phossm1	-1.7189	0.1292
glucose6phossm2	-1.5900	0.2581
h2osm	-0.1631	0.0702
pism1	-0.6805	-0.0331
pism2	-0.5544	0.0930

A longer list of standard molar entropies is provided in reference (8).

15.6 Calculation of Standard Transformed Entropies of Formation of Reactants

There are two ways to calculate the standard transformed entropy of formation of a reactant: (1) It can be calculated from $\Delta_f G_i'^{\circ}$ and $\Delta_f H_i'^{\circ}$ using

$$\Delta_f S_i'^{\circ} = (\Delta_f H_i'^{\circ} - \Delta_f G_i'^{\circ})/T \tag{15.6-1}$$

(2) It can be calculated from the $\Delta_f S_j^{\circ}$ of species given in the preceding section. This can be done starting with equation 3.3-7 for the Legendre transform for the entropy of a system at specified pH:

$$S' = S - n_H(H)\overline{S}(H^+) \tag{15.6-2}$$

This is a rather lengthy derivation, but there is an easier way since the function of temperature, pH, and ionic strength for the standard transformed Gibbs energy of a species contains all the thermodynamic information about the species (see Section 3.4). The equation for the standard transformed entropy of formation for a species can be obtained by use of $\Delta_f S_j'^{\circ} = -\partial \Delta_f G_j'^{\circ}/\partial T$. The standard transformed Gibbs energy of formation of a species is given by equation 3.6-1, which is given by

$$\Delta_f G_j'^{\circ}(I) = \Delta_f G_j^{\circ}(I=0) + N_H(j) RT\ln(10)\,pH - RT\alpha(z_i^2 - N_H(j))\,I^{1/2}/(1 + 1.6\,I^{1/2}) \tag{15.6-3}$$

Taking the indicated partial derivative yields

$$\Delta_f S_j'^{\circ}(I) = \Delta_f S_j^{\circ}(I=0) - N_H(j) R\ln(10)\,pH + R(\alpha + T\,\partial\alpha/\partial T)\{z_i^2 - N_H(j)\}\,I^{1/2}/(1 + 1.6\,I^{1/2}) \tag{15.6-4}$$

The standard transformed entropy of formation of a pseudoisomer group is given by

$$\Delta_f S_i'^{\circ} = \sum r_j \Delta_f S_j'^{\circ} - R\sum r_j \ln r_j \tag{15.6-5}$$

This equation can be derived using equation 15.6-1 by substituting the expression for the standard transformed enthalpy of a reactant

$$\Delta_f H_i'^{\circ} = \sum r_j \Delta_f H_j'^{\circ} \tag{15.6-6}$$

and the expression for the standard transformed Gibbs energy of a reactant

$$\Delta_f G_i'^{\circ} = \sum r_j \Delta_f G_j'^{\circ} + RT\sum r_j \ln r_j \tag{15.6-7}$$

Thus the function of pH and ionic strength at 298.15 K that yields $\Delta_f S_i'^{\circ}$ can be obtained for a reactant using equation 15.6-5 when $\Delta_f S_j^{\circ}$ is known for all the species and $\Delta_f G_j^{\circ}$ are available so that the equilibrium mole fractions r_j can be calculated for the species. These functions of pH and ionic strength at 298.15 K can be derived using the program **calc-Sformreactant** (8). In this program the coefficient for the ionic strength adjustment for the standard transformed entropy of formation of a species is given by

$$\{RT^2\,(\partial\alpha/\partial T) - (-RT\alpha)\}/T = R\,(\alpha + \partial\alpha/\partial T) \tag{15.6-8}$$

Values of this coefficient are given in Table 1.1 for a number of temperatures, and at 25 °C it is given by $(2.91482+1.4775)/298.15 = 0.0147319$.

To obtain a table $\Delta_f S_j{}^\circ$ as a function of pH from $S_m{}^\circ$ of the species the twelve reactants, we first have to use **calcentropylist** and then use **calcSformreactant** .

```
In[49]:=
  calcentropylist[speciesmat_] :=
   Module[{dGzero, dHzero, zi, nH}, (*This program is used to assemble
      the list of standard entropies of formation of the species of a
      reactant in kJ K^-1 mol^-1 at 298.15 K and zero ionic strength.*)
     {dGzero, dHzero, zi, nH} = Transpose[speciesmat];
     (dHzero - dGzero) / 298.15]
```

```
In[50]:=
  acetaldehydeSlist = calcentropylist[acetaldehydesp]
```

```
Out[50]= {-0.245615}
```

```
In[51]:=
  adenineSlist = calcentropylist[adeninesp]
```

```
Out[51]= {-0.616804, -0.603824}
```

```
In[52]:=
  adenosineSlist = calcentropylist[adenosinesp];
```

```
In[53]:=
  ammoniaSlist = calcentropylist[ammoniasp];
```

```
In[54]:=
  adpSlist = calcentropylist[adpsp];
```

```
In[55]:=
  ampSlist = calcentropylist[ampsp];
```

```
In[56]:=
  atpSlist = calcentropylist[atpsp];
```

```
In[57]:=
  co2totSlist = calcentropylist[co2totsp];
```

```
In[58]:=
  ethanolSlist = calcentropylist[ethanolsp];
```

```
In[59]:=
  formateSlist = calcentropylist[formatesp];
```

```
In[60]:=
  glucoseSlist = calcentropylist[glucosesp];
```

```
In[61]:=
  glucose6phosSlist = calcentropylist[glucose6phossp];
```

```
In[62]:=
  h2oSlist = calcentropylist[h2osp];
```

```
In[63]:=
  piSlist = calcentropylist[pisp];
```

```
In[64]:=
  calctrSformreactant[speciesmat_, entropylist_] :=
   Module[{dGzero, dHzero, zi, nH, pHterm, isterm, gpfnsp,
     dGreactant, ri, isentropy, dSfnsp, pHtermS, avgentropy, entropymix},
    (*This program derives the function of pH and ionic strength (is) that gives
      the standard transformed entropy of formation of a reactant (sum of species)
      at 298.15 K.  The first input is a matrix that gives the standard Gibbs
      energy of formation,the standard enthalpy of formation, the electric charge,
     and the number of hydrogen atoms in the species in the reactant.  There
      is a row in the matrix for each species of the reactant.  dSfnsp is a list of
      the functions for the species.  Entropies are expressed in kJ K^-1 mol^-1.*)
    {dGzero, dHzero, zi, nH} = Transpose[speciesmat];
      (*Calculate the functions for the
      standard Gibbs energies of formation of the species.*)
    pHterm = nH * 8.31451 * .29815 * Log[10^-pH];
    isterm = 2.91482 * ((zi^2) - nH) * (is^.5) / (1 + 1.6 * is^.5);
    gpfnsp = dGzero - pHterm - isterm;
    (*Calculate the standard
      transformed Gibbs energy of formation for the reactant.*)
    dGreactant = -8.31451 * .29815 *
      Log[Apply[Plus, Exp[-1 * gpfnsp / (8.31451 * .29815)]]];
    (*Calculate the equilibrium mole fractions of the species in the reactant.*)
    ri = Exp[(dGreactant - gpfnsp) / (8.31451 * .29815)];
    (*Calculate the standard transfomed entropies of formation of the species and
       then calculate the mole fraction-weighted average entropy of the reactant.*)
    pHtermS = nH * 8.31451 * 10^-3 * Log[10^-pH];
    isentropy = .0147319 * ((zi^2) - nH) * (is^.5) / (1 + 1.6 * is^.5);
    dSfnsp = entropylist + pHtermS + isentropy;
    avgentropy = ri.dSfnsp;
    (*Calculate the entropy of mixing of the species.*)
    entropymix = 8.31451 * 10^-3 * ri.Log[ri];
    avgentropy - entropymix]
```

This program is used to derive the functions of pH and ionic strength that give the standard transformed entropies of formation of all the reactants in Table 15.2. In the next section these functions are used to calculate the standard transformed entropies of reaction for four enzyme-catalyzed reactions.

```
In[65]:=
  acetaldehydeentropy =
    calctrSformreactant[acetaldehydesp, acetaldehydeSlist] /. is → .25 /.
     pH → {5, 6, 7, 8, 9};

In[66]:=
  adenosineentropy =
    calctrSformreactant[adenosinesp, adenosineSlist] /. is → .25 /. pH → {5, 6, 7, 8, 9};

In[67]:=
  adpentropy = calctrSformreactant[adpsp, adpSlist] /. is → .25 /. pH → {5, 6, 7, 8, 9};

In[68]:=
  ampentropy = calctrSformreactant[ampsp, ampSlist] /. is → .25 /. pH → {5, 6, 7, 8, 9};

In[69]:=
  atpentropy = calctrSformreactant[atpsp, atpSlist] /. is → .25 /. pH → {5, 6, 7, 8, 9};
```

```
In[70]:=
  co2totentropy =
    calctrSformreactant[co2totsp, co2totSlist] /. is → .25 /. pH → {5, 6, 7, 8, 9};

In[71]:=
  ethanolentropy =
    calctrSformreactant[ethanolsp, ethanolSlist] /. is → .25 /. pH → {5, 6, 7, 8, 9};

In[72]:=
  formateentropy =
    calctrSformreactant[formatesp, formateSlist] /. is → .25 /. pH → {5, 6, 7, 8, 9};

In[73]:=
  glucoseentropy =
    calctrSformreactant[glucosesp, glucoseSlist] /. is → .25 /. pH → {5, 6, 7, 8, 9};

In[74]:=
  glucose6phosentropy =
    calctrSformreactant[glucose6phossp, glucose6phosSlist] /. is → .25 /.
      pH → {5, 6, 7, 8, 9};

In[75]:=
  h2oentropy = calctrSformreactant[h2osp, h2oSlist] /. is → .25 /. pH → {5, 6, 7, 8, 9};

In[76]:=
  pientropy = calctrSformreactant[pisp, piSlist] /. is → .25 /. pH → {5, 6, 7, 8, 9};
```

Table 15.3 $\Delta_f S_i{'}^\circ$ in kJ K^{-1} mol^{-1} for reactants at 298.15 K, 0.25 M ionic strength, and five pHs

```
In[77]:=
  PaddedForm[TableForm[{acetaldehydeentropy, adenosineentropy, adpentropy,
    ampentropy, atpentropy, co2totentropy, ethanolentropy, formateentropy,
    glucoseentropy, glucose6phosentropy, h2oentropy, pientropy},
    TableHeadings → {{"acetaldehyde", "adenosine", "adp", "amp", "atp",
      "co2tot", "ethanol", "formate", "glucose", "glucose6phos", "h2o", "pi"},
    {"pH 5", "pH 6", "pH 7", "pH 8", "pH 9"}}], {5, 4}]
```

Out[77]//PaddedForm=

	pH 5	pH 6	pH 7	pH 8	pH 9
acetaldehyde	-0.6449	-0.7215	-0.7980	-0.8746	-0.9512
adenosine	-2.7304	-2.9781	-3.2269	-3.4758	-3.7246
adp	-3.5445	-3.7906	-4.0333	-4.2661	-4.4961
amp	-3.1447	-3.3918	-3.6336	-3.8657	-4.0957
atp	-3.9519	-4.1980	-4.4420	-4.6756	-4.9059
co2tot	-0.4535	-0.4767	-0.4889	-0.5052	-0.5158
ethanol	-0.9566	-1.0715	-1.1864	-1.3012	-1.4161
formate	-0.3458	-0.3649	-0.3841	-0.4032	-0.4223
glucose	-2.3593	-2.5890	-2.8187	-3.0485	-3.2782
glucose6phos	-2.7827	-3.0064	-3.2211	-3.4323	-3.6429
h2o	-0.3628	-0.4011	-0.4394	-0.4776	-0.5159
pi	-0.7494	-0.7834	-0.8046	-0.8218	-0.8406

There is an easier way to make this table. BasicBiochemData3 provides the functions of temperature, pH, and ionic strength for $\Delta_f S_i{'}^\circ$ for the 94 reactants for which $\Delta_f H^\circ$ values are known for all the species. These functions can be used to make tables or plots that show these dependencies. Table 15.3 at 298.15 K can be calculated as follows, but it can also be prepared for any temperature in the range 273.15 K to about 313.15 K and other ionic strengths.

```
In[78]:=
PaddedForm[
  TableForm[{acetaldehydeST, adenosineST, adpST, ampST, atpST, co2totST, ethanolST,
      formateST, glucoseST, glucose6phosST, h2oST, piST} /. t → 298.15 /. is → .25 /.
    pH → {5, 6, 7, 8, 9}, TableHeadings → {{"acetaldehyde", "adenosine", "adp", "amp",
      "atp", "co2tot", "ethanol", "formate", "glucose", "glucose6phos", "h2o", "pi"},
    {"pH 5", "pH 6", "pH 7", "pH 8", "pH 9"}}], {5, 4}];
```

These values can be calculated using equation 15.6-1, and the fact that exactly the same values are obtained is a test of three programs, **calcGmat**, **calcHmat**, and **calctrSformreactant**.

Table 15.3 is for 0.25 M ionic strength, but the effect of ionic strength can be studied by making tables like the following:

Table 15.4 Effect of ionic strength on $\Delta_f S_i'^\circ$ (phosphate) in kJ K^{-1} mol^{-1} at 298.15 K and five pHs

```
In[79]:=
PaddedForm[
  TableForm[piST /. t → 298.15 /. is → {0, .1, .25} /. pH → {5, 6, 7, 8, 9}, TableHeadings →
    {{"I=0", "I=0.10", "I=0.25"}, {"pH 5", "pH 6", "pH 7", "pH 8", "pH 9"}}], {5, 4}]
```

```
Out[79]//PaddedForm=
```

	pH 5	pH 6	pH 7	pH 8	pH 9
I=0	-0.7457	-0.7830	-0.8140	-0.8341	-0.8529
I=0.10	-0.7486	-0.7837	-0.8075	-0.8248	-0.8436
I=0.25	-0.7494	-0.7834	-0.8046	-0.8218	-0.8406

The rate of change of the standard transformed entropy of formation of a reactant with pH is given by (10)

$$(\partial \Delta_f S_i'^\circ / \partial pH) = - R\ln(10)\{\partial(T\bar{N}_{Hi})/\partial T\} \qquad (15.6\text{-}9)$$

where $\bar{N}_{Hi}$ is the average number of hydrogen ions bound by the reactant.

15.7 Calculation of Standard Transformed Entropies of Reaction

The standard transformed entropy of reaction at a specified pH can be calculated in two ways:

$$\Delta_r S_i'^\circ = \Sigma v_i' \Delta_f S_i'^\circ = \Sigma v_i' S_{mi}'^\circ \qquad (15.7\text{-}1)$$

When differences between $\Delta_f S_i'^\circ$ of reactants and products are taken the entropies of the elements cancel. In the calculations here we will use the first form of equation 15.7-1. Since the functions $\Delta_f S_i'^\circ$ for reactants in an enzyme-catalyzed reaction can be added and subtracted, the complicated function for the reaction can be used to calculate $\Delta_r S'^\circ$ at specified temperature, pH, and ionic strength.

```
In[80]:=
rx1 = (adpST + piST - (atpST + h2oST)) /. t → 298.15 /. is → .25 /. pH → {5, 6, 7, 8, 9};
```

```
In[81]:=
rx2 = (ampST + piST - (adpST + h2oST)) /. t → 298.15 /. is → .25 /. pH → {5, 6, 7, 8, 9};
```

```
In[82]:=
rx3 = (adenosineST + piST - (ampST + h2oST)) /. t → 298.15 /. is → .25 /. pH → {5, 6, 7, 8, 9};
```

In[83]:=
rx4 = (glucoseST + piST - (glucose6phosST + h2oST)) /. t → 298.15 /. is → .25 /. pH → {5, 6, 7, 8, 9};

Table 15.5 gives the changes in standard transformed entropies for four reactions.

Table 15.5 $\Delta_r S'^\circ$ in kJ K^{-1} mol^{-1} for reactions at 298.15 K, 0.25 M ionic strength, and pHs 5, 6, 7, 8, and 9

In[84]:=
  PaddedForm[TableForm[{rx1, rx2, rx3, rx4},
    TableHeadings → {{"atp+h2o=adp+pi", "adp+h2o=amp+pi",
      "amp+h2o=adenosine+pi", "glucose6phos+h2o=glucose+pi"}, {" pH 5",
      " pH 6", " pH 7", " pH 8", " pH 9"}}, TableSpacing → {1, .2}], {6, 4}]

Out[84]//PaddedForm=

	pH 5	pH 6	pH 7	pH 8	pH 9
atp+h2o=adp+pi	0.0207	0.0252	0.0435	0.0654	0.0850
adp+h2o=amp+pi	0.0132	0.0165	0.0344	0.0562	0.0758
amp+h2o=adenosine+pi	0.0276	0.0313	0.0414	0.0458	0.0464
glucose6phos+h2o=glucose+pi	0.0368	0.0351	0.0371	0.0396	0.0400

Since these values are all positive, the change in standard transformed entropy favors reaction.
The rate of change of the standard transformed reaction entropy with pH is given by (10)

$$(\partial\Delta_r S'^\circ/\partial pH) = - R\ln(10)\{\partial(T\Delta_r N_H)/\partial T\} \tag{15.7-2}$$

where $\Delta_r N_H$ is the change in binding of hydrogen ions in the reaction. For the hydrolysis of ATP at high pH, $\Delta_r N_H = -1$ so that $(\partial T\Delta_r N_H)/\partial T) = -1$. Thus $(\partial\Delta_r S_i'^\circ/\partial pH) = R\ln(10) = 0.0191471$ kJ K^{-1}mol^{-1}.

15.8 Discussion

Use of the third law is not the only way to get large molecules of biochemical reactants into tables of species properties. If apparent equilibrium constants and heats of reaction can be determined for a pathway of reactions from smaller molecules (for which $\Delta_f G^\circ$ and $\Delta_f H^\circ$ are known with respect to the elements) to form the large molecule, then the properties of the species of the large molecule can be determined relative to the elements in their reference states. This method has its problems in that it is very difficult to determine apparent equilibrium constants greater than about 10^3 to 10^4 and the number of reactions in the path may be large and some of the reactants may not be readily available in pure form. Thus it is fortunate that the third law method is available.

Quantum mechanical calculations can be used to obtain estimates of molar entropies of species. An example is the study of the conversion of chorismate to prephenate by Kast and coworkers (11). They used quantum mechanics to estimate $\Delta_r S^\circ$ for chorismate^{2-}(aq) = prephenate^{2-}(aq). Since $\Delta_r H^\circ$ was obtained experimentally, this made it possible to estimate the equilibrium constant for this reaction.

References

1. R. N. Goldberg and Y. Tewari, Thermodynamic and transport properties of carbohydrates and their monophosphates: The pentoses and hexoses, J. Phys. Chem. Ref. Data 18, 809-880 (1989).
2. R. A. Alberty and R. N. Goldberg, Calorimetric determination of the standard transformed enthalpy of a biochemical reaction at specified pH and pMg, Biophys. Chem. 47, 213-223 (1993).
3. D. D. Wagman, W. H. Evans, V. B. Parker, R. H. Schumm, I. Halow, S. M. Bailey, K. L. Churney, and R. L. Nuttall, The NBS tables of chemical thermodynamic properties, J. Phys. Chem. Ref. Data, 11, Supplement 2 (1982).
4. J. Boerio-Goates, M. R. Francis, R. N. Goldberg, M. A. V. Ribeiro da Silva, M. D. M. C. Ribeiro da Silva, and Y. Tewari, Thermochemistry of adenosine, J. Chem. Thermodyn. 33, 929-947 (2001).

5. J. S. Boyer, M. R. Francis, and J. Boeiro-Goates, Heat-capacity measurements and thermodynamic funnctions of crystal-line adenine: Revised thermodynamic properties of aqueous adenine, J. Chem. Thermo. 35, 1917-1928 (2003).

6. J. Boeiro-Goates, S. D. Hopkins, R. A. R. Monteiro, M. D. M. C. Riberio da Silva, M. A. V. Riberio da Silva, and R. N. Goldberg, Thermochemistry of inosine, J. Chem. Thermodyn., .37, 1239-1249 (2005).

7. R. A. Alberty, BasicBiochemData3, 2005.

In[85]:= **http : // library.wolfram.com / infocenter / MathSource / 5704**

8. R. A. Alberty, Standard molar entropies, standard entropics of formation, and standard transformed entropies of formation in the thermodynamics of enzyme-catalyzed reactions, J. Chem. Thermodyn., in press.

9. Merck Index, Merck and Co. Inc., 1983.

10. R. A. Alberty, Effect of temperature on standard transformed thermodynamic properties of bichemical reactions with emphasis on the Maxwell equations, J. Phys. Chem. 197 B, 3631-3635 (2003).

11. P. Kast, Y. B. Tewari, O. Wiest, D. Hilvert, K. N. Houk, and R.N. Goldberg, Thermodynamics of the conversion of chorismate to prephenate: Experimental results and theoretical predicitions, J. Phys. Chem. B 101, 10976-10982 (1997).

Appendix

1. BasicBiochemData3.nb

2. Tables of Transformed Thermodynamic Properties

3. Glossary of Names of Reactants

4. Glossary of Symbols for Thermodynamic Properties

5. List ofMathematica Programs

6. Sources of Biochemical Thermodynamic Information on the Web

```
BeginPackage["BiochemThermo`BasicBiochemData3`"];
```

BasicBiochemData3.nb

Robert A. Alberty
Department of Chemistry 6-215
Massachusetts Institute of Technology
Cambridge, MA 02139
alberty@mit.edu

Abstract: The most efficient way to store thermodynamic data on enzyme-catalyzed reactions is to use matrices of species properties. Since equilibrium in enzyme-catalyzed reactions is reached at specified pH values, the thermodynamics of the reactions is discussed in terms of transformed thermodynamic properties. These transformed thermodynamic properties are complicated functions of temperature, pH, and ionic strength that can be calculated from the matrices of species values. The most important of these transformed thermodynamic properties is the standard transformed Gibbs energy of formation of a reactant (sum of species). It is the most important because when this function of temperature, pH, and ionic strength is known, all the other standard transformed properties can be calculated by taking partial derivatives. The species database in this package contains data matrices for 199 reactants. For 94 of these reactants, standard enthalpies of formation of species are known, and so standard transformed Gibbs energies, standard transformed enthalpies, standard transformed entropies, and average numbers of hydrogen atoms can be calculated as functions of temperature, pH, and ionic strength. For reactions between these 94 reactants, the changes in these properties can be calculated over a range of temperatures, pHs, and ionic strengths, and so can apparent equilibrium constants. For the other 105 reactants, only standard transformed Gibbs energies of formation and average numbers of hydrogen atoms at 298.15 K can be calculated. The loading of this package provides functions of pH and ionic strength at 298.15 K for standard transformed Gibbs energies of formation and average numbers of hydrogen atoms for 199 reactants. It also provides functions of temperature, pH, and ionic strength for the standard transformed Gibbs energies of formation, standard transformed enthalpies of formation, standard transformed entropies of formation, and average numbers of hydrogen atoms for 94 reactants. Thus loading this package makes available 774 mathematical functions for these properties. These functions can be added and subtracted to obtain changes in these properties in biochemical reactions and apparent equilibrium constants.

1 Introduction

Chemical reaction systems are discussed in terms of species, and many chemical thermodynamic properties can be calculated from the species properties given in the next section, for example pKs. However, in making calculations on enzyme-catalyzed reactions it is useful to take the pH as an independent variable. When this is done the principal thermodynamic properties of a reactant are the standard transformed Gibbs energy of formation $\Delta_f G'°$, the standard transformed enthalpy of formation $\Delta_f H'°$, the standard transformed entropy of formation $\Delta_f S'°$, and the average number of hydrogen atoms in the reactant $\overline{N}_H$. These properties are related by the following equations:

$$\Delta_f H_i'° = -T^2 \frac{\partial(\Delta_f G_i'°/T)}{\partial T} \tag{1}$$

$$\Delta_f S_i'° = - \frac{\partial \Delta_f G_i'°}{\partial T} \tag{2}$$

$$\overline{N}_H\,(i) = \frac{1}{RT\ln\,(10)}\,\frac{\partial\Delta_f\,G_i{}^{'\,\circ}}{\partial pH} \tag{3}$$

These equations emphasize the importance of being able to express $\Delta_f\,G\,'\,^{\circ}$ as a function of temperature, pH, and ionic strength because the other three properties can be calculated by taking partial derivatives.

The calculation of $\Delta_f\,G^{\circ}$ and $\Delta_f\,H^{\circ}$ of species from experimental data on apparent equilibrium constants and transformed enthalpies of reaction is described in R. A. Alberty, Thermodynamics of Biochemical Reactions, Wiley, Hoboken, NJ (2003) and a number of places in the literature. That is not discussed here because this package is oriented toward the derivation of mathematical functions to calculate thermodynamic properties at specified T, pH, and ionic strength. There are two types of biochemical reactants in the database:

(1) reactants for which $\Delta_f\,G^{\,\circ}$ values are known for all species that are significant in the range pH 5 - 9

(2) reactants for which $\Delta_f\,G^{\,\circ}$ and $\Delta_f\,H^{\circ}$ are known for all species that are significant in the range pH 5 - 9

There are 199 reactants of the first type and 94 reactants of the second type. For the first type of reactant, it is possible to calculate $\Delta_f\,G\,'\,^{\circ}$ and $\overline{N}_H$ as functions of pH and ionic strength at 298.15 K. For the second type of reactant, it is possible to calculate $\Delta_f\,G\,'\,^{\circ}$, $\Delta_f\,H\,'\,^{\circ}$, $\Delta_f\,S\,'\,^{\circ}$, and $\overline{N}_H$ as functions of temperature, pH , and ionic strength. The equations for calculating these transformed properties from species properties are quite complicated, and so it is fortunate that they can be derived by using Mathematica's symbolic capabilities. The program **calcdGmat** is used with the first type of reactants to derive the functions of pH and ionic strength at 298.15 K. The program **derivetrGT** is used with the second type of reactants to derive the functions of temperature, pH, and ionic strength.

The transformed thermodynamic properties of biochemical reactions ($K\,'$, $\Delta_r\,G\,'\,^{\circ}$, $\Delta_r\,H\,'\,^{\circ}$, $\Delta_r\,S\,'\,^{\circ}$, and $\Delta_r\,N_H$) are related to properties of reactants by the following equations:

$$\Delta_r\,G^{'\,o} = \sum\,\nu_i{}'\Delta_f\,G_i{}^{'\,o} = -\,RT\ln K\,' \tag{4}$$

$$\Delta_r\,H^{'\,o} = \sum\,\nu_i{}'\Delta_f\,H_i{}^{'\,o} \tag{5}$$

$$\Delta_r\,S^{'\,o} = \sum\,\nu_i{}'\Delta_f\,S_i{}^{'\,o} \tag{6}$$

$$\Delta_r\,N_H = \sum\,\nu_i{}'\overline{N}_H \tag{7}$$

The $\nu_i{}'$ are the stoichiometric numbers of reactants in the biochemical equation (positive for reactants on the right side of the equation and negative for reactants on the left side). The prime is needed on the stoichiometric numbers to distinguish them from the stoichiomeric numbers in the underlying chemical reactions.

This package provides mathematical functions for the following properties of reactants:

(1) Functions of pH and ionic strength at 298.15 K for $\Delta_f\,G_i{}^{'\,o}$ for 199 reactants, for example, atp.

(2) Functions of pH and ionic strength at 298.15 K for $\overline{N}_H$ for 199 reactants, for example, atpNH.

(3) Functions of temperature, pH and ionic strength for $\Delta_f\,G_i{}^{'\,\,o}$ for 94 reactants, for example, atpGT.

(4) Functions of temperature, pH and ionic strength for $\Delta_f\,H_i{}^{'\,o}$ for 94 reactants, for example, atpHT

(5) Functions of temperature, pH and ionic strength for $\Delta_f\,S_i{}^{'\,\,o}$ for 94 reactants, for example, atpST

(6) Functions of temperature, pH and ionic strength for $\overline{N}_H$ for 94 reactants, for example, atpNHT

These functions can be evaluated at desired temperatures, pHs, and ionic strengths by use of the operation ReplaceAll: for example, atpGT/.t->313.15/.pH->7/.is->0.25. They can also be used to make tables and plots.

The functions for biochemical reactants can be added and subtracted to obtain standard transformed thermodynamic properties for enzyme-catalyzed reactions and their apparent equilibrium constants. The standard transformed Gibbs energy of reaction for atp + h2o = adp + pi is given as a function of temperature, pH, and ionic strength by adpGT+h2oGT-(atpGT+h2oGT). The other properties can also be calculated by changing the suffixes. These functions can also be used to produce plots and tables.

Instructions for the use of the package BasicBiochemData3.m

When the notebook BasicBiochemData3.nb was made, a package version BasicBiochemData3.m was made automatically. The package consists of the *Mathematica* input without the text. When the 199 small matrices of species data or the 774 functions are needed, the command   <<BiochemThermo`BasicBiochem-Data3`
is used to make this information available. It is necessary for the user of this book to put BasicBiochemData3.m in their computer so that
<<BiochemThermo`BasicBiochemData3`
will work. To load this package properly, it is first necessary to create a folder named BiochemThermo and put BasicBiochemData3.m into this folder. Then it is necessary to find out where to put this folder as follows: Open a Mathematica session and evaluate $UserBaseDirectory. On my Mac computer, this yields /Users/robertalberty/Library/*Mathematica*. In this *Mathematica* file, you will find Applications. Put the folder BiochemThermo (with BasicBiochemData3.m in it) into this Applications directory. Now this package becomes available whenever you load it using
<<BiochemThermo`BasicBiochemData3`
Each chapter should be opened in a fresh workspace. When this package is loaded, all the species data on 199 reactants and the 774 functions under this package will be available for calculations. The properties for adenosine triphosphate are named atpsp, atp, atpNH, atpGT, atpHT, atpST, and atpNHT. These functions are all protected; that is, none of them can be changed without unprotecting them.

```
Off[General::spell1];
Off[General::spell];
```

Usage statements for the programs used to derive functions of temperature, pH, and ionic strength for reactants:

```
calcdGmat::usage="calcdGmat[species_] produces the function of pH and ionic
strength (is) that gives the standard transformed Gibbs energy of formation of a
reactant (sum of species) at 298.15 K.  The input speciesmat is a matrix that
gives the standard Gibbs energy of formation, the standard enthalpy of formation,
the electric charge, and the number of hydrogen atoms in each species. There is
a row in the matrix for each species of the reactant. gpfnsp is a list of the
functions for the species.  Energies are expressed in kJ mol^-1.";

calcNHmat::usage="calcNHmat[speciesmat_]
produces the function of pH and ionic strength (is) that gives the average
number of hydrogen atoms in a reactant (sum of species) at 298.15 K.
The input speciesmat is a matrix that gives the standard Gibbs energy of
formation, the standard enthalpy of formation, the electric charge, and
the number of hydrogen atoms in each species. There is a row in the matrix
for each species of the reactant. gpfnsp is a list of the functions for
the species.";

derivetrGibbsT::usage="derivetrGibbsT[speciesmat_] derives the function of T
(in Kelvin), pH, and ionic strength (is) that gives
the standard transformed Gibbs energy of
formation of a reactant (sum of species).  The input speciesmat is a matrix
that gives the standard Gibbs energy of formation in kJ mol^-1 at 298.15 K and
zero ionic strength, the standard enthalpy of formation in kJ mol^-1 at 298.15 K
and zero ionic strength, the electric charge, and the number of hydrogen atoms in
each species.  There is a row in the matrix for each species of the reactant.
gpfnsp is a list of the functions for the standard transformed Gibbs energies
of the species. The corresponding functions for other transformed properties
can be obtained by taking partial derivatives.  The standard transformed Gibbs
energy of formation of a reactant in kJ mol^-1 can be calculated at any temperature
in the range 273.15 K to 313.15 K, any pH in the range 5 to 9, and any ionic
strength in the range 0 to 0.35 M by use of ReplaceAll (/.).";
```

derivetrHT::usage="derivetrHT[speciesmat_] derives the function of T (in Kelvin), pH, and ionic strength (is) that gives the standard transformed enthalpy of formation of a reactant (sum of species). The input speciesmat is a matrix that gives the standard Gibbs energy of formation in kJ mol^-1 at 298.15 K and zero ionic strength, the standard enthalpy of formation in kJ mol^-1 at 298.15 K and zero ionic strength, the electric charge, and the number of hydrogen atoms in each species. There is a row in the matrix for each species of the reactant. gpfnsp is a list of the functions for the standard transformed Gibbs energies of the species. This program applies the Gibbs-Helmholtz equation to the function for the standard transformed Gibbs energy of formation of a reactant. The standard transformed enthalpy of formation of a reactant in kJ mol^-1 can be calculated at any temperature in the range 273.15 K to 313.15 K, any pH in the range 5 to 9, and any ionic strength in the range 0 to 0.35 M by use of ReplaceAll (/.)

derivetrST::usage="derivetrST[speciesmat_] derives the function of T (in Kelvin), pH, and ionic strength (is) that gives the stndard transformed entropy of formation of a reactant (sum of species). The input speciesmat is a matrix that gives the standard Gibbs energy of formation in kJ mol^-1 at 298.15 K and zero ionic strength, the standard enthalpy of formation in kJ mol^-1 at 298.15 K and zero ionic strength, the electric charge, and the number of hydrogen atoms in each species. There is a row in the matrix for each species of the reactant. gpfnsp is a list of the functions for the standard transformed Gibbs energies of the species. This program applies the equation for the entropy to the function for the standard transformed Gibbs energy of formation of a reactant. The standard transformed entropy of formation of a reactant in kJ K^-1 mol^-1 can be calculated at any temperature in the range 273.15 K to 313.15 K, any pH in the range 5 to 9, and any ionic strength in the range 0 to 0.35 M by use of ReplaceAll (/.).";

deriveNHT::usage="deriveNHT[speciesmat_] derives the function of T (in Kelvin), pH, and ionic strength (is) that gives the average number of hydrogen atoms in a reactant (sum of species). The input speciesmat is a matrix that gives the standard Gibbs energy of formation in kJ mol^-1 at 298.15 K and zero ionic strength, the standard enthalpy of formation in kJ mol^-1 at 298.15 K and zero ionic strength, the electric charge, and the number of hydrogen atoms in each species. There is a row in the matrix for each species of the reactant. gpfnsp is a list of the functions for the standard transformed Gibbs energies of the species. The average number can be calculated at any temperature in the range 273.15 K to 313.15 K, any pH in the range 5 to 9, and any ionic strength in the range 0 to 0.35 M by use of ReplaceAll (/.).";

Remove all the names to be used.

Remove [acetaldehydesp,acetatesp,acetoacetatesp,acetoacetylcoAsp,acetonesp,
acetylcoAsp,acetylphossp,aconitatecissp,adeninesp,adenosinesp,
adenosinephosphosulfatesp,adpsp,alaninesp,ammoniasp,ampsp,arabinosesp,
arabinose5phossp,asparagineLsp,aspartatesp,atpsp,bpgsp,butanalsp,butanolnsp,
butyratesp,cellobiosesp,citratesp,citrateisosp,co2gsp,co2totsp,coAsp,
coAglutathionesp,coaqsp,cogsp,creatinesp,creatininesp,cyclicampsp,cysteineLsp,
cystineLsp,cytochromecoxsp,cytochromecredsp,deoxyadenosinesp,deoxyadpsp,
deoxyampsp,deoxyatpsp,deoxyribosesp,deoxyribose1phossp,deoxyribose5phossp,
dihydroxyacetonesp,dihydroxyacetonephossp,ethaneaqsp,ethanolsp,
ethylacetatesp,fadenzoxsp,fadenzredsp,fadoxsp,fadredsp,ferredoxinoxsp,
ferredoxinredsp,ferricsp,ferroussp,fmnoxsp,fmnredsp,formatesp,fructosesp,
fructose16phossp,fructose6phossp,fumaratesp,galactono14lactonesp,
galactosesp,galactose1phossp,galactose6phossp,gluconolactonesp,
gluconolactone6phossp,glucosesp,glucose1phossp,glucose6phossp,glutamatesp,
glutaminesp,glutathioneoxsp,glutathioneredsp,glyceraldehydesp,
glyceraldehydephossp,glyceratesp,glycerolsp,glycerol3phossp,glycinesp,
glycolatesp,glycylglycinesp,glyoxylatesp,h2aqsp,h2gsp,h2osp,h2o2aqsp,
h2saqsp,hydroxy2oxoglutaratesp,hydroxyglutaratesp,hydroxypropionatebsp,
hydroxypyruvatesp,hypoxanthinesp,i2crsp,iditolsp,idpsp,impsp,indolesp,
inosinesp,iodideionsp,isomaltosesp,itpsp,ketoglutaramatesp,ketoglutaratesp,

```
lactatesp,lactosesp,lactulosesp,leucineisoLsp,leucineLsp,lyxosesp,malatesp,
maleatesp,maltosesp,malylcoAsp,mannitol1phossp,mannitolDsp,mannosesp,
mannose1phossp,mannose6phossp,methaneaqsp,methanegsp,methanolsp,methionineLsp,
methyl2oxopentanoatesp,methylmalatesp,methylmaleatesp,methylmalonylcoAsp,
n2aqsp,n2gsp,n2oaqsp,nadoxsp,nadpoxsp,nadpredsp,nadredsp,
nicotinamideribonucleotidesp,nitratesp,nitritesp,noaqsp,o2aqsp,o2gsp,
oxalatesp,oxaloacetatesp,oxalosuccinatesp,oxalylcoAsp,oxoprolinesp,
palmitatesp,pepsp,phenylalanineLsp,phosphoglycerate2sp,phosphoglycerate3sp,
phosphohydroxypyruvatesp,phosphoserinesp,pisp,ppisp,propanol2sp,propanolnsp,
propanoylcoAsp,prppsp,pyruvatesp,retinalsp,retinolsp,ribitolsp,ribosesp,
ribose1phossp,ribose5phossp,ribulosesp,ribulose5phossp,serineLsp,sorbitolsp,
sorbitol6phossp,sorbosesp,succinatesp,succinylcoAsp,sucrosesp,sulfatesp,
sulfitesp,sulfurcrsp,thioredoxinoxsp,thioredoxinredsp,threoninesp,
transcinnamatesp,trehalosesp,tryptophanLsp,tyrosineLsp,ubiquinoneoxsp,
ubiquinoneredsp,uratesp,ureasp,uricacidsp,valineLsp,xylitolsp,xylosesp,
xylulosesp]

Remove[acetaldehyde,acetate,acetoacetate,acetoacetylcoA,acetone,acetylcoA,
acetylphos,aconitatecis,adenine,adenosine,adenosinephosphosulfate,adp,
alanine,ammonia,amp,arabinose,arabinose5phos,asparagineL,aspartate,atp,bpg,
butanal,butanoln,butyrate,cellobiose,citrate,citrateiso,co2g,co2tot,coA,
coAglutathione,coaq,cog,creatine,creatinine,cyclicamp,cysteineL,cystineL,
cytochromecox,cytochromecred,deoxyadenosine,deoxyadp,deoxyamp,deoxyatp,
deoxyribose,deoxyribose1phos,deoxyribose5phos,dihydroxyacetone,
dihydroxyacetonephos,ethaneaq,ethanol,ethylacetate,fadenzox,fadenzred,fadox,
fadred,ferredoxinox,ferredoxinred,ferric,ferrous,fmnox,fmnred,formate,
fructose,fructose16phos,fructose6phos,fumarate,galactono14lactone,galactose,
galactose1phos,galactose6phos,gluconolactone,gluconolactone6phos,glucose,
glucose1phos,glucose6phos,glutamate,glutamine,glutathioneox,glutathionered,
glyceraldehyde,glyceraldehydephos,glycerate,glycerol,glycerol3phos,glycine,
glycolate,glycylglycine,glyoxylate,h2aq,h2g,h2o,h2o2aq,h2saq,hydroxy2oxoglutarate,
hydroxyglutarate,hydroxypropionateb,hydroxypyruvate,hypoxanthine,i2cr,iditol,
idp,imp,indole,inosine,iodideion,isomaltose,itp,ketoglutaramate,ketoglutarate,
lactate,lactose,lactulose,leucineisoL,leucineL,lyxose,malate,maleate,maltose,
malylcoA,mannitol1phos,mannitolD,mannose,mannose1phos,mannose6phos,methaneaq,
methaneg,methanol,methionineL,methyl2oxopentanoate,methylmalate,methylmaleate,
methylmalonylcoA,n2aq,n2g,n2oaq,nadox,nadpox,nadpred,nadred,
nicotinamideribonucleotide,nitrate,nitrite,noaq,o2aq,o2g,oxalate,oxaloacetate,
oxalosuccinate,oxalylcoA,oxoproline,palmitate,pep,phenylalanineL,
phosphoglycerate2,phosphoglycerate3,phosphohydroxypyruvate,phosphoserine,pi,
ppi,propanol2,propanoln,propanoylcoA,prpp,pyruvate,retinal,retinol,ribitol,
ribose,ribose1phos,ribose5phos,ribulose,ribulose5phos,serineL,sorbitol,
sorbitol6phos,sorbose,succinate,succinylcoA,sucrose,sulfate,sulfite,sulfurcr,
thioredoxinox,thioredoxinred,threonine,transcinnamate,trehalose,tryptophanL,
tyrosineL,ubiquinoneox,ubiquinonered,urate,urea,uricacid,valineL,xylitol,
xylose,xylulose]

Remove[acetaldehydeNH,acetateNH,acetoacetateNH,acetoacetylcoANH,acetoneNH,
acetylcoANH,acetylphosNH,aconitatecisNH,adenineNH,adenosineNH,
adenosinephosphosulfateNH,adpNH,alanineNH,ammoniaNH,ampNH,arabinoseNH,
arabinose5phosNH,asparagineLNH,aspartateNH,atpNH,bpgNH,butanalNH,butanolnNH,
butyrateNH,cellobioseNH,citrateNH,citrateisoNH,co2gNH,co2totNH,coANH,
coAglutathioneNH,coaqNH,cogNH,creatineNH,creatinineNH,cyclicampNH,cysteineLNH,
cystineLNH,cytochromecoxNH,cytochromecredNH,deoxyadenosineNH,deoxyadpNH,
deoxyampNH,deoxyatpNH,deoxyriboseNH,deoxyribose1phosNH,deoxyribose5phosNH,
dihydroxyacetoneNH,dihydroxyacetonephosNH,ethaneaqNH,ethanolNH,
ethylacetateNH,fadenzoxNH,fadenzredNH,fadoxNH,fadredNH,ferredoxinoxNH,
ferredoxinredNH,ferricNH,ferrousNH,fmnoxNH,fmnredNH,formateNH,fructoseNH,
fructose16phosNH,fructose6phosNH,fumarateNH,galactono14lactoneNH,
galactoseNH,galactose1phosNH,galactose6phosNH,gluconolactoneNH,
gluconolactone6phosNH,glucoseNH,glucose1phosNH,glucose6phosNH,glutamateNH,
glutamineNH,glutathioneoxNH,glutathioneredNH,glyceraldehydeNH,
glyceraldehydephosNH,glycerateNH,glycerolNH,glycerol3phosNH,glycineNH,
glycolateNH,glycylglycineNH,glyoxylateNH,h2aqNH,h2gNH,h2oNH,h2o2aqNH,
h2saqNH,hydroxy2oxoglutarateNH,hydroxyglutarateNH,hydroxypropionatebNH,
```

hydroxypyruvateNH,hypoxanthineNH,i2crNH,iditolNH,idpNH,impNH,indoleNH,
inosineNH,iodideionNH,isomaltoseNH,itpNH,ketoglutaramateNH,
ketoglutarateNH,lactateNH,lactoseNH,lactuloseNH,leucineisoLNH,leucineLNH,
lyxoseNH,malateNH,maleateNH,maltoseNH,malylcoANH,mannitol1phosNH,mannitolDNH,
mannoseNH,mannose1phosNH,mannose6phosNH,methaneaqNH,methanegNH,methanolNH,
methionineLNH,methyl2oxopentanoateNH,methylmalateNH,methylmaleateNH,
methylmalonylcoANH,n2aqNH,n2gNH,n2oaqNH,nadoxNH,nadpoxNH,nadpredNH,nadredNH,
nicotinamideribonucleotideNH,nitrateNH,nitriteNH,noaqNH,o2aqNH,o2gNH,
oxalateNH,oxaloacetateNH,oxalosuccinateNH,oxalylcoANH,oxoprolineNH,
palmitateNH,pepNH,phenylalanineLNH,phosphoglycerate2NH,phosphoglycerate3NH,
phosphohydroxypyruvateNH,phosphoserineNH,piNH,ppiNH,propanol2NH,propanolnNH,
propanoylcoANH,prppNH,pyruvateNH,retinalNH,retinolNH,ribitolNH,riboseNH,
ribose1phosNH,ribose5phosNH,ribuloseNH,ribulose5phosNH,serineLNH,sorbitolNH,
sorbitol6phosNH,sorboseNH,succinateNH,succinylcoANH,sucroseNH,sulfateNH,
sulfiteNH,sulfurcrNH,thioredoxinoxNH,thioredoxinredNH,threonineNH,
transcinnamateNH,trehaloseNH,tryptophanLNH,tyrosineLNH,ubiquinoneoxNH,
ubiquinoneredNH,urateNH,ureaNH,uricacidNH,valineLNH,xylitolNH,xyloseNH,
xyluloseNH]

Remove[acetaldehydeGT,acetateGT,acetoneGT,adenineGT,adenosineGT,adpGT,
alanineGT,ammoniaGT,ampGT,arabinoseGT,asparagineLGT,aspartateGT,atpGT,
citrateGT,co2gGT,co2totGT,coaqGT,cogGT,ethaneaqGT,ethanolGT,ethylacetateGT,
ferricGT,ferrousGT,formateGT,fructoseGT,fructose16phosGT,fructose6phosGT,
fumarateGT,galactoseGT,glucoseGT,glucose6phosGT,glutamateGT,glutamineGT,
glycerolGT,glycerol3phosGT,glycineGT,glycylglycineGT,h2aqGT,h2gGT,h2oGT,
h2o2aqGT,h2saqGT,i2crGT,idpGT,impGT,indoleGT,inosineGT,iodideionGT,
isomaltoseGT,itpGT,ketoglutarateGT,lactateGT,lactoseGT,leucineLGT,malateGT,
maltoseGT,mannoseGT,mannose6phosGT,methaneaqGT,methanegGT,methanolGT,n2aqGT,
n2gGT,n2oaqGT,nadoxGT,nadpoxGT,nadpredGT,nadredGT,nitrateGT,nitriteGT,
noaqGT,o2aqGT,o2gGT,oxaloacetateGT,pepGT,piGT,ppiGT,propanol2GT,pyruvateGT,
riboseGT,ribose1phosGT,ribose5phosGT,ribuloseGT,sorboseGT,succinateGT,
sucroseGT,sulfateGT,sulfiteGT,sulfurcrGT,tryptophanLGT,ureaGT,valineLGT,
xyloseGT,xyluloseGT]

Remove[acetaldehydeHT,acetateHT,acetoneHT,adenineHT,adenosineHT,adpHT,alanineHT,
ammoniaHT,ampHT,arabinoseHT,asparagineLHT,aspartateHT,atpHT,citrateHT,co2gHT,
co2totHT,coaqHT,cogHT,ethaneaqHT,ethanolHT,ethylacetateHT,ferricHT,ferrousHT,
formateHT,fructoseHT,fructose16phosHT,fructose6phosHT,fumarateHT,galactoseHT,
glucoseHT,glucose6phosHT,glutamateHT,glutamineHT,glycerolHT,glycerol3phosHT,
glycineHT,glycylglycineHT,h2aqHT,h2gHT,h2oHT,h2o2aqHT,h2saqHT,i2crHT,idpHT,
impHT,indoleHT,inosineHT,iodideionHT,isomaltoseHT,itpHT,ketoglutarateHT,
lactateHT,lactoseHT,leucineLHT,malateHT,maltoseHT,mannoseHT,mannose6phosHT,
methaneaqHT,methanegHT,methanolHT,n2aqHT,n2gHT,n2oaqHT,nadoxHT,nadpoxHT,
nadpredHT,nadredHT,nitrateHT,nitriteHT,noaqHT,o2aqHT,o2gHT,oxaloacetateHT,
pepHT,piHT,ppiHT,propanol2HT,pyruvateHT,riboseHT,ribose1phosHT,ribose5phosHT,
ribuloseHT,sorboseHT,succinateHT,sucroseHT,sulfateHT,sulfiteHT,sulfurcrHT,
tryptophanLHT,ureaHT,valineLHT,xyloseHT,xyluloseHT]

Remove[acetaldehydeST,acetateST,acetoneST,adenineST,adenosineST,adpST,alanineST,
ammoniaST,ampST,arabinoseST,asparagineLST,aspartateST,atpST,citrateST,co2gST,
co2totST,coaqST,cogST,ethaneaqST,ethanolST,ethylacetateST,ferricST,ferrousST,
formateST,fructoseST,fructose16phosST,fructose6phosST,fumarateST,galactoseST,
glucoseST,glucose6phosST,glutamateST,glutamineST,glycerolST,glycerol3phosST,
glycineST,glycylglycineST,h2aqST,h2gST,h2oST,h2o2aqST,h2saqST,i2crST,idpST,
impST,indoleST,inosineST,iodideionST,isomaltoseST,itpST,ketoglutarateST,
lactateST,lactoseST,leucineLST,malateST,maltoseST,mannoseST,mannose6phosST,
methaneaqST,methanegST,methanolST,n2aqST,n2gST,n2oaqST,nadoxST,nadpoxST,
nadpredST,nadredST,nitrateST,nitriteST,noaqST,o2aqST,o2gST,oxaloacetateST,
pepST,piST,ppiST,propanol2ST,pyruvateST,riboseST,ribose1phosST,ribose5phosST,
ribuloseST,sorboseST,succinateST,sucroseST,sulfateST,sulfiteST,sulfurcrST,
tryptophanLST,ureaST,valineLST,xyloseST,xyluloseST]

Remove[acetaldehydeNHT,acetateNHT,acetoneNHT,adenineNHT,adenosineNHT,adpNHT,
alanineNHT,ammoniaNHT,ampNHT,arabinoseNHT,asparagineLNHT,aspartateNHT,atpNHT,
citrateNHT,co2gNHT,co2totNHT,coaqNHT,cogNHT,ethaneaqNHT,ethanolNHT,

```
    ethylacetateNHT,ferricNHT,ferrousNHT,formateNHT,fructoseNHT,fructose16phosNHT,
    fructose6phosNHT,fumarateNHT,galactoseNHT,glucoseNHT,glucose6phosNHT,
    glutamateNHT,glutamineNHT,glycerolNHT,glycerol3phosNHT,glycineNHT,
    glycylglycineNHT,h2aqNHT,h2gNHT,h2oNHT,h2o2aqNHT,h2saqNHT,i2crNHT,idpNHT,
    impNHT,indoleNHT,inosineNHT,iodideionNHT,isomaltoseNHT,itpNHT,
    ketoglutarateNHT,lactateNHT,lactoseNHT,leucineLNHT,malateNHT,maltoseNHT,
    mannoseNHT,mannose6phosNHT,methaneaqNHT,methanegNHT,methanolNHT,n2aqNHT,
    n2gNHT,n2oaqNHT,nadoxNHT,nadpoxNHT,nadpredNHT,nadredNHT,nitrateNHT,nitriteNHT,
    noaqNHT,o2aqNHT,o2gNHT,oxaloacetateNHT,pepNHT,piNHT,ppiNHT,propanol2NHT,
    pyruvateNHT,riboseNHT,ribose1phosNHT,ribose5phosNHT,ribuloseNHT,sorboseNHT,
    succinateNHT,sucroseNHT,sulfateNHT,sulfiteNHT,sulfurcrNHT,tryptophanLNHT,
    ureaNHT,valineLNHT,xyloseNHT,xyluloseNHT]
```

Unprotect all the names to be used.

```
    Unprotect [t,pH, is,
    acetaldehydesp,acetatesp,acetoacetatesp,acetoacetylcoAsp,acetonesp,
    acetylcoAsp,acetylphossp,aconitatecissp,adeninesp,adenosinesp,
    adenosinephosphosulfatesp,adpsp,alaninesp,ammoniasp,ampsp,arabinosesp,
    arabinose5phossp,asparagineLsp,aspartatesp,atpsp,bpgsp,butanalsp,butanolnsp,
    butyratesp,cellobiosesp,citratesp,citrateisosp,co2gsp,co2totsp,coAsp,
    coAglutathionesp,coaqsp,cogsp,creatinesp,creatininesp,cyclicampsp,cysteineLsp,
    cystineLsp,cytochromecoxsp,cytochromecredsp,deoxyadenosinesp,deoxyadpsp,
    deoxyampsp,deoxyatpsp,deoxyribosesp,deoxyribose1phossp,deoxyribose5phossp,
    dihydroxyacetonesp,dihydroxyacetonephossp,ethaneaqsp,ethanolsp,
    ethylacetatesp,fadenzoxsp,fadenzredsp,fadoxsp,fadredsp,ferredoxinoxsp,
    ferredoxinredsp,ferricsp,ferroussp,fmnoxsp,fmnredsp,formatesp,fructosesp,
    fructose16phossp,fructose6phossp,fumaratesp,galactono14lactonesp,
    galactosesp,galactose1phossp,galactose6phossp,gluconolactonesp,
    gluconolactone6phossp,glucosesp,glucose1phossp,glucose6phossp,glutamatesp,
    glutaminesp,glutathioneoxsp,glutathioneredsp,glyceraldehydesp,
    glyceraldehydephossp,glyceratesp,glycerolsp,glycerol3phossp,glycinesp,
    glycolatesp,glycylglycinesp,glyoxylatesp,h2aqsp,h2gsp,h2osp,h2o2aqsp,
    h2saqsp,hydroxy2oxoglutaratesp,hydroxyglutaratesp,hydroxypropionatebsp,
    hydroxypyruvatesp,hypoxanthinesp,i2crsp,iditolsp,idpsp,impsp,indolesp,
    inosinesp,iodideionsp,isomaltosesp,itpsp,ketoglutaramatesp,
    ketoglutaratesp,lactatesp,lactosesp,lactulosesp,leucineisoLsp,leucineLsp,
    lyxosesp,malatesp,maleatesp,maltosesp,malylcoAsp,mannitol1phossp,mannitolDsp,
    mannosesp,mannose1phossp,mannose6phossp,methaneaqsp,methanegsp,methanolsp,
    methionineLsp,methyl2oxopentanoatesp,methylmalatesp,methylmaleatesp,
    methylmalonylcoAsp,n2aqsp,n2gsp,n2oaqsp,nadoxsp,nadpoxsp,nadpredsp,nadredsp,
    nicotinamideribonucleotidesp,nitratesp,nitritesp,noaqsp,o2aqsp,o2gsp,
    oxalatesp,oxaloacetatesp,oxalosuccinatesp,oxalylcoAsp,oxoprolinesp,
    palmitatesp,pepsp,phenylalanineLsp,phosphoglycerate2sp,phosphoglycerate3sp,
    phosphohydroxypyruvatesp,phosphoserinesp,pisp,ppisp,propanol2sp,propanolnsp,
    propanoylcoAsp,prppsp,pyruvatesp,retinalsp,retinolsp,ribitolsp,ribosesp,
    ribose1phossp,ribose5phossp,ribulosesp,ribulose5phossp,serineLsp,sorbitolsp,
    sorbitol6phossp,sorbosesp,succinatesp,succinylcoAsp,sucrosesp,sulfatesp,
    sulfitesp,sulfurcrsp,thioredoxinoxsp,thioredoxinredsp,threoninesp,
    transcinnamatesp,trehalosesp,tryptophanLsp,tyrosineLsp,ubiquinoneoxsp,
    ubiquinoneredsp,uratesp,ureasp,uricacidsp,valineLsp,xylitolsp,xylosesp,
    xylulosesp];
```

```
    Unprotect[acetaldehyde,acetate,acetoacetate,acetoacetylcoA,acetone,acetylcoA,
    acetylphos,aconitatecis,adenine,adenosine,adenosinephosphosulfate,adp,alanine,
    ammonia,amp,arabinose,arabinose5phos,asparagineL,aspartate,atp,bpg,butanal,
    butanoln,butyrate,cellobiose,citrate,citrateiso,co2g,co2tot,coA,
    coAglutathione,coaq,cog,creatine,creatinine,cyclicamp,cysteineL,cystineL,
    cytochromecox,cytochromecred,deoxyadenosine,deoxyadp,deoxyamp,deoxyatp,
    deoxyribose,deoxyribose1phos,deoxyribose5phos,dihydroxyacetone,
    dihydroxyacetonephos,ethaneaq,ethanol,ethylacetate,fadenzox,fadenzred,fadox,
    fadred,ferredoxinox,ferredoxinred,ferric,ferrous,fmnox,fmnred,formate,
    fructose,fructose16phos,fructose6phos,fumarate,galactono14lactone,galactose,
    galactose1phos,galactose6phos,gluconolactone,gluconolactone6phos,glucose,
    glucose1phos,glucose6phos,glutamate,glutamine,glutathioneox,glutathionered,
```

glyceraldehyde,glyceraldehydephos,glycerate,glycerol,glycerol3phos,glycine,
glycolate,glycylglycine,glyoxylate,h2aq,h2g,h2o,h2o2aq,h2saq,
hydroxy2oxoglutarate,hydroxyglutarate,hydroxypropionateb,hydroxypyruvate,
hypoxanthine,i2cr,iditol,idp,imp,indole,inosine,iodideion,isomaltose,itp,
ketoglutaramate,ketoglutarate,lactate,lactose,lactulose,leucineisoL,
leucineL,lyxose,malate,maleate,maltose,malylcoA,mannitol1phos,mannitolD,
mannose,mannose1phos,mannose6phos,methaneaq,methaneg,methanol,methionineL,
methyl2oxopentanoate,methylmalate,methylmaleate,methylmalonylcoA,n2aq,n2g,
n2oaq,nadox,nadpox,nadpred,nadred,nicotinamideribonucleotide,nitrate,
nitrite,noaq,o2aq,o2g,oxalate,oxaloacetate,oxalosuccinate,oxalylcoA,
oxoproline,palmitate,pep,phenylalanineL,phosphoglycerate2,phosphoglycerate3,
phosphohydroxypyruvate,phosphoserine,pi,ppi,propanol2,propanoln,
propanoylcoA,prpp,pyruvate,retinal,retinol,ribitol,ribose,ribose1phos,
ribose5phos,ribulose,ribulose5phos,serineL,sorbitol,sorbitol6phos,sorbose,
succinate,succinylcoA,sucrose,sulfate,sulfite,sulfurcr,thioredoxinox,
thioredoxinred,threonine,transcinnamate,trehalose,tryptophanL,tyrosineL,
ubiquinoneox,ubiquinonered,urate,urea,uricacid,valineL,xylitol,xylose,
xylulose];

Unprotect[acetaldehydeNH,acetateNH,acetoacetateNH,acetoacetylcoANH,acetoneNH,
acetylcoANH,acetylphosNH,aconitatecisNH,adenineNH,adenosineNH,
adenosinephosphosulfateNH,adpNH,alanineNH,ammoniaNH,ampNH,arabinoseNH,
arabinose5phosNH,asparagineLNH,aspartateNH,atpNH,bpgNH,butanalNH,butanolnNH,
butyrateNH,cellobioseNH,citrateNH,citrateisoNH,co2gNH,co2totNH,coANH,
coAglutathioneNH,coaqNH,cogNH,creatineNH,creatinineNH,cyclicampNH,cysteineLNH,
cystineLNH,cytochromecoxNH,cytochromecredNH,deoxyadenosineNH,deoxyadpNH,
deoxyampNH,deoxyatpNH,deoxyriboseNH,deoxyribose1phosNH,deoxyribose5phosNH,
dihydroxyacetoneNH,dihydroxyacetonephosNH,ethaneaqNH,ethanolNH,
ethylacetateNH,fadenzoxNH,fadenzredNH,fadoxNH,fadredNH,ferredoxinoxNH,
ferredoxinredNH,ferricNH,ferrousNH,fmnoxNH,fmnredNH,formateNH,fructoseNH,
fructose16phosNH,fructose6phosNH,fumarateNH,galactono14lactoneNH,
galactoseNH,galactose1phosNH,galactose6phosNH,gluconolactoneNH,
gluconolactone6phosNH,glucoseNH,glucose1phosNH,glucose6phosNH,glutamateNH,
glutamineNH,glutathioneoxNH,glutathioneredNH,glyceraldehydeNH,
glyceraldehydephosNH,glycerateNH,glycerolNH,glycerol3phosNH,glycineNH,
glycolateNH,glycylglycineNH,glyoxylateNH,h2aqNH,h2gNH,h2oNH,h2o2aqNH,
h2saqNH,hydroxy2oxoglutarateNH,hydroxyglutarateNH,hydroxypropionatebNH,
hydroxypyruvateNH,hypoxanthineNH,i2crNH,iditolNH,idpNH,impNH,indoleNH,
inosineNH,iodideionNH,isomaltoseNH,itpNH,ketoglutaramateNH,
ketoglutarateNH,lactateNH,lactoseNH,lactuloseNH,leucineisoLNH,leucineLNH,
lyxoseNH,malateNH,maleateNH,maltoseNH,malylcoANH,mannitol1phosNH,mannitolDNH,
mannoseNH,mannose1phosNH,mannose6phosNH,methaneaqNH,methanegNH,methanolNH,
methionineLNH,methyl2oxopentanoateNH,methylmalateNH,methylmaleateNH,
methylmalonylcoANH,n2aqNH,n2gNH,n2oaqNH,nadoxNH,nadpoxNH,nadpredNH,nadredNH,
nicotinamideribonucleotideNH,nitrateNH,nitriteNH,noaqNH,o2aqNH,o2gNH,
oxalateNH,oxaloacetateNH,oxalosuccinateNH,oxalylcoANH,oxoprolineNH,
palmitateNH,pepNH,phenylalanineLNH,phosphoglycerate2NH,phosphoglycerate3NH,
phosphohydroxypyruvateNH,phosphoserineNH,piNH,ppiNH,propanol2NH,propanolnNH,
propanoylcoANH,prppNH,pyruvateNH,retinalNH,retinolNH,ribitolNH,riboseNH,
ribose1phosNH,ribose5phosNH,ribuloseNH,ribulose5phosNH,serineLNH,sorbitolNH,
sorbitol6phosNH,sorboseNH,succinateNH,succinylcoANH,sucroseNH,sulfateNH,
sulfiteNH,sulfurcrNH,thioredoxinoxNH,thioredoxinredNH,threonineNH,
transcinnamateNH,trehaloseNH,tryptophanLNH,tyrosineLNH,ubiquinoneoxNH,
ubiquinoneredNH,urateNH,ureaNH,uricacidNH,valineLNH,xylitolNH,xyloseNH,
xyluloseNH];

Unprotect[acetaldehydeGT,acetateGT,acetoneGT,adenineGT,adenosineGT,adpGT,
alanineGT,ammoniaGT,ampGT,arabinoseGT,asparagineLGT,aspartateGT,atpGT,
citrateGT,co2gGT,co2totGT,coaqGT,cogGT,ethaneaqGT,ethanolGT,ethylacetateGT,
ferricGT,ferrousGT,formateGT,fructoseGT,fructose16phosGT,fructose6phosGT,
fumarateGT,galactoseGT,glucoseGT,glucose6phosGT,glutamateGT,glutamineGT,
glycerolGT,glycerol3phosGT,glycineGT,glycylglycineGT,h2aqGT,h2gGT,h2oGT,
h2o2aqGT,h2saqGT,i2crGT,idpGT,impGT,indoleGT,inosineGT,iodideionGT,
isomaltoseGT,itpGT,ketoglutarateGT,lactateGT,lactoseGT,leucineLGT,malateGT,
maltoseGT,mannoseGT,mannose6phosGT,methaneaqGT,methanegGT,methanolGT,n2aqGT,

```
n2gGT,n2oaqGT,nadoxGT,nadpoxGT,nadpredGT,nadredGT,nitrateGT,nitriteGT,
noaqGT,o2aqGT,o2gGT,oxaloacetateGT,pepGT,piGT,ppiGT,propanol2GT,pyruvateGT,
riboseGT,ribose1phosGT,ribose5phosGT,ribuloseGT,sorboseGT,succinateGT,
sucroseGT,sulfateGT,sulfiteGT,sulfurcrGT,tryptophanLGT,ureaGT,valineLGT,
xyloseGT,xyluloseGT];

Unprotect[acetaldehydeHT,acetateHT,acetoneHT,adenineHT,adenosineHT,adpHT,alanineHT,
ammoniaHT,ampHT,arabinoseHT,asparagineLHT,aspartateHT,atpHT,citrateHT,co2gHT,
co2totHT,coaqHT,cogHT,ethaneaqHT,ethanolHT,ethylacetateHT,ferricHT,ferrousHT,
formateHT,fructoseHT,fructose16phosHT,fructose6phosHT,fumarateHT,galactoseHT,
glucoseHT,glucose6phosHT,glutamateHT,glutamineHT,glycerolHT,glycerol3phosHT,
glycineHT,glycylglycineHT,h2aqHT,h2gHT,h2oHT,h2o2aqHT,h2saqHT,i2crHT,idpHT,
impHT,indoleHT,inosineHT,iodideionHT,isomaltoseHT,itpHT,ketoglutarateHT,
lactateHT,lactoseHT,leucineLHT,malateHT,maltoseHT,mannoseHT,mannose6phosHT,
methaneaqHT,methanegHT,methanolHT,n2aqHT,n2gHT,n2oaqHT,nadoxHT,nadpoxHT,
nadpredHT,nadredHT,nitrateHT,nitriteHT,noaqHT,o2aqHT,o2gHT,oxaloacetateHT,
pepHT,piHT,ppiHT,propanol2HT,pyruvateHT,riboseHT,ribose1phosHT,ribose5phosHT,
ribuloseHT,sorboseHT,succinateHT,sucroseHT,sulfateHT,sulfiteHT,sulfurcrHT,
tryptophanLHT,ureaHT,valineLHT,xyloseHT,xyluloseHT];

Unprotect[acetaldehydeST,acetateST,acetoneST,adenineST,adenosineST,adpST,alanineST,
ammoniaST,ampST,arabinoseST,asparagineLST,aspartateST,atpST,citrateST,co2gST,
co2totST,coaqST,cogST,ethaneaqST,ethanolST,ethylacetateST,ferricST,ferrousST,
formateST,fructoseST,fructose16phosST,fructose6phosST,fumarateST,galactoseST,
glucoseST,glucose6phosST,glutamateST,glutamineST,glycerolST,glycerol3phosST,
glycineST,glycylglycineST,h2aqST,h2gST,h2oST,h2o2aqST,h2saqST,i2crST,idpST,
impST,indoleST,inosineST,iodideionST,isomaltoseST,itpST,ketoglutarateST,
lactateST,lactoseST,leucineLST,malateST,maltoseST,mannoseST,mannose6phosST,
methaneaqST,methanegST,methanolST,n2aqST,n2gST,n2oaqST,nadoxST,nadpoxST,
nadpredST,nadredST,nitrateST,nitriteST,noaqST,o2aqST,o2gST,oxaloacetateST,
pepST,piST,ppiST,propanol2ST,pyruvateST,riboseST,ribose1phosST,ribose5phosST,
ribuloseST,sorboseST,succinateST,sucroseST,sulfateST,sulfiteST,sulfurcrST,
tryptophanLST,ureaST,valineLST,xyloseST,xyluloseST];

Unprotect[acetaldehydeNHT,acetateNHT,acetoneNHT,adenineNHT,adenosineNHT,adpNHT,
alanineNHT,ammoniaNHT,ampNHT,arabinoseNHT,asparagineLNHT,aspartateNHT,atpNHT,
citrateNHT,co2gNHT,co2totNHT,coaqNHT,cogNHT,ethaneaqNHT,ethanolNHT,
ethylacetateNHT,ferricNHT,ferrousNHT,formateNHT,fructoseNHT,fructose16phosNHT,
fructose6phosNHT,fumarateNHT,galactoseNHT,glucoseNHT,glucose6phosNHT,
glutamateNHT,glutamineNHT,glycerolNHT,glycerol3phosNHT,glycineNHT,
glycylglycineNHT,h2aqNHT,h2gNHT,h2oNHT,h2o2aqNHT,h2saqNHT,i2crNHT,idpNHT,
impNHT,indoleNHT,inosineNHT,iodideionNHT,isomaltoseNHT,itpNHT,
ketoglutarateNHT,lactateNHT,lactoseNHT,leucineLNHT,malateNHT,maltoseNHT,
mannoseNHT,mannose6phosNHT,methaneaqNHT,methanegNHT,methanolNHT,n2aqNHT,
n2gNHT,n2oaqNHT,nadoxNHT,nadpoxNHT,nadpredNHT,nadredNHT,nitrateNHT,nitriteNHT,
noaqNHT,o2aqNHT,o2gNHT,oxaloacetateNHT,pepNHT,piNHT,ppiNHT,propanol2NHT,
pyruvateNHT,riboseNHT,ribose1phosNHT,ribose5phosNHT,ribuloseNHT,sorboseNHT,
succinateNHT,sucroseNHT,sulfateNHT,sulfiteNHT,sulfurcrNHT,tryptophanLNHT,
ureaNHT,valineLNHT,xyloseNHT,xyluloseNHT];
```

2 Basic data on the species that make up a reactant

The rows in the data matrices are of the form $\{\Delta_f G°, \Delta_f H°, z, N_H\}$, where the energies are in kJ mol^{-1} at 298.15 K and zero ionic strength, z is the charge number, and N_H is the number of hydrogen atoms in the species.

```
acetaldehydesp={{-139.,-212.23,0,4}};
acetatesp={{-369.31,-486.01,-1,3},{-396.45,-485.76,0,4}};
acetoacetatesp={{-482.49,_,0,5}};
acetoacetylcoAsp={{-285.32,_,0,5}};
```

```
acetonesp={{-159.7,-221.71,0,6}};
acetylcoAsp={{-180.36,_,0,3}};
acetylphossp={{-1219.49,_,-2,3},{-1269.08,_,-1,4},{-1298.26,_,0,5}};
aconitatecissp={{-917.13,_,-3,3}};
adeninesp={{313.40,129.5,0,5},{289.43,109.4,1,6}};
adenosinesp={{-194.5,-621.3,0,13},{-214.28,-637.7,1,14}};
adenosinephosphosulfatesp={{-1541.98,_,-1,12}};
adpsp={{-1906.13,-2626.54,-3,12},{-1947.1,-2620.94,-2,
        13},{-1971.98,-2638.54,-1,14}};
alaninesp={{-371.,-554.8,0,7}};
ammoniasp={{-26.5,-80.29,0,3},{-79.31,-132.51,1,4}};
ampsp={{-1040.45,-1635.37,-2,12},{-1078.86,-1629.97,-1,
        13},{-1101.63,-1648.07,0,14}};
arabinosesp={{-742.23,-1043.79,0,10}};
arabinose5phossp={{-1598.34,_,-2,9},{-1636.53,_,-1,10}};
asparagineLsp={{-525.93,-766.09,0,8}};
aspartatesp={{-695.88,-943.41,-1,6}};
atpsp={{-2768.1,-3619.21,-4,12},{-2811.48,-3612.91,-3,
        13},{-2838.18,-3627.91,-2,14}};
bpgsp={{-2356.14,_,-4,4},{-2401.58,_,-3,5}};
butanalsp={{-129.87,_,0,8}};
butanolnsp={{-171.84,_,0,10}};
butyratesp={{-352.63,_,-1,7}};
cellobiosesp={{-1585.94,_,0,22}};
citratesp={{-1162.69,-1515.11,-3,5},{-1199.18,-1518.46,-2,
        6},{-1226.33,-1520.88,-1,7}};
citrateisosp={{-1156.04,_,-3,5},{-1192.57,_,-2,6},{-1219.47,_,-1,7}};
co2gsp={{-394.36,-393.5,0,0}};
co2totsp={{-527.81,-677.14,-2,0},{-586.77,-691.99,-1,1},{-623.11,-699.63,0,
        2}};
coAsp={{0,_,-1,0},{-47.83,_,0,1}};
coAglutathionesp={{-35.85,_,-1,15}};
coaqsp={{-119.9,-120.96,0,0}};
cogsp={{-137.17,-110.53,0,0}};
creatinesp={{-259.2,_,0,9}};
creatininesp={{-23.14,_,0,7}};
cyclicampsp={{-629.68,_,-1,7}};
cysteineLsp={{-291.,_,-1,6},{-338.82,_,0,7}};
cystineLsp={{-666.51,_,0,12}};
cytochromecoxsp={{0,_,3,0}};
cytochromecredsp={{-24.51,_,2,0}};
deoxyadenosinesp={{-46.64,_,0,13},{-66.42,_,1,14}};
deoxyadpsp={{-1758.27,_,-3,12},{-1799.24,_,-2,13},{-1824.12,_,-1,14}};
deoxyampsp={{-892.59,_,-2,12},{-931.0,_,-1,13},{-953.77,_,0,14}};
deoxyatpsp={{-2620.24,_,-4,12},{-2663.62,_,-3,13},{-2690.32,_,-2,14}};
deoxyribosesp={{-604.14,_,0,10}};
deoxyribose1phossp={{-1439.84,_,-2,9},{-1478.02,_,-1,10}};
deoxyribose5phossp={{-1447.92,_,-2,9},{-1486.11,_,-1,10}};
dihydroxyacetonesp={{-451.0,_,0,6}};
dihydroxyacetonephossp={{-1296.26,_,-2,5},{-1328.8,_,-1,6}};
ethaneaqsp={{-17.01,-102.09,0,6}};
ethanolsp={{-181.64,-288.3,0,6}};
ethylacetatesp={{-337.65,-482.,0,8}};
fadenzoxsp={{0,_,-2,31}};
fadenzredsp={{-88.6,_,-2,33}};
fadoxsp={{0,_,-2,31}};
fadredsp={{-38.88,_,-2,33}};
ferredoxinoxsp={{0,_,1,0}};
ferredoxinredsp={{38.07,_,0,0}};
ferricsp={{-4.70,-48.5,3,0}};
ferroussp={{-78.90,-89.10,2,0}};
fmnoxsp={{0,_,-2,19}};
fmnredsp={{-38.88,_,-2,21}};
formatesp={{-351.,-425.55,-1,1}};
fructosesp={{-915.51,-1259.38,0,12}};
```

```
fructose16phossp={{-2601.4,-3343.25,-4,10},{-2639.36,-3341.45,-3,11},{-2673.\
\
89,-3339.65,-2,12}};
fructose6phossp={{-1760.8,-2265.17,-2,11},{-1796.6,-2263.37,-1,12}};
fumaratesp={{-601.87,-777.39,-2,2},{-628.14,-774.46,-1,3},{-645.8,-774.88,0,
      4}};
galactono14lactonesp={{-905.34,_,0,10}};
galactosesp={{-908.93,-1255.2,0,12}};
galactose1phossp={{-1756.69,_,-2,11},{-1791.77,_,-1,12}};
galactose6phossp={{-1756.81,_,-2,11},{-1793.57,_,-1,12}};
gluconolactonesp={{-903.50,_,0,10}};
gluconolactone6phossp={{-1746.08,_,-2,9},{-1782.73,_,-1,10}};
glucosesp={{-915.9,-1262.19,0,12}};
glucose1phossp={{-1756.87,_,-2,11},{-1793.98,_,-1,12}};
glucose6phossp={{-1763.94,-2276.44,-2,11},{-1800.59,-2274.64,-1,12}};
glutamatesp={{-697.47,-979.89,-1,8}};
glutaminesp={{-528.02,-805.,0,10}};
glutathioneoxsp={{0,_,-2,30}};
glutathioneredsp={{34.17,_,-2,15},{-13.44,_,-1,16}};
glyceraldehydesp={{-440.02,_,0,6}};
glyceraldehydephossp={{-1288.6,_,-2,5},{-1321.14,_,-1,6}};
glyceratesp={{-661.01,_,-1,5}};
glycerolsp={{-497.48,-676.55,0,8}};
glycerol3phossp={{-1358.96,-1724.59,-2,7},{-1397.04,-1722.79,-1,8}};
glycinesp={{-379.91,-523.,0,5}};
glycolatesp={{-530.95,_,-1,3}};
glycylglycinesp={{-520.2,-734.25,0,8}};
glyoxylatesp={{-468.6,_,-1,1}};
h2aqsp={{17.6,-4.2,0,2}};
h2gsp={{0,0,0,2}};
h2osp={{-237.19,-285.83,0,2}};
h2o2aqsp={{-134.03,-191.17,0,2}};
h2saqsp={{85.8,33.1,-2,0},{12.08,-17.6,-1,1},{-27.83,-39.7,0,2}};
hydroxy2oxoglutaratesp={{-951.78,_,-2,4}};
hydroxyglutaratesp={{-840.18,_,-2,6}};
hydroxypropionatebsp={{-518.4,_,-1,5}};
hydroxypyruvatesp={{-608.55,_,-1,3}};
hypoxanthinesp={{89.5,_,0,4}};
i2crsp={{0,0,0,0}};
iditolsp={{-944.73,_,0,14}};
idpsp={{-2066.44,-2797.18,-4,10},{-2121.01,-2825.88,-3,
      11},{-2161.99,-2820.28,-2,12}};
impsp={{-1200.43,-1799.61,-3,10},{-1255.42,-1834.71,-2,
      11},{-1293.84,-1829.31,-1,12}};
indolesp={{223.8,97.5,0,7}};
inosinesp={{-358.02,-792.74,-1,11},{-409.15,-819.84,0,12}};
iodideionsp={{-51.57,-55.19,-1,0}};
isomaltosesp={{-1587.71,-2244.48,0,22}};
itpsp={{-2925.31,-3794.65,-5,10},{-2982.90,-3818.60,-4,
      11},{-3026.28,-3812.25,-3,12}};
ketoglutaramatesp={{-595.88,_,-1,5}};
ketoglutaratesp={{-793.41,-1044.06,-2,4}};
lactatesp={{-516.72,-686.64,-1,5}};
lactosesp={{-1567.33,-2233.08,0,22}};
lactulosesp={{-1575.22,_,0,22}};
leucineisoLsp={{-343.9,_,0,13}};
leucineLsp={{-352.25,-643.37,0,13}};
lyxosesp={{-749.14,_,0,10}};
malatesp={{-842.66,-1079.79,-2,4},{-872.68,-1079.95,-1,5}};
maleatesp={{-592.09,_,-2,2}};
maltosesp={{-1574.69,-2238.06,0,22}};
malylcoAsp={{-663.44,_,-1,4},{-687.47,_,0,5}};
mannitol1phossp={{-1793.58,_,-2,13},{-1830.68,_,-1,14}};
mannitolDsp={{-942.61,_,0,14}};
mannosesp={{-910.,-1258.66,0,12}};
```

```
mannose1phossp={{-1754.56,_,-2,11},{-1791.32,_,-1,12}};
mannose6phossp={{-1759.87,-2273.71,-2,11},{-1796.63,-2271.90,-1,12}};
methaneaqsp={{-34.33,-89.04,0,4}};
methanegsp={{-50.72,-74.81,0,4}};
methanolsp={{-175.31,-245.93,0,4}};
methionineLsp={{-502.92,_,0,11}};
methyl2oxopentanoatesp={{-445.14,_,-1,9}};
methylmalatesp={{-843.90,_,-2,6}};
methylmaleatesp={{-600.7,_,-2,4},{-636.49,_,-1,5}};
methylmalonylcoAsp={{-502.48,_,-1,4},{-526.51,_,0,5}};
n2aqsp={{18.7,-10.54,0,0}};
n2gsp={{0,0,0,0}};
n2oaqsp={{104.2,82.05,0,0}};
nadoxsp={{0,0,-1,26}};
nadpoxsp={{-835.18,-1007.48,-3,25}};
nadpredsp={{-809.71,-1036.66,-4,26}};
nadredsp={{22.65,-31.94,-2,27}};
nicotinamideribonucleotidesp={{840.08,_,-2,14},{803.32,_,-1,15}};
nitratesp={{-108.74,-205.0,-1,0},{-111.25,-207.36,0,1}};
nitritesp={{-32.2,-104.6,-1,0},{-50.6,-119.2,0,1}};
noaqsp={{86.55,90.25,0,0}};
o2aqsp={{16.4,-11.7,0,0}};
o2gsp={{0,0,0,0}};
oxalatesp={{-673.9,825.1,-2,0},{-698.33,_,-1,1}};
oxaloacetatesp={{-793.29,-959.90,-2,2}};
oxalosuccinatesp={{-1138.88,_,-2,4}};
oxalylcoAsp={{-509.96,_,-1,0}};
oxoprolinesp={{-469.15,_,-1,6}};
palmitatesp={{-259.4,_,-1,31}};
pepsp={{-1263.65,-1621.38,-3,2},{-1303.61,-1619.58,-2,3}};
phenylalanineLsp={{-207.1,_,0,11}};
phosphoglycerate2sp={{-1496.38,_,-3,4},{-1539.99,_,-2,5}};
phosphoglycerate3sp={{-1502.54,_,-3,4},{-1545.52,_,-2,5}};
phosphohydroxypyruvatesp={{-1448.67,_,-3,2}};
phosphoserinesp={{-1360.75,_,-2,6},{-1397.51,_,-1,7}};
pisp={{-1096.1,-1299.,-2,1},{-1137.3,-1302.6,-1,2}};
ppisp={{-1919.86,-2293.47,-4,0},{-1973.86,-2294.87,-3,
        1},{-2012.21,-2295.37,-2,2},{-2025.11,-2290.37,-1,
        3},{-2029.85,-2281.17,0,4}};
propanol2sp={{-185.23,-330.83,0,8}};
propanolnsp={{-175.81,_,0,8}};
propanoylcoAsp={{-179.14,_,0,5}};
prppsp={{-3284.25,_,-5,8},{-3325.23,_,-4,9},{-3363.42,_,-3,10}};
pyruvatesp={{-472.27,-596.22,-1,3}};
retinalsp={{0,_,0,28}};
retinolsp={{-27.91,_,0,30}};
ribitolsp={{-774.8,_,0,12}};
ribosesp={{-752.00,-1034.00,0,10}};
ribose1phossp={{-1587.70,-2041.48,-2,9},{-1625.88,-2030.18,-1,10}};
ribose5phossp={{-1595.78,-2041.48,-2,9},{-1633.96,-2030.18,-1,10}};
ribulosesp={{-735.94,-1023.02,0,10}};
ribulose5phossp={{-1595.32,_,-2,9},{-1633.50,_,-1,10}};
serineLsp={{-510.87,_,0,7}};
sorbitolsp={{-946.36,_,0,14}};
sorbitol6phossp={{-1793.10,_,-2,13},{-1829.75,_,-1,14}};
sorbosesp={{-911.95,-1263.3,0,12}};
succinatesp={{-690.44,-908.68,-2,4},{-722.62,-908.84,-1,5},{-746.64,-912.2,0,
        6}};
succinylcoAsp={{-509.72, _, -1, 4}, {-533.76, _, 0,
    5}};
sucrosesp={{-1564.7,-2199.87,0,22}};
sulfatesp={{-744.53,-909.27,-2,0},{-755.91,-887.34,-1,1}};
sulfitesp={{-486.5,-635.5,-2,0},{-527.78,-626.22,-1,1}};
sulfurcrsp={{0,0,0,0}};
thioredoxinoxsp={{0,_,0,0}};
```

```
thioredoxinredsp={{69.88,_,-2,0},{20.56,_,-1,1},{-25.37,_,0,2}};
threoninesp={{-528.89,_,0,9}};
transcinnamatesp={{-128.75,_,-1,7}};
trehalosesp={{-1582.76,_,0,22}};
tryptophanLsp={{-114.7,-405.2,0,12}};
tyrosineLsp={{-370.7,_,0,11}};
ubiquinoneoxsp={{0,_,0,90}};
ubiquinoneredsp={{-89.92,_,0,92}};
uratesp={{-325.9,_,-1,3}};
ureasp={{-202.8,-317.65,0,4}};
uricacidsp={{-356.9,_,0,4}};
valineLsp={{-358.65,-611.99,0,11}};
xylitolsp={{-784.09,_,0,12}};
xylosesp={{-750.49,-1045.94,0,10}};
xylulosesp={{-746.15,-1029.65,0,10}};


Begin["Private`"];
```

---

3 Programs that are used

```
calcdGmat[speciesmat_] :=
  Module[{dGzero, dHzero, zi, nH, pHterm, isterm,
    gpfnsp},(*This program produces the function of pH and ionic
strength (is) that gives the standard transformed Gibbs energy of formation of a
reactant (sum of species) at 298.15 K.  The input speciesmat is a matrix that
gives the standard Gibbs energy of formation, the standard enthalpy of formation,
the electric charge, and the number of hydrogen atoms in each species. There is
a row in the matrix for each species of the reactant. gpfnsp is a list of the
functions for the species.  Energies are expressed in kJ mol^-1.*)
    {dGzero, dHzero, zi, nH} =
     Transpose[speciesmat];
     pHterm = nH*8.31451*0.29815*Log[10^(-pH)];
     isterm = 2.91482*(zi^2 - nH)*(is^0.5/(1 + 1.6*is^0.5));
    gpfnsp = dGzero - pHterm - isterm;
    -8.31451*0.29815*Log[Plus @@
      Exp[-(gpfnsp/(8.31451*0.29815))]]]]

calcNHmat[speciesmat_] :=
Module[{dGzero, dHzero,zi, nH, pHterm, isterm,gpfnsp,trgef},(*This program
produces the function of pH and ionic strength (is) that gives the average
number of hydrogen atoms in a reactant (sum of species) at 298.15 K.
The input speciesmat is a matrix that gives the standard Gibbs energy of
formation, the standard enthalpy of formation, the electric charge, and
the number of hydrogen atoms in each species. There is a row in the matrix
for each species of the reactant. gpfnsp is a list of the functions for
the species.*)
{dGzero,dHzero,zi,nH}=Transpose[speciesmat];
pHterm = nH*8.31451*.29815*Log[10^-pH];
isterm = 2.91482*((zi^2) - nH)*(is^.5)/(1 + 1.6*is^.5);
gpfnsp=dGzero - pHterm - isterm;
trgef=-8.31451*.29815*Log[Apply[Plus,Exp[-1*gpfnsp/(8.31451*.29815)]]];
D[trgef,pH]/(8.31451*.29815*Log[10])]

derivetrGibbsT[speciesmat_]:=Module[{dGzero,dGzeroT,dHzero,zi,nH,gibbscoeff
,pHterm, isterm,gpfnsp},(*This program derives the function of T (in Kelvin),
pH, and ionic strength (is) that gives the standard transformed Gibbs energy
of formation of a reactant (sum of species).  The input speciesmat is a matrix
that gives the standard Gibbs energy of formation in kJ mol^-1 at 298.15 K and
zero ionic strength, the standard enthalpy of formation in kJ mol^-1 at 298.15 K
```

and zero ionic strength, the electric charge, and the number of hydrogen atoms in
each species. There is a row in the matrix for each species of the reactant.
gpfnsp is a list of the functions for the standard transformed Gibbs energies
of the species. The corresponding functions for other transformed properties
can be obtained by taking partial derivatives. The standard transformed Gibbs
energy of formation of a reactant in kJ mol^-1 can be calculated at any temperature
in the range 273.15 K to 313.15 K, any pH in the range 5 to 9, and any ionic
strength in the range 0 to 0.35 M by use of ReplaceAll (/.).*)
{dGzero,dHzero,zi,nH}=Transpose[speciesmat];
gibbscoeff=(9.20483*t)/10^3-(1.284668*t^2)/10^5+(4.95199*t^3)/10^8;
dGzeroT=(dGzero*t)/298.15+dHzero*(1-t/298.15);
pHterm=(nH*8.31451*t*Log[10^(-pH)])/1000;
istermG=(gibbscoeff*(zi^2-nH)*is^0.5)/(1+1.6*is^0.5);
gpfnsp=dGzeroT-pHterm-istermG;
-((8.31451*t*Log[Plus@@(E^(-(gpfnsp/((8.31451*t)/1000))))])/1000)]

derivetrHT[speciesmat_]:=Module[{dGzero,dGzeroT,dHzero,zi,nH,gibbscoeff,
pHterm,isterm,gpfnsp,gibbsereactant},(*This program derives the function of T (in
Kelvin), pH, and ionic strength (is) that gives the stndard transformed enthalpy of
formation of a reactant (sum of species). The input speciesmat is a matrix
that gives the standard Gibbs energy of formation in kJ mol^-1 at 298.15 K and
zero ionic strength, the standard enthalpy of formation in kJ mol^-1 at 298.15 K
and zero ionic strength, the electric charge, and the number of hydrogen atoms in
each species. There is a row in the matrix for each species of the reactant.
gpfnsp is a list of the functions for the standard transformed Gibbs energies
of the species. This program applies the Gibbs-Helmholtz equation to the function
for the standard transformed Gibbs energy of formation of a reactant. The
standard transformed enthalpy of formation of a reactant in kJ mol^-1 can be
calculated at any temperature in the range 273.15 K to 313.15 K, any pH in the
range 5 to 9, and any ionic strength in the range 0 to 0.35 M by use of ReplaceAll (/.)
{dGzero,dHzero,zi,nH}=Transpose[speciesmat];
gibbscoeff=(9.20483*t)/10^3-(1.284668*t^2)/10^5+(4.95199*t^3)/10^8;
dGzeroT=(dGzero*t)/298.15+dHzero*(1-t/298.15);
pHterm=(nH*8.31451*t*Log[10^(-pH)])/1000;
istermG=(gibbscoeff*(zi^2-nH)*is^0.5)/(1+1.6*is^0.5);
gpfnsp=dGzeroT-pHterm-istermG;
gibbsereactant=-((8.31451*t*Log[Plus@@(E^(-(gpfnsp/((8.31451*t)/1000))))])/1000);
-t^2*D[gibbsereactant/t,t]]]

derivetrST[speciesmat_]:=Module[{dGzero,dGzeroT,dHzero,zi,nH,gibbscoeff,pHterm,
isterm,gpfnsp,gibbsereactant},(*This program derives the function of T
(in Kelvin), pH, and ionic strength (is) that gives the stndard transformed
entropy of formation of a reactant (sum of species). The input speciesmat
is a matrix that gives the standard Gibbs energy of formation in kJ mol^-1
at 298.15 K and zero ionic strength, the standard enthalpy of formation in
kJ mol^-1 at 298.15 K and zero ionic strength, the electric charge, and the
number of hydrogen atoms in each species. There is a row in the matrix for
each species of the reactant. gpfnsp is a list of the functions for the
standard transformed Gibbs energies of the species. This program applies
the equation for the entropy to the function for the standard transformed
Gibbs energy of formation of a reactant. The standard transformed entropy
of formation of a reactant in kJ K^-1 mol^-1 can be calculated at any
temperature in the range 273.15 K to 313.15 K, any pH in the range 5 to 9,
and any ionic strength in the range 0 to 0.35 M by use of ReplaceAll (/.).*)
{dGzero,dHzero,zi,nH}=Transpose[speciesmat];
gibbscoeff=(9.20483*t)/10^3-(1.284668*t^2)/10^5+(4.95199*t^3)/10^8;
dGzeroT=(dGzero*t)/298.15+dHzero*(1-t/298.15);
pHterm=(nH*8.31451*t*Log[10^(-pH)])/1000;
istermG=(gibbscoeff*(zi^2-nH)*is^0.5)/(1+1.6*is^0.5);
gpfnsp=dGzeroT-pHterm-istermG;
gibbsereactant=-((8.31451*t*Log[Plus@@(E^(-(gpfnsp/((8.31451*t)/1000))))])/1000);
-D[gibbsereactant,t]]]

deriveNHT[speciesmat_]:=Module[{dGzero,dGzeroT,dHzero,zi,nH,gibbscoeff,pHterm,
isterm,gpfnsp,trgefT},(*This program derives the function of T (in Kelvin), pH,

and ionic strength (is) that gives the average number of hydrogen atoms in a reactant (sum of species). The input speciesmat is a matrix that gives the standard Gibbs energy of formation in kJ mol^-1 at 298.15 K and zero ionic strength, the standard enthalpy of formation in kJ mol^-1 at 298.15 K and zero ionic strength, the electric charge, and the number of hydrogen atoms in each species. There is a row in the matrix for each species of the reactant. gpfnsp is a list of the functions for the standard transformed Gibbs energies of the species. The average number can be calculated at any temperature in the range 273.15 K to 313.15 K, any pH in the range 5 to 9, and any ionic strength in the range 0 to 0.35 M by use of ReplaceAll (/.).*)

```
{dGzero,dHzero,zi,nH}=Transpose[speciesmat];
gibbscoeff=(9.20483*t)/10^3-(1.284668*t^2)/10^5+(4.95199*t^3)/10^8;
dGzeroT=(dGzero*t)/298.15+dHzero*(1-t/298.15);
pHterm=(nH*8.31451*t*Log[10^(-pH)])/1000;
istermG=(gibbscoeff*(zi^2-nH)*is^0.5)/(1+1.6*is^0.5);
gpfnsp=dGzeroT-pHterm-istermG;
trgefT=-((8.31451*t*Log[Plus@@(E^(-(gpfnsp/((8.31451*t)/1000))))]));
D[trgefT,pH]/(8.31451*t*Log[10])]
```

4 Calculation of the functions of pH and ionic strength for the standard transformed Gibbs energies of formation of reactants at 298.15 K and desired pHs and ionic strengths

These functions are given the following *Mathematica* names of reactants, which have to be cleared first.

```
Clear[acetaldehyde,acetate,acetoacetate,acetoacetylcoA,acetone,acetylcoA,
acetylphos,aconitatecis,adenine,adenosine,adenosinephosphosulfate,adp,alanine,
ammonia,amp,arabinose,arabinose5phos,asparagineL,aspartate,atp,bpg,butanal,
butanoln,butyrate,cellobiose,citrate,citrateiso,co2g,co2tot,coA,
coAglutathione,coaq,cog,creatine,creatinine,cyclicamp,cysteineL,cystineL,
cytochromecox,cytochromecred,deoxyadenosine,deoxyadp,deoxyamp,deoxyatp,
deoxyribose,deoxyribose1phos,deoxyribose5phos,dihydroxyacetone,
dihydroxyacetonephos,ethaneaq,ethanol,ethylacetate,fadenzox,fadenzred,fadox,
fadred,ferredoxinox,ferredoxinred,ferric,ferrous,fmnox,fmnred,formate,
fructose,fructose16phos,fructose6phos,fumarate,galactono14lactone,galactose,
galactose1phos,galactose6phos,gluconolactone,gluconolactone6phos,glucose,
glucose1phos,glucose6phos,glutamate,glutamine,glutathioneox,glutathionered,
glyceraldehyde,glyceraldehydephos,glycerate,glycerol,glycerol3phos,glycine,
glycolate,glycylglycine,glyoxylate,h2aq,h2g,h2o,h2o2aq,h2saq,
hydroxy2oxoglutarate,hydroxyglutarate,hydroxypropionateb,hydroxypyruvate,
hypoxanthine,i2cr,iditol,idp,imp,indole,inosine,iodideion,isomaltose,itp,
ketoglutaramate,ketoglutarate,lactate,lactose,lactulose,leucineisoL,
leucineL,lyxose,malate,maleate,maltose,malylcoA,mannitol1phos,mannitolD,
mannose,mannose1phos,mannose6phos,methaneaq,methaneg,methanol,methionineL,
methyl2oxopentanoate,methylmalate,methylmaleate,methylmalonylcoA,n2aq,n2g,
n2oaq,nadox,nadpox,nadpred,nadred,nicotinamideribonucleotide,nitrate,
nitrite,noaq,o2aq,o2g,oxalate,oxaloacetate,oxalosuccinate,oxalylcoA,
oxoproline,palmitate,pep,phenylalanineL,phosphoglycerate2,phosphoglycerate3,
phosphohydroxypyruvate,phosphoserine,pi,ppi,propanol2,propanoln,
propanoylcoA,prpp,pyruvate,retinal,retinol,ribitol,ribose,ribose1phos,
ribose5phos,ribulose,ribulose5phos,serineL,sorbitol,sorbitol6phos,sorbose,
succinate,succinylcoA,sucrose,sulfate,sulfite,sulfurcr,thioredoxinox,
thioredoxinred,threonine,transcinnamate,trehalose,tryptophanL,tyrosineL,
ubiquinoneox,ubiquinonered,urate,urea,uricacid,valineL,xylitol,xylose,
xylulose]
```

The program **calcdGmat** is applied in a single step by use of Map.

```
Evaluate[{acetaldehyde,acetate,acetoacetate,acetoacetylcoA,acetone,acetylcoA,
acetylphos,aconitatecis,adenine,adenosine,adenosinephosphosulfate,adp,alanine,
```

```
ammonia,amp,arabinose,arabinose5phos,asparagineL,aspartate,atp,bpg,butanal,
butanoln,butyrate,cellobiose,citrate,citrateiso,co2g,co2tot,coA,
coAglutathione,coaq,cog,creatine,creatinine,cyclicamp,cysteineL,cystineL,
cytochromecox,cytochromecred,deoxyadenosine,deoxyadp,deoxyamp,deoxyatp,
deoxyribose,deoxyribose1phos,deoxyribose5phos,dihydroxyacetone,
dihydroxyacetonephos,ethaneaq,ethanol,ethylacetate,fadenzox,fadenzred,fadox,
fadred,ferredoxinox,ferredoxinred,ferric,ferrous,fmnox,fmnred,formate,
fructose,fructose16phos,fructose6phos,fumarate,galactono14lactone,galactose,
galactose1phos,galactose6phos,gluconolactone,gluconolactone6phos,glucose,
glucose1phos,glucose6phos,glutamate,glutamine,glutathioneox,glutathionered,
glyceraldehyde,glyceraldehydephos,glycerate,glycerol,glycerol3phos,glycine,
glycolate,glycylglycine,glyoxylate,h2aq,h2g,h2o,h2o2aq,h2saq,
hydroxy2oxoglutarate,hydroxyglutarate,hydroxypropionateb,hydroxypyruvate,
hypoxanthine,i2cr,iditol,idp,imp,indole,inosine,iodideion,isomaltose,itp,
ketoglutaramate,ketoglutarate,lactate,lactose,lactulose,leucineisoL,
leucineL,lyxose,malate,maleate,maltose,malylcoA,mannitol1phos,mannitolD,
mannose,mannose1phos,mannose6phos,methaneaq,methaneg,methanol,methionineL,
methyl2oxopentanoate,methylmalate,methylmaleate,methylmalonylcoA,n2aq,n2g,
n2oaq,nadox,nadpox,nadpred,nadred,nicotinamideribonucleotide,nitrate,
nitrite,noaq,o2aq,o2g,oxalate,oxaloacetate,oxalosuccinate,oxalylcoA,
oxoproline,palmitate,pep,phenylalanineL,phosphoglycerate2,phosphoglycerate3,
phosphohydroxypyruvate,phosphoserine,pi,ppi,propanol2,propanoln,
propanoylcoA,prpp,pyruvate,retinal,retinol,ribitol,ribose,ribose1phos,
ribose5phos,ribulose,ribulose5phos,serineL,sorbitol,sorbitol6phos,sorbose,.
succinate,succinylcoA,sucrose,sulfate,sulfite,sulfurcr,thioredoxinox,
thioredoxinred,threonine,transcinnamate,trehalose,tryptophanL,tyrosineL,
ubiquinoneox,ubiquinonered,urate,urea,uricacid,valineL,xylitol,xylose,
xylulose}]=Map[\
calcdGmat,{acetaldehydesp,acetatesp,acetoacetatesp,acetoacetylcoAsp,acetonesp,
acetylcoAsp,acetylphossp,aconitatecissp,adeninesp,adenosinesp,
adenosinephosphosulfatesp,adpsp,alaninesp,ammoniasp,ampsp,arabinosesp,
arabinose5phossp,asparagineLsp,aspartatesp,atpsp,bpgsp,butanalsp,butanolnsp,
butyratesp,cellobiosesp,citratesp,citrateisosp,co2gsp,co2totsp,coAsp,
coAglutathionesp,coaqsp,cogsp,creatinesp,creatininesp,cyclicampsp,cysteineLsp,
cystineLsp,cytochromecoxsp,cytochromecredsp,deoxyadenosinesp,deoxyadpsp,
deoxyampsp,deoxyatpsp,deoxyribosesp,deoxyribose1phossp,deoxyribose5phossp,
dihydroxyacetonesp,dihydroxyacetonephossp,ethaneaqsp,ethanolsp,
ethylacetatesp,fadenzoxsp,fadenzredsp,fadoxsp,fadredsp,ferredoxinoxsp,
ferredoxinredsp,ferricsp,ferroussp,fmnoxsp,fmnredsp,formatesp,fructosesp,
fructose16phossp,fructose6phossp,fumaratesp,galactono14lactonesp,
galactosesp,galactose1phossp,galactose6phossp,gluconolactonesp,
gluconolactone6phossp,glucosesp,glucose1phossp,glucose6phossp,glutamatesp,
glutaminesp,glutathioneoxsp,glutathioneredsp,glyceraldehydesp,
glyceraldehydephossp,glyceratesp,glycerolsp,glycerol3phossp,glycinesp,
glycolatesp,glycylglycinesp,glyoxylatesp,h2aqsp,h2gsp,h2osp,h2o2aqsp,
h2saqsp,hydroxy2oxoglutaratesp,hydroxyglutaratesp,hydroxypropionatebsp,
hydroxypyruvatesp,hypoxanthinesp,i2crsp,iditolsp,idpsp,impsp,indolesp,
inosinesp,iodideionsp,isomaltosesp,itpsp,ketoglutaramatesp,
ketoglutaratesp,lactatesp,lactosesp,lactulosesp,leucineisoLsp,leucineLsp,
lyxosesp,malatesp,maleatesp,maltosesp,malylcoAsp,mannitol1phossp,mannitolDsp,
mannosesp,mannose1phossp,mannose6phossp,methaneaqsp,methanegsp,methanolsp,
methionineLsp,methyl2oxopentanoatesp,methylmalatesp,methylmaleatesp,
methylmalonylcoAsp,n2aqsp,n2gsp,n2oaqsp,nadoxsp,nadpoxsp,nadpredsp,nadredsp,
nicotinamideribonucleotidesp,nitratesp,nitritesp,noaqsp,o2aqsp,o2gsp,
oxalatesp,oxaloacetatesp,oxalosuccinatesp,oxalylcoAsp,oxoprolinesp,
palmitatesp,pepsp,phenylalanineLsp,phosphoglycerate2sp,phosphoglycerate3sp,
phosphohydroxypyruvatesp,phosphoserinesp,pisp,ppisp,propanol2sp,propanolnsp,
propanoylcoAsp,prppsp,pyruvatesp,retinalsp,retinolsp,ribitolsp,ribosesp,
ribose1phossp,ribose5phossp,ribulosesp,ribulose5phossp,serineLsp,sorbitolsp,
sorbitol6phossp,sorbosesp,succinatesp,succinylcoAsp,sucrosesp,sulfatesp,
sulfitesp,sulfurcrsp,thioredoxinoxsp,thioredoxinredsp,threoninesp,
transcinnamatesp,trehalosesp,tryptophanLsp,tyrosineLsp,ubiquinoneoxsp,
ubiquinoneredsp,uratesp,ureasp,uricacidsp,valineLsp,xylitolsp,xylosesp,
xylulosesp}];
```

■ 5 Calculation of the functions of pH and ionic strength for the average number of hydrogen atoms in the reactants at 298.15 K and desired pHs and ionic strengths

The function of pH and ionic stength that gives the average number of hydrogen atoms $\overline{N}_H$ in a reactant can be derived by differentiating the function for the standard transformed Gibbs energy of formation. The names to be assigned to these functions arc given by the following list:

```
Clear[acetaldehydeNH,acetateNH,acetoacetateNH,acetoacetylcoANH,acetoneNH,
acetylcoANH,acetylphosNH,aconitatecisNH,adenineNH,adenosineNH,
adenosinephosphosulfateNH,adpNH,alanineNH,ammoniaNH,ampNH,arabinoseNH,
arabinose5phosNH,asparagineLNH,aspartateNH,atpNH,bpgNH,butanalNH,butanolnNH,
butyrateNH,cellobioseNH,citrateNH,citrateisoNH,co2gNH,co2totNH,coANH,
coAglutathioneNH,coaqNH,cogNH,creatineNH,creatinineNH,cyclicampNH,cysteineLNH,
cystineLNH,cytochromecoxNH,cytochromecredNH,deoxyadenosineNH,deoxyadpNH,
deoxyampNH,deoxyatpNH,deoxyriboseNH,deoxyribose1phosNH,deoxyribose5phosNH,
dihydroxyacetoneNH,dihydroxyacetonephosNH,ethaneaqNH,ethanolNH,
ethylacetateNH,fadenzoxNH,fadenzredNH,fadoxNH,fadredNH,ferredoxinoxNH,
ferredoxinredNH,ferricNH,ferrousNH,fmnoxNH,fmnredNH,formateNH,fructoseNH,
fructose16phosNH,fructose6phosNH,fumarateNH,galactono14lactoneNH,
galactoseNH,galactose1phosNH,galactose6phosNH,gluconolactoneNH,
gluconolactone6phosNH,glucoseNH,glucose1phosNH,glucose6phosNH,glutamateNH,
glutamineNH,glutathioneoxNH,glutathioneredNH,glyceraldehydeNH,
glyceraldehydephosNH,glycerateNH,glycerolNH,glycerol3phosNH,glycineNH,
glycolateNH,glycylglycineNH,glyoxylateNH,h2aqNH,h2gNH,h2oNH,h2o2aqNH,
h2saqNH,hydroxy2oxoglutarateNH,hydroxyglutarateNH,hydroxypropionatebNH,
hydroxypyruvateNH,hypoxanthineNH,i2crNH,iditolNH,idpNH,impNH,indoleNH,
inosineNH,iodideionNH,isomaltoseNH,itpNH,ketoglutaramateNH,
ketoglutarateNH,lactateNH,lactoseNH,lactuloseNH,leucineisoLNH,leucineLNH,
lyxoseNH,malateNH,maleateNH,maltoseNH,malylcoANH,mannitol1phosNH,mannitolDNH,
mannoseNH,mannose1phosNH,mannose6phosNH,methaneaqNH,methanegNH,methanolNH,
methionineLNH,methyl2oxopentanoateNH,methylmalateNH,methylmaleateNH,
methylmalonylcoANH,n2aqNH,n2gNH,n2oaqNH,nadoxNH,nadpoxNH,nadpredNH,nadredNH,
nicotinamideribonucleotideNH,nitrateNH,nitriteNH,noaqNH,o2aqNH,o2gNH,
oxalateNH,oxaloacetateNH,oxalosuccinateNH,oxalylcoANH,oxoprolineNH,
palmitateNH,pepNH,phenylalanineLNH,phosphoglycerate2NH,phosphoglycerate3NH,
phosphohydroxypyruvateNH,phosphoserineNH,piNH,ppiNH,propanol2NH,propanolnNH,
propanoylcoANH,prppNH,pyruvateNH,retinalNH,retinolNH,ribitolNH,riboseNH,
ribose1phosNH,ribose5phosNH,ribuloseNH,ribulose5phosNH,serineLNH,sorbitolNH,
sorbitol6phosNH,sorboseNH,succinateNH,succinylcoANH,sucroseNH,sulfateNH,
sulfiteNH,sulfurcrNH,thioredoxinoxNH,thioredoxinredNH,threonineNH,
transcinnamateNH,trehaloseNH,tryptophanLNH,tyrosineLNH,ubiquinoneoxNH,
ubiquinoneredNH,urateNH,ureaNH,uricacidNH,valineLNH,xylitolNH,xyloseNH,
xyluloseNH]
```

The program calcNHmat is applied in a single step by use of Map.

```
Evaluate[{acetaldehydeNH,acetateNH,acetoacetateNH,acetoacetylcoANH,acetoneNH,
acetylcoANH,acetylphosNH,aconitatecisNH,adenineNH,adenosineNH,
adenosinephosphosulfateNH,adpNH,alanineNH,ammoniaNH,ampNH,arabinoseNH,
arabinose5phosNH,asparagineLNH,aspartateNH,atpNH,bpgNH,butanalNH,butanolnNH,
butyrateNH,cellobioseNH,citrateNH,citrateisoNH,co2gNH,co2totNH,coANH,
coAglutathioneNH,coaqNH,cogNH,creatineNH,creatinineNH,cyclicampNH,cysteineLNH,
cystineLNH,cytochromecoxNH,cytochromecredNH,deoxyadenosineNH,deoxyadpNH,
deoxyampNH,deoxyatpNH,deoxyriboseNH,deoxyribose1phosNH,deoxyribose5phosNH,
dihydroxyacetoneNH,dihydroxyacetonephosNH,ethaneaqNH,ethanolNH,
ethylacetateNH,fadenzoxNH,fadenzredNH,fadoxNH,fadredNH,ferredoxinoxNH,
ferredoxinredNH,ferricNH,ferrousNH,fmnoxNH,fmnredNH,formateNH,fructoseNH,
fructose16phosNH,fructose6phosNH,fumarateNH,galactono14lactoneNH,
```

```
galactoseNH,galactose1phosNH,galactose6phosNH,gluconolactoneNH,
gluconolactone6phosNH,glucoseNH,glucose1phosNH,glucose6phosNH,glutamateNH,
glutamineNH,glutathioneoxNH,glutathioneredNH,glyceraldehydeNH,
glyceraldehydephosNH,glycerateNH,glycerolNH,glycerol3phosNH,glycineNH,
glycolateNH,glycylglycineNH,glyoxylateNH,h2aqNH,h2gNH,h2oNH,h2o2aqNH,
h2saqNH,hydroxy2oxoglutarateNH,hydroxyglutarateNH,hydroxypropionatebNH,
hydroxypyruvateNH,hypoxanthineNH,i2crNH,iditolNH,idpNH,impNH,indoleNH,
inosineNH,iodideionNH,isomaltoseNH,itpNH,ketoglutaramateNH,
ketoglutarateNH,lactateNH,lactoseNH,lactuloseNH,leucineisoLNH,leucineLNH,
lyxoseNH,malateNH,maleateNH,maltoseNH,malylcoANH,mannitol1phosNH,mannitolDNH,
mannoseNH,mannose1phosNH,mannose6phosNH,methaneaqNH,methanegNH,methanolNH,
methionineLNH,methyl2oxopentanoateNH,methylmalateNH,methylmaleateNH,
methylmalonylcoANH,n2aqNH,n2gNH,n2oaqNH,nadoxNH,nadpoxNH,nadpredNH,nadredNH,
nicotinamideribonucleotideNH,nitrateNH,nitriteNH,noaqNH,o2aqNH,o2gNH,
oxalateNH,oxaloacetateNH,oxalosuccinateNH,oxalylcoANH,oxoprolineNH,
palmitateNH,pepNH,phenylalanineLNH,phosphoglycerate2NH,phosphoglycerate3NH,
phosphohydroxypyruvateNH,phosphoserineNH,piNH,ppiNH,propanol2NH,propanolnNH,
propanoylcoANH,prppNH,pyruvateNH,retinalNH,retinolNH,ribitolNH,riboseNH,
ribose1phosNH,ribose5phosNH,ribuloseNH,ribulose5phosNH,serineLNH,sorbitolNH,
sorbitol6phosNH,sorboseNH,succinateNH,succinylcoANH,sucroseNH,sulfateNH,
sulfiteNH,sulfurcrNH,thioredoxinoxNH,thioredoxinredNH,threonineNH,
transcinnamateNH,trehaloseNH,tryptophanLNH,tyrosineLNH,ubiquinoneoxNH,
ubiquinoneredNH,urateNH,ureaNH,uricacidNH,valineLNH,xylitolNH,xyloseNH,
xyluloseNH}]=Map[calcNHmat,{acetaldehydesp,acetatesp,acetoacetatesp,
acetoacetylcoAsp,acetonesp,
acetylcoAsp,acetylphossp,aconitatecissp,adeninesp,adenosinesp,
adenosinephosphosulfatesp,adpsp,alaninesp,ammoniasp,ampsp,arabinosesp,
arabinose5phossp,asparagineLsp,aspartatesp,atpsp,bpgsp,butanalsp,butanolnsp,
butyratesp,cellobiosesp,citratesp,citrateisosp,co2gsp,co2totsp,coAsp,
coAglutathionesp,coaqsp,cogsp,creatinesp,creatininesp,cyclicampsp,cysteineLsp,
cystineLsp,cytochromecoxsp,cytochromecredsp,deoxyadenosinesp,deoxyadpsp,
deoxyampsp,deoxyatpsp,deoxyribosesp,deoxyribose1phossp,deoxyribose5phossp,
dihydroxyacetonesp,dihydroxyacetonephossp,ethaneaqsp,ethanolsp,
ethylacetatesp,fadenzoxsp,fadenzredsp,fadoxsp,fadredsp,ferredoxinoxsp,
ferredoxinredsp,ferricsp,ferroussp,fmnoxsp,fmnredsp,formatesp,fructosesp,
fructose16phossp,fructose6phossp,fumaratesp,galactono14lactonesp,
galactosesp,galactose1phossp,galactose6phossp,gluconolactonesp,
gluconolactone6phossp,glucosesp,glucose1phossp,glucose6phossp,glutamatesp,
glutaminesp,glutathioneoxsp,glutathioneredsp,glyceraldehydesp,
glyceraldehydephossp,glyceratesp,glycerolsp,glycerol3phossp,glycinesp,
glycolatesp,glycylglycinesp,glyoxylatesp,h2aqsp,h2gsp,h2osp,h2o2aqsp,
h2saqsp,hydroxy2oxoglutaratesp,hydroxyglutaratesp,hydroxypropionatebsp,
hydroxypyruvatesp,hypoxanthinesp,i2crsp,iditolsp,idpsp,impsp,indolesp,
inosinesp,iodideionsp,isomaltosesp,itpsp,ketoglutaramatesp,
ketoglutaratesp,lactatesp,lactosesp,lactulosesp,leucineisoLsp,leucineLsp,
lyxosesp,malatesp,maleatesp,maltosesp,malylcoAsp,mannitol1phossp,mannitolDsp,
mannosesp,mannose1phossp,mannose6phossp,methaneaqsp,methanegsp,methanolsp,
methionineLsp,methyl2oxopentanoatesp,methylmalatesp,methylmaleatesp,
methylmalonylcoAsp,n2aqsp,n2gsp,n2oaqsp,nadoxsp,nadpoxsp,nadpredsp,nadredsp,
nicotinamideribonucleotidesp,nitratesp,nitritesp,noaqsp,o2aqsp,o2gsp,
oxalatesp,oxaloacetatesp,oxalosuccinatesp,oxalylcoAsp,oxoprolinesp,
palmitatesp,pepsp,phenylalanineLsp,phosphoglycerate2sp,phosphoglycerate3sp,
phosphohydroxypyruvatesp,phosphoserinesp,pisp,ppisp,propanol2sp,propanolnsp,
propanoylcoAsp,prppsp,pyruvatesp,retinalsp,retinolsp,ribitolsp,ribosesp,
ribose1phossp,ribose5phossp,ribulosesp,ribulose5phossp,serineLsp,sorbitolsp,
sorbitol6phossp,sorbosesp,succinatesp,succinylcoAsp,sucrosesp,sulfatesp,
sulfitesp,sulfurcrsp,thioredoxinoxsp,thioredoxinredsp,threoninesp,
transcinnamatesp,trehalosesp,tryptophanLsp,tyrosineLsp,ubiquinoneoxsp,
ubiquinoneredsp,uratesp,ureasp,uricacidsp,valineLsp,xylitolsp,xylosesp,
xylulosesp}];
```

6 Calculation of the functions of temperature, pH and ionic strength for the standard transformed Gibbs energies of formation for reactants for which standard enthalpies of formation of the species are known

When $\Delta_f H°$ are known for all species of a reactant, $\Delta_f G'°$, $\overline{N}_H$, $\Delta_f H'°$, and $\Delta_f S'°$ can be calculated at temperatures in the range 273.15 - 313.15 K, pHs in the range 5 - 9, and ionic strengths in the range zero to 0.35 M. The names of the 94 reactants for which all $\Delta_f H°$ are known are given by

```
Clear[acetaldehydeGT,acetateGT,acetoneGT,adenineGT,adenosineGT,adpGT,
alanineGT,ammoniaGT,ampGT,arabinoseGT,asparagineLGT,aspartateGT,atpGT,
citrateGT,co2gGT,co2totGT,coaqGT,cogGT,ethaneaqGT,ethanolGT,ethylacetateGT,
ferricGT,ferrousGT,formateGT,fructoseGT,fructose16phosGT,fructose6phosGT,
fumarateGT,galactoseGT,glucoseGT,glucose6phosGT,glutamateGT,glutamineGT,
glycerolGT,glycerol3phosGT,glycineGT,glycylglycineGT,h2aqGT,h2gGT,h2oGT,
h2o2aqGT,h2saqGT,i2crGT,idpGT,impGT,indoleGT,inosineGT,iodideionGT,
isomaltoseGT,itpGT,ketoglutarateGT,lactateGT,lactoseGT,leucineLGT,malateGT,
maltoseGT,mannoseGT,mannose6phosGT,methaneaqGT,methanegGT,methanolGT,n2aqGT,
n2gGT,n2oaqGT,nadoxGT,nadpoxGT,nadpredGT,nadredGT,nitrateGT,nitriteGT,
noaqGT,o2aqGT,o2gGT,oxaloacetateGT,pepGT,piGT,ppiGT,propanol2GT,pyruvateGT,
riboseGT,ribose1phosGT,ribose5phosGT,ribuloseGT,sorboseGT,succinateGT,
sucroseGT,sulfateGT,sulfiteGT,sulfurcrGT,tryptophanLGT,ureaGT,valineLGT,
xyloseGT,xyluloseGT]
```

```
Evaluate[{acetaldehydeGT,acetateGT,acetoneGT,adenineGT,adenosineGT,adpGT,
alanineGT,ammoniaGT,ampGT,arabinoseGT,asparagineLGT,aspartateGT,atpGT,
citrateGT,co2gGT,co2totGT,coaqGT,cogGT,ethaneaqGT,ethanolGT,ethylacetateGT,
ferricGT,ferrousGT,formateGT,fructoseGT,fructose16phosGT,fructose6phosGT,
fumarateGT,galactoseGT,glucoseGT,glucose6phosGT,glutamateGT,glutamineGT,
glycerolGT,glycerol3phosGT,glycineGT,glycylglycineGT,h2aqGT,h2gGT,h2oGT,
h2o2aqGT,h2saqGT,i2crGT,idpGT,impGT,indoleGT,inosineGT,iodideionGT,
isomaltoseGT,itpGT,ketoglutarateGT,lactateGT,lactoseGT,leucineLGT,malateGT,
maltoseGT,mannoseGT,mannose6phosGT,methaneaqGT,methanegGT,methanolGT,n2aqGT,
n2gGT,n2oaqGT,nadoxGT,nadpoxGT,nadpredGT,nadredGT,nitrateGT,nitriteGT,
noaqGT,o2aqGT,o2gGT,oxaloacetateGT,pepGT,piGT,ppiGT,propanol2GT,pyruvateGT,
riboseGT,ribose1phosGT,ribose5phosGT,ribuloseGT,sorboseGT,succinateGT,
sucroseGT,sulfateGT,sulfiteGT,sulfurcrGT,tryptophanLGT,ureaGT,valineLGT,
xyloseGT,xyluloseGT}]=Map[derivetrGibbsT,{acetaldehydesp,acetatesp,acetonesp,
adeninesp,adenosinesp,adpsp,alaninesp,ammoniasp,ampsp,arabinosesp,
asparagineLsp,aspartatesp,atpsp,citratesp,co2gsp,co2totsp,coaqsp,cogsp,
ethaneaqsp,ethanolsp,ethylacetatesp,ferricsp,ferroussp,formatesp,fructosesp,
fructose16phossp,fructose6phossp,fumaratesp,galactosesp,glucosesp,
glucose6phossp,glutamatesp,glutaminesp,glycerolsp,glycerol3phossp,glycinesp,
glycylglycinesp,h2aqsp,h2gsp,h2osp,h2o2aqsp,h2saqsp,i2crsp,idpsp,impsp,
indolesp,inosinesp,iodideionsp,isomaltosesp,itpsp,ketoglutaratesp,lactatesp,
lactosesp,leucineLsp,malatesp,maltosesp,mannosesp,mannose6phossp,
methaneaqsp,methanegsp,methanolsp,n2aqsp,n2gsp,n2oaqsp,nadoxsp,nadpoxsp,
nadpredsp,nadredsp,nitratesp,nitritesp,noaqsp,o2aqsp,o2gsp,oxaloacetatesp,
pepsp,pisp,ppisp,propanol2sp,pyruvatesp,ribosesp,ribose1phossp,ribose5phossp,
ribulosesp,sorbosesp,succinatesp,sucrosesp,sulfatesp,sulfitesp,sulfurcrsp,
tryptophanLsp,ureasp,valineLsp,xylosesp,xylulosesp}];
```

7 Calculation of the functions of temperature, pH and ionic strength for the standard transformed enthalpies of formation of reactants for which standard enthalpies of formation of the species are known

The function of T, pH, and ionic strength that gives the standard transformed enthalpy of formation $\Delta_f H'°$ for a reactant can be derived by differentiating the function for the standard transformed Gibbs energy of formation. This is done by the program **derivetrHT**.

```
Clear[acetaldehydeHT,acetateHT,acetoneHT,adenineHT,adenosineHT,adpHT,alanineHT,
 ammoniaHT,ampHT,arabinoseHT,asparagineLHT,aspartateHT,atpHT,citrateHT,co2gHT,
 co2totHT,coaqHT,cogHT,ethaneaqHT,ethanolHT,ethylacetateHT,ferricHT,ferrousHT,
 formateHT,fructoseHT,fructose16phosHT,fructose6phosHT,fumarateHT,galactoseHT,
 glucoseHT,glucose6phosHT,glutamateHT,glutamineHT,glycerolHT,glycerol3phosHT,
 glycineHT,glycylglycineHT,h2aqHT,h2gHT,h2oHT,h2o2aqHT,h2saqHT,i2crHT,idpHT,
 impHT,indoleHT,inosineHT,iodideionHT,isomaltoseHT,itpHT,ketoglutarateHT,
 lactateHT,lactoseHT,leucineLHT,malateHT,maltoseHT,mannoseHT,mannose6phosHT,
 methaneaqHT,methanegHT,methanolHT,n2aqHT,n2gHT,n2oaqHT,nadoxHT,nadpoxHT,
 nadpredHT,nadredHT,nitrateHT,nitriteHT,noaqHT,o2aqHT,o2gHT,oxaloacetateHT,
 pepHT,piHT,ppiHT,propanol2HT,pyruvateHT,riboseHT,ribose1phosHT,ribose5phosHT,
 ribuloseHT,sorboseHT,succinateHT,sucroseHT,sulfateHT,sulfiteHT,sulfurcrHT,
 tryptophanLHT,ureaHT,valineLHT,xyloseHT,xyluloseHT]

Evaluate[{acetaldehydeHT,acetateHT,acetoneHT,adenineHT,adenosineHT,adpHT,alanineHT,
 ammoniaHT,ampHT,arabinoseHT,asparagineLHT,aspartateHT,atpHT,citrateHT,co2gHT,
 co2totHT,coaqHT,cogHT,ethaneaqHT,ethanolHT,ethylacetateHT,ferricHT,ferrousHT,
 formateHT,fructoseHT,fructose16phosHT,fructose6phosHT,fumarateHT,galactoseHT,
 glucoseHT,glucose6phosHT,glutamateHT,glutamineHT,glycerolHT,glycerol3phosHT,
 glycineHT,glycylglycineHT,h2aqHT,h2gHT,h2oHT,h2o2aqHT,h2saqHT,i2crHT,idpHT,
 impHT,indoleHT,inosineHT,iodideionHT,isomaltoseHT,itpHT,ketoglutarateHT,
 lactateHT,lactoseHT,leucineLHT,malateHT,maltoseHT,mannoseHT,mannose6phosHT,
 methaneaqHT,methanegHT,methanolHT,n2aqHT,n2gHT,n2oaqHT,nadoxHT,nadpoxHT,
 nadpredHT,nadredHT,nitrateHT,nitriteHT,noaqHT,o2aqHT,o2gHT,oxaloacetateHT,
 pepHT,piHT,ppiHT,propanol2HT,pyruvateHT,riboseHT,ribose1phosHT,ribose5phosHT,
 ribuloseHT,sorboseHT,succinateHT,sucroseHT,sulfateHT,sulfiteHT,sulfurcrHT,
 tryptophanLHT,ureaHT,valineLHT,xyloseHT,xyluloseHT}]=Map[derivetrHT,{acetaldehydesp,
 acetatesp,acetonesp,adeninesp,adenosinesp,adpsp,alaninesp,
 ammoniasp,ampsp,arabinosesp,asparagineLsp,aspartatesp,atpsp,citratesp,co2gsp,
 co2totsp,coaqsp,cogsp,ethaneaqsp,ethanolsp,ethylacetatesp,ferricsp,ferroussp,
 formatesp,fructosesp,fructose16phossp,fructose6phossp,fumaratesp,galactosesp,
 glucosesp,glucose6phossp,glutamatesp,glutaminesp,glycerolsp,glycerol3phossp,
 glycinesp,glycylglycinesp,h2aqsp,h2gsp,h2osp,h2o2aqsp,h2saqsp,i2crsp,idpsp,
 impsp,indolesp,inosinesp,iodideionsp,isomaltosesp,itpsp,ketoglutaratesp,
 lactatesp,lactosesp,leucineLsp,malatesp,maltosesp,mannosesp,mannose6phossp,
 methaneaqsp,methanegsp,methanolsp,n2aqsp,n2gsp,n2oaqsp,nadoxsp,nadpoxsp,
 nadpredsp,nadredsp,nitratesp,nitritesp,noaqsp,o2aqsp,o2gsp,oxaloacetatesp,
 pepsp,pisp,ppisp,propanol2sp,pyruvatesp,ribosesp,ribose1phossp,ribose5phossp,
 ribulosesp,sorbosesp,succinatesp,sucrosesp,sulfatesp,sulfitesp,sulfurcrsp,
 tryptophanLsp,ureasp,valineLsp,xylosesp,xylulosesp}];
```

8 Calculation of the functions of temperature, pH and ionic strength for the standard transformed entropies of formation of reactants for which standard enthalpies of formation of the species are known

The function of T, pH, and ionic strength that gives the standard transformed entropy of formation $\Delta_f S'°$ for a reactant can be derived by differentiating the function for the standard transformed Gibbs energy of formation. This is done by the program **derivetrST**.

```
Clear[acetaldehydeST,acetateST,acetoneST,adenineST,adenosineST,adpST,alanineST,
ammoniaST,ampST,arabinoseST,asparagineLST,aspartateST,atpST,citrateST,co2gST,
co2totST,coaqST,cogST,ethaneaqST,ethanolST,ethylacetateST,ferricST,ferrousST,
formateST,fructoseST,fructose16phosST,fructose6phosST,fumarateST,galactoseST,
glucoseST,glucose6phosST,glutamateST,glutamineST,glycerolST,glycerol3phosST,
glycineST,glycylglycineST,h2aqST,h2gST,h2oST,h2o2aqST,h2saqST,i2crST,idpST,
impST,indoleST,inosineST,iodideionST,isomaltoseST,itpST,ketoglutarateST,
lactateST,lactoseST,leucineLST,malateST,maltoseST,mannoseST,mannose6phosST,
methaneaqST,methanegST,methanolST,n2aqST,n2gST,n2oaqST,nadoxST,nadpoxST,
nadpredST,nadredST,nitrateST,nitriteST,noaqST,o2aqST,o2gST,oxaloacetateST,
pepST,piST,ppiST,propanol2ST,pyruvateST,riboseST,ribose1phosST,ribose5phosST,
ribuloseST,sorboseST,succinateST,sucroseST,sulfateST,sulfiteST,sulfurcrST,
tryptophanLST,ureaST,valineLST,xyloseST,xyluloseST]

Evaluate[{acetaldehydeST,acetateST,acetoneST,adenineST,adenosineST,adpST,alanineST,
ammoniaST,ampST,arabinoseST,asparagineLST,aspartateST,atpST,citrateST,co2gST,
co2totST,coaqST,cogST,ethaneaqST,ethanolST,ethylacetateST,ferricST,ferrousST,
formateST,fructoseST,fructose16phosST,fructose6phosST,fumarateST,galactoseST,
glucoseST,glucose6phosST,glutamateST,glutamineST,glycerolST,glycerol3phosST,
glycineST,glycylglycineST,h2aqST,h2gST,h2oST,h2o2aqST,h2saqST,i2crST,idpST,
impST,indoleST,inosineST,iodideionST,isomaltoseST,itpST,ketoglutarateST,
lactateST,lactoseST,leucineLST,malateST,maltoseST,mannoseST,mannose6phosST,
methaneaqST,methanegST,methanolST,n2aqST,n2gST,n2oaqST,nadoxST,nadpoxST,
nadpredST,nadredST,nitrateST,nitriteST,noaqST,o2aqST,o2gST,oxaloacetateST,
pepST,piST,ppiST,propanol2ST,pyruvateST,riboseST,ribose1phosST,ribose5phosST,
ribuloseST,sorboseST,succinateST,sucroseST,sulfateST,sulfiteST,sulfurcrST,
tryptophanLST,ureaST,valineLST,xyloseST,xyluloseST}]=Map[derivetrST,{acetaldehydesp,
acetatesp,acetonesp,adeninesp,adenosinesp,adpsp,alaninesp,
ammoniasp,ampsp,arabinosesp,asparagineLsp,aspartatesp,atpsp,citratesp,co2gsp,
co2totsp,coaqsp,cogsp,ethaneaqsp,ethanolsp,ethylacetatesp,ferricsp,ferroussp,
formatesp,fructosesp,fructose16phossp,fructose6phossp,fumaratesp,galactosesp,
glucosesp,glucose6phossp,glutamatesp,glutaminesp,glycerolsp,glycerol3phossp,
glycinesp,glycylglycinesp,h2aqsp,h2gsp,h2osp,h2o2aqsp,h2saqsp,i2crsp,idpsp,
impsp,indolesp,inosinesp,iodideionsp,isomaltosesp,itpsp,ketoglutaratesp,
lactatesp,lactosesp,leucineLsp,malatesp,maltosesp,mannosesp,mannose6phossp,
methaneaqsp,methanegsp,methanolsp,n2aqsp,n2gsp,n2oaqsp,nadoxsp,nadpoxsp,
nadpredsp,nadredsp,nitratesp,nitritesp,noaqsp,o2aqsp,o2gsp,oxaloacetatesp,
pepsp,pisp,ppisp,propanol2sp,pyruvatesp,ribosesp,ribose1phossp,ribose5phossp,
ribulosesp,sorbosesp,succinatesp,sucrosesp,sulfatesp,sulfitesp,sulfurcrsp,
tryptophanLsp,ureasp,valineLsp,xylosesp,xylulosesp}];
```

9 Calculation of the functions of temperature, pH and ionic strength for the average numbers of hydrogen atoms in reactants for which standard enthalpies of formation of the species are known

The function of temperature, pH. and ionic stength that gives the average number of hydrogen atoms $\overline{N}_H$ in a reactant can be derived by differentiating the function for the standard transformed Gibbs energy of formation. This is done by the program **deriveNHT**.

```
Clear[acetaldehydeNHT,acetateNHT,acetoneNHT,adenineNHT,adenosineNHT,adpNHT,
alanineNHT,ammoniaNHT,ampNHT,arabinoseNHT,asparagineLNHT,aspartateNHT,atpNHT,
citrateNHT,co2gNHT,co2totNHT,coaqNHT,cogNHT,ethaneaqNHT,ethanolNHT,
ethylacetateNHT,ferricNHT,ferrousNHT,formateNHT,fructoseNHT,fructose16phosNHT,
fructose6phosNHT,fumarateNHT,galactoseNHT,glucoseNHT,glucose6phosNHT,
glutamateNHT,glutamineNHT,glycerolNHT,glycerol3phosNHT,glycineNHT,
glycylglycineNHT,h2aqNHT,h2gNHT,h2oNHT,h2o2aqNHT,h2saqNHT,i2crNHT,idpNHT,
impNHT,indoleNHT,inosineNHT,iodideionNHT,isomaltoseNHT,itpNHT,
ketoglutarateNHT,lactateNHT,lactoseNHT,leucineLNHT,malateNHT,maltoseNHT,
mannoseNHT,mannose6phosNHT,methaneaqNHT,methanegNHT,methanolNHT,n2aqNHT,
n2gNHT,n2oaqNHT,nadoxNHT,nadpoxNHT,nadpredNHT,nadredNHT,nitrateNHT,nitriteNHT,
noaqNHT,o2aqNHT,o2gNHT,oxaloacetateNHT,pepNHT,piNHT,ppiNHT,propanol2NHT,
pyruvateNHT,riboseNHT,ribose1phosNHT,ribose5phosNHT,ribuloseNHT,sorboseNHT,
succinateNHT,sucroseNHT,sulfateNHT,sulfiteNHT,sulfurcrNHT,tryptophanLNHT,
ureaNHT,valineLNHT,xyloseNHT,xyluloseNHT]

Evaluate[{acetaldehydeNHT,acetateNHT,acetoneNHT,adenineNHT,adenosineNHT,adpNHT,
alanineNHT,ammoniaNHT,ampNHT,arabinoseNHT,asparagineLNHT,aspartateNHT,atpNHT,
citrateNHT,co2gNHT,co2totNHT,coaqNHT,cogNHT,ethaneaqNHT,ethanolNHT,
ethylacetateNHT,ferricNHT,ferrousNHT,formateNHT,fructoseNHT,fructose16phosNHT,
fructose6phosNHT,fumarateNHT,galactoseNHT,glucoseNHT,glucose6phosNHT,
glutamateNHT,glutamineNHT,glycerolNHT,glycerol3phosNHT,glycineNHT,
glycylglycineNHT,h2aqNHT,h2gNHT,h2oNHT,h2o2aqNHT,h2saqNHT,i2crNHT,idpNHT,
impNHT,indoleNHT,inosineNHT,iodideionNHT,isomaltoseNHT,itpNHT,
ketoglutarateNHT,lactateNHT,lactoseNHT,leucineLNHT,malateNHT,maltoseNHT,
mannoseNHT,mannose6phosNHT,methaneaqNHT,methanegNHT,methanolNHT,n2aqNHT,
n2gNHT,n2oaqNHT,nadoxNHT,nadpoxNHT,nadpredNHT,nadredNHT,nitrateNHT,nitriteNHT,
noaqNHT,o2aqNHT,o2gNHT,oxaloacetateNHT,pepNHT,piNHT,ppiNHT,propanol2NHT,
pyruvateNHT,riboseNHT,ribose1phosNHT,ribose5phosNHT,ribuloseNHT,sorboseNHT,
succinateNHT,sucroseNHT,sulfateNHT,sulfiteNHT,sulfurcrNHT,tryptophanLNHT,
ureaNHT,valineLNHT,xyloseNHT,xyluloseNHT}]=Map[deriveNHT,{acetaldehydesp,
acetatesp,acetonesp,adeninesp,adenosinesp,adpsp,alaninesp,
ammoniasp,ampsp,arabinosesp,asparagineLsp,aspartatesp,atpsp,citratesp,co2gsp,
co2totsp,coaqsp,cogsp,ethaneaqsp,ethanolsp,ethylacetatesp,ferricsp,ferroussp,
formatesp,fructosesp,fructose16phossp,fructose6phossp,fumaratesp,galactosesp,
glucosesp,glucose6phossp,glutamatesp,glutaminesp,glycerolsp,glycerol3phossp,
glycinesp,glycylglycinesp,h2aqsp,h2gsp,h2osp,h2o2aqsp,h2saqsp,i2crsp,idpsp,
impsp,indolesp,inosinesp,iodideionsp,isomaltosesp,itpsp,ketoglutaratesp,
lactatesp,lactosesp,leucineLsp,malatesp,maltosesp,mannosesp,mannose6phossp,
methaneaqsp,methanegsp,methanolsp,n2aqsp,n2gsp,n2oaqsp,nadoxsp,nadpoxsp,
nadpredsp,nadredsp,nitratesp,nitritesp,noaqsp,o2aqsp,o2gsp,oxaloacetatesp,
pepsp,pisp,ppisp,propanol2sp,pyruvatesp,ribosesp,ribose1phossp,ribose5phossp,
ribulosesp,sorbosesp,succinatesp,sucrosesp,sulfatesp,sulfitesp,sulfurcrsp,
tryptophanLsp,ureasp,valineLsp,xylosesp,xylulosesp}];

End[]

Private`
```

Now we protect all the names that have been assigned so that they cannot be changed.

```
Protect [t,pH, is,
acetaldehydesp,acetatesp,acetoacetatesp,acetoacetylcoAsp,acetonesp,
acetylcoAsp,acetylphossp,aconitatecissp,adeninesp,adenosinesp,
adenosinephosphosulfatesp,adpsp,alaninesp,ammoniasp,ampsp,arabinosesp,
arabinose5phossp,asparagineLsp,aspartatesp,atpsp,bpgsp,butanalsp,butanolnsp,
butyratesp,cellobiosesp,citratesp,citrateisosp,co2gsp,co2totsp,coAsp,
coAglutathionesp,coaqsp,cogsp,creatinesp,creatininesp,cyclicampsp,cysteineLsp,
cystineLsp,cytochromecoxsp,cytochromecredsp,deoxyadenosinesp,deoxyadpsp,
deoxyampsp,deoxyatpsp,deoxyribosesp,deoxyribose1phossp,deoxyribose5phossp,
dihydroxyacetonesp,dihydroxyacetonephossp,ethaneaqsp,ethanolsp,
ethylacetatesp,fadenzoxsp,fadenzredsp,fadoxsp,fadredsp,ferredoxinoxsp,
ferredoxinredsp,ferricsp,ferroussp,fmnoxsp,fmnredsp,formatesp,fructosesp,
fructose16phossp,fructose6phossp,fumaratesp,galactono14lactonesp,
```

galactosesp,galactose1phossp,galactose6phossp,gluconolactonesp,
gluconolactone6phossp,glucosesp,glucose1phossp,glucose6phossp,glutamatesp,
glutaminesp,glutathioneoxsp,glutathioneredsp,glyceraldehydesp,
glyceraldehydephossp,glyceratesp,glycerolsp,glycerol3phossp,glycinesp,
glycolatesp,glycylglycinesp,glyoxylatesp,h2aqsp,h2gsp,h2osp,h2o2aqsp,
h2saqsp,hydroxy2oxoglutaratesp,hydroxyglutaratesp,hydroxypropionatebsp,
hydroxypyruvatesp,hypoxanthinesp,i2crsp,iditolsp,idpsp,impsp,indolesp,
inosinesp,iodideionsp,isomaltosesp,itpsp,ketoglutaramatesp,
ketoglutaratesp,lactatesp,lactosesp,lactulosesp,leucineisoLsp,leucineLsp,
lyxosesp,malatesp,maleatesp,maltosesp,malylcoAsp,mannitol1phossp,mannitolDsp,
mannosesp,mannose1phossp,mannose6phossp,methaneaqsp,methanegsp,methanolsp,
methionineLsp,methyl2oxopentanoatesp,methylmalatesp,methylmaleatesp,
methylmalonylcoAsp,n2aqsp,n2gsp,n2oaqsp,nadoxsp,nadpoxsp,nadpredsp,nadredsp,
nicotinamideribonucleotidesp,nitratesp,nitritesp,noaqsp,o2aqsp,o2gsp,
oxalatesp,oxaloacetatesp,oxalosuccinatesp,oxalylcoAsp,oxoprolinesp,
palmitatesp,pepsp,phenylalanineLsp,phosphoglycerate2sp,phosphoglycerate3sp,
phosphohydroxypyruvatesp,phosphoserinesp,pisp,ppisp,propanol2sp,propanolnsp,
propanoylcoAsp,prppsp,pyruvatesp,retinalsp,retinolsp,ribitolsp,ribosesp,
ribose1phossp,ribose5phossp,ribulosesp,ribulose5phossp,serineLsp,sorbitolsp,
sorbitol6phossp,sorbosesp,succinatesp,succinylcoAsp,sucrosesp,sulfatesp,
sulfitesp,sulfurcrsp,thioredoxinoxsp,thioredoxinredsp,threoninesp,
transcinnamatesp,trehalosesp,tryptophanLsp,tyrosineLsp,ubiquinoneoxsp,
ubiquinoneredsp,uratesp,ureasp,uricacidsp,valineLsp,xylitolsp,xylosesp,
xylulosesp];

Protect[acetaldehyde,acetate,acetoacetate,acetoacetylcoA,acetone,acetylcoA,
acetylphos,aconitatecis,adenine,adenosine,adenosinephosphosulfate,adp,alanine,
ammonia,amp,arabinose,arabinose5phos,asparagineL,aspartate,atp,bpg,butanal,
butanoln,butyrate,cellobiose,citrate,citrateiso,co2g,co2tot,coA,
coAglutathione,coaq,cog,creatine,creatinine,cyclicamp,cysteineL,cystineL,
cytochromecox,cytochromecred,deoxyadenosine,deoxyadp,deoxyamp,deoxyatp,
deoxyribose,deoxyribose1phos,deoxyribose5phos,dihydroxyacetone,
dihydroxyacetonephos,ethaneaq,ethanol,ethylacetate,fadenzox,fadenzred,fadox,
fadred,ferredoxinox,ferredoxinred,ferric,ferrous,fmnox,fmnred,formate,
fructose,fructose16phos,fructose6phos,fumarate,galactono14lactone,galactose,
galactose1phos,galactose6phos,gluconolactone,gluconolactone6phos,glucose,
glucose1phos,glucose6phos,glutamate,glutamine,glutathioneox,glutathionered,
glyceraldehyde,glyceraldehydephos,glycerate,glycerol,glycerol3phos,glycine,
glycolate,glycylglycine,glyoxylate,h2aq,h2g,h2o,h2o2aq,h2saq,
hydroxy2oxoglutarate,hydroxyglutarate,hydroxypropionateb,hydroxypyruvate,
hypoxanthine,i2cr,iditol,idp,imp,indole,inosine,iodideion,isomaltose,itp,
ketoglutaramate,ketoglutarate,lactate,lactose,lactulose,leucineisoL,
leucineL,lyxose,malate,maleate,maltose,malylcoA,mannitol1phos,mannitolD,
mannose,mannose1phos,mannose6phos,methaneaq,methaneg,methanol,methionineL,
methyl2oxopentanoate,methylmalate,methylmaleate,methylmalonylcoA,n2aq,n2g,
n2oaq,nadox,nadpox,nadpred,nadred,nicotinamideribonucleotide,nitrate,
nitrite,noaq,o2aq,o2g,oxalate,oxaloacetate,oxalosuccinate,oxalylcoA,
oxoproline,palmitate,pep,phenylalanineL,phosphoglycerate2,phosphoglycerate3,
phosphohydroxypyruvate,phosphoserine,pi,ppi,propanol2,propanoln,
propanoylcoA,prpp,pyruvate,retinal,retinol,ribitol,ribose,ribose1phos,
ribose5phos,ribulose,ribulose5phos,serineL,sorbitol,sorbitol6phos,sorbose,
succinate,succinylcoA,sucrose,sulfate,sulfite,sulfurcr,thioredoxinox,
thioredoxinred,threonine,transcinnamate,trehalose,tryptophanL,tyrosineL,
ubiquinoneox,ubiquinonered,urate,urea,uricacid,valineL,xylitol,xylose,
xylulose];

Protect[acetaldehydeNH,acetateNH,acetoacetateNH,acetoacetylcoANH,acetoneNH,
acetylcoANH,acetylphosNH,aconitatecisNH,adenineNH,adenosineNH,
adenosinephosphosulfateNH,adpNH,alanineNH,ammoniaNH,ampNH,arabinoseNH,
arabinose5phosNH,asparagineLNH,aspartateNH,atpNH,bpgNH,butanalNH,butanolnNH,
butyrateNH,cellobioseNH,citrateNH,citrateisoNH,co2gNH,co2totNH,coANH,
coAglutathioneNH,coaqNH,cogNH,creatineNH,creatinineNH,cyclicampNH,cysteineLNH,
cystineLNH,cytochromecoxNH,cytochromecredNH,deoxyadenosineNH,deoxyadpNH,
deoxyampNH,deoxyatpNH,deoxyriboseNH,deoxyribose1phosNH,deoxyribose5phosNH,
dihydroxyacetoneNH,dihydroxyacetonephosNH,ethaneaqNH,ethanolNH,

```
ethylacetateNH,fadenzoxNH,fadenzredNH,fadoxNH,fadredNH,ferredoxinoxNH,
ferredoxinredNH,ferricNH,ferrousNH,fmnoxNH,fmnredNH,formateNH,fructoseNH,
fructose16phosNH,fructose6phosNH,fumarateNH,galactono14lactoneNH,
galactoseNH,galactose1phosNH,galactose6phosNH,gluconolactoneNH,
gluconolactone6phosNH,glucoseNH,glucose1phosNH,glucose6phosNH,glutamateNH,
glutamineNH,glutathioneoxNH,glutathioneredNH,glyceraldehydeNH,
glyceraldehydephosNH,glycerateNH,glycerolNH,glycerol3phosNH,glycineNH,
glycolateNH,glycylglycineNH,glyoxylateNH,h2aqNH,h2gNH,h2oNH,h2o2aqNH,
h2saqNH,hydroxy2oxoglutarateNH,hydroxyglutarateNH,hydroxypropionatebNH,
hydroxypyruvateNH,hypoxanthineNH,i2crNH,iditolNH,idpNH,impNH,indoleNH,
inosineNH,iodideionNH,isomaltoseNH,itpNH,ketoglutaramateNH,
ketoglutarateNH,lactateNH,lactoseNH,lactuloseNH,leucineisoLNH,leucineLNH,
lyxoseNH,malateNH,maleateNH,maltoseNH,malylcoANH,mannitol1phosNH,mannitolDNH,
mannoseNH,mannose1phosNH,mannose6phosNH,methaneaqNH,methanegNH,methanolNH,
methionineLNH,methyl2oxopentanoateNH,methylmalateNH,methylmaleateNH,
methylmalonylcoANH,n2aqNH,n2gNH,n2oaqNH,nadoxNH,nadpoxNH,nadpredNH,nadredNH,
nicotinamideribonucleotideNH,nitrateNH,nitriteNH,noaqNH,o2aqNH,o2gNH,
oxalateNH,oxaloacetateNH,oxalosuccinateNH,oxalylcoANH,oxoprolineNH,
palmitateNH,pepNH,phenylalanineLNH,phosphoglycerate2NH,phosphoglycerate3NH,
phosphohydroxypyruvateNH,phosphoserineNH,piNH,ppiNH,propanol2NH,propanolnNH,
propanoylcoANH,prppNH,pyruvateNH,retinalNH,retinolNH,ribitolNH,riboseNH,
ribose1phosNH,ribose5phosNH,riboseNH,ribulose5phosNH,serineLNH,sorbitolNH,
sorbitol6phosNH,sorboseNH,succinateNH,succinylcoANH,sucroseNH,sulfateNH,
sulfiteNH,sulfurcrNH,thioredoxinoxNH,thioredoxinredNH,threonineNH,
transcinnamateNH,trehaloseNH,tryptophanLNH,tyrosineLNH,ubiquinoneoxNH,
ubiquinoneredNH,urateNH,ureaNH,uricacidNH,valineLNH,xylitolNH,xyloseNH,
xyluloseNH];

Protect[acetaldehydeGT,acetateGT,acetoneGT,adenineGT,adenosineGT,adpGT,
alanineGT,ammoniaGT,ampGT,arabinoseGT,asparagineLGT,aspartateGT,atpGT,
citrateGT,co2gGT,co2totGT,coaqGT,cogGT,ethaneaqGT,ethanolGT,ethylacetateGT,
ferricGT,ferrousGT,formateGT,fructoseGT,fructose16phosGT,fructose6phosGT,
fumarateGT,galactoseGT,glucoseGT,glucose6phosGT,glutamateGT,glutamineGT,
glycerolGT,glycerol3phosGT,glycineGT,glycylglycineGT,h2aqGT,h2gGT,h2oGT,
h2o2aqGT,h2saqGT,i2crGT,idpGT,impGT,indoleGT,inosineGT,iodideionGT,
isomaltoseGT,itpGT,ketoglutarateGT,lactateGT,lactoseGT,leucineLGT,malateGT,
maltoseGT,mannoseGT,mannose6phosGT,methaneaqGT,methanegGT,methanolGT,n2aqGT,
n2gGT,n2oaqGT,nadoxGT,nadpoxGT,nadpredGT,nadredGT,nitrateGT,nitriteGT,
noaqGT,o2aqGT,o2gGT,oxaloacetateGT,pepGT,piGT,ppiGT,propanol2GT,pyruvateGT,
riboseGT,ribose1phosGT,ribose5phosGT,ribuloseGT,sorboseGT,succinateGT,
sucroseGT,sulfateGT,sulfiteGT,sulfurcrGT,tryptophanLGT,ureaGT,valineLGT,
xyloseGT,xyluloseGT];

Protect[
acetaldehydeHT,acetateHT,acetoneHT,adenineHT,adenosineHT,adpHT,alanineHT,
ammoniaHT,ampHT,arabinoseHT,asparagineLHT,aspartateHT,atpHT,citrateHT,co2gHT,
co2totHT,coaqHT,cogHT,ethaneaqHT,ethanolHT,ethylacetateHT,ferricHT,ferrousHT,
formateHT,fructoseHT,fructose16phosHT,fructose6phosHT,fumarateHT,galactoseHT,
glucoseHT,glucose6phosHT,glutamateHT,glutamineHT,glycerolHT,glycerol3phosHT,
glycineHT,glycylglycineHT,h2aqHT,h2gHT,h2oHT,h2o2aqHT,h2saqHT,i2crHT,idpHT,
impHT,indoleHT,inosineHT,iodideionHT,isomaltoseHT,itpHT,ketoglutarateHT,
lactateHT,lactoseHT,leucineLHT,malateHT,maltoseHT,mannoseHT,mannose6phosHT,
methaneaqHT,methanegHT,methanolHT,n2aqHT,n2gHT,n2oaqHT,nadoxHT,nadpoxHT,
nadpredHT,nadredHT,nitrateHT,nitriteHT,noaqHT,o2aqHT,o2gHT,oxaloacetateHT,
pepHT,piHT,ppiHT,propanol2HT,pyruvateHT,riboseHT,ribose1phosHT,ribose5phosHT,
ribuloseHT,sorboseHT,succinateHT,sucroseHT,sulfateHT,sulfiteHT,sulfurcrHT,
tryptophanLHT,ureaHT,valineLHT,xyloseHT,xyluloseHT];

Protect[acetaldehydeST,acetateST,acetoneST,adenineST,adenosineST,adpST,alanineST,
ammoniaST,ampST,arabinoseST,asparagineLST,aspartateST,atpST,citrateST,co2gST,
co2totST,coaqST,cogST,ethaneaqST,ethanolST,ethylacetateST,ferricST,ferrousST,
formateST,fructoseST,fructose16phosST,fructose6phosST,fumarateST,galactoseST,
glucoseST,glucose6phosST,glutamateST,glutamineST,glycerolST,glycerol3phosST,
glycineST,glycylglycineST,h2aqST,h2gST,h2oST,h2o2aqST,h2saqST,i2crST,idpST,
impST,indoleST,inosineST,iodideionST,isomaltoseST,itpST,ketoglutarateST,
```

```
  lactateST,lactoseST,leucineLST,malateST,maltoseST,mannoseST,mannose6phosST,
methaneaqST,methanegST,methanolST,n2aqST,n2gST,n2oaqST,nadoxST,nadpoxST,
nadpredST,nadredST,nitrateST,nitriteST,noaqST,o2aqST,o2gST,oxaloacetateST,
pepST,piST,ppiST,propanol2ST,pyruvateST,riboseST,ribose1phosST,ribose5phosST,
riburoseST,sorboseST,succinateST,sucroseST,sulfateST,sulfiteST,sulfurcrST,
tryptophanLST,ureaST,valineLST,xyloseST,xyluloseST];

Protect[acetaldehydeNHT,acetateNHT,acetoneNHT,adenineNHT,adenosineNHT,adpNHT,
alanineNHT,ammoniaNHT,ampNHT,arabinoseNHT,asparagineLNHT,aspartateNHT,atpNHT,
citrateNHT,co2gNHT,co2totNHT,coaqNHT,cogNHT,ethaneaqNHT,ethanolNHT,
ethylacetateNHT,ferricNHT,ferrousNHT,formateNHT,fructoseNHT,fructose16phosNHT,
fructose6phosNHT,fumarateNHT,galactoseNHT,glucoseNHT,glucose6phosNHT,
glutamateNHT,glutamineNHT,glycerolNHT,glycerol3phosNHT,glycineNHT,
glycylglycineNHT,h2aqNHT,h2gNHT,h2oNHT,h2o2aqNHT,h2saqNHT,i2crNHT,idpNHT,
impNHT,indoleNHT,inosineNHT,iodideionNHT,isomaltoseNHT,itpNHT,
ketoglutarateNHT,lactateNHT,lactoseNHT,leucineLNHT,malateNHT,maltoseNHT,
mannoseNHT,mannose6phosNHT,methaneaqNHT,methanegNHT,methanolNHT,n2aqNHT,
n2gNHT,n2oaqNHT,nadoxNHT,nadpoxNHT,nadpredNHT,nadredNHT,nitrateNHT,nitriteNHT,
noaqNHT,o2aqNHT,o2gNHT,oxaloacetateNHT,pepNHT,piNHT,ppiNHT,propanol2NHT,
pyruvateNHT,riboseNHT,ribose1phosNHT,ribose5phosNHT,ribuloseNHT,sorboseNHT,
succinateNHT,sucroseNHT,sulfateNHT,sulfiteNHT,sulfurcrNHT,tryptophanLNHT,
ureaNHT,valineLNHT,xyloseNHT,xyluloseNHT];

EndPackage[]
```

Appendix 2: Tables of Transformed Thermodynamic Properties

The functions of pH and ionic strength and functions of temperature, pH, and ionic strength that yield standard transformed thermodynamic properties are produced in the package BasicBiochemData3. These functions are used here to make the following four tables:

Table A1 $\Delta_f G'°$ at 298.15 K, ionic strength 0.25 M, and five pHs for 199 reactants
Table A2 $\overline{N}_H$ at 298.15 K, ionic strength 0.25 M, and five pHs for 199 reactants
Table A3 $\Delta_f H'°$ at 298.15 K, ionic strength 0.25 M, and five pHs for 94 reactants
Table A4 $\Delta_f S'°$ at 298.15 K, ionic strength 0.25 M, and five pHs for 94 reactants

Tables A1 and A2 can be recalculated at other pHs and ionic strengths. Tables A3 and A4 can be recalculated at other temperatures, pHs, and ionic strengths, and tables for $\Delta_f G'°$ and $\overline{N}_H$ for 94 reactants can also be calculated at other temperatures, pHs, and ionic strengths.

The following lists of functions are used to make the tables

```
In[2]:=  Off[General::"spell"];
         Off[General::"spell1"];

In[4]:=  << BiochemThermo`BasicBiochemData3`

In[5]:=  listGibbsfnpHis = {acetaldehyde, acetate, acetoacetate, acetoacetylcoA, acetone,
            acetylcoA, acetylphos, aconitatecis, adenine, adenosine, adenosinephosphosulfate,
            adp, alanine, ammonia, amp, arabinose, arabinose5phos, asparagineL, aspartate,
            atp, bpg, butanal, butanoln, butyrate, cellobiose, citrate, citrateiso, co2g,
            co2tot, coA, coAglutathione, coaq, cog, creatine, creatinine, cyclicamp, cysteineL,
            cystineL, cytochromecox, cytochromecred, deoxyadenosine, deoxyadp, deoxyamp,
            deoxyatp, deoxyribose, deoxyribose1phos, deoxyribose5phos, dihydroxyacetone,
            dihydroxyacetonephos, ethaneaq, ethanol, ethylacetate, fadenzox, fadenzred, fadox,
            fadred, ferredoxinox, ferredoxinred, ferric, ferrous, fmnox, fmnred, formate,
            fructose, fructose16phos, fructose6phos, fumarate, galactono14lactone, galactose,
            galactose1phos, galactose6phos, gluconolactone, gluconolactone6phos, glucose,
            glucose1phos, glucose6phos, glutamate, glutamine, glutathioneox, glutathionered,
            glyceraldehyde, glyceraldehydephos, glycerate, glycerol, glycerol3phos,
            glycine, glycolate, glycylglycine, glyoxylate, h2aq, h2g, h2o, h2o2aq, h2saq,
            hydroxy2oxoglutarate, hydroxyglutarate, hydroxypropionateb, hydroxypyruvate,
            hypoxanthine, i2cr, iditol, idp, imp, indole, inosine, iodideion, isomaltose,
            itp, ketoglutaramate, ketoglutarate, lactate, lactose, lactulose, leucineisoL,
            leucineL, lyxose, malate, maleate, maltose, malylcoA, mannitol1phos, mannitolD,
            mannose, mannose1phos, mannose6phos, methaneaq, methaneg, methanol, methionineL,
            methyl2oxopentanoate, methylmalate, methylmaleate, methylmalonylcoA, n2aq, n2g,
            n2oaq, nadox, nadpox, nadpred, nadred, nicotinamideribonucleotide, nitrate,
            nitrite, noaq, o2aq, o2g, oxalate, oxaloacetate, oxalosuccinate, oxalylcoA,
            oxoproline, palmitate, pep, phenylalanineL, phosphoglycerate2, phosphoglycerate3,
            phosphohydroxypyruvate, phosphoserine, pi, ppi, propanol2, propanoln, propanoylcoA,
            prpp, pyruvate, retinal, retinol, ribitol, ribose, ribose1phos, ribose5phos,
            ribulose, ribulose5phos, serineL, sorbitol, sorbitol6phos, sorbose, succinate,
            succinylcoA, sucrose, sulfate, sulfite, sulfurcr, thioredoxinox, thioredoxinred,
            threonine, transcinnamate, trehalose, tryptophanL, tyrosineL, ubiquinoneox,
            ubiquinonered, urate, urea, uricacid, valineL, xylitol, xylose, xylulose};
```

```
In[6]:= listreactantsNH = {acetaldehydeNH, acetateNH, acetoacetateNH, acetoacetylcoANH,
        acetoneNH, acetylcoANH, acetylphosNH, aconitatecisNH, adenineNH, adenosineNH,
        adenosinephosphosulfateNH, adpNH, alanineNH, ammoniaNH, ampNH, arabinoseNH,
        arabinose5phosNH, asparagineLNH, aspartateNH, atpNH, bpgNH, butanalNH, butanolnNH,
        butyrateNH, cellobioseNH, citrateNH, citrateisoNH, co2gNH, co2totNH, coANH,
        coAglutathioneNH, coaqNH, cogNH, creatineNH, creatinineNH, cyclicampNH, cysteineLNH,
        cystineLNH, cytochromecoxNH, cytochromecredNH, deoxyadenosineNH, deoxyadpNH,
        deoxyampNH, deoxyatpNH, deoxyriboseNH, deoxyribose1phosNH, deoxyribose5phosNH,
        dihydroxyacetoneNH, dihydroxyacetonephosNH, ethaneaqNH, ethanolNH, ethylacetateNH,
        fadenzoxNH, fadenzredNH, fadoxNH, fadredNH, ferredoxinoxNH, ferredoxinredNH,
        ferricNH, ferrousNH, fmnoxNH, fmnredNH, formateNH, fructoseNH, fructose16phosNH,
        fructose6phosNH, fumarateNH, galactono14lactoneNH, galactoseNH, galactose1phosNH,
        galactose6phosNH, gluconolactoneNH, gluconolactone6phosNH, glucoseNH,
        glucose1phosNH, glucose6phosNH, glutamateNH, glutamineNH, glutathioneoxNH,
        glutathioneredNH, glyceraldehydeNH, glyceraldehydephosNH, glycerateNH, glycerolNH,
        glycerol3phosNH, glycineNH, glycolateNH, glycylglycineNH, glyoxylateNH, h2aqNH,
        h2gNH, h2oNH, h2o2aqNH, h2saqNH, hydroxy2oxoglutarateNH, hydroxyglutarateNH,
        hydroxypropionatebNH, hydroxypyruvateNH, hypoxanthineNH, i2crNH, iditolNH, idpNH,
        impNH, indoleNH, inosineNH, iodideionNH, isomaltoseNH, itpNH, ketoglutaramateNH,
        ketoglutarateNH, lactateNH, lactoseNH, lactuloseNH, leucineisoLNH, leucineLNH,
        lyxoseNH, malateNH, maleateNH, maltoseNH, malylcoANH, mannitol1phosNH, mannitolDNH,
        mannoseNH, mannose1phosNH, mannose6phosNH, methaneaqNH, methanegNH, methanolNH,
        methionineLNH, methyl2oxopentanoateNH, methylmalateNH, methylmaleateNH,
        methylmalonylcoANH, n2aqNH, n2gNH, n2oaqNH, nadoxNH, nadpoxNH, nadpredNH, nadredNH,
        nicotinamideribonucleotideNH, nitrateNH, nitriteNH, noaqNH, o2aqNH, o2gNH, oxalateNH,
        oxaloacetateNH, oxalosuccinateNH, oxalylcoANH, oxoprolineNH, palmitateNH, pepNH,
        phenylalanineLNH, phosphoglycerate2NH, phosphoglycerate3NH, phosphohydroxypyruvateNH,
        phosphoserineNH, piNH, ppiNH, propanol2NH, propanolnNH, propanoylcoANH,
        prppNH, pyruvateNH, retinalNH, retinolNH, ribitolNH, riboseNH, ribose1phosNH,
        ribose5phosNH, ribuloseNH, ribulose5phosNH, serineLNH, sorbitolNH, sorbitol6phosNH,
        sorboseNH, succinateNH, succinylcoANH, sucroseNH, sulfateNH, sulfiteNH,
        sulfurcrNH, thioredoxinoxNH, thioredoxinredNH, threonineNH, transcinnamateNH,
        trehaloseNH, tryptophanLNH, tyrosineLNH, ubiquinoneoxNH, ubiquinoneredNH,
        urateNH, ureaNH, uricacidNH, valineLNH, xylitolNH, xyloseNH, xyluloseNH};

In[7]:= listreactantsHT = {acetaldehydeHT, acetateHT, acetoneHT, adenineHT, adenosineHT,
        adpHT, alanineHT, ammoniaHT, ampHT, arabinoseHT, asparagineLHT, aspartateHT,
        atpHT, citrateHT, co2gHT, co2totHT, coaqHT, cogHT, ethaneaqHT, ethanolHT,
        ethylacetateHT, ferricHT, ferrousHT, formateHT, fructoseHT, fructose16phosHT,
        fructose6phosHT, fumarateHT, galactoseHT, glucoseHT, glucose6phosHT, glutamateHT,
        glutamineHT, glycerolHT, glycerol3phosHT, glycineHT, glycylglycineHT, h2aqHT,
        h2gHT, h2oHT, h2o2aqHT, h2saqHT, i2crHT, idpHT, impHT, indoleHT, inosineHT,
        iodideionHT, isomaltoseHT, itpHT, ketoglutarateHT, lactateHT, lactoseHT,
        leucineLHT, malateHT, maltoseHT, mannoseHT, mannose6phosHT, methaneaqHT,
        methanegHT, methanolHT, n2aqHT, n2gHT, n2oaqHT, nadoxHT, nadpoxHT, nadpredHT,
        nadredHT, nitrateHT, nitriteHT, noaqHT, o2aqHT, o2gHT, oxaloacetateHT,
        pepHT, piHT, ppiHT, propanol2HT, pyruvateHT, riboseHT, ribose1phosHT,
        ribose5phosHT, ribuloseHT, sorboseHT, succinateHT, sucroseHT, sulfateHT,
        sulfiteHT, sulfurcrHT, tryptophanLHT, ureaHT, valineLHT, xyloseHT, xyluloseHT};
```

```
In[8]:= listreactantsST = {acetaldehydeST, acetateST, acetoneST, adenineST, adenosineST,
           adpST, alanineST, ammoniaST, ampST, arabinoseST, asparagineLST, aspartateST,
           atpST, citrateST, co2gST, co2totST, coaqST, cogST, ethaneaqST, ethanolST,
           ethylacetateST, ferricST, ferrousST, formateST, fructoseST, fructose16phosST,
           fructose6phosST, fumarateST, galactoseST, glucoseST, glucose6phosST, glutamateST,
           glutamineST, glycerolST, glycerol3phosST, glycineST, glycylglycineST, h2aqST,
           h2gST, h2oST, h2o2aqST, h2saqST, i2crST, idpST, impST, indoleST, inosineST,
           iodideionST, isomaltoseST, itpST, ketoglutarateST, lactateST, lactoseST,
           leucineLST, malateST, maltoseST, mannoseST, mannose6phosST, methaneaqST,
           methanegST, methanolST, n2aqST, n2gST, n2oaqST, nadoxST, nadpoxST, nadpredST,
           nadredST, nitrateST, nitriteST, noaqST, o2aqST, o2gST, oxaloacetateST,
           pepST, piST, ppiST, propanol2ST, pyruvateST, riboseST, ribose1phosST,
           ribose5phosST, ribuloseST, sorboseST, succinateST, sucroseST, sulfateST,
           sulfiteST, sulfurcrST, tryptophanLST, ureaST, valineLST, xyloseST, xyluloseST};
```

The following list of 199 names is use to make Tables 1 and 2.

```
In[9]:= listnames = {"acetaldehyde", "acetate", "acetoacetate", "acetoacetylcoA",
           "acetone", "acetylcoA", "acetylphos", "aconitatecis", "adenine", "adenosine",
           "adenosinephosphosulfate", "adp", "alanine", "ammonia", "amp", "arabinose",
           "arabinose5phos", "asparagineL", "aspartate", "atp", "bpg", "butanal", "butanoln",
           "butyrate", "cellobiose", "citrate", "citrateiso", "co2g", "co2tot", "coA",
           "coAglutathione", "coaq", "cog", "creatine", "creatinine", "cyclicamp", "cysteineL",
           "cystineL", "cytochromecox", "cytochromecred", "deoxyadenosine", "deoxyadp",
           "deoxyamp", "deoxyatp", "deoxyribose", "deoxyribose1phos", "deoxyribose5phos",
           "dihydroxyacetone", "dihydroxyacetonephos", "ethaneaq", "ethanol", "ethylacetate",
           "fadenzox", "fadenzred", "fadox", "fadred", "ferredoxinox", "ferredoxinred",
           "ferric", "ferrous", "fmnox", "fmnred", "formate", "fructose", "fructose16phos",
           "fructose6phos", "fumarate", "galactono14lactone", "galactose", "galactose1phos",
           "galactose6phos", "gluconolactone", "gluconolactone6phos", "glucose",
           "glucose1phos", "glucose6phos", "glutamate", "glutamine", "glutathioneox",
           "glutathionered", "glyceraldehyde", "glyceraldehydephos", "glycerate", "glycerol",
           "glycerol3phos", "glycine", "glycolate", "glycylglycine", "glyoxylate", "h2aq",
           "h2g", "h2o", "h2o2aq", "h2saq", "hydroxy2oxoglutarate", "hydroxyglutarate",
           "hydroxypropionateb", "hydroxypyruvate", "hypoxanthine", "i2cr", "iditol", "idp",
           "imp", "indole", "inosine", "iodideion", "isomaltose", "itp", "ketoglutaramate",
           "ketoglutarate", "lactate", "lactose", "lactulose", "leucineisoL", "leucineL",
           "lyxose", "malate", "maleate", "maltose", "malylcoA", "mannitol1phos", "mannitolD",
           "mannose", "mannose1phos", "mannose6phos", "methaneaq", "methaneg", "methanol",
           "methionineL", "methyl2oxopentanoate", "methylmalate", "methylmaleate",
           "methylmalonylcoA", "n2aq", "n2g", "n2oaq", "nadox", "nadpox", "nadpred", "nadred",
           "nicotinamideribonucleotide", "nitrate", "nitrite", "noaq", "o2aq", "o2g", "oxalate",
           "oxaloacetate", "oxalosuccinate", "oxalylcoA", "oxoproline", "palmitate", "pep",
           "phenylalanineL", "phosphoglycerate2", "phosphoglycerate3", "phosphohydroxypyruvate",
           "phosphoserine", "pi", "ppi", "propanol2", "propanoln", "propanoylcoA",
           "prpp", "pyruvate", "retinal", "retinol", "ribitol", "ribose", "ribose1phos",
           "ribose5phos", "ribulose", "ribulose5phos", "serineL", "sorbitol", "sorbitol6phos",
           "sorbose", "succinate", "succinylcoA", "sucrose", "sulfate", "sulfite",
           "sulfurcr", "thioredoxinox", "thioredoxinred", "threonine", "transcinnamate",
           "trehalose", "tryptophanL", "tyrosineL", "ubiquinoneox", "ubiquinonered",
           "urate", "urea", "uricacid", "valineL", "xylitol", "xylose", "xylulose"};
```

The following list of 94 names is used to make Tables 3 and 4.

```
In[10]:= nameswithH = {"acetaldehyde", "acetate", "acetone", "adenine", "adenosine",
        "adp", "alanine", "ammonia", "amp", "arabinose", "asparagineL", "aspartate",
        "atp", "citrate", "co2g", "co2tot", "coaq", "cog", "ethaneaq", "ethanol",
        "ethylacetate", "ferric", "ferrous", "formate", "fructose", "fructose16phos",
        "fructose6phos", "fumarate", "galactose", "glucose", "glucose6phos", "glutamate",
        "glutamine", "glycerol", "glycerol3phos", "glycine", "glycylglycine", "h2aq",
        "h2g", "h2o", "h2o2aq", "h2saq", "i2cr", "idp", "imp", "indole", "inosine",
        "iodideion", "isomaltose", "itp", "ketoglutarate", "lactate", "lactose",
        "leucineL", "malate", "maltose", "mannose", "mannose6phos", "methaneaq",
        "methaneg", "methanol", "n2aq", "n2g", "n2oaq", "nadox", "nadpox", "nadpred",
        "nadred", "nitrate", "nitrite", "noaq", "o2aq", "o2g", "oxaloacetate",
        "pep", "pi", "ppi", "propanol2", "pyruvate", "ribose", "ribose1phos",
        "ribose5phos", "ribulose", "sorbose", "succinate", "sucrose", "sulfate",
        "sulfite", "sulfurcr", "tryptophanL", "urea", "valineL", "xylose", "xylulose"};
```

Table A1 Standard Transformed Gibbs Energies of Formation of Reactants in kJ mol^{-1} at 298.15 K, Ionic Strength 0.25 M, and pHs 5, 6, 7, 8, and 9

```
In[11]:= table1 = PaddedForm[TableForm[listGibbsfnpHis /. is → .25 /. pH → {5, 6, 7, 8, 9},
        TableHeadings → {listnames, {"  pH 5", "  pH 6", "  pH 7", "  pH 8", "pH 9"}},
        TableSpacing → {1, 1}], {6, 2}]
```

Out[11]//PaddedForm=

	pH 5	pH 6	pH 7	pH 8	pH 9
acetaldehyde	-21.60	1.23	24.06	46.90	69.73
acetate	-282.71	-265.02	-247.83	-230.70	-213.57
acetoacetate	-335.74	-307.20	-278.66	-250.12	-221.58
acetoacetylcoA	-138.57	-110.03	-81.49	-52.95	-24.41
acetone	16.40	50.65	84.90	119.14	153.39
acetylcoA	-92.31	-75.19	-58.06	-40.94	-23.81
acetylphos	-1153.77	-1129.84	-1107.02	-1085.39	-1066.49
aconitatecis	-836.37	-819.24	-802.12	-785.00	-767.87
adenine	459.79	488.65	517.23	545.77	574.31
adenosine	186.98	261.25	335.46	409.66	483.87
adenosinephosphosulfate	-1190.59	-1122.09	-1053.60	-985.10	-916.61
adp	-1569.05	-1495.55	-1424.70	-1355.78	-1287.24
alanine	-165.55	-125.59	-85.64	-45.68	-5.73
ammonia	37.28	60.11	82.93	105.64	127.51
amp	-698.40	-625.22	-554.83	-486.04	-417.51
arabinose	-448.73	-391.65	-334.57	-277.49	-220.41
arabinose5phos	-1344.02	-1288.15	-1234.99	-1183.35	-1131.94
asparagineL	-291.13	-245.47	-199.80	-154.14	-108.47
aspartate	-520.59	-486.34	-452.09	-417.85	-383.60
atp	-2437.46	-2363.76	-2292.50	-2223.44	-2154.88
bpg	-2262.15	-2233.92	-2207.30	-2183.36	-2160.38
butanal	104.93	150.59	196.26	241.92	287.59
butanoln	121.66	178.74	235.82	292.90	349.98
butyrate	-147.99	-108.03	-68.08	-28.12	11.83
cellobiose	-940.24	-814.67	-689.09	-563.51	-437.93
citrate	-1027.23	-995.44	-966.23	-937.62	-909.07
citrateiso	-1020.58	-988.80	-959.58	-930.97	-902.42
co2g	-394.36				
co2tot	-564.61	-554.49	-547.10	-541.18	-535.80
coA	-18.48	-12.79	-7.26	-2.82	-1.10
coAglutathione	403.59	489.21	574.83	660.45	746.07

coaq	-119.90				
cog	-137.17				
creatine	4.95	56.32	107.69	159.07	210.44
creatinine	182.31	222.27	262.22	302.18	342.13
cyclicamp	-425.04	-385.08	-345.13	-305.17	-265.22
cysteineL	-133.37	-93.43	-53.65	-14.97	20.99
cystineL	-314.31	-245.81	-177.32	-108.82	-40.33
cytochromecox	-7.29				
cytochromecred	-27.75				
deoxyadenosine	334.84	409.11	483.32	557.52	631.73
deoxyadp	-1421.19	-1347.69	-1276.84	-1207.92	-1139.38
deoxyamp	-550.54	-477.36	-406.97	-338.18	-269.65
deoxyatp	-2289.60	-2215.90	-2144.64	-2075.58	-2007.02
deoxyribose	-310.64	-253.56	-196.48	-139.40	-82.32
deoxyribose1phos	-1185.51	-1129.65	-1076.49	-1024.85	-973.44
deoxyribose5phos	-1193.60	-1137.73	-1084.57	-1032.93	-981.52
dihydroxyacetone	-274.90	-240.65	-206.40	-172.16	-137.91
dihydroxyacetonephos	-1154.88	-1124.53	-1095.70	-1067.13	-1038.59
ethaneaq	159.09	193.34	227.59	261.83	296.08
ethanol	-5.54	28.71	62.96	97.20	131.45
ethylacetate	-102.85	-57.19	-11.52	34.14	79.81
fadenzox	906.61	1083.56	1260.51	1437.46	1614.40
fadenzred	876.71	1065.07	1253.44	1441.80	1630.17
fadox	906.61	1083.56	1260.51	1437.46	1614.40
fadred	926.43	1114.79	1303.16	1491.52	1679.89
ferredoxinox	-0.81				
ferredoxinred	38.07				
ferric	-11.99				
ferrous	-82.14				
fmnox	554.41	662.86	771.31	879.77	988.22
fmnred	574.23	694.10	813.97	933.84	1053.70
formate	-322.46	-316.75	-311.04	-305.34	-299.63
fructose	-563.31	-494.81	-426.32	-357.82	-289.33
fructose16phos	-2326.42	-2264.57	-2206.78	-2149.62	-2092.54
fructose6phos	-1445.66	-1379.42	-1315.74	-1252.84	-1190.04
fumarate	-546.67	-535.02	-523.58	-512.16	-500.74
galactono14lactone	-611.84	-554.76	-497.68	-440.60	-383.52
galactose	-556.73	-488.23	-419.74	-351.24	-282.75
galactose1phos	-1440.96	-1375.09	-1311.60	-1248.72	-1185.93
galactose6phos	-1442.49	-1375.79	-1311.80	-1248.85	-1186.05
gluconolactone	-610.00	-552.92	-495.84	-438.76	-381.68
gluconolactone6phos	-1490.37	-1435.13	-1382.60	-1331.07	-1279.68
glucose	-563.70	-495.20	-426.71	-358.21	-289.72
glucose1phos	-1442.86	-1376.01	-1311.89	-1248.92	-1186.11
glucose6phos	-1449.53	-1382.88	-1318.92	-1255.98	-1193.18
glutamate	-463.48	-417.82	-372.15	-326.49	-280.82
glutamine	-234.52	-177.44	-120.36	-63.28	-6.20
glutathioneox	877.26	1048.50	1219.74	1390.98	1562.22
glutathionered	455.34	546.64	637.62	726.89	813.52
glyceraldehyde	-263.92	-229.67	-195.42	-161.18	-126.93
glyceraldehydephos	-1147.22	-1116.87	-1088.04	-1059.47	-1030.93
glycerate	-515.07	-486.53	-457.99	-429.45	-400.91
glycerol	-262.68	-217.02	-171.35	-125.69	-80.02
glycerol3phos	-1163.24	-1118.83	-1077.13	-1036.91	-996.93

glycine	-233.16	-204.62	-176.08	-147.54	-119.00
glycolate	-443.71	-426.59	-409.46	-392.34	-375.21
glycylglycine	-285.40	-239.74	-194.07	-148.41	-102.74
glyoxylate	-440.06	-434.35	-428.64	-422.94	-417.23
h2aq	76.30	87.72	99.13	110.55	121.96
h2g	58.70	70.12	81.53	92.95	104.36
h2o	-178.49	-167.07	-155.66	-144.24	-132.83
h2o2aq	-75.33	-63.91	-52.50	-41.08	-29.67
h2saq	30.82	41.84	51.01	57.62	63.44
hydroxy2oxoglutarate	-837.62	-814.79	-791.95	-769.12	-746.29
hydroxyglutarate	-667.32	-633.07	-598.82	-564.57	-530.33
hydroxypropionateb	-372.46	-343.92	-315.38	-286.84	-258.30
hydroxypyruvate	-521.31	-504.19	-487.06	-469.94	-452.81
hypoxanthine	206.90	229.73	252.56	275.40	298.23
i2cr	0.00				
iditol	-533.83	-453.92	-374.01	-294.09	-214.18
idp	-1813.14	-1745.49	-1680.43	-1617.91	-1558.16
imp	-942.62	-875.25	-810.61	-747.85	-687.07
indole	429.25	469.21	509.16	549.12	589.07
inosine	-56.95	11.54	79.99	148.06	214.21
iodideion	-52.38				
isomaltose	-942.01	-816.44	-690.86	-565.28	-439.70
itp	-2681.45	-2613.60	-2548.10	-2485.18	-2424.69
ketoglutaramate	-449.94	-421.40	-392.86	-364.32	-335.78
ketoglutarate	-679.25	-656.42	-633.58	-610.75	-587.92
lactate	-370.78	-342.24	-313.70	-285.16	-256.62
lactose	-921.63	-796.06	-670.48	-544.90	-419.32
lactulose	-929.52	-803.95	-678.37	-552.79	-427.21
leucineisoL	37.65	111.85	186.06	260.26	334.47
leucineL	29.30	103.50	177.71	251.91	326.12
lyxose	-455.64	-398.56	-341.48	-284.40	-227.32
malate	-729.49	-705.79	-682.85	-660.00	-637.17
maleate	-536.63	-525.21	-513.80	-502.38	-490.96
maltose	-928.99	-803.42	-677.84	-552.26	-426.68
malylcoA	-547.05	-524.04	-501.19	-478.35	-455.52
mannitol1phos	-1420.87	-1342.60	-1267.06	-1192.68	-1118.45
mannitolD	-531.71	-451.80	-371.89	-291.97	-212.06
mannose	-557.80	-489.30	-420.81	-352.31	-283.82
mannose1phos	-1440.24	-1373.54	-1309.55	-1246.60	-1183.80
mannose6phos	-1445.55	-1378.85	-1314.86	-1251.91	-1189.11
methaneaq	83.07	105.90	128.73	151.57	174.40
methaneg	66.68	89.51	112.34	135.18	158.01
methanol	-57.91	-35.08	-12.25	10.59	33.42
methionineL	-180.07	-117.28	-54.49	8.29	71.08
methyl2oxopentanoate	-181.80	-130.43	-79.06	-27.68	23.69
methylmalate	-671.04	-636.79	-602.54	-568.29	-534.05
methylmaleate	-491.00	-464.72	-441.00	-418.06	-395.21
methylmalonylcoA	-386.09	-363.08	-340.23	-317.39	-294.56
n2aq	18.70				
n2g	0.00				
n2oaq	104.20				
nadox	762.29	910.70	1059.11	1207.51	1355.92
nadpox	-108.72	33.98	176.68	319.38	462.08
nadpred	-59.57	88.84	237.25	385.66	534.07

nadred	811.86	965.98	1120.09	1274.21	1428.33
nicotinamideribonucleotide	1242.45	1326.27	1407.39	1487.46	1567.39
nitrate	-109.55	-109.55	-109.55	-109.55	-109.55
nitrite	-33.03	-33.01	-33.01	-33.01	-33.01
noaq	86.55				
o2aq	16.40				
o2g	0.00				
oxalate	-677.26	-677.15	-677.14	-677.14	-677.14
oxaloacetate	-737.83	-726.41	-715.00	-703.58	-692.16
oxalosuccinate	-1024.72	-1001.89	-979.05	-956.22	-933.39
oxalylcoA	-510.77				
oxoproline	-293.86	-259.61	-225.36	-191.12	-156.87
palmitate	649.64	826.59	1003.54	1180.48	1357.43
pep	-1218.97	-1203.00	-1189.73	-1178.02	-1166.58
phenylalanineL	115.75	178.54	241.33	304.11	366.90
phosphoglycerate2	-1396.52	-1368.31	-1341.79	-1317.92	-1294.95
phosphoglycerate3	-1402.06	-1373.94	-1347.73	-1324.05	-1301.11
phosphohydroxypyruvate	-1397.26	-1385.84	-1374.43	-1363.01	-1351.59
phosphoserine	-1193.18	-1155.02	-1119.57	-1085.16	-1050.90
pi	-1079.46	-1068.49	-1059.49	-1052.97	-1047.17
ppi	-1957.07	-1947.46	-1940.66	-1935.64	-1933.29
propanol2	49.57	95.23	140.90	186.56	232.23
propanoln	58.99	104.65	150.32	195.98	241.65
propanoylcoA	-32.39	-3.85	24.69	53.23	81.77
prpp	-3077.91	-3025.46	-2978.51	-2932.71	-2887.04
pyruvate	-385.03	-367.91	-350.78	-333.66	-316.53
retinal	821.80	981.62	1141.45	1301.27	1461.10
retinol	852.59	1023.83	1195.07	1366.31	1537.55
ribitol	-422.60	-354.10	-285.61	-217.11	-148.62
ribose	-458.50	-401.42	-344.34	-287.26	-230.18
ribose1phos	-1333.37	-1277.51	-1224.35	-1172.71	-1121.30
ribose5phos	-1341.45	-1285.59	-1232.43	-1180.79	-1129.38
ribulose	-442.44	-385.36	-328.28	-271.20	-214.12
ribulose5phos	-1340.99	-1285.13	-1231.97	-1180.33	-1128.92
serineL	-305.42	-265.46	-225.51	-185.55	-145.60
sorbitol	-535.46	-455.55	-375.64	-295.72	-215.81
sorbitol6phos	-1419.99	-1341.92	-1266.55	-1192.19	-1117.97
sorbose	-559.75	-491.25	-422.76	-354.26	-285.77
succinate	-578.32	-553.72	-530.64	-507.79	-484.95
succinylcoA	-393.33	-370.32	-347.47	-324.63	-301.80
sucrose	-919.00	-793.43	-667.85	-542.27	-416.69
sulfate	-747.77	-747.77	-747.77	-747.77	-747.77
sulfite	-499.29	-494.02	-490.68	-489.85	-489.75
sulfurcr	0.00				
thioredoxinox	0.00				
thioredoxinred	33.33	44.70	55.74	64.03	66.35
threonine	-264.74	-213.37	-162.00	-110.62	-59.25
transcinnamate	75.89	115.85	155.80	195.76	235.71
trehalose	-937.06	-811.49	-685.91	-560.33	-434.75
tryptophanL	237.50	306.00	374.49	442.99	511.48
tyrosineL	-47.85	14.94	77.73	140.51	203.30
ubiquinoneox	2641.49	3155.21	3668.94	4182.66	4696.38
ubiquinonered	2610.27	3135.41	3660.55	4185.69	4710.83
urate	-238.66	-221.54	-204.41	-187.29	-170.16

urea	-85.40	-62.57	-39.74	-16.90	5.93
uricacid	-239.50	-216.67	-193.84	-171.00	-148.17
valineL	-35.80	26.99	89.78	152.56	215.35
xylitol	-431.89	-363.39	-294.90	-226.40	-157.91
xylose	-456.99	-399.91	-342.83	-285.75	-228.67
xylulose	-452.65	-395.57	-338.49	-281.41	-224.33

Table A2 Average Number of Hydrogen Atoms in a Reactant at 298.15 K, Ionic Strength 0.25 M, and pHs 5, 6, 7, 8, and 9

In[12]:= **table2 = PaddedForm[TableForm[listreactantsNH /. is → .25 /. pH → {5, 6, 7, 8, 9},**
      TableHeadings → {listnames, {" pH 5", " pH 6", " pH 7", " pH 8", " pH 9"}},
      TableSpacing → {1, 1}] // N, {4, 2}]

Out[12]//PaddedForm=

	pH 5	pH 6	pH 7	pH 8	pH 9
acetaldehyde	4.00				
acetate	3.23	3.03	3.00	3.00	3.00
acetoacetate	5.00				
acetoacetylcoA	5.00				
acetone	6.00				
acetylcoA	3.00				
acetylphos	4.40	4.06	3.94	3.57	3.12
aconitatecis	3.00				
adenine	5.14	5.02	5.00	5.00	5.00
adenosine	13.03	13.00	13.00	13.00	13.00
adenosinephosphosulfate	12.00				
adp	13.01	12.69	12.18	12.02	12.00
alanine	7.00				
ammonia	4.00	4.00	3.99	3.95	3.64
amp	12.98	12.60	12.13	12.01	12.00
arabinose	10.00				
arabinose5phos	9.93	9.57	9.12	9.01	9.00
asparagineL	8.00				
aspartate	6.00				
atp	13.03	12.75	12.23	12.03	12.00
bpg	4.99	4.87	4.40	4.06	4.01
butanal	8.00				
butanoln	10.00				
butyrate	7.00				
cellobiose	22.00				
citrate	5.91	5.27	5.03	5.00	5.00
citrateiso	5.90	5.27	5.03	5.00	5.00
co2g	0.00				
co2tot	1.92	1.55	1.11	0.99	0.85
coA	1.00	0.99	0.93	0.55	0.11
coAglutathione	15.00				
coaq	0.00				
cog	0.00				
creatine	9.00				
creatinine	7.00				
cyclicamp	7.00				
cysteineL	7.00	6.99	6.93	6.55	6.11
cystineL	12.00				
cytochromecox	0.00				

cytochromecred	0.00			
deoxyadenosine	13.03	13.00	13.00	13.00	13.00
deoxyadp	13.01	12.69	12.18	12.02	12.00
deoxyamp	12.98	12.60	12.13	12.01	12.00
deoxyatp	13.03	12.75	12.23	12.03	12.00
deoxyribose	10.00				
deoxyribose1phos	9.93	9.57	9.12	9.01	9.00
deoxyribose5phos	9.93	9.57	9.12	9.01	9.00
dihydroxyacetone	6.00				
dihydroxyacetonephos	5.58	5.12	5.01	5.00	5.00
ethaneaq	6.00				
ethanol	6.00				
ethylacetate	8.00				
fadenzox	31.00				
fadenzred	33.00				
fadox	31.00				
fadred	33.00				
ferredoxinox	0.00				
ferredoxinred	0.00				
ferric	0.00				
ferrous	0.00				
fmnox	19.00				
fmnred	21.00				
formate	1.00				
fructose	12.00				
fructose16phos	11.44	10.31	10.03	10.00	10.00
fructose6phos	11.84	11.34	11.05	11.01	11.00
fumarate	2.10	2.01	2.00	2.00	2.00
galactono14lactone	10.00				
galactose	12.00				
galactose1phos	11.79	11.27	11.04	11.00	11.00
galactose6phos	11.88	11.43	11.07	11.01	11.00
gluconolactone	10.00				
gluconolactone6phos	9.88	9.42	9.07	9.01	9.00
glucose	12.00				
glucose1phos	11.90	11.46	11.08	11.01	11.00
glucose6phos	11.88	11.42	11.07	11.01	11.00
glutamate	8.00				
glutamine	10.00				
glutathioneox	30.00				
glutathionered	16.00	15.98	15.86	15.37	15.06
glyceraldehyde	6.00				
glyceraldehydephos	5.58	5.12	5.01	5.00	5.00
glycerate	5.00				
glycerol	8.00				
glycerol3phos	7.93	7.56	7.11	7.01	7.00
glycine	5.00				
glycolate	3.00				
glycylglycine	8.00				
glyoxylate	1.00				
h2aq	2.00				
h2g	2.00				
h2o	2.00				
h2o2aq	2.00				

h2saq	1.98	1.84	1.34	1.05	1.00
hydroxy2oxoglutarate	4.00				
hydroxyglutarate	6.00				
hydroxypropionateb	5.00				
hydroxypyruvate	3.00				
hypoxanthine	4.00				
i2cr	0.00				
iditol	14.00				
idp	11.96	11.68	11.14	10.75	10.21
imp	11.94	11.59	11.11	10.87	10.38
indole	7.00				
inosine	12.00	12.00	11.98	11.83	11.32
iodideion	0.00				
isomaltose	22.00				
itp	11.97	11.74	11.21	10.85	10.32
ketoglutaramate	5.00				
ketoglutarate	4.00				
lactate	5.00				
lactose	22.00				
lactulose	22.00				
leucineisoL	13.00				
leucineL	13.00				
lyxose	10.00				
malate	4.33	4.05	4.00	4.00	4.00
maleate	2.00				
maltose	22.00				
malylcoA	4.08	4.01	4.00	4.00	4.00
mannitol1phos	13.90	13.46	13.08	13.01	13.00
mannitolD	14.00				
mannose	12.00				
mannose1phos	11.88	11.43	11.07	11.01	11.00
mannose6phos	11.88	11.43	11.07	11.01	11.00
methaneaq	4.00				
methaneg	4.00				
methanol	4.00				
methionineL	11.00				
methyl2oxopentanoate	9.00				
methylmalate	6.00				
methylmaleate	4.83	4.34	4.05	4.01	4.00
methylmalonylcoA	4.08	4.01	4.00	4.00	4.00
n2aq	0.00				
n2g	0.00				
n2oaq	0.00				
nadox	26.00				
nadpox	25.00				
nadpred	26.00				
nadred	27.00				
nicotinamideribonucleotide	14.88	14.43	14.07	14.01	14.00
nitrate	0.00	1.43×10^{-6}	1.43×10^{-7}	1.43×10^{-8}	1.43×10^{-9}
nitrite	0.01	0.00	0.00	8.71×10^{-6}	8.71×10^{-7}
noaq	0.00				
o2aq	0.00				
o2g	0.00				

oxalate	0.05	0.01	0.00	0.00	5.16×10^{-6}
oxaloacetate	2.00				
oxalosuccinate	4.00				
oxalylcoA	0.00				
oxoproline	6.00				
palmitate	31.00				
pep	2.93	2.59	2.12	2.01	2.00
phenylalanineL	11.00				
phosphoglycerate2	4.98	4.86	4.38	4.06	4.01
phosphoglycerate3	4.98	4.83	4.32	4.05	4.00
phosphohydroxypyruvate	2.00				
phosphoserine	6.88	6.43	6.07	6.01	6.00
pi	1.98	1.82	1.31	1.04	1.00
ppi	1.88	1.42	1.02	0.69	0.17
propanol2	8.00				
propanoln	8.00				
propanoylcoA	5.00				
prpp	9.72	8.56	8.06	8.01	8.00
pyruvate	3.00				
retinal	28.00				
retinol	30.00				
ribitol	12.00				
ribose	10.00				
ribose1phos	9.93	9.57	9.12	9.01	9.00
ribose5phos	9.93	9.57	9.12	9.01	9.00
ribulose	10.00				
ribulose5phos	9.93	9.57	9.12	9.01	9.00
serineL	7.00				
sorbitol	14.00				
sorbitol6phos	13.88	13.42	13.07	13.01	13.00
sorbose	12.00				
succinate	4.60	4.11	4.01	4.00	4.00
succinylcoA	4.08	4.01	4.00	4.00	4.00
sucrose	22.00				
sulfate	0.00	0.00	2.67×10^{-6}	2.67×10^{-7}	2.67×10^{-8}
sulfite	0.98	0.82	0.32	0.04	0.00
sulfurcr	0.00				
thioredoxinox	0.00				
thioredoxinred	2.00	1.98	1.83	0.89	0.12
threonine	9.00				
transcinnamate	7.00				
trehalose	22.00				
tryptophanL	12.00				
tyrosineL	11.00				
ubiquinoneox	90.00				
ubiquinonered	92.00				
urate	3.00				
urea	4.00				
uricacid	4.00				
valineL	11.00				
xylitol	12.00				
xylose	10.00				
xylulose	10.00				

Table A3 Standard Transformed Enthalpies of Formation of Reactants in kJ mol^{-1} at 298.15 K, Ionic Strength 0.25 M, and pHs 5, 6, 7, 8, and 9 (This table can also be calculated for temperatures from 273.15 K to about 313.15 K.)

```
In[13]:=  table3 =
           PaddedForm[TableForm[listreactantsHT /. t → 298.15 /. is → .25 /. pH → {5, 6, 7, 8, 9},
             TableHeadings → {nameswithH, {"  pH 5", "  pH 6", "  pH 7", "  pH 8", "  pH 9"}},
             TableSpacing → {1, 1}] // N, {6, 2}]
```

Out[13]//PaddedForm=

	pH 5	pH 6	pH 7	pH 8	pH 9
acetaldehyde	-213.88	-213.88	-213.88	-213.88	-213.88
acetate	-486.96	-486.85	-486.84	-486.83	-486.83
acetone	-224.18	-224.18	-224.18	-224.18	-224.18
adenine	124.69	127.13	127.41	127.44	127.44
adenosine	-627.12	-626.70	-626.66	-626.66	-626.66
adp	-2625.85	-2625.73	-2627.23	-2627.71	-2627.77
alanine	-557.68	-557.68	-557.68	-557.68	-557.68
ammonia	-133.74	-133.72	-133.45	-130.98	-115.00
amp	-1636.00	-1636.50	-1638.19	-1638.61	-1638.66
arabinose	-1047.91	-1047.91	-1047.91	-1047.91	-1047.91
asparagineL	-769.39	-769.39	-769.39	-769.39	-769.39
aspartate	-945.47	-945.47	-945.47	-945.47	-945.47
atp	-3615.72	-3615.41	-3616.89	-3617.48	-3617.55
citrate	-1518.56	-1515.00	-1513.66	-1513.48	-1513.46
co2g	-393.50				
co2tot	-699.81	-696.62	-692.88	-691.81	-689.57
coaq	-120.96				
cog	-110.53				
ethaneaq	-104.56	-104.56	-104.56	-104.56	-104.56
ethanol	-290.77	-290.77	-290.77	-290.77	-290.77
ethylacetate	-485.30	-485.30	-485.30	-485.30	-485.30
ferric	-44.79				
ferrous	-87.45				
formate	-425.55	-425.55	-425.55	-425.55	-425.55
fructose	-1264.32	-1264.32	-1264.32	-1264.32	-1264.32
fructose16phos	-3342.48	-3341.21	-3340.83	-3340.78	-3340.78
fructose6phos	-2267.93	-2268.00	-2268.05	-2268.05	-2268.05
fumarate	-776.44	-776.55	-776.56	-776.57	-776.57
galactose	-1260.14	-1260.14	-1260.14	-1260.14	-1260.14
glucose	-1267.13	-1267.13	-1267.13	-1267.13	-1267.13
glucose6phos	-2279.19	-2279.26	-2279.31	-2279.32	-2279.32
glutamate	-982.77	-982.77	-982.77	-982.77	-982.77
glutamine	-809.12	-809.12	-809.12	-809.12	-809.12
glycerol	-679.85	-679.85	-679.85	-679.85	-679.85
glycerol3phos	-1725.68	-1725.74	-1725.81	-1725.82	-1725.83
glycine	-525.06	-525.06	-525.06	-525.06	-525.06
glycylglycine	-737.55	-737.55	-737.55	-737.55	-737.55
h2aq	-5.02	-5.02	-5.02	-5.02	-5.02
h2g	-0.82	-0.82	-0.82	-0.82	-0.82
h2o	-286.65	-286.65	-286.65	-286.65	-286.65
h2o2aq	-191.99	-191.99	-191.99	-191.99	-191.99
h2saq	-40.08	-36.77	-25.35	-18.71	-17.69
i2cr	0.00				
idp	-2823.72	-2824.54	-2825.21	-2818.05	-2801.45
imp	-1834.08	-1835.34	-1836.59	-1832.29	-1814.21
indole	94.62	94.62	94.62	94.62	94.62
inosine	-824.78	-824.72	-824.20	-819.90	-805.81
iodideion	-54.78				
isomaltose	-2253.54	-2253.54	-2253.54	-2253.54	-2253.54
itp	-3813.59	-3814.25	-3815.41	-3811.65	-3797.44
ketoglutarate	-1044.06	-1044.06	-1044.06	-1044.06	-1044.06
lactate	-688.29	-688.29	-688.29	-688.29	-688.29

lactose	-2242.14	-2242.14	-2242.14	-2242.14	-2242.14
leucineL	-648.73	-648.73	-648.73	-648.73	-648.73
malate	-1080.39	-1079.87	-1079.80	-1079.79	-1079.79
maltose	-2247.12	-2247.12	-2247.12	-2247.12	-2247.12
mannose	-1263.60	-1263.60	-1263.60	-1263.60	-1263.60
mannose6phos	-2276.45	-2276.52	-2276.58	-2276.59	-2276.59
methaneaq	-90.69	-90.69	-90.69	-90.69	-90.69
methaneg	-76.46	-76.46	-76.46	-76.46	-76.46
methanol	-247.58	-247.58	-247.58	-247.58	-247.58
n2aq	-10.54				
n2g	0.00				
n2oaq	82.05				
nadox	-10.30	-10.30	-10.30	-10.30	-10.30
nadpox	-1014.07	-1014.07	-1014.07	-1014.07	-1014.07
nadpred	-1040.78	-1040.78	-1040.78	-1040.78	-1040.78
nadred	-41.41	-41.41	-41.41	-41.41	-41.41
nitrate	-204.59	-204.59	-204.59	-204.59	-204.59
nitrite	-104.32	-104.20	-104.19	-104.19	-104.19
noaq	90.25				
o2aq	-11.70				
o2g	0.00				
oxaloacetate	-959.08	-959.08	-959.08	-959.08	-959.08
pep	-1619.12	-1618.89	-1618.58	-1618.51	-1618.50
pi	-1302.90	-1302.05	-1299.39	-1297.99	-1297.79
ppi	-2294.19	-2292.82	-2291.57	-2290.09	-2287.70
propanol2	-334.13	-334.13	-334.13	-334.13	-334.13
pyruvate	-597.04	-597.04	-597.04	-597.04	-597.04
ribose	-1038.12	-1038.12	-1038.12	-1038.12	-1038.12
ribose1phos	-2034.57	-2038.04	-2042.41	-2043.41	-2043.53
ribose5phos	-2034.57	-2038.04	-2042.41	-2043.41	-2043.53
ribulose	-1027.14	-1027.14	-1027.14	-1027.14	-1027.14
sorbose	-1268.24	-1268.24	-1268.24	-1268.24	-1268.24
succinate	-909.87	-908.88	-908.70	-908.68	-908.68
sucrose	-2208.93	-2208.93	-2208.93	-2208.93	-2208.93
sulfate	-907.62	-907.62	-907.62	-907.62	-907.62
sulfite	-626.38	-627.58	-631.44	-633.52	-633.82
sulfurcr	0.00				
tryptophanL	-410.14	-410.14	-410.14	-410.14	-410.14
urea	-319.30	-319.30	-319.30	-319.30	-319.30
valineL	-616.52	-616.52	-616.52	-616.52	-616.52
xylose	-1050.06	-1050.06	-1050.06	-1050.06	-1050.06
xylulose	-1033.77	-1033.77	-1033.77	-1033.77	-1033.77

Table A4 Standard Transformed Entropies of Formation of Reactants in kJ K^{-1} mol^{-1} at 298.15 K, Ionic Strength 0.25 M, and pHs 5, 6, 7, 8, and 9 (This table can also be calculated for temperatures of 273.15 K to about 313.15 K.)

```
In[14]:= table4 =
         PaddedForm[TableForm[listreactantsST /. t → 298.15 /. is → .25 /. pH → {5, 6, 7, 8, 9},
            TableHeadings → {nameswithH, {"  pH 5", "  pH 6", "  pH 7", "  pH 8", "  pH 9"}},
            TableSpacing → {1, 1}], {6, 4}]

Out[14]//PaddedForm=
```

	pH 5	pH 6	pH 7	pH 8	pH 9
acetaldehyde	-0.6449	-0.7215	-0.7981	-0.8746	-0.9512
acetate	-0.6851	-0.7440	-0.8016	-0.8591	-0.9165

acetone	-0.8069	-0.9218	-1.0367	-1.1515	-1.2664
adenine	-1.1239	-1.2126	-1.3075	-1.4031	-1.4988
adenosine	-2.7305	-2.9782	-3.2270	-3.4758	-3.7247
adp	-3.5445	-3.7906	-4.0333	-4.2661	-4.4962
alanine	-1.3152	-1.4492	-1.5832	-1.7173	-1.8513
ammonia	-0.5736	-0.6501	-0.7258	-0.7936	-0.8134
amp	-3.1447	-3.3918	-3.6336	-3.8658	-4.0958
arabinose	-2.0097	-2.2011	-2.3926	-2.5840	-2.7754
asparagineL	-1.6041	-1.7572	-1.9104	-2.0636	-2.2167
aspartate	-1.4251	-1.5399	-1.6548	-1.7697	-1.8845
atp	-3.9519	-4.1981	-4.4420	-4.6756	-4.9058
citrate	-1.6479	-1.7426	-1.8361	-1.9315	-2.0272
co2g	0.0029				
co2tot	-0.4535	-0.4767	-0.4889	-0.5052	-0.5158
coaq	-0.0036				
cog	0.0894				
ethaneaq	-0.8843	-0.9992	-1.1140	-1.2289	-1.3438
ethanol	-0.9567	-1.0715	-1.1864	-1.3013	-1.4161
ethylacetate	-1.2827	-1.4359	-1.5890	-1.7422	-1.8954
ferric	-0.1100				
ferrous	-0.0178				
formate	-0.3458	-0.3649	-0.3841	-0.4032	-0.4223
fructose	-2.3512	-2.5809	-2.8107	-3.0404	-3.2702
fructose16phos	-3.4079	-3.6111	-3.8036	-3.9952	-4.1866
fructose6phos	-2.7579	-2.9803	-3.1941	-3.4051	-3.6157
fumarate	-0.7707	-0.8101	-0.8485	-0.8868	-0.9251
galactose	-2.3593	-2.5890	-2.8187	-3.0485	-3.2782
glucose	-2.3593	-2.5891	-2.8188	-3.0485	-3.2783
glucose6phos	-2.7827	-3.0065	-3.2212	-3.4323	-3.6430
glutamate	-1.7417	-1.8949	-2.0480	-2.2012	-2.3544
glutamine	-1.9272	-2.1187	-2.3101	-2.5016	-2.6930
glycerol	-1.3992	-1.5523	-1.7055	-1.8587	-2.0118
glycerol3phos	-1.8865	-2.0356	-2.1757	-2.3106	-2.4447
glycine	-0.9790	-1.0748	-1.1705	-1.2662	-1.3619
glycylglycine	-1.5165	-1.6697	-1.8228	-1.9760	-2.1291
h2aq	-0.2728	-0.3111	-0.3493	-0.3876	-0.4259
h2g	-0.1996	-0.2379	-0.2762	-0.3145	-0.3528
h2o	-0.3628	-0.4011	-0.4394	-0.4777	-0.5159
h2o2aq	-0.3913	-0.4296	-0.4679	-0.5062	-0.5445
h2saq	-0.2378	-0.2637	-0.2561	-0.2560	-0.2721
i2cr	0.0000				
idp	-3.3895	-3.6191	-3.8396	-4.0253	-4.1700
imp	-2.9900	-3.2202	-3.4412	-3.6372	-3.7804
indole	-1.1224	-1.2564	-1.3904	-1.5244	-1.6584
inosine	-2.5753	-2.8048	-3.0327	-3.2466	-3.4212
iodideion	-0.0080				
isomaltose	-4.3989	-4.8201	-5.2413	-5.6625	-6.0836
itp	-3.7972	-4.0270	-4.2505	-4.4490	-4.6042
ketoglutarate	-1.2236	-1.3002	-1.3767	-1.4533	-1.5299
lactate	-1.0649	-1.1606	-1.2564	-1.3521	-1.4478
lactose	-4.4290	-4.8502	-5.2714	-5.6926	-6.1138
leucineL	-2.2741	-2.5230	-2.7719	-3.0208	-3.2696
malate	-1.1769	-1.2547	-1.3314	-1.4080	-1.4846
maltose	-4.4210	-4.8422	-5.2634	-5.6846	-6.1058

mannose	-2.3673	-2.5970	-2.8267	-3.0565	-3.2862
mannose6phos	-2.7868	-3.0108	-3.2256	-3.4368	-3.6474
methaneaq	-0.5828	-0.6594	-0.7359	-0.8125	-0.8891
methaneg	-0.4801	-0.5567	-0.6332	-0.7098	-0.7864
methanol	-0.6361	-0.7127	-0.7893	-0.8659	-0.9425
n2aq	-0.0981				
n2g	0.0000				
n2oaq	-0.0743				
nadox	-2.5913	-3.0890	-3.5868	-4.0846	-4.5823
nadpox	-3.0366	-3.5152	-3.9938	-4.4724	-4.9510
nadpred	-3.2910	-3.7888	-4.2865	-4.7843	-5.2821
nadred	-2.8619	-3.3788	-3.8957	-4.4126	-4.9295
nitrate	-0.3188	-0.3188	-0.3188	-0.3188	-0.3188
nitrite	-0.2391	-0.2388	-0.2387	-0.2387	-0.2387
noaq	0.0124				
o2aq	-0.0942				
o2g	0.0000				
oxaloacetate	-0.7421	-0.7804	-0.8186	-0.8569	-0.8952
pep	-1.3421	-1.3949	-1.4384	-1.4774	-1.5158
pi	-0.7494	-0.7834	-0.8046	-0.8218	-0.8406
ppi	-1.1307	-1.1583	-1.1770	-1.1888	-1.1887
propanol2	-1.2869	-1.4401	-1.5932	-1.7464	-1.8996
pyruvate	-0.7111	-0.7685	-0.8260	-0.8834	-0.9408
ribose	-1.9440	-2.1355	-2.3269	-2.5184	-2.7098
ribose1phos	-2.3518	-2.5509	-2.7438	-2.9204	-3.0932
ribose5phos	-2.3247	-2.5238	-2.7167	-2.8933	-3.0661
ribulose	-1.9611	-2.1525	-2.3440	-2.5354	-2.7269
sorbose	-2.3763	-2.6060	-2.8358	-3.0655	-3.2952
succinate	-1.1121	-1.1912	-1.2680	-1.3446	-1.4212
sucrose	-4.3264	-4.7476	-5.1688	-5.5900	-6.0112
sulfate	-0.5361	-0.5361	-0.5362	-0.5362	-0.5362
sulfite	-0.4263	-0.4480	-0.4721	-0.4819	-0.4832
sulfurcr	0.0000				
tryptophanL	-2.1722	-2.4019	-2.6317	-2.8614	-3.0912
urea	-0.7845	-0.8611	-0.9377	-1.0142	-1.0908
valineL	-1.9477	-2.1583	-2.3689	-2.5795	-2.7901
xylose	-1.9892	-2.1806	-2.3721	-2.5635	-2.7550
xylulose	-1.9491	-2.1405	-2.3320	-2.5234	-2.7149

Appendix 3: Glossary of Names of Reactants

All are aqueous reactants (sums of species) except when labelled g for gas or cr for crystalline. The name that is used in *Mathematica* is given first

acetaldehyde, acetaldehyde
acetate, acetate
acetoacetate, acetoacetate
acetoacetylcoA, acetoacetyl-CoA
acetone, acetone
acetylcoA, acetyl-CoA
acetylphos, acetyl phosphate
aconitatecis, cis-aconitate
adenine, adenine
adenosine, adenosine
adenosinephosphosulfate, adenylsulfate
adp, ADP
alanine, L-alanine
ammonia, NH_3 and NH_4^+
amp, AMP
arabinose, arabinose
arabinose5phos, arabinose 5-phosphate
asparagineL, L-asparagine
aspartate, L-aspartate
atp, ATP
bpg, 2,3-bisphospho-D-glycerate
butanal, butanal
butanoln, n-butanol
butyrate, butyrate
cellobiose, cellobiose
citrate, citrate
citrateiso, isocitrate
co2g, CO_2 gas
co2tot, H_2CO_3, CO_2, HCO_3^-, and CO_3^{2-}
coA, CoA
coAglutathione, CoA-glutathione
coaq, CO
cog, CO gas
creatine, creatine
creatinine, creatinine
cyclicamp, 3',5'-cyclicAMP
cysteineL, L-cysteine
cystineL, L-cystine
cytochromecox, ferricytochrome
cytochromecred, ferrocytochrome
deoxyadenosine, deoxyadenosine
deoxyadp, dADP
deoxyamp, dAMP
deoxyatp, dATP

deoxyribose, 2-deoxy-D-ribose
deoxyribose1phos, 2-deoxy-D-ribose 1-phosphate
deoxyribose5phos, 2-deoxy-D-ribose 5-phosphate
dihydroxyacetone, glycerone
dihydroxyacetonephos, glycerone phosphate
ethaneaq, ethane
ethanol, ethanol
ethylacetate, ethyl acetate
fadenzox, FAD bound by enzyme
fadenzred, $FADH_2$ bound by enzyme
fadox, FAD
fadred, $FADH_2$
ferredoxinox, oxidized ferredoxin
ferredoxinred, reduced ferredoxin
ferric, Fe^{3+}
ferrous, Fe^{2+}
fmnox, FMN
fmnred, $FMNH_2$
formate, formate
fructose, D-fructose
fructose16phos, D-fructose 1,6-bisphosphate
fructose6phos, D-fructose 6-phosphate
fumarate, fumarate
galactono14lactone, D-galactono-1,4-lactone
galactose, D-galactose
galactose1phos, galactose 1-phosphate
galactose6phos, galactose 6-phosphate
gluconolactone, D-glucono-1,5-lactone
gluconolactone6phos, D-glucono-1,5-lactone 6-phosphate
glucose, D-glucose
glucose1phos, D-glucose 1-phosphate
glucose6phos, D-glucose 6-phosphate
glutamate, L-glutamate
glutamine, L-glutamine
glutathioneox, GSSG
glutathionered, GSH
glyceraldehyde, D-glyceraldehyde
glyceraldehydephos, D-glyceraldehyde-3-phosphate
glycerate, D-glycerate
glycerol, glycerol
glycerol3phos, Sn-glycerol 3-phosphate
glycine, glycine
glycolate, glycolate
glycylglycine, glycylglycine
glyoxylate, glyoxylate
h2aq, H_2
h2g, H_2 gas
h2o, H_2O
h2o2aq, H_2O_2

h2saq, H_2 S
hydroxy2oxoglutarate, 4-hydroxy-2-oxoglutarate
hydroxyglutarate, (S)-2-hydroxyglutarate
hydroxypropionateb, 3-hydroxypropionate
hydroxypyruvate, 3-hydroxypyruvate
hypoxanthine, hypoxanthine
i2cr, crystalline iodine
iditol, L-iditol
idp, inosine diphosphate
imp, inosine monophosphate (5'-inosineate)
indole, indole
inosine, inosine
iodideion, iodide ion
isomaltose, isomaltose
itp, inosine triphosphate
ketoglutaramate, ketoglutaramate
ketoglutarate, 2-oxoglutarate
lactate, (S)-lactate
lactose, lactose
lactulose, lactulose
leucineisoL, L-isoleucine
leucineL, L-leucine
lyxose, D-lyxose
malate, (S)-malate
maleate, maleate
maltose, maltose
malylcoA, malyl-CoA
mannitol1phos, D-mannitol 1-phosphate
mannitolD, D-mannitol
mannose, D-mannose
mannose1phos, D-mannose 1-phosphate
mannose6phos, D-mannose 6-phosphate
methaneaq, methane
methaneg, methane gas
methanol, methanol
methionineL, L-methionine
methyl2oxopentanoate, 4-methyl-2-oxopentanoate
methylmalate, (R)-2-methylmalate
methylmaleate, 2-methylmaleate
methylmalonylcoA, (S)methylmalonyl-CoA
n2aq, N_2
n2g, N_2 gas
n2oaq, nitrous oxide
nadox, NAD^+
nadpox, $NAPD^+$
nadpred, NADPH
nadred, NADH
nicotinamideribonucleotide, nicotinamide ribonucleotide
nitrate, NO_3^-

nitrite, NO_2^-
noaq, nitric oxide
o2aq, O_2
o2g, O_2 gas
oxalate, oxalate
oxaloacetate, oxaloacetate
oxalosuccinate, oxalosuccinate
oxalylcoA, oxalyl-CoA
oxoproline, 4-oxoproline
palmitate, palmitate
pep, phosphoenolphosphate
phenylalanineL, L-phenylalanine
phosphoglycerate2, 2-phospho-D-glycerate
phosphoglycerate3, 3-phospho-D-glycerate
phosphohydroxypyruvate, 3-phosphohydroxypyruvate
phosphoserine, D-phospho-L-serine
pi, phosphate (orthophosphate)
ppi, diphosphate (pyrophosphate)
propanol2, propan-2-ol
propanoln, propan-1-ol
propanoylcoA, propanoyl-CoA
prpp, 5-phosphoribosyl-α-pyrophosphate
pyruvate, pyruvate
retinal, retinal
retinol, retinol
ribitol, ribitol
ribose, D-ribose
ribose1phos, D-ribose 1-phoshate
ribose5phos, D-ribose 5-phoshate
ribulose, D-ribulose
ribulose5phos, D-ribulose 5-phosphate
serineL, L-serine
sorbitol, D-sorbitol
sorbitol6phos, D-sorbitol 6-phosphate
sorbose, L-sorbose
succinate, succinate
succinylcoA, succinyl-CoA
sucrose, sucrose
sulfate, SO_4^{2-}
sulfite, SO_3^-
sulfurcr, crystalline sulfur
thioredoxinox, thioredoxin disulfide
thioredoxinred, thioredoxin
threonine, L-threonine
transcinnamate, transcinnamate
trehalose, trehalose
tryptophanL, L-tryptophan
tyrosineL, L-tyrosine
ubiquinoneox, ubiquinone

ubiquinonered, ubiquinole
urate, urate
urea, urea
uricacid, uric acid
valineL, L-valine
xylitol, L-xylitol
xylose, D-xylose
xylulose, D-xyulose

Appendix 4: Glossary of symbols for thermodynamic properties

SI units and sizes of matrices are indicated in parentheses.

a_j activity of species j

A conservation matrix for a system of chemical reactions ($C \times N$)

A' conservation matrix for a system of enzyme-catalyzed reactions at specified pH ($C' \times N'$)

A'' conservation matrix for a system of enzyme-catalyzed reactions at specified pH and specified availability of oxygen atoms or coenzymes ($C'' \times N''$)

B empirical constant in the extended Debye-Huckel equation ($1.6 \text{ kg}^{1/2} \text{ mol}^{-1/2}$)

c_j concentration of species j (mol L^{-1})

$c°$ standard concentration (1 mol L^{-1})

C number of components in a system of chemical reactions

C' number of components in a system of enzyme-catalyzed reactions at specified pH

C'' number of components in a system of enzyme-catalyzed reactions at specified pH and specified availability of oxygen atoms or coenzymes

$C_{\text{Pm}}°$ standard molar heat capacity of a species (J K^{-1} mol^{-1})

$\Delta_f C_P'°$ standard transformed molar heat capacity of formation of a reactant (J K^{-1} mol^{-1})

$\Delta_r C_P'°$ standard transformed heat capacity of reaction (J K^{-1} mol^{-1})

D number of variables needed to describe the extensive state of a system

D' number of variables needed to describe the extensive state of a system when the concentrations of one or more species have been specified

D'' number of variables needed to describe the extensive state of a system when the concentrations of one or more species and one or more reactants have been specified

E electromotive force (electric potential difference) or reduction potential (V)

$E°$ standard electromotive force of a cell or standard reduction potential (V)

E' apparent electromotive force or apparent reduction potential at a specified pH (V)

$E'°$ apparent standard electromotive force of a cell or apparent standard reduction potential (V)

F Faraday constant (96,485 C mol^{-1})

G Gibbs energy of a system at specified T, P, and ionic strength (J)

G' transformed Gibbs energy of a system at specified T, P, ionic strength, and specified concentrations of one or more species (J)

G'' further transformed Gibbs energy of a system at specified T, P, ionic strength, and specified concentrations of one or more species and one or more reactants (J)

$\Delta_f G_j$ Gibbs energy of formation of species j at specified T, P, and ionic strength (J mol^{-1})

$\Delta_f G_j°$ standard Gibbs energy of formation of species j at specified T, P, and ionic strength (J mol^{-1})

$\Delta_r G$ Gibbs energy of chemical reaction (J mol^{-1})

$\Delta_r G°$ standard Gibbs energy of chemical reaction (J mol^{-1})

$\Delta_f G_j'$ transformed Gibbs energy of formation of species j at specified T, P, ionic strength, and specified concentrations of one or more species (J mol^{-1})

$\Delta_f G_i'$ transformed Gibbs energy of formation of reactant i at specified T, P, ionic strength, and specified concentrations of one or more species (J mol^{-1})

$\Delta_f G_i'°$ standard transformed Gibbs energy of formation of reactant i at specified T, P, ionic strength, and specified concentrations of one or more species (J mol^{-1})

$\Delta_r G'$ transformed Gibbs energy of a biochemical reaction at a specified concentration of a species (J mol^{-1})

$\Delta_r G'°$ standard transformed Gibbs energy of reaction at a specified concentration of a species (J mol^{-1})

$\Delta_f G''$ further transformed Gibbs energy of formation at specified concentrations of one or more reactants (J mol^{-1})

$\Delta_f G_i''°$ standard further transformed Gibbs energy of formation of a pseudoisomer group of reactants at specified T, P, ionic strength, and specified concentrations of one or more species and one or more reactants (J mol^{-1})

$\Delta_r G''$	further transformed Gibbs energy of a biochemical reaction at a specified concentration of one or more reactants (J mol^{-1})
$\Delta_r G''^{\circ}$	standard further transformed Gibbs energy of reaction at specified concentrations of one or more reactants (J mol^{-1})
H	enthalpy of a system at specified T, P, and ionic strength (J)
H'	transformed enthalpy of a system at specified T, P, ionic strength, and specified concentrations of one or more species (J)
H''	further transformed enthalpy of a system at specified T, P, ionic strength, and specified concentrations of one or more species and one or more reactants (J)
$\Delta_r H(cal)$	enthalpy change in a calorimetric experiment (J mol^{-1})
$\Delta_f H_j$	enthalpy of formation of species j at specified T, P, and ionic strength (J mol^{-1})
$\Delta_f H_j^{\circ}$	standard enthalpy of formation of species j at specified T, P, and ionic strength (J mol^{-1})
$\Delta_r H$	enthalpy of chemical reaction (J mol^{-1})
$\Delta_r H^{\circ}$	standard enthalpy of chemical reaction (J mol^{-1})
$\Delta_f H_j'$	transformed enthalpy of formation of species j at specified T, P, ionic strength, and specified concentrations of one or more species (J mol^{-1})
$\Delta_f H_i'$	transformed enthalpy of formation of reactant i at specified T, P, ionic strength, and specified concentrations of one or more species (J mol^{-1})
$\Delta_f H_i'^{\circ}$	standard transformed enthalpy of formation of reactant i at specified T, P, ionic strength, and specified concentrations of one or more species (J mol^{-1})
$\Delta_r H'$	transformed enthalpy of a biochemical reaction at a specified concentration of a species (J mol^{-1})
$\Delta_r H'^{\circ}$	standard transformed enthalpy of reaction at a specified concentration of a species (J mol^{-1})
$\Delta_f H''$	further transformed enthalpy of formation at specified concentrations of one or more reactants (J mol^{-1})
$\Delta_f H_i''^{\circ}$	standard further transformed enthalpy of formation of a pseudoisomer group of reactants at specified T, P, ionic strength, and specified concentrations of one or more species and one or more reactants (J mol^{-1})
$\Delta_r H''$	further transformed enthalpy of a biochemical reaction at a specified concentration of one or more reactants (J mol^{-1})
$\Delta_r H''^{\circ}$	standard further transformed enthalpy of reaction at specified concentrations of one or more reactants (J mol^{-1})
I	ionic strength (mol L^{-1})
K	equilibrium constant written in terms of concentrations of species at specified T, P, and ionic strength
K'	apparent equilibrium constant written in terms of concentrations of reactants (sums of species) at specified T, P, ionic strength and concentrations of one or more species
K''	apparent equilibrium constant written in terms of concentrations of pseudoisomer groups (sums of reactants) at specified T, P, ionic strength, and concentrations of one or more species and one or more reactants
K_H	Henry's law constant
K_H'	apparent Henry's law constant at a specified pH
K_{Mg}	dissociation constant of a magnesium complex ion
n	total amount in a system (mol)
n_i	amount of species i in a system (mol)
$\{n_i\}$	set of amounts of species in a system (mol)
n_i'	amount of reactant i (sum of species) (mol)
n_i''	amount of pseudoisomer group i (sum of reactants) (mol)
$n_c(i)$	amount of component i (mol)
$n_c(i)'$	amount of apparent component i at specified concentrations of one or more species (mol)
$n_c(i)''$	amount of apparent component i at specified concentrations of one or more species and one or more reactants (mol)
$\boldsymbol{n}$	column vector of amounts of species (Nx1) (mol)
$\boldsymbol{n'}$	column vector of amounts of reactants (sum of species) (N'x1) (mol)

n''	column vector of amounts of pseudoisomer groups of reactants (N''x1) (mol)
n_c	column vector of amounts of components (Cx1) (mol)
n_{nc}	column vector of amounts of noncomponents ((N-C)x1) (mol)
n_c'	column vector of amounts of apparent components at specified concentrations of one or more species (C'x1) (mol)
n_c''	column vector of amounts of apparent components at specified concentrations of one or more species and one or more reactants (C''x1) (mol)
N_s	number of different kinds of species in a system
N'	number of different reactants (sums of species) in a system
N''	number of different pseudoisomer groups of reactants in a system
N_{iso}	number of isomers in an isomer group or pseudoisomers in a pseudoisomer group
$N_H(j)$	number of hydrogen atoms in species j
$N_{ATP}(i)$	number of ATP in reactant i (can be positive or negative)
$N_{Mg}(j)$	number of magnesium atoms in a molecule of j
$N_{O2}(i)$	number of oxygen molecules bound in i
N_c	matrix of numbers of specified components in the N_s species (CxN_s)
$\overline{N}_H(i)$	average number of hydrogen atoms bound by a molecule of i
$\overline{N}_{Mg}(i)$	average number of magnesium atoms bound by a molecule of i
$\Delta_r N_H$	change in binding of hydrogen ions in a biochemical reaction at specified T, P, ionic strength and concentrations of one or more species
$\Delta_r N_{Mg}$	change in the binding of magnesium ions in a biochemical reaction at specified T, P, pH, and ionic strength
P	pressure (bar)
pH_a	- $\log\{a(H^+)\}$
$pH = pH_c$	- $\log[H^+]$ at specified T, P, and ionic strength
pMg	- $\log[Mg^{2+}]$ at specified T,P, and ionic strength
pK	- $\log K$ for the dissociation of an acid at specified T, P, and ionic strength
q	heat flow into a system (J)
Q	reaction quotient at T and P
Q'	reaction quotient at specified T, P, pH, pMg, and ionic strength
r_i	mole fraction of isomer i within an isomer group or pseudoisomer i within a pseudoisomer group
R	gas constant (8.31451 J K^{-1} mol^{-1})
R	number of independent reactions in a system described in terms of species
R'	number of independent reactions in a system described in terms of reactants (sums of species)
R''	number of independent reactions in a system described in terms of pseudoisomer groups of reactants
S	entropy of a system at specified T, P, and ionic strength (J K^{-1})
S'	transformed entropy of a system at specified T, P, ionic strength, and specified concentrations of one or more species (J K^{-1})
S''	further transformed entropy of a system at specified T, P, ionic strength, and specified concentrations of one or more species and one or more reactants (J K^{-1})
S_m	molar entropy of a species (J K^{-1} mol^{-1})
S_m'	molar transformed entropy of a species (J K^{-1} mol^{-1})
$S_m^{\,\circ}$	standard molar entropy of a species (J K^{-1} mol^{-1})
$S_m'^{\,\circ}$	standard transformed molar entropy of a species (J K^{-1} mol^{-1})
$\Delta_f S_j$	entropy of formation of species j at specified T, P, and ionic strength (J K^{-1} mol^{-1})
$\Delta_f S_j^{\,\circ}$	standard entropy of formation of species j at specified T, P, and ionic strength (J K^{-1} mol^{-1})
$\Delta_r S$	entropy of chemical reaction (J K^{-1} mol^{-1})
$\Delta_r S^{\circ}$	standard entropy of chemical reaction (J K^{-1} mol^{-1})
$\Delta_f S_j'$	transformed entropy of formation of species j at specified T, P, ionic strength, and specified concentrations of one or more species (J K^{-1} mol^{-1})

$\Delta_f S_i{}'$	transformed entropy of formation of reactant i at specified T, P, ionic strength, and specified concentrations of one or more species (J K^{-1} mol^{-1})
$\Delta_f S_i{}'{}^\circ$	standard transformed entropy of formation of reactant i at specified T, P, ionic strength, and specified concentrations of one or more species (J K^{-1} mol^{-1})
$\Delta_r S'$	transformed entropy of a biochemical reaction at a specified concentration of a species (J K^{-1} mol^{-1})
$\Delta_r S'{}^\circ$	standard transformed entropy of reaction at a specified concentration of a species (J K^{-1} mol^{-1})
$\Delta_f S''$	further transformed entropy of formation at specified concentrations of one or more reactants (J K^{-1} mol^{-1})
$\Delta_f S_i{}''{}^\circ$	standard further transformed entropy of formation of a pseudoisomer group of reactants at specified T, P, ionic strength, and specified concentrations of one or more species and one or more reactants (J K^{-1} mol^{-1})
$\Delta_r S''$	further transformed entropy of a biochemical reaction at a specified concentration of one or more reactants (J K^{-1} mol^{-1})
$\Delta_r S''{}^\circ$	standard further transformed entropy of reaction at specified concentrations of one or more reactants (J K^{-1} mol^{-1})
$s_i{}'$	stoichiometric number of step I
s'	pathway matrix (Rx1)
T	temperature (K)
t	Celsius temperature (°C)
U	internal energy (J)
U'	transformed internal energy (J)
V	volume (m^3)
V_m	molar volume (m^3 mol^{-1})
w	work done on a system (J)
Y	fractional saturation
z_j	charge number of ion j

α	Debye-Hückel constant (1.17582 kg$^{1/2}$ mol$^{-1/2}$ at 298.15 K)
γ_i	activity coefficient of species i
γ	surface tension (N m^{-1})
μ_j	chemical potential of species j at specified T, P, and ionic strength (J mol^{-1})
$\{\mu_i\}$	set of chemical potentials (J mol^{-1})
$\mu_i{}'$	transformed chemical potential of reactant i at specified T, P, ionic strength, and concentrations of one or more species (J mol^{-1})
$\mu_i{}''$	further transformed chemical potential of pseudoisomer group i at specified T, P, ionic strength, and concentrations of one or more species and one or more reactants (J mol^{-1})
$\mu_j{}^\circ$	standard chemical potential of species j at specified T, P, and ionic strength (J mol^{-1})
$\mu_i{}'{}^\circ$	standard transformed chemical potential of reactant i (J mol^{-1})
$\mu_i{}''{}^\circ$	standard further transformed chemical potential of pseudoisomer group i at specified T, P, ionic strength, and concentrations of one or more species and one or more reactants (J mol^{-1})
$\boldsymbol{\mu}$	vector of chemical potentials of species at specified T, P, and ionic strength (1xN) (J mol^{-1})
$\boldsymbol{\mu}'$	vector of transformed chemical potentials of reactants at specified T, P, ionic strength, and concentrations of one or more species (1xN') (J mol^{-1})
$\boldsymbol{\mu}''$	vector of further transformed chemical potentials of pseudoisomer groups of reactants at specified T, P, ionic strength, and concentrations of one or more species and one or more reactants (1xN'') (J mol^{-1})
μ_{ci}	chemical potential of component i at specified T, P, and ionic strength (J mol^{-1})
$\mu_{ci}{}'$	transformed chemical potential of component i at specified T, P, ionic strength, and concentrations of one or more species (J mol^{-1})

μ_{ci}''	further transformed chemical potential of component i at specified T, P, ionic strength, and concentrations of one or more species and one or more reactants (J mol^{-1})
μ_c	vector of chemical potentials of components at specified T, P, and ionic strength $(1 \times C)$ (J mol^{-1})
μ_c'	vector of transformed chemical potentials of components at specified T, P, ionic strength, and concentrations of one or more species $(1 \times C')$ (J mol^{-1})
μ_{nc}'	vector of chemical potentials of noncomponents at specified T, P, and ionic strength $(1 \times (N-C))$ (J mol^{-1})
μ_c''	vector of further transformed chemical potentials of components at specified T, P, ionic strength, and concentrations of one or more species and one or more reactants $(1 \times C'')$ (J mol^{-1})
ν_i	stoichiometric number of species i in a chemical reaction
ν_{ij}	stoichiometric number of species i in reaction j
ν_i'	stoichiometric number of reactant i in a biochemical reaction
ν_{ij}'	stoichiometric number of reactant i in reaction j
ν_{ij}''	stoichiometric number of pseudoisomer group i in reaction j
$\boldsymbol{\nu}$	stoichiometric number matrix in terms of species $(N \times R)$
$\boldsymbol{\nu}'$	stoichiometric number matrix in terms of reactants $(N' \times R')$
$\boldsymbol{\nu}''$	stoichiometric number matrix in terms of pseudoisomer groups of reactants $(N'' \times R'')$
ν_i	stoichiometric number of species i
ν_i'	apparent stoichiometric number of reactant i (sum of species)
ν_i''	apparent stoichiometric number of reactant i (sum of species) when the concentration of a reactant has been specified
$|\nu_e|$	number of electrons in a half reaction
ξ	extent of chemical reaction (mol)
ξ'	extent of biochemical reaction (mol)
$\boldsymbol{\xi}$	extent of reaction column vector at specified T, P, and ionic strength $(R \times 1)$ (mol)
$\boldsymbol{\xi}'$	extent of reaction column vector at specified T, P, ionic strength, and concentrations of one or more species $(R' \times 1)$ (mol)
$\boldsymbol{\xi}''$	extent of reaction column vector at specified T, P, ionic strength, and concentrations of one or more species and one or more reactants $(R'' \times 1)$ (mol)
L	binding potential (J mol^{-1})
κ	acid dissociation constant of an independent acidic group

Appendix 5: List of *Mathematica* Programs

All the *Mathematica* programs used in each chapter are listed. Programs are not duplicated, but references are given as to where they can be found.

Chapter 1 Thermodynamics of the Dissociation of Weak Acids

Section 1.6

```
calcpK[speciesmat_, no_, is_] := Module[{lnkzero, sigmanuzsq, lnK},
  (*Calculates pKs for a weak acid at 298.15 K at specified ionic
    strengths (is) when the number no of the pK is specified. pKs
    are numbered 1,2,3,... from the highest pK to the lowest pK,
  but the highest pK for a weak acid may be omitted if it is outside
  of the range 5 t0 9. For H3PO4,pK1=calcpK[pisp,1,{0}]=7.22.*)
  lnkzero = (speciesmat[[no + 1, 1]] - speciesmat[[no, 1]]) / (8.31451 * 0.29815);
  sigmanuzsq = speciesmat[[no, 3]]^2 - speciesmat[[no + 1, 3]]^2 + 1;
  lnK = lnkzero + (1.17582 * is^0.5 * sigmanuzsq) / (1 + 1.6 * is^0.5);
  N[-(lnK / Log[10])]]
```

```
calcpK298is[speciesmat_] := Module[{glist, hlist, zlist, nHlist, glistis, ghydionis},
  (*This program derives the functions of ionic strength that yields the pKs
    at 298.15 K for weak acids. The first function of ionic strength
    is for the acid with the fewest hydrogen atoms. The program has a
    single argument so that it can be used with Map. The functions can be
    evaluated by use of calcpK298is[atpsp]/.is→{0,.1,.25}, for example.*)
  {glist, hlist, zlist, nHlist} = Transpose[speciesmat];
glistis = Table[glist[[i]] - 2.91482 * zlist[[i]]^2 * is^.5 / (1 + 1.6 * is^.5),
  {i, 1, Length[zlist]}];
  ghydionis = -2.91482 * is^.5 / (1 + 1.6 * is^.5);
  Table[((glistis[[i - 1]] - glistis[[i]] + ghydionis) / (8.31451 * .29815 * Log[10])),
  {i, 2, Length[zlist]}]]
```

Section 1.7

```
calcpKT[speciesmat_, n_, is_, t_] := Module[
  {basicspeciesG, acidicspeciesG, coeff, basicspeciesI, acidicspeciesI, hydionI},
  (*Calculates pKs for a weak acid at temperatures (t) in the range 273.15-
    313.15 K and ionic strengths (is) in the rangw 0-
    0.35 M when the number (n) of the pK is specified. pKs are numbered 1,2,
    3,... from the highest pK to the lowest pK, but the highest pK for a weak
    acid may be omitted if it is outside of the range 5 to 9. The first step is
    to calculate the standard Gibbs energies of formation at zero ionic strength
    as a function of temperature. The second step is to adjust these values to
    the desired ionic strength. A list of temperatures can be used. For example,
  pK1=calcpKT[atpsp,1,0,{273.15,298.15,313.15}]={7.50,7.60,7,65}.*)
  basicspeciesG = (t / 298.15) * speciesmat[[n, 1]] + (1 - t / 298.15) * speciesmat[[n, 2]];
  acidicspeciesG =
    (t / 298.15) * speciesmat[[n + 1, 1]] + (1 - t / 298.15) * speciesmat[[n + 1, 2]];
  coeff = (9.20483 * 10^-3) * t - (1.28467 * 10^-5) * t^2 + (4.95199 * 10^-8) * t^3;
  basicspeciesI = basicspeciesG - coeff * speciesmat[[n, 3]]^2 * is^.5 / (1 + 1.6 * is^.5)
  acidicspeciesI =
    acidicspeciesG - coeff * speciesmat[[n + 1, 3]]^2 * is^.5 / (1 + 1.6 * is^.5);
  hydionI = -coeff * is^.5 / (1 + 1.6 * is^.5);
  (1 / (8.31451 * (t / 1000) * Log[10])) * (hydionI + basicspeciesI - acidicspeciesI)]
```

```
calcpKTfn[speciesmat_] :=
 Module[{glist, hlist, zlist, nHlist, coeff, speciesGT, spceiesGTis, hydionis},
  (*Derives the function of temperature, pH,
    and ionic strength that gives the pKs for a weak acid.  pKs are numbered 1,
    2,3,... from the highest pK to the lowest pK,
    but the highest pK for a weak acid may be omitted if it is outside
     of the range 5 to 9. The first step is to calculate the standard
     Gibbs energies of formation at zero ionic strength as a function of
     temperature.  The second step is to adjust these values to the desired
     ionic strength. The output is a list of functions, with as many functions
     as pKs.  The third step is to make a table of the pKs.  For example,
     calcpKTfn[atpsp]/.t→{273.15,298.15,313.15}/.is→{0,.1,.25}*)
  {glist, hlist, zlist, nHlist} = Transpose[speciesmat];
  (*Calculate functions of temperature for the Gibbs energies of all species.*)
  speciesGT = (t / 298.15) * glist + (1 - t / 298.15) * hlist;
  (*Adjust these functions of temperature
    to make them functions of ionic strength as well.*)
  coeff = (9.20483 * 10^-3) * t - (1.28467 * 10^-5) * t^2 + (4.95199 * 10^-8) * t^3;
  speciesGTis = speciesGT - coeff * zlist^2 * is^.5 / (1 + 1.6 * is^.5);
  hydionis = -coeff * is^.5 / (1 + 1.6 * is^.5);
  (*Make a list of the Gibbs energies of
    dissociation for all weak acids and convert them to pKs.*)
  Table[((speciesGTis[[i - 1]] - speciesGTis[[i]] + hydionis) /
      (8.31451 * (t / 1000) * Log[10])), {i, 2, Length[zlist]}]]
```

Section 1.8

```
calcGHSdiss[speciesmat_, no_, is_] :=
 Module[{lnkzero, sigmanuzsq, lnK, dGI, dHzero, dHI, dSI},
  (*Calculates {dGI,dHI,dSI} for a weak acid at 298.15 K at specified ionic
    strengths (is) when the number no of the pK is specified. pKs are numbered 1,
    2,3,... from the highest pK to the lowest pK,
    but the highest pK for a weak acid may be omitted if it is outside of the
     range 5 t0 9. The Gibbs energy and enthalpy are given in kJ mol^-1,
    and the entropy is given in J K^-1 mol^-1.  For H3PO4,
    pK1=calcGHSdiss[pisp,1,{0}]={{41.2},{3.6},{-126.111}}.*)
  lnkzero = (speciesmat[[no + 1, 1]] - speciesmat[[no, 1]]) / (8.31451 * 0.29815);
  sigmanuzsq = speciesmat[[no, 3]]^2 - speciesmat[[no + 1, 3]]^2 + 1;
  lnK = lnkzero + (1.17582 * is^0.5 * sigmanuzsq) / (1 + 1.6 * is^0.5);
  (*Calculate the Gibbs energy of acid dissociation.*)
  dGI = -8.31451 * .29815 * lnK;
  (*Calculate the enthalpy of dissociation.*)
  dHzero = speciesmat[[no, 2]] - speciesmat[[no + 1, 2]];
  sigmanuzsq = speciesmat[[no, 3]]^2 - speciesmat[[no + 1, 3]]^2 + 1;
  dHI = dHzero - (1.4775 * is^0.5 * sigmanuzsq) / (1 + 1.6 * is^0.5);
  (*Calculate the entropy of dissociation.*)
  dSI = (dHI - dGI) / .29815;
  Transpose[{dGI, dHI, dSI}]]
```

```
calcGHSdissfn[speciesmat_] :=
 Module[{glist, hlist, zlist, nHlist, glistis, ghydionis, gibbs, hlistis,
    hhydionis, enthalpy, entropy}, (*This program derives the functions of
      ionic strength that yield {G,H,S} functions of ionic strength at
      298.15 K for weak acids.  The first function of ionic strength is for
      the acid with the fewest hydrogen atoms.  The program has a single
      argument so that it can be used with Map.  The functions can be
      evaluated by use of calcGHSdissfn[atpsp]/.is→{0,.1,.25}, for example.*)
   {glist, hlist, zlist, nHlist} = Transpose[speciesmat];
glistis = Table[glist[[i]] - 2.91482 * zlist[[i]]^2 * is^.5 / (1 + 1.6 * is^.5),
    {i, 1, Length[zlist]}];
   ghydionis = -2.91482 * is^.5 / (1 + 1.6 * is^.5);
   gibbs =
    Table[((glistis[[i - 1]] - glistis[[i]] + ghydionis)), {i, 2, Length[zlist]}];
   hlistis = Table[hlist[[i]] - 1.4775 * zlist[[i]]^2 * is^.5 / (1 + 1.6 * is^.5),
     {i, 1, Length[zlist]}];
   hhydionis = -1.4775 * is^.5 / (1 + 1.6 * is^.5);
   enthalpy =
    Table[((hlistis[[i - 1]] - hlistis[[i]] + hhydionis)), {i, 2, Length[zlist]}];
   entropy = (enthalpy - gibbs) / .29815;
   Transpose[{gibbs, enthalpy, entropy}]]]
```

Chapter 3 Biochemical Reactions at Specified Temperature and Various pKs

Section 3.6

```
calcdGmat[speciesmat_] :=
Module[{dGzero,dHzero, zi, nH, pHterm, isterm,gpfnsp},(*This program derives the
function of pH and ionic strength (is) that gives the standard transformed Gibbs
energy of formation of a reactant (sum of species) at 298.15 K.  The input
speciesmat is a matrix that gives the standard Gibbs energy of formation, the
standard enthalpy of formation, the electric charge, and the number of hydrogen
atoms in each species. There is a row in the matrix for each species of the
reactant. gpfnsp is a list of the functions for the species.  Energies are
expressed in kJ mol^-1.*)
{dGzero,dHzero,zi,nH}=Transpose[speciesmat];
pHterm = nH*8.31451*.29815*Log[10^-pH];
isterm = 2.91482*((zi^2) - nH)*(is^.5)/(1 + 1.6*is^.5);
gpfnsp=dGzero - pHterm - isterm;
-8.31451*.29815*Log[Apply[Plus,Exp[-1*gpfnsp/(8.31451*.29815)]]]]]]
```

```
calcdHmat[speciesmat_] :=
 Module[{dGzero, dHzero, zi, nH, dhfnsp, pHterm, isenth, dgfnsp, dGreactant, ri},
   (*This program derives the function of ionic strength (is) that gives the
     standard transformed enthalpy of formation of a reactant (sum of species)
     at 298.15 K.  The input is a matrix that gives the standard Gibbs energy
     of formation,the standard enthalpy of formation, the electric charge,
   and the number of hydrogen atoms in the species in the reactant.  There
     is a row in the matrix for each species of the reactant.  dhfnsp is a list
     of the functions for the species.  Energies are expressed in kJ mol^-1.*)
     {dGzero, dHzero, zi, nH} = Transpose[speciesmat];
     isenth = 1.4775 * ((zi^2) - nH) * (is^.5) / (1 + 1.6 * is^.5);
     dhfnsp = dHzero + isenth;
   (*Calculate the functions for the
     standard Gibbs energies of formation of the species.*)
   pHterm = nH * 8.31451 * .29815 * Log[10^-pH];
   gpfnsp = dGzero - pHterm - isenth * 2.91482 / 1.4775;
   (*Calculate the standard
     transformed Gibbs energy of formation for the reactant.*)
   dGreactant = -8.31451 * .29815 *
     Log[Apply[Plus, Exp[-1 * gpfnsp / (8.31451 * .29815)]]]];
   (*Calculate the equilibrium mole fractions of the species in the
     reactant and the mole fraction-weighted average of the functions
     for the standard transformed enthalpies of the species.*)
   ri = Exp[(dGreactant - gpfnsp) / (8.31451 * .29815)];
   ri.dhfnsp]

derivetrS[speciesmat_] :=
 Module[{dG, dH}, (*This program derives the function of pH and ionic strength
     (is) that gives the standard transformed entropy of formation of a reactant
     (sum of species) at 298.15 K.  The entropy is given in J K^-1 mol^-1.*)
   dG = calcdGmat[speciesmat];
   dH = calcdHmat[speciesmat];
   (dH - dG) / .29815]

calcavHbound[speciesmat_] :=
 Module[{gfn}, (*This program derives the function of pH
     and ionic strength that gives the average number of hydrogen
     atoms bound by a biochemical reactant at 298.15 K.*)
   gfn = calcdGmat[speciesmat];
   (1 / (8.31451 * .29815 * Log[10])) * D[gfn, pH]]
```

    Section 3.8

```
deriverxfn[eq_] := Module[{function},
  (*Derives the function of pH and ionic strength that gives the thermodynamic
     properties of a biochemical reaction typed in the form atpG+h2oG+de==
   adpG+piG.  Other suffixes can be used for H, S, and NH.*)
   function = Solve[eq, de]; function[[1, 1, 2]]]
```

    Section 3.12

```
calcpK[speciesmat_, no_, is_] (See Section 1.6)
```

```
derivetrGspecies[speciesmat_] :=
Module[{dGzero,dHzero, zi, nH, pHterm, isterm},(*This program derives the
functions of pH and ionic strength (is) that gives the standard transformed Gibbs
energies of formation of the species of a reactant at 298.15 K.  The input
speciesmat is a matrix that gives the standard Gibbs energy of formation, the
standard enthalpy of formation, the electric charge, and the number of hydrogen
atoms in each species. There is a row in the matrix for each species of the
reactant. The output is a list of the functions for the species of the reactant.
Energies are expressed in kJ mol^-1.*)
{dGzero,dHzero,zi,nH}=Transpose[speciesmat];
pHterm = nH*8.31451*.29815*Log[10^-pH];
isterm = 2.91482*((zi^2) - nH)*(is^.5)/(1 + 1.6*is^.5);
dGzero - pHterm - isterm]
```

Chapter 4 Biochemical Reactiions at Various pHs and Various Temperatures

Section 4.1


```
derivetrGibbsT[speciesmat_]:=Module[{dGzero,dGzeroT,dHzero,zi,nH,gibbscoeff,pHterm
, isterm,gpfnsp},(*This program derives the function of T (in Kelvin), pH, and
ionic strength (is) that gives the standard transformed Gibbs energy of formation
of a reactant (sum of species).  The input speciesmat is a matrix that gives the
standard Gibbs energy of formation in kJ mol^-1 at 298.15 K and zero ionic
strength, the standard enthalpy of formation in kJ mol^-1 at 298.15 K and zero
ionic strength, the electric charge, and the number of hydrogen atoms in each
species.  There is a row in the matrix for each species of the reactant.  gpfnsp
is a list of the functions for the transformed Gibbs energies of the species.
The corresponding functions for other transformed properties can be obtained by
taking partial derivatives.  The standard transformed Gibbs energy of formation
of a reactant in kJ mol^-1 can be calculated at any temperature in the range
273.15 K to 313.15 K, any pH in the range 5 to 9, and any ionic strength in the
range 0 to 0.35 M by use of the assignment operator (/.).*)
{dGzero,dHzero,zi,nH}=Transpose[speciesmat];
gibbscoeff=(9.20483*t)/10^3-(1.284668*t^2)/10^5+(4.95199*t^3)/10^8;
dGzeroT=(dGzero*t)/298.15+dHzero*(1-t/298.15);
pHterm=(nH*8.31451*t*Log[10^(-pH)])/1000;
istermG=(gibbscoeff*(zi^2-nH)*is^0.5)/(1+1.6*is^0.5);
gpfnsp=dGzeroT-pHterm-istermG;
-((8.31451*t*Log[Plus@@(E^(-(gpfnsp/((8.31451*t)/1000))))])/1000)]
```


Section 4.4


```
derivefnGHSNHrx[eq_] :=
 Module[{function, functionG, functionH, functionS, functionNH},
   (*Derives the functions of temperature, pH,
    and ionic strength that give the standard transformed reaction Gibbs energy,
    standard transformed reaction enthalpy, standard transformed reaction entropy,
    and the change in binding of hydrogen ions of a biochemical reaction
      or half reaction typed in the form atpGT+h2oGT+de==adpGT+piGT.*)
   function = Solve[eq, de];
   functionG = function[[1, 1, 2]];
   functionH = -t^2 * D[function[[1, 1, 2]] / t, t];
   functionS = -1000 * D[function[[1, 1, 2]], t];
   functionNH = (1 / (8.31451 * (t / 1000) * Log[10])) * D[function[[1, 1, 2]], pH];
   {functionG, functionH, functionS, functionNH}]
```

Chapter 5 Biochemical Reactions at Specified pH, pMg, and Various Temperatures

Section 5.2

```
deriveGMgT[speciesmat_] :=
Module[{dGzero,dHzero,dGzeroT,zi,nH,nMg,pHterm,stdGMg,
pMgterm,coeffis,isterm,gpfnsp},(*This program derives the function of T, pH, pMg
and ionic strength (is) that gives the standard transformed Gibbs energy of
formation of a reactant (sum of species).  The input speciesmat is a matrix that
gives the standard Gibbs energy of formation, the standard enthalpy of formation,
the electric charge, the number of hydrogen atoms, and the number of magnesium
atoms in each species. There is a row in the matrix for each species of the
reactant. gpfnsp is a list of the functions for the species.  Energies in the
output are in kJ mol^-1.*)
{dGzero,dHzero,zi,nH,nMg} = Transpose[speciesmat];
pHterm = nH*8.31451*t*Log[10]*pH/1000;
stdGMg=(t/298.15)*(-455.3)+(1-t/298.15)*(-467.00);
pMgterm = nMg*(-stdGMg+8.31451*(t/1000)*Log[10]*pMg);
coeffis=(9.20483*t)/10^3-(1.284668*t^2)/10^5+(4.95199*t^3)/10^8;
dGzeroT=(t/298.15)*dGzero+(1-t/298.15)*dHzero;
isterm = coeffis*((zi^2)-nH-4*nMg)*(is^.5)/(1 + 1.6*is^.5);
gpfnsp=dGzeroT+pHterm+pMgterm-isterm;
-8.31451*(t/1000)*Log[Apply[Plus,Exp[-1*gpfnsp/((8.31451*(t/1000)))]]]]];
```

Section 5.3

```
deriveGHSNHNMg[speciesmat_] :=
Module[{dGzero,dHzero,dGzeroT,zi,nH,nMg,pHterm,stdGMg,
pMgterm,coeffis,isterm,gpfnsp,gibbsfn,enthalfn,entfn,nHfn,nMgfn},(*This program
derives the function of T, pH, pMg and ionic strength (is) that gives the
standard transformed Gibbs energy of formation of a reactant (sum of species).
Then partial differentiation is used to derive the functions that yield the
standard transformed enthalp of formation, standard transformed entropy of
formation, average numbe of hydrogen ions bound and average number of magnesium
ions bound.  The input speciesmat is a matrix that gives the standard Gibbs
energy of formation, the standard enthalpy of formation, the electric charge, the
number of hydrogen atoms, and the number of magnesium atoms in each species.
There is a row in the matrix for each species of the reactant. gpfnsp is a list
of the functions for the species.  Energies are expressed in kJ mol^-1 in the
program until gibbsfn is needed, but the output is changed to J mol^-1 so that
differentiations can be made with respect to t in K. The output is a list of five
functions.  Energies in the output are in kJ mol^-1 and entropies in J K^-1
mol^-1.*)
{dGzero,dHzero,zi,nH,nMg} = Transpose[speciesmat];
pHterm = nH*8.31451*t*Log[10]*pH/1000;
stdGMg=(t/298.15)*(-455.3)+(1-t/298.15)*(-467.00);
pMgterm = nMg*(-stdGMg+8.31451*(t/1000)*Log[10]*pMg);
coeffis=(9.20483*t)/10^3-(1.284668*t^2)/10^5+(4.95199*t^3)/10^8;
dGzeroT=(t/298.15)*dGzero+(1-t/298.15)*dHzero;
isterm = coeffis*((zi^2)-nH-4*nMg)*(is^.5)/(1 + 1.6*is^.5);
gpfnsp=dGzeroT+pHterm+pMgterm-isterm;
gibbsfn=-8.31451*(t/1000)*Log[Apply[Plus,Exp[-1*gpfnsp/((8.31451*(t/1000)))]]];
enthalfn=-t^2*D[gibbsfn/t,t];
entfn=-1000*D[gibbsfn,t];
nHfn=(1/(8.31451*(t/1000)*Log[10]))*D[gibbsfn,pH];
nMgfn=(1/(8.31451*(t/1000)*Log[10]))*D[gibbsfn,pMg];
{gibbsfn,enthalfn,entfn,nHfn,nMgfn}]
```

Chapter 6 Development of a Database on Species

Section 6.3

```
calcGef1sp[equat_, pHc_, ionstr_, z1_, nH1_] :=
  Module[{energy, trGereactant},(*This program uses ∑viΔfGi'°=-RTlnK' to
calculate the standard Gibbs energy of formation of the species of a reactant
that does not have a pK in the range 4 to 10. The equation is of the form
pyruvate+atp-x-adp==-8.31451*.29815*Log[K'], where K' is the apparent equilibrium
constant at 298.15 K, pHc, and ionic strength is.  The reactant has charge number
z1 and hydrogen atom number nH1.  The output is the species vector without the
standard enthalpy of formation.*)
    energy = Solve[equat, x] /. pH -> pHc /. is -> ionstr;
    trGereactant = energy[[1,1,2]];
    gef1 = trGereactant - nH1*8.31451*0.29815*Log[10]*pHc +
      (2.91482*(z1^2 - nH1)*ionstr^0.5)/
      (1 + 1.6*ionstr^0.5);
      {{gef1, _, z1, nH1}}]

calcGef2sp[equat_, pHc_, ionstr_, z1_, nH1_, pK0_] :=
  Module[{energy, trGereactant, pKe, trgefpHis,gef1, gef2},(*This program uses ∑
viΔfGi'°=-RTlnK' to calculate the standard Gibbs energies of formation of the two
species of a reactant for which the pK at zero ionic strength is pK0. The
equation is of the form pyruvate+atp-x-adp==-8.31451*.29815*Log[K'], where K' is
the apparent equilibrium constant at 298.15 K, pHc, and ionic strength is.  The
more basic form of the reactant has charge number z1 and hydrogen atom number
nH1.  The output is the species matrix without the standard enthalpies of
formation.*)
  energy = Solve[equat, x] /. pH -> pHc /. is -> ionstr;
    trGereactant = energy[[1,1,2]];
    pKe = pK0 + (0.510651*ionstr^0.5*2*z1)/
      (1 + 1.6*ionstr^0.5); trgefpHis =
     trGereactant + 8.31451*0.29815*Log[1 + 10^(pKe - pHc)];
    gef1 = trgefpHis - nH1*8.31451*0.29815*Log[10]*pHc +
      (2.91482*(z1^2 - nH1)*ionstr^0.5)/
      (1 + 1.6*ionstr^0.5);
    gef2 = gef1 + 8.31451*0.29815*Log[10^(-pK0)];
    {{gef1, _, z1, nH1}, {gef2, _, z1 + 1, nH1 + 1}}]

calcGef3sp[equat_, pHc_, ionstr_, z1_, nH1_, pK10_,
    pK20_] := Module[{energy, trGereactant, pKe, trgefpHis,
      gef1, gef2, gef3, pK1e, pK2e},(*This program uses ∑viΔfGi'°=-RTlnK' to
calculate the standard Gibbs energies of formation of the three species of a
reactant for which the pKs at zero ionic strength is pK10 and pK20. The equation
is of the form pyruvate+atp-x-adp==-8.31451*.29815*Log[K'], where K' is the
apparent equilibrium constant at 298.15 K, pHc, and ionic strength is.  The more
basic form of the reactant has charge number z1 and hydrogen atom number nH1.
The output is the species matrix without the standard enthalpies of formation of
the three species.*)
    energy = Solve[equat, x] /. pH -> pHc /. is -> ionstr;
    trGereactant = energy[[1,1,2]];
    pK1e = pK10 + (0.510651*ionstr^0.5*2*z1)/
      (1 + 1.6*ionstr^0.5);
    pK2e = pK20 + (0.510651*ionstr^0.5*(2*z1 + 2))/
      (1 + 1.6*ionstr^0.5); trgefpHis =
     trGereactant + 8.31451*0.29815*
      Log[1 + 10^(pK1e - pHc) + 10^(pK1e + pK2e - 2*pHc)];
    gef1 = trgefpHis - nH1*8.31451*0.29815*Log[10]*pHc +
      (2.91482*(z1^2 - nH1)*ionstr^0.5)/
      (1 + 1.6*ionstr^0.5);
    gef2 = gef1 + 8.31451*0.29815*Log[10^(-pK10)];
    gef3 = gef2 + 8.31451*0.29815*Log[10^(-pK20)];
    {{gef1, _, z1, nH1}, {gef2, _, z1 + 1, nH1 + 1},
     {gef3, _, z1 + 2, nH1 + 2}}]
```

```
calcHf1sp[equat_, spmat_, pHc_, ionstr_] :=
 Module[{energy, trHreactant, enthf1, gef1, dHzero1, z1, nH1},
  (*This program uses ∑viΔfHi'°=ΔrH'° (298.15 K) to calculate the standard
      enthalpy of formation (I=0) of the single species of a reactant for which
      the species matrix (spmat) contains ΔfG° at zero ionic strength.  The
      reaction equation (equat) is of the form x+nadredh-malateh-nadoxh==89.5,
    where 89.5 kJ mol^-1 is the heat of reaction and x is oxaloacetate.  The
      species matrix (spmat) is that for oxaloacetate.  The calorimetric
      experiment is at pHc and ionic strength ionstr.  The reactant
      x has charge number z1 and hydrogen atom number nH1.  The
      output is the complete species matrix for x. 11-21-04*)
  {gef1, dHzero1, z1, nH1} = Transpose[spmat];
  energy = Solve[equat, x] /. pH -> pHc /. is → ionstr;
  trHreactant = energy[[1, 1, 2]];
  enthf1 = trHreactant - 1.4775 * (z1^2 - nH1) * ionstr^0.5 / (1 + 1.6 * ionstr^0.5);
  Flatten[{gef1, enthf1, z1, nH1}]]


calcHf2sp[equat_, spmat_, pHc_, ionstr_, dHdisszero_] :=
 Module[{dGzero, dHzero, zi, nHi, pHterm, isterm, gpfnsp,
   energy, trHreactant, stdtrGereactant, r1, r2, solution,
   dH1zero, dH2zero, dH1, dH2}, (*This program uses ∑viΔfH'°=
    ΔrH'° (298.15 K) to calculate the standard enthalpy of formation (I=0) of the
      two species of a reactant for which the species matrix (spmat) contains
      ΔfG° at zero ionic strength for the two species of the reactant.  The
      reaction equation (equat) is of the form mannoseh+pih-x-h2oh==1.7,
    where 1.7 kJ mol^-1 is the heat of reaction and x is mannnose6phosh.  The
      species matrix (spmat) is that for mannose6phos.  The calorimetric
      experiment is at pHc and ionic strength ionstr.  The first step
      in the calculation is to use the information on the standard
      Gibbs energies of formation of the species of the reactant
      of interest to calculate the equilibrium mole fractions r1
      (base form) and r2 (acidform) of the two species of the reactant of
      interest.  The final output is the complete species matrix for x.*)
  {dGzero, dHzero, zi, nHi} = Transpose[spmat];
  pHterm = nHi * 8.31451 * .29815 * Log[10^-pH] /. pH -> pHc;
 isterm = 2.91482 * ((zi^2) - nHi) * (is^.5) / (1 + 1.6 * is^.5) /. is → ionstr;
 gpfnsp = dGzero - pHterm - isterm;
  stdtrGereactant =
   -8.31451 * .29815 * Log[Apply[Plus, Exp[-1 * gpfnsp / (8.31451 * .29815)]]];
  r1 = Exp[(stdtrGereactant - gpfnsp[[1]]) / (8.31451 * .29815)];
  r2 = Exp[(stdtrGereactant - gpfnsp[[2]]) / (8.31451 * .29815)];
 (*Now calculate dfH'° (reactant) from ΔrH'° (298.15 K) for the reaction.*)
  energy = Solve[equat, x] /. pH -> pHc /. is → ionstr;
 trHreactant = energy[[1, 1, 2]];
  (*dH1xero is given by the following equation. dH2zero is
    calculated from the equation for the enthalpy of dissociation.*)
  solution = Solve[trHreactant ==
     r1 * (dH1zero + 1.4775 * (zi[[1]]^2 - nHi[[1]]) * ionstr^0.5 / (1 + 1.6 * ionstr^0.5))
      r2 * (dH1zero - dHdisszero + 1.4775 * (zi[[2]]^2 - nHi[[2]]) *
          ionstr^0.5 / (1 + 1.6 * ionstr^0.5)), dH1zero];
  dH1 = solution[[1, 1, 2]];
  dH2 = dH1 - dHdisszero;
  Transpose[{dGzero, {dH1, dH2}, zi, nHi}]]
```

```
calcHf3sp[equat_, spmat_, pHc_, ionstr_, dHdisszero1_, dHdisszero2_] :=
 Module[{dGzero, dHzero, zi, nHi, pHterm, isterm, gpfnsp, energy,
   trHreactant, stdtrGereactant, r1, r2, r3, solution, dH1zero,
   dH2zero, dH3zero, dH1expt, dH2expt, dH3expt, dH1, dH2, dH3},
   (*This program uses ∑viΔfHi'°=ΔrH'° (298.15 K) to calculate the standard
       enthalpy of formation (I=0) of the three species of a reactant
       for which the species matrix (spmat) contains ΔfG° at zero ionic
       strength for the three species of the reactant.  The reaction
       equation (equat) is of the form adph+x-fructose6phos-atph=-84.2,
     where 84.2 kJ mol^-1 is the heat of reaction and x is fructose16phosh.  The
       species matrix (spmat) is that for fructose16phos.  The calorimetric
       experiment is at pHc and ionic strength ionstr.  The first step
       in the calculation is to use the information on the standard
       Gibbs energies of formation of the species of the reactant
       of interest to calculate the equilibrium mole fractions r1
       (base form) and r2 (acidform) of the two species of the reactant of
       interest.  The final output is the complete species matrix for x.*)
   {dGzero, dHzero, zi, nHi} = Transpose[spmat];
   pHterm = nHi * 8.31451 * .29815 * Log[10^-pH] /. pH -> pHc;
isterm = 2.91482 * ((zi^2) - nHi) * (is^.5) / (1 + 1.6 * is^.5) /. is → ionstr;
gpfnsp = dGzero - pHterm - isterm;
   stdtrGereactant =
     -8.31451 * .29815 * Log[Apply[Plus, Exp[-1*gpfnsp / (8.31451 * .29815)]]];
   r1 = Exp[(stdtrGereactant - gpfnsp[[1]]) / (8.31451 * .29815)];
   r2 = Exp[(stdtrGereactant - gpfnsp[[2]]) / (8.31451 * .29815)];
   r3 = Exp[(stdtrGereactant - gpfnsp[[3]]) / (8.31451 * .29815)];
(*Now calculate dfH'° (reactant) from ΔrH'° (298.15 K) for the reaction.*)
   energy = Solve[equat, x] /. pH -> pHc /. is → ionstr;
trHreactant = energy[[1, 1, 2]];
   (*The standard transformed enthalpies of formation of the three species are
       given by the following six equations. dH1zero, dH2zero,  and dH3zero
       are also related by the equations for the enthalpy of dissociation.*)
   solution = Solve[{dH1expt == dH1zero + 1.4775 * (zi[[1]]^2 - nHi[[1]]) *
         ionstr^0.5 / (1 + 1.6 * ionstr^0.5), dH2expt == dH2zero +
        1.4775 * (zi[[2]]^2 - nHi[[2]]) * ionstr^0.5 / (1 + 1.6 * ionstr^0.5), dH3expt ==
       dH3zero + 1.4775 * (zi[[3]]^2 - nHi[[3]]) * ionstr^0.5 / (1 + 1.6 * ionstr^0.5),
       trHreactant == r1 * dH1expt + r2 * dH2expt + r3 * dH3expt,
       dH2zero == dH1zero - dHdisszero1, dH3zero == dH2zero - dHdisszero2},
      {dH1zero, dH2zero, dH3zero}, {dH1expt, dH2expt, dH3expt}];
   dHzerocalc = {solution[[1, 1, 2]], solution[[1, 2, 2]],
     solution[[1, 3, 2]]};
   Transpose[{dGzero, dHzerocalc, zi, nHi}]]


calcdGmat[speciesmat_]    (See Section 3.6)


calckprime[eq_, pHlist_, islist_] := Module[{energy, dG},(*Calculates the
apparent equilibrium constant at specified pHs and ionic strengths for a
biochemical reaction typed in the form atp+h2o+de==adp+pi.  The names of
reactants call the appropriate functions of pH and ionic strength.  pHlist and is
list can be lists.*)
   energy = Solve[eq, de];
    dG = energy[[1,1,2]] /. pH -> pHlist /. is -> islist;
    E^(-(dG/(8.31451*0.29815)))]
```

Section 6.4

```
calcdGmat313[speciesmat_] :=
Module[{dGzero, dHzero,zi, nH, pHterm, isterm,gpfnsp},(*This program produces the
function of pH and ionic strength (is) that gives the standard transformed Gibbs
energy of formation of a reactant (sum of species) at 313.15 K.  The input
speciesmat is a matrix that gives the standard Gibbs energy of formation, the
standard enthalpy of formation, the electric charge, and the number of hydrogen
atoms in each species. There is a row in the matrix for each species of the
reactant. gpfnsp is a list of the functions for the species.  Energies are
expressed in kJ mol^-1.*)
{dGzero,dHzero,zi,nH}=Transpose[speciesmat];
pHterm = nH*8.31451*.31315*Log[10^-pH];
isterm = 3.14338*((zi^2) - nH)*(is^.5)/(1 + 1.6*is^.5);
gpfnsp=dGzero - pHterm - isterm;
-8.31451*.31315*Log[Apply[Plus,Exp[-1*gpfnsp/(8.31451*.31315)]]]]]


calcdHmat313[speciesmat_] :=
 Module[{dHzero, zi, nH, dhfnsp, dGzero, pHterm, isenth, dgfnsp, dGreactant, ri},
   (*This program produces the function of ionic strength (is) that gives the
     standard transformed enthalpy of formation of a reactant (sum of species)
     at 298.15 K.  The input is a matrix that gives the standard Gibbs energy
     of formation,the standard enthalpy of formation, the electric charge,
   and the number of hydrogen atoms in the species in the reactant.  There
     is a row in the matrix for each species of the reactant.  dhfnsp is a list
     of the functions for the species.  Energies are expressed in kJ mol^-1.*)
     {dGzero, dHzero, zi, nH} = Transpose[speciesmat];
     isenth = 1.78158 * ((zi^2) - nH) * (is^.5) / (1 + 1.6 * is^.5);
     dhfnsp = dHzero + isenth;
   (*Now calculate the functions for the
     standard Gibbs energies of formation of the species.*)
    dGzero = speciesmat[[All, 1]];
   pHterm = nH * 8.31451 * .31315 * Log[10^-pH];
   gpfnsp = dGzero - pHterm - isenth * 3.14338 / 1.78158;
   (*Now calculate the standard
     transformed Gibbs energy of formation for the reactant.*)
   dGreactant = -8.31451 * .31315 *
     Log[Apply[Plus, Exp[-1 * gpfnsp / (8.31451 * .31315)]]];
   (*Now calculate the equilibrium mole fractions of the species in
      the reactant and the mole fraction-weighted average of the
      functions for the standard transformed enthalpies of the species.*)
  ri = Exp[ (dGreactant - gpfnsp) / (8.31451 * .31315)];
  ri.dhfnsp]


calconespeciesprops[trGT_, trHT_, nH_, z_, t_, pH_, is_] :=
 Module[{gcoeff, hcoeff, gadjust, hadjust},
   (*This program calculates the standard Gibbs energy of formation of a single
     species and its standard enthalpy of formation at 298.15 K and zero ionic
     strength. The input data are the standard transformed Gibbs energy of
     formation (trGT) and standard transformed enthalpy of formation (trHT)
     at a specified temperature (t in K), pH, and ionic strength (is).  The
     energies are in kJ mol^-1. The output is the usual data matrix for 298.15 K,
   including the charge number and the number of hydrogen atoms in the species.*)
   gcoeff = 9.20483 * 10^-3 * t - 1.28467 * 10^-5 * t^2 + 4.95199 * 10^-8 * t^3;
   hcoeff = -1.28466 * 10^-5 * t^2 + 9.90399 * 10^-8 * t^3;
   gadjust = -nH * 8.31451 * (t / 1000) * Log[10] * pH - gcoeff * nH * is^.5 / (1 + 1.6 * is^.5);
   hadjust = hcoeff * nH * is^.5 / (1 + 1.6 * is^.5);
   htab = trHT + hadjust;
   gspT = trGT + gadjust;
   gtab = (298.15 / t) * gspT - ((298.15 / t) - 1) * htab;
   {{gtab, htab, z, nH}}]
```

Chapter 7 Uses of Matrices in Biochemical Thermodynamics

Section 7.7

```
equcalcc[as_, lnk_, no_] :=
 Module[{l, x, b, ac, m, n, e, k}, (*as=conservation matrix. lnk=-(1/RT)
        (Gibbs energy of formation vector at T). no=initial composition vector.*)
  (*Setup*)
  {m, n} = Dimensions[as];
  b = as.no;
  ac = as;
  (*Initialize*)l = LinearSolve[as.Transpose[as], -as.(lnk + Log[n])];
  (*Solve*)Do[e = b - ac.(x = E^(lnk + l.as));
   If[(10^-10) > Max[Abs[e]], Break[]];
   l = l + LinearSolve[ac.Transpose[as*Table[x, {m}]], e], {k, 100}];
  If[k = 100, Return["Algorithm Failed"]];
  Return[x]]

equcalcrx[nt_, lnkr_, no_] :=
 Module[{as, lnk}, (*nt=transposed stoichiometric number matrix. lnkr=ln of
        equilibrium constants of rxs (vector). no=initial composition vector.*)
  (*Setup*)
  lnk = LinearSolve[nt, lnkr];
  as = NullSpace[nt];
  equcalcc[as, lnk, no]]
```

Appendix 7

```
mkeqm[c_List,s_List]:=(*c_List is the list of stoichiometric numbers for a
reaction. s_List is a list of the names of species or reactants. These names
have to be put in quotation marks.*)Map[Max[#,0]&,-c].s->Map[Max[#,0]&,c].s

nameMatrix[m_List,s_List]:=(*m_List is the transposed stoichiometric number
matrix for the system of reactions. s_List is a list of the names of species or
reactants. These names have to be put in quotation marks.*)Map[mkeqm[#,s]&,m]
```

Chapter 8 Oxidoreductase Reactions (Class 1) at 298.15 K

Section 8.2

```
calcappredpot[eq_, nu_, pHlist_, islist_] :=
 Module[{energy},(*Calculates the standard apparent reduction potential of a
half reaction in volts at specified pHs and ionic strengths for a biochemical
half reaction typed in the form nadox+de==nadred. The names of the reactants
call the corresponding functions of pH and ionic strength. nu is the number of
electrons involved. pHlist and islist can be lists.*)
  energy = Solve[eq, de];
    -(energy[[1,1,2]]/(nu*96.485)) /. pH -> pHlist /.
    is -> islist]
```

```
redpotsinorder[evaldata_] :=
 Module[{}, (*This program sorts the apparent reduction potentials in decreasing
        order and produces a table.  evaldata is in the following form: evaldata=
        {{"nadox+2e=nadred",calcappredpot[nadox+de==nadred,2,pH,is]},
            {"co2tot+pyruvate+2e=malate+h2o",
             calcappredpot[co2tot+pyruvate+de==malate+h2o,2,pH,is]},
                {"n2aq+8e=2ammonia+h2aq",calcappredpot[n2aq+de==2*ammonia+h2aq,
                 8,pH,is]}}/.pH->7/.is->.25*)
    TableForm[evaldata[[Reverse[Ordering[Transpose[evaldata][[2]]]]]]]]

redpotsinalphaborder[evaldata_] :=
 Module[{}, (*This program sorts the apparent reduction potentials in alphabetical
        order and produces a table.  evaldata is in the following form: evaldata=
        {{"nadox+2e=nadred",calcappredpot[nadox+de==nadred,2,pH,is]},
            {"co2tot+pyruvate+2e=malate+h2o",
             calcappredpot[co2tot+pyruvate+de==malate+h2o,2,pH,is]},
                {"n2aq+8e=2ammonia+h2aq",calcappredpot[n2aq+de==2*ammonia+h2aq,
                 8,pH,is]}}/.pH->7/.is->.25*)
    TableForm[evaldata[[Ordering[Transpose[evaldata][[1]]]]]]]
```

Section 8.3

```
derivefnGNHEMFrx[eq_, nu_] :=
 Module[{function}, (*Derives the functions of pH and ionic strength that give
        the standard transformed Gibbs energy of reaction, change in number
        of hydrogen ions bound, and the standard apparent reduction potential
        at 298.15 K for a biochemical half reaction typed in as, for example,
        n2aq+de==2*ammonia+h2aq. nu is the number of formal electrons.  The
        standard transformed Gibbs energy of reaction is in kJ mol^-1.*)
    function = Solve[eq, de];
    functionG = function[[1, 1, 2]];
    functionNH = (1 / (8.31451 * .29815 * Log[10])) * D[functionG, pH];
    functionEMF = -functionG / (nu * 96.485);
    {functionG, functionNH, functionEMF}]
```

Section 8.5

```
derivefnGNHKprimerx[eq_] :=
 Module[{function, functionG, functionNH, functionlogKprime},
   (*Derives the functions of pH and ionic strength that give the standard
     transformed Gibbs energy of reaction, change in number of hydrogen ions
     bound, and the base 10 log of the apparent equilibrium constant at 298.15
     K for a biochemical reaction typed in as, for example, atp+h2o+de==adp+
     pi. The standard transformed Gibbs energy of reaction is in kJ mol^-1.*)
    function = Solve[eq, de];
    functionG = function[[1, 1, 2]];
    functionNH = (1 / (8.31451 * .29815 * Log[10])) * D[functionG, pH];
    functionlogKprime = -functionG / (8.31451 * .29815 * Log[10]);
    {functionG, functionNH, functionlogKprime}];
```

```
calctrGerx[eq_, pHlist_, islist_] :=
 Module[{energy}, (*Calculates the standard transformed Gibbs energy
      of reaction in kJ mol^-1 at specified pHs and ionic strengths
      for a biochemical reaction typed in the form atp+h2o+de==
   adp+pi.  The names of reactants call the appropriate functions
      of pH and ionic strength.  pHlist and is list can be lists.*)
   energy = Solve[eq, de];
   energy[[1, 1, 2]] /. pH -> pHlist /. is -> islist]

calcNHrx[eq_, pHlist_, islist_] := Module[{energy},(*This program calculates the
change in the binding of hydrogen ions in a biochemical reaction at specified pHs
and ionic strengths.  The reaction is entered in the form atp+h2o+de==adp+pi.*)
   energy = Solve[eq, de];
     D[energy[[1,1,2]], pH]/(8.31451*0.29815*Log[10]) /.
      pH -> pHlist /. is -> islist]
```

Chapter 9 Transferase Reactions (Class 2) at 298.15 K

Section 9.2

```
derivefnGNHKprimerx[eq_]  (See Section 8.5)
```

Section 9.3

```
calctrGerx[eq_, pHlist_, islist_]  (See Section 8.5)
```

```
calcNHrx[eq_, pHlist_, islist_]    (See Section 8.5)
```

Chapter 10 Hydrolase Reactions (Class 3) at 298.15 K

Section 10.2

```
derivefnGNHKprimerx[eq_]  (See Section 8.5)
```

Section 10.3

```
calctrGerx[eq_, pHlist_, islist_]  (See Section 8.5)
```

```
calcNHrx[eq_, pHlist_, islist_] := Module[{energy},(*This program calculates the
change in the binding of hydrogen ions in a biochemical reaction at specified pHs
and ionic strengths.  The reaction is entered in the form atp+h2o+de==adp+pi.*)
   energy = Solve[eq, de];
     D[energy[[1,1,2]], pH]/(8.31451*0.29815*Log[10]) /.
      pH -> pHlist /. is -> islist]
```

Section 10.4

```
calcpK1is0[speciesmat_] :=
 Module[{lnkzero}, (*This program calculates pK1 at 298.15
   K and zero ionic stregth for a weak acid.*)
   lnkzero = (speciesmat[[2, 1]] - speciesmat[[1, 1]]) / (8.31451 * .29815);
   N[-lnkzero / Log[10]]]
```

Section 10.5

```
calcdGzerohighpH[estersp_, sugarsp_] :=
 Module[{}, (*estersp is the species matrix of the ester. sugar
    is the species matrix of the sugar.  Δ_rG° is in kJ mol⁻¹.*)
   (sugarsp[[1]] + pisp[[1]] - (estersp[[1]] + h2osp[[1]])) [[1]]]
```

Chapter 11 Lyase Reactions (Class 4), Isomerase Reactions Class 4), and Ligase Reactions (Class 6) at 298.15 K

Section 11.2

```
derivefnGNHKprimerx[eq_]   (See Section 8.5)
```

Section 11.3

```
calctrGerx[eq_, pHlist_, islist_]  (See Section 8.5)
```

```
calcNHrx[eq_, pHlist_, islist_]    (See Section 8.5)
```

Chapter 12 Survey of Reactions at 298.15 K

```
round[vec_, params_:{6, 2}] :=(*When a list of numbers has more digits to the
right of the decimal point than you want, say 6, you can request 2 by using
round[vec,{6,2}],*)
  Flatten[Map[NumberForm[#1, params] & , {vec}, {2}]]
```

```
trGibbsRxSummary[eq_, title_, reaction_, pHlist_, islist_] :=
 Module[{functiom, trGibbse, dvtNH, vectorNH},
   (*When this program is given a reaction equation in the form acetaldehyde+
      nadred+de==ethanol+nadox,it calculates the standard transformed Gibbs
      energies of reaction in kJ mol^-1 and the change in binding of hydrogen
      ions in the reaction at the desired pHs and ionic strengths at 298.15
      K.  title_ is in the form "EC 1.1.1.1 Alcohol dehydrogenase".  reaction_
      is in the form "acetaldehyde+nadred=ethanol+nadox".*)
   function = Solve[eq, de];
   trGibbse = round[function[[1, 1, 2]] /. pH -> pHlist /. is -> islist, {4, 2}];
   dvtNH = (1 / (8.31451 * .29815 * Log[10])) * D[function[[1, 1, 2]], pH];
   vectorNH = round[dvtNH /. pH -> pHlist /. is -> islist, {4, 2}];
   Print[title]; Print[reaction]; Print[trGibbse]; Print[vectorNH]]
```

Chapter 13 Survey of Reactions at Various Temperatures

Section 13

```
rxSummaryT[title_, reaction_, mathfunct_, temp_, pHlist_, islist_] :=
 Module[{trGibbse, trenthalpy, trentropy, logK, vectorNH, table},
   (*title_ is in the form "EC 1.1.1.1 Alcohol dehydrogenase".
       reaction_ is in the form "acetaldehyde+nadred=ethanol+nadox".
       mathfunct is of the form ethanolGT+nadoxGT-(acetaldehydeGT+nadredGT).
       temp is of the form 298.15 or other in the range 273.15 to 313.15.
       pHlist is of the form {5,6,7,8,9} or other.
       islist is of the form 0.25 or other.
       This program uses the mathfunction of T, pH, and ionic strength to calculate
     the standard transformed Gibbs energies of reaction in kJ mol^-1,
     the standard transformed enthalpies of formation in kJ mol^-1,
     the standard transformed entropy of reaction in J K^-1 mol^-1,
     the change in the binding of hydrogen ions,
     and logK' at the desired temperature, pHs, and ionic strength.*)
   trGibbse = mathfunct /. t → temp /. pH -> pHlist /. is -> islist;
   trenthalpy = -t^2 * D[mathfunct / t, t] /. t → temp /. pH -> pHlist /. is -> islist;
   trentropy = -D[mathfunct, t] * 1000 /. t → temp /. pH -> pHlist /. is -> islist;
   vectorNH = (1 / (8.31451 * (t / 1000) * Log[10])) * D[mathfunct, pH] /. t → temp /.
     pH -> pHlist /. is -> islist;
   logK = -mathfunct / (8.31452 * temp * Log[10] / 1000) /. t → temp /. pH -> pHlist /.
     is -> islist;
   table = PaddedForm[TableForm[{trGibbse, trenthalpy, trentropy, vectorNH, logK},
     TableHeadings → {{"Δr G ' °", "Δr H ' °", "Δr S ' °", "Δr NH", "logK '"},
       {"  pH 5", "  pH 6", "  pH 7", "  pH 8", "  pH 9"}}], {4, 2}];
   Print[title]; Print[reaction]; Print[table]]
```

Chapter 14 Protein-Ligand Binding

Section 14.4

```
calcstdfrtrGe := Module[{pK1M, pK2M, pK1MO2, pK2MO2, k, pM, pMO2, kprimeO2},
   (*This program derives the function of temperature, pH, and concentration
     of molecular oxygen that yields the standard further transformed Gibbs
     energy of formation of HavMO2av.  Energies are in joules per mole.*)
   pK1M = 7.85 - (37.7 * 10^3 / (8.3145 * Log[10])) * (1 / 293.15 - 1 / t);
   pK2M = 5.46 - (-6.3 * 10^3 / (8.3145 * Log[10])) * (1 / 293.15 - 1 / t);
   pK1MO2 = 6.67 - (37.7 * 10^3 / (8.3145 * Log[10])) * (1 / 293.15 - 1 / t);
   pK2MO2 = 6.04 - (-6.3 * 10^3 / (8.3145 * Log[10])) * (1 / 293.15 - 1 / t);
   k = 10^-5 * Exp[(60.7 * 10^3 / 8.3145) * (1 / 293.15 - 1 / t)];
   pM = (1 + 10^ (-pH + pK1M) + 10^ (-2 * pH + pK1M + pK2M));
   pMO2 = (1 + 10^ (-pH + pK1MO2) + 10^ (-2 * pH + pK1MO2 + pK2MO2));
   kprimeO2 = k * pM / pMO2;
   -8.3145 * t * Log[pM] - 8.3145 * t * Log[1 + co2 / kprimeO2]]
```

Chapter 15 Calorimetry

Section 15.4

```
calcSform1sp[reactantname_, speciesmat_] :=
 Module[{dGzero, dHzero, zi, nH}, (*This program is used to calculate the
     standard entropy of formation of the single species of a reactant in
     kJ K^-1 mol^-1 at 298.15 K and zero ionic strength.  The reactant
     name should be in quotation marks.  The output is a 1 x 4 matrix.*)
   {dGzero, dHzero, zi, nH} = Transpose[speciesmat];
   Transpose[{{reactantname}, (dHzero - dGzero) / 298.15, zi, nH}]]
```

```
calcSform2sp[reactantname_, speciesmat_] :=
 Module[{dGzero, dHzero, zi, nH},
   (*This program is used to calculate the standard entropies of
     formation of the two species of a reactant in kJ K^-1 mol^-1
     at 298.15 K and zero ionic strength.  The reactant name should
     be put in quotation marks.  The output is a 2 x4 matrix.*)
   {dGzero, dHzero, zi, nH} = Transpose[speciesmat];
   Transpose[{{reactantname, ""}, (dHzero - dGzero) / 298.15, zi, nH}]]

calcSform3sp[reactantname_, speciesmat_] :=
 Module[{dGzero, dHzero, zi, nH},
   (*This program is used to calculate the standard entropies of
     formation of the three species of a reactant in kJ K^-1 mol^-1
     at 298.15 K and zero ionic strength.  The reactant name should
     be put in quotation marks.  The output is a 3 x4 matrix.*)
   {dGzero, dHzero, zi, nH} = Transpose[speciesmat];
   Transpose[{{reactantname, "", ""}, (dHzero - dGzero) / 298.15, zi, nH}]]

calcentropylist[speciesmat_] :=
 Module[{dGzero, dHzero, zi, nH}, (*This program is used to assemble the
     list of standard entropies of formation of the species of a
     reactant in kJ K^-1 mol^-1 at 298.15 K and zero ionic strength.*)
   {dGzero, dHzero, zi, nH} = Transpose[speciesmat];
   (dHzero - dGzero) / 298.15]
```

Section 15.5

```
calctrSformreactant[speciesmat_, entropylist_] :=
 Module[{dGzero, dHzero, zi, nH, pHterm, isterm, gpfnsp, dGreactant,
   ri, isentropy, dSfnsp, pHtermS, avgentropy, entropymix},
   (*This program derives the function of pH and ionic strength
     (is) that gives the standard transformed entropy of formation
     of a reactant (sum of species) at 298.15 K.  The first input
     is a matrix that gives the standard Gibbs energy of formation,
    the standard enthalpy of formation, the electric charge,
    and the number of hydrogen atoms in the species in the
    reactant.  There is a row in the matrix for each species
    of the reactant.  dSfnsp is a list of the functions for
    the species.  Entropies are expressed in kJ K^-1 mol^-1.*)
   {dGzero, dHzero, zi, nH} = Transpose[speciesmat];
    (*Calculate the functions for the
    standard Gibbs energies of formation of the species.*)
   pHterm = nH * 8.31451 * .29815 * Log[10^-pH];
   isterm = 2.91482 * ((zi^2) - nH) * (is^.5) / (1 + 1.6 * is^.5);
   gpfnsp = dGzero - pHterm - isterm;
   (*Calculate the standard transformed
     Gibbs energy of formation for the reactant.*)
   dGreactant = -8.31451 * .29815 *
     Log[Apply[Plus, Exp[-1 * gpfnsp / (8.31451 * .29815)]]];
   (*Calculate the equilibrium mole fractions of
     the species in the reactant.*)
   ri = Exp[(dGreactant - gpfnsp) / (8.31451 * .29815)];
   (*Calculate the standard transfomed entropies of formation
      of the species and then calculate the mole fraction-
     weighted average entropy of the reactant.*)
   pHtermS = nH * 8.31451 * 10^-3 * Log[10^-pH];
   isentropy = .0147319 * ((zi^2) - nH) * (is^.5) / (1 + 1.6 * is^.5);
   dSfnsp = entropylist + pHtermS + isentropy;
   avgentropy = ri.dSfnsp;
   (*Calculate the entropy of mixing of the species.*)
   entropymix = 8.31451 * 10^-3 * ri.Log[ri];
   avgentropy - entropymix]
```

BasicBiochemData3

```
calcdGmat[speciesmat_]     (See Section 3.6)
```

```
calcNHmat[speciesmat_]     (See Section 3.6)
```

```
calctrGibbsT[speciesmat_]     (See Section 4.1)
```

```
derivetrHT[speciesmat_] :=
 Module[{dGzero, dGzeroT, dHzero, zi, nH, gibbscoeff, pHterm, isterm, gpfnsp,
   gibbsereactant}, (*This program derives the function of T (in Kelvin),
    pH,and ionic strength (is) that gives the stndard transformed enthalpy
    of formation of a reactant (sum of species).The input speciesmat is a
    matrix that gives the standard Gibbs energy of formation in kJ mol^-1
    at 298.15 K and zero ionic strength,the standard enthalpy of formation
    in kJ mol^-1 at 298.15 K and zero ionic strength,the electric charge,
    and the number of hydrogen atoms in each species.There is a row in
```

```
     the matrix for each species of the reactant.gpfnsp is a list of the
     functions for the standard transformed Gibbs energies of the species.This
     program applies the Gibbs-Helmholtz equation to the function for
     the standard transformed Gibbs energy of formation of a reactant.The
     standard transformed enthalpy of formation of a reactant in kJ mol^-1
     can be calculated at any temperature in the range 273.15 K to 313.15 K,
   any pH in the range 5 to 9,and any ionic strength in the range
     0 to 0.35 M by use of the assignment operator (/.).*)
   {dGzero, dHzero, zi, nH} = Transpose[speciesmat];
   gibbscoeff = (9.20483 * t) / 10^3 - (1.284668 * t^2) / 10^5 + (4.95199 * t^3) / 10^8;
   dGzeroT = (dGzero * t) / 298.15 + dHzero * (1 - t / 298.15);
   pHterm = (nH * 8.31451 * t * Log[10^(-pH)]) / 1000;
   istermG = (gibbscoeff * (zi^2 - nH) * is^0.5) / (1 + 1.6 * is^0.5);
   gpfnsp = dGzeroT - pHterm - istermG;
   gibbsereactant =
     - ((8.31451 * t * Log[Plus @@ (E^(-(gpfnsp / ((8.31451 * t) / 1000)))))]) / 1000);
   -t^2 * D[gibbsereactant / t, t]]

derivetrST[speciesmat_] :=
 Module[{dGzero, dGzeroT, dHzero, zi, nH, gibbscoeff, pHterm, isterm, gpfnsp,
   gibbsereactant}, (*This program derives the function of T (in Kelvin),
     pH,and ionic strength (is) that gives the stndard transformed entropy
     of formation of a reactant (sum of species).The input speciesmat is a
     matrix that gives the standard Gibbs energy of formation in kJ mol^-1
     at 298.15 K and zero ionic strength,the standard enthalpy of formation
     in kJ mol^-1 at 298.15 K and zero ionic strength,the electric charge,
   and the number of hydrogen atoms in each species.There is a row in the
     matrix for each species of the reactant.gpfnsp is a list of the functions for
     the standard transformed Gibbs energies of the species.This program applies
     the equation for the entropy to the function for the standard transformed
     Gibbs energy of formation of a reactant.The standard transformed entropy
     of formation of a reactant in kJ K^-1 mol^-1 can be calculated at any
     temperature in the range 273.15 K to 313.15 K,any pH in the range 5 to 9,
     and any ionic strength in the range 0 to 0.35 M by use of the assignment
     operator (/.).*){dGzero, dHzero, zi, nH} = Transpose[speciesmat];
   gibbscoeff = (9.20483 * t) / 10^3 - (1.284668 * t^2) / 10^5 + (4.95199 * t^3) / 10^8;
   dGzeroT = (dGzero * t) / 298.15 + dHzero * (1 - t / 298.15);
   pHterm = (nH * 8.31451 * t * Log[10^(-pH)]) / 1000;
   istermG = (gibbscoeff * (zi^2 - nH) * is^0.5) / (1 + 1.6 * is^0.5);
   gpfnsp = dGzeroT - pHterm - istermG;
   gibbsereactant =
     - ((8.31451 * t * Log[Plus @@ (E^(-(gpfnsp / ((8.31451 * t) / 1000)))))]) / 1000);
   -D[gibbsereactant, t]]

deriveNHT[speciesmat_] := Module[
   {dGzero, dGzeroT, dHzero, zi, nH, gibbscoeff, pHterm, isterm, gpfnsp, trgefT},
   (*This program derives the function of T (in Kelvin),pH,
     and ionic strength (is) that gives the average number of hydrogen
     atoms in a reactant (sum of species).The input speciesmat is a matrix
     that gives the standard Gibbs energy of formation in kJ mol^-1 at
     298.15 K and zero ionic strength,the standard enthalpy of formation
     in kJ mol^-1 at 298.15 K and zero ionic strength,the electric charge,
   and the number of hydrogen atoms in each species.There is a row in the
     matrix for each species of the reactant.gpfnsp is a list of the functions for
     the standard transformed Gibbs energies of the species.The average number
     can be calculated at any temperature in the range 273.15 K to 313.15 K,
   any pH in the range 5 to 9,and any ionic strength in the range
     0 to 0.35 M by use of the assignment operator (/.).*)
```

```
{dGzero, dHzero, zi, nH} = Transpose[speciesmat];
gibbscoeff = (9.20483 * t) / 10^3 - (1.284668 * t^2) / 10^5 + (4.95199 * t^3) / 10^8;
dGzeroT = (dGzero * t) / 298.15 + dHzero * (1 - t / 298.15);
pHterm = (nH * 8.31451 * t * Log[10^(-pH)]) / 1000;
istermG = (gibbscoeff * (zi^2 - nH) * is^0.5) / (1 + 1.6 * is^0.5);
gpfnsp = dGzeroT - pHterm - istermG;
trgefT = - ((8.31451 * t * Log[Plus @@ (E^(-(gpfnsp / ((8.31451 * t) / 1000))))]));
D[trgefT, pH] / (8.31451 * t * Log[10])]
```

Appendix 6: Sources of Biochemical Thermodynamic Information on the Web

A web address (URL: Uniform Resource Locator) can be made live in the sense that when it is clicked in *Mathematica*, the resource appears on the screen. Starting with a web address in text, it can be made live in the following six steps:

1. Select a web address in text.
2. Change style to input.
3. Copy web address.
4. Choose Input->Create Hyperlink
5. Select Notebook or URL.
6. Paste in URL.

There are several types of information on the web that are useful in making thermodynamic calculations:

(1) The first is experimental data on apparent equilibrium constants and transformed enthalpies of enzyme-catalyzed reactions. R. N. Goldberg and Y. B. Tewara have evaluated these data in the literature and have published six review articles in J. Phys. Chem. Ref. Data. In addition R. N. Goldberg, Y. B. Tewara, and T. N. Bhat have put up a web site to assist in use these data.

http : // xpdb.nist.gov / enzyme_thermodynamics /

(2) The second type is *Mathematica* packages that provide data and programs for making calculations. The following three packages are available at *MathSource*:

R. A. Alberty, BasicBioChemData2: Data and Programs for Biochemical Thermodynamics (2003).

http : // library.wolfram.com / infocenter / MathSource / 797

This package provides data on the species of 131 reactants at 298.15 K and programs for calculating various transformed thermodynamic properties. Programs are given for the calculation of apparent equilibrium constants and other transformed thermodynamic properties of enzyme-catalyzed reactions by simply typing in the reaction.

R. A. Alberty, ProteinLigandProg (2003).

http : // library.wolfram.com / infocenter / MathSource / 4808

This package shows how to calculate various thermodynamic properties of a protein-ligand binding. There are three ways to discuss the thermodynamics of the formation of $H_{av} MO_{2\,av}$.

R. A. Alberty, BasicBiochemData3.nb (2005).

http : // library.wolfram.com / infocenter / MathSource / 5704

This package provides data on the species of 199 reactants at 298.15 K and programs for calculating various transformed thermodynamic properties. Loading this package provides functions of pH and ionic strength at 298.15 K for standard transformed Gibbs energies of formation and average numbers of hydrogen atoms for 199 reactants. It also provides functions of temperature, pH, and ionic strength for more properties of 94 reactants. Thus loading this package makes available 774 mathematical functions for these properties.

(3) The third type of URLs provides lists of enzyme-catalyzed reactions and certain information about the enzymes:

E. C. Webb, Enzyme Nomenclature 1992, Academic Press, New York (1992).

http : // www.chem.qmw.ac.uk / iubmb / enzyme /

This URL is an update of the IUBMB recommendations of names for enzymes. This site is the responsibility of the Nomenclature Committee of NC-IUBMB. This section gives balanced equations for enzyme-catalyzed reactions and certain references and information, arranged by EC number. Links are provided to BRENDA, EXPASY, GDT, KEGG, UM-BBD, ERGO, and PDP.

EC-PDP Enzyme Structure Database

http : // www.ebi.acuk / thornton - srv / databases / enzymes /

This gives enzyme-catalyzed reactions and additional information.

Swissprot Enzyme (Enzyme nomenclature database)

http : // us.expasy.org / enzyme /

This is a repository relative to the nomenclature of enzymes. It accepts EC numbers. It provides connections with BRENDA, EMR, KEGG, IUBMB, and BioCarti.

(4) A fourth type of URLs are international recommendations on biochemical thermodynamics:

R. A. Alberty, A. Cornish-Bowden, Q. H. Gibson, R. N. Goldberg, G. G. Hammes, W. Jencks, K. F. Tipton, R. Veech, H. V. Westerhoff, and E. C. Webb, Recommendations for nomenclature and tables in biochemical thermodynamics, Pure Appl. Chem. 66, 1641-1666 (1994). Reprinted in Europ. J. Biochem. 240, 1-14 (1996).

http://www.chem.qmw.ac.uk/iubmb/thermod/

(5) A fifth type of URL provides a *Mathematica* package for treating complex equilibria in aqueous solution.

D. L. Akers and R. N. Goldberg; "BioEqCalc: A package for performing equilibrium calculations on biochemical reactions," Mathematica J., 8, 86-113 (2001).

http://www.mathematica-journal.com/issue/v8i1/

This package also shows how to input a chemical equation and obtain the corresponding stoichiometric matrix.

Index